T0212726

Lecture Notes in Artificial Intelligence 9336

Subseries of Lecture Notes in Computer Science

More information about this series at http://www.springer.com/series/1244

Marco Gavanelli · Evelina Lamma
Fabrizio Riguzzi (Eds.)

AI*IA 2015 Advances in Artificial Intelligence

XIVth International Conference
of the Italian Association for Artificial Intelligence
Ferrara, Italy, September 23–25, 2015
Proceedings

 Springer

Editors
Marco Gavanelli
Dipartimento di Ingegneria
Università di Ferrara
Ferrara
Italy

Fabrizio Riguzzi
Dipartimento di Matematica e Informatica
Universita di Ferrara
Ferrara
Italy

Evelina Lamma
Dipartimento di Ingegneria
Università di Ferrara
Ferrara
Italy

ISSN 0302-9743   ISSN 1611-3349   (electronic)
Lecture Notes in Artificial Intelligence
ISBN 978-3-319-24308-5   ISBN 978-3-319-24309-2   (eBook)
DOI 10.1007/978-3-319-24309-2

Library of Congress Control Number: 2015950944

LNCS Sublibrary: SL7 – Artificial Intelligence

Springer Cham Heidelberg New York Dordrecht London

Printed on acid-free paper

Springer International Publishing AG Switzerland is part of Springer Science+Business Media
(www.springer.com)

Preface

This book collects the contributions accepted for *AI*IA 2015, the 14th Conference of the Italian Association for Artificial Intelligence,* held in Ferrara, Italy, September 23–25, 2015. The conference is organized by AI*IA (the Italian Association for Artificial Intelligence) and it is held every other year.

The Program Committee (PC) received 44 valid submissions. Each paper was carefully reviewed by at least three members of the PC, who selected the 35 papers that are presented in these proceedings.

Following the 2013 edition of the conference, we adopted a "social" model: the papers were made available to conference participants in advance, each paper was shortly presented at the conference and was assigned a time slot and a reserved table where the authors were available for discussing their work with the interested audience. In this way, we aim at fostering discussion and facilitating idea exchange, community creation, and collaboration.

AI*IA 2015 featured exciting keynotes by Laurent Perron, Technical Leader of the Operations Research Team at Google; Kristian Kersting, Head of the Statistical Relational Activity Mining Group, Fraunhofer IAIS, Technical University of Dortmund; Oren Etzioni, Director of the Allen Institute for Artificial Intelligence; and Kevin Warwick, Deputy Vice Chancellor (Research) at Coventry University.

The program of the conference also included six workshops: the *First Workshop on Artificial Intelligence and Design (AIDE 2015),* the *Second Workshop on Artificial Intelligence and Robotics (AIRO 2015),* the *Sixth Italian Workshop on Planning and Scheduling (IPS 2015),* the *First Workshop on Intelligent Techniques at Libraries and Archives (IT@LIA 2015),* the *Fourth Italian Workshop on Machine Learning and Data Mining (MLDM.it 2015),* and the *22nd RCRA International Workshop on Experimental Evaluation of Algorithms for Solving Problems with Combinatorial Explosion (RCRA 2015),* plus a doctoral consortium.

The chairs wish to thank the Program Committee members and the anonymous reviewers for their careful work in the selection of the best papers; the chairs of the workshops and of the doctoral consortium for organizing the respective events, as well as Elena Bellodi, Giuseppe Cota, Andrea Peano, and Riccardo Zese for their help during the organization of the conference.

July 2015

Marco Gavanelli
Evelina Lamma
Fabrizio Riguzzi

Organization

AI*IA 2015 was organized by AI*IA (Associazione Italiana per l'Intelligenza Artificiale), in cooperation with the Departments of Engineering and of Mathematics and Computer Science of the University of Ferrara (Italy).

Executive Committee

Conference Chairs

Marco Gavanelli	Università di Ferrara, Italy
Evelina Lamma	Università di Ferrara, Italy
Fabrizio Riguzzi	Università di Ferrara, Italy

Doctoral Consortium Chairs

Elena Bellodi	Università di Ferrara, Italy
Alessio Bonfietti	Università di Bologna, Italy

Panel Chair

Piero Poccianti	Consorzio Operativo Gruppo MPS, Italy

Program Committee

Matteo Baldoni	Università di Torino, Italy
Stefania Bandini	Università Milano-Bicocca, Italy
Roberto Basili	Università di Roma Tor Vergata, Italy
Nicola Basilico	Università di Milano, Italy
Elena Bellodi	Università di Ferrara, Italy
Federico Bergenti	Università di Parma, Italy
Stefano Bistarelli	Università di Perugia, Italy
Luciana Bordoni	ENEA, Italy
Francesco Buccafurri	Università Mediterranea di Reggio Calabria, Italy
Stefano Cagnoni	Università di Parma, Italy
Diego Calvanese	Free University of Bozen-Bolzano, Italy
Amedeo Cappelli	CNR, Italy
Luigia Carlucci Aiello	Sapienza Università di Roma, Italy
Amedeo Cesta	CNR, Italy
Antonio Chella	Università di Palermo, Italy
Carlo Combi	Università di Verona, Italy
Gabriella Cortellessa	CNR, Italy
Stefania Costantini	Università di L'Aquila, Italy

Giuseppe De Giacomo	Sapienza Università di Roma, Italy
Francesco Donini	CNR, Italy
Agostino Dovier	Università di Udine, Italy
Floriana Esposito	Università di Bari, Italy
Stefano Ferilli	Università di Bari, Italy
Marco Gavanelli	Università di Ferrara, Italy
Nicola Guarino	CNR, Italy
Luca Iocchi	Sapienza Università di Roma, Italy
Evelina Lamma	Università di Ferrara, Italy
Nicola Leone	Università della Calabria, Italy
Chendong Li	Dell, USA
Francesca Alessandra Lisi	Università di Bari, Italy
Bernardo Magnini	FBK, Italy
Sara Manzoni	Università Milano-Bicocca, Italy
Alberto Martelli	Università di Torino, Italy
Paola Mello	Università di Bologna, Italy
Alessio Micheli	Università di Pisa, Italy
Alfredo Milani	Università di Perugia, Italy
Michela Milano	Università di Bologna, Italy
Stefania Montani	Università del Piemonte Orientale, Italy
Alessandro Moschitti	Università di Trento, Italy
Roberto Navigli	Sapienza Università di Roma, Italy
Angelo Oddi	CNR, Italy
Andrea Omicini	Università di Bologna, Italy
Maria Teresa Pazienza	Università di Roma Tor Vergata, Italy
Roberto Pirrone	Università di Palermo, Italy
Piero Poccianti	Consorzio Operativo Gruppo MPS, Italy
Gian Luca Pozzato	Università di Torino, Italy
Luca Pulina	Università di Sassari, Italy
Daniele P. Radicioni	Università di Torino, Italy
Francesco Ricca	Università della Calabria, Italy
Fabrizio Riguzzi	Università di Ferrara, Italy
Andrea Roli	Università di Bologna, Italy
Salvatore Ruggieri	Università di Pisa, Italy
Fabio Sartori	Università Milano-Bicocca, Italy
Ken Satoh	National Institute of Informatics and Sokendai, Japan
Andrea Schaerf	Università di Udine, Italy
Floriano Scioscia	Politecnico di Bari, Italy
Giovanni Semeraro	Università di Bari, Italy
Roberto Serra	Università di Modena e Reggio Emilia, Italy
Francesca Toni	Imperial College, UK
Pietro Torasso	Università di Torino, Italy
Eloisa Vargiu	Barcelona Digital Technology Center, Spain
Marco Villani	Università di Modena e Reggio Emilia, Italy
Giuseppe Vizzari	Università Milano-Bicocca, Italy

Additional Reviewers

Bacciu, Davide	De Benedictis, Riccardo	Mencar, Corrado
Barlacchi, Gianni	Degeler, Viktoriya	Patti, Viviana
Basile, Pierpaolo	Franzoni, Valentina	Poggioni, Valentina
Bellandi, Andrea	Furletti, Barbara	Portinale, Luigi
Benotto, Giulia	Fuscà, Davide	Sato, Taisuke
Bloisi, Domenico Daniele	Gallicchio, Claudio	Takahashi, Kazuko
Cauteruccio, Francesco	Georgievski, Ilche	Tesconi, Maurizio
D'Amato, Claudia	Lieto, Antonio	
Dal Palù, Alessandro	Manna, Marco	

Sponsoring Institutions

AI*IA 2015 was partially funded by the Artificial Intelligence Journal, by the Departments of Engineering and of Mathematics and Computer Science of the University of Ferrara, by the Istituto Nazionale di Alta Matematica "F. Severi" - Gruppo Nazionale per il Calcolo Scientifico, by Dario Flaccovio Editore, and by the Italian Association for Artificial Intelligence.

Contents

Knowledge Representation and Reasoning

Machine Learning

Swarm Intelligence and Genetic Algorithms

Collective Self-Awareness and Self-Expression for Efficient Network Exploration

Michele Amoretti$^{(\boxtimes)}$ and Stefano Cagnoni

Department of Information Engineering, Universita degli Studi di Parma,
Parma, Italy
{michele.amoretti,stefano.cagnoni}@unipr.it

Abstract. Message broadcasting and topology discovery are classical problems for distributed systems, both of which are related to the concept of network exploration. Typical decentralized approaches assume that network nodes are provided with traditional routing tables. In this paper we propose a novel network exploration approach based on collective self-awareness and self-expression, resulting from the simultaneous application of two strategies, namely hierarchy and recursion, which imply the adoption of unusual routing tables. We show how the proposed approach may provide distributed systems with improved efficiency and scalability, with respect to traditional approaches.

Keywords: Collective self-awareness · Collective self-expression · Hierarchy and recursion · Network exploration

1 Introduction

Network exploration is the bottom line of several problems for distributed systems, *i.e.*, systems consisting of multiple autonomous nodes that communicate through a network. Example of such problems are message broadcasting [1,2] and topology discovery [3,4]. Centralized solutions are not scalable and highly inefficient. Thus, decentralized approaches are usually adopted, assuming that network nodes are provided with traditional routing tables — *i.e.*, data tables that list the routes to particular network destinations and, in some cases, metrics associated to those routes.

In this paper we propose a novel network exploration approach based on *collective self-awareness and self-expression* [5], resulting from the simultaneous application of two strategies, namely *hierarchy and recursion*, which imply the adoption of unusual routing tables. With respect to traditional approaches, the one we propose may provide distributed systems with improved efficiency and scalability.

The paper is organized as follows. In Section 2, related work on network exploration and self-aware computing is discussed. In Section 3, the concepts of collective self-awareness and self-expression are summarized, with particular focus on their implementation based on hierarchy and recursion (HR). In Section 4,

© Springer International Publishing Switzerland 2015
M. Gavanelli et al. (Eds.): AI*IA 2015, LNAI 9336, pp. 3–16, 2015.
DOI: 10.1007/978-3-319-24309-2_1

our HR-based network exploration algorithm is illustrated. In Section 5, simulation results are presented and discussed. The last section concludes the paper presenting future research lines. As an appendix, a short survey on consciousness, self-Awareness and self-Expression in psychology and cognitive science is proposed.

2   Related Work

Network exploration is a necessary task in several contexts. Among others, multiple-message broadcasting is particularly important for wireless sensor networks (WSNs), since it is a basic operation in many applications, such as updating of routing tables and several kinds of data aggregation functions in WSNs. Classical Flooding (CF) is the simplest way of implementing multi-hop broadcast: when a node receives a broadcast packet for the first time, it forwards the packet to its neighbors (duplicates are detected and dropped). However, CF can be very costly in terms of wasted bandwidth, and also inefficient, because of broadcast storms that may be generated by concomitant packet retransmissions [6]. Considering also that in Low power and Lossy Networks (LLNs) nodes are extremely energy-constrained, a finite message budget is a realistic assumption [1]. The optimal solution consists of building a minimum Connected Dominating Set (CDS), defined as the minimum set of relays that guarantees network connectivity. However, finding a minimum CDS is known as an NP-hard problem [7] to which several authors proposed distributed approximate solutions. Recently, the IETF ROLL working group standardized RPL, the routing protocol for LLNs [8], and also proposed Multicast Protocol for Low power and Lossy Networks (MPL), a forwarding mechanism for LLN networks [9]. Later, La et al. [10] have introduced Beacon-based Forwarding Tree (BFT), an energy-efficient multi-hop broadcasting scheme that achieves performance similar to MPL, although it fits better the case of nodes with low radio duty cycling[1] MAC layers of the type of beacon-enabled IEEE 802.15.4. More generally, Yu et al. [2] have proposed a distributed algorithm for multiple-message broadcasting in unstructured wireless networks under a global interference model, as well as a lower bound for randomized distributed multiple-message broadcast algorithms under the assumed network model.

Another domain where network exploration plays a prominent role is topology discovery. Several authors have proposed agent-oriented approaches based on learning [3,11,12]. Agents explore the network to i) acquire information, on their own, about visited nodes (first-hand knowledge); ii) collect information from other agents, by means of direct or implicit communication. Not using agents, Li et al. [4] have recently proposed an IPv6 network router-level topology discovery method, combining the topology information obtained by *traceroute* and OSPF [13].

All the aforementioned network exploration strategies assume that network nodes are provided with traditional routing tables. Our approach, instead,

---

[1] Radio duty cycling is the proportion between the periods nodes are on and off.

implies the adoption of unusual routing tables. Before introducing them, we need to recall some background concepts.

Self-aware computing systems and applications proactively maintain information about their own environments and internal states [14]. In detail, self-expression refers to i) goal revision and ii) self-adaptive behavior, which derives from reasoning about the knowledge associated with the system's self-awareness.

According to Faniyi *et al.* [14], a self-aware node "possesses information about its internal state and has sufficient knowledge of its environment to determine how it is perceived by other parts of the system." Self-awareness produces behavioral models of the node. Self-expression encompasses goal revision and self-adaptive behavior deriving from reasoning about such models. The same authors [14] have also developed a computational translation of the layered self-awareness model proposed by psychologist Neisser [15] (a short survey on consciousness, self-awareness and self-expression in psychology and cognitive science is proposed at the end of this paper). From the bottom up to the top of the stack, there are five levels of self-awareness, with increasing complexity: stimulus awareness, interaction awareness, time awareness, goal awareness and meta-self-awareness. This latter layer is the node's awareness of its own self-awareness capabilities (or its lack). The conceptual framework is illustrated in Figure 1. Self-awareness implies processing information collected from the internal and external sensors of the node. Data provided by the internal sensors contribute to *private* self-awareness construction. Information about the node's interactions with the physical environment and other nodes, instead, contributes to *public* self-awareness construction.

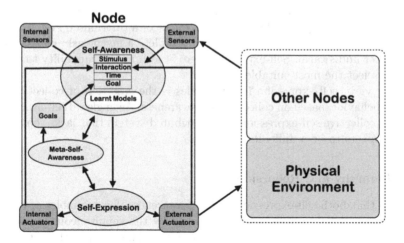

Fig. 1. Representation of a self-aware and self-expressive node, according to the models defined by Faniyi *et al.* [14].

Our recent contribution to self-aware computing is the definition of collective self-awareness and self-expression [5], which is the baseline for the network exploration approach we present in this paper. In the following section, we illustrate the main principles of collective self-awareness and self-expression.

3   Collective Self-Awareness and Self-expression

According to Mitchell [16], complex systems are dynamic systems composed of interconnected parts that, as a whole, exhibit properties that could not be inferred from the properties of the individual parts. Ant colonies, immune systems and humans are examples of complex systems, where collective self-awareness is an emergent effect [14].

In computing/networking systems, the adaptive mechanisms that can be implemented within a single node coincide with self-expression, if they are based on the node's self-awareness capabilities. What about self-awareness and self-expression of a distributed system as a whole? Can actual global self-awareness be achieved only by providing the distributed system with a centralized omniscient monitor? Luckily, the answer is no.

A computing node exhibits self-expression if it is able to assert its behavior upon either itself or other nodes, the relevance of such a behavior being proportional to a notion of authority in the network [14]. The behavior of the node is affected by its state, context, goals and constraints.

Self-expression for ensembles of cooperating computational entities is the ability to deploy run-time changes of the coordination pattern, according to Cabri *et al.* [17]. In other words, the distributed system expresses itself (meaning that it still does what it is supposed to do) independently of unexpected situations and, to accomplish this, it can modify its original internal organization. For example, suppose that each component of a distributed system knows three different collaborative approaches to complete a given task: master-slave, peer-to-peer and swarm. Self-expression here is seen as the capability to collaboratively select the most suitable strategy.

In our view, self-expression for ensembles is the assertion of collective self-adaptive behavior, based on collective self-awareness. As in global self-awareness, achieving collective self-expression in a distributed system that lacks centralized control appears to be a difficult task.

3.1   Hierarchy and Recursion

We claim that both self-expression and self-awareness, for ensembles of cooperating computational entities, can be achieved by the simultaneous application of two strategies, namely *hierarchy* and *recursion*. Hierarchy is the categorization of a group of nodes according to their capability or status. Recursion is the repeated use of a single, flexible functional unit for different capabilities over different scopes of a distributed system.

A possible implementation of this principle is *recursive networking*, developed to describe multi-layer virtual networks that embed networks as nodes inside other networks. In the last decade, recursive networking has evolved to become a possible architecture for the future Internet [18]. In particular, it is a prominent approach to designing quantum networks [19].

Moreover, in the context of Content-Centric Networking (CCN) [20], an emerging approach which is particularly promising to face the scalability issues that the Internet of Things is raising, there is room for both static and dynamic hierarchies.

For example, consider the network illustrated in Figure 2. The routing table at node 4.2 contains information on how to reach any other node in the network. For scalability purposes, the table has more precise information about nearby destinations (node 4.4 and node 4.7), and vague information about more remote destinations (NET9), obtained using hierarchy and recursion. Routing tables are initialized and updated with information exchanged between directly attached neighbors.

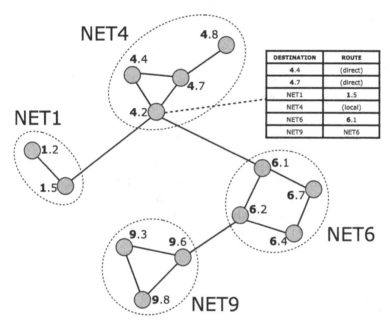

DESTINATION	ROUTE
4.4	(direct)
4.7	(direct)
NET1	1.5
NET4	(local)
NET6	6.1
NET9	NET6

Fig. 2. Hierarchy and recursion: the routing table at node 4.2 contains information on how to reach any other node in the network.

By means of hierarchy and recursion, global self-awareness is available at every node and enables global self-expression. In the network example, packets are forwarded according to routing tables. Forwarding too many packets to the same neighbor may cause congestion on that node. Having a feedback about such a negative effect of the forwarding node's behavior (*i.e.*, having local self-awareness) may lead to a modification of that behavior, supported by the routing table (which is a way of building global self-awareness). Namely, an alternative destination for packets may be chosen. This local self-expression process may also trigger a routing table update. On the other hand, a routing table update may also derive from the exchange of routing information with known nodes. The simultaneous and collaborative update of HR-based routing tables, which may also take into account the possibility of changing hierarchies, is actually a global self-expression process.

More precisely, the networking example in Figure 2 (and further developed in the subsequent sections) fits the conceptual framework in the following way:

- stimuli-awareness: the network is stimuli-aware, meaning that nodes know how to manage messages;
- interaction-awareness: nodes and subnets interact by exchanging messages and control information, i.e., routing table updates; they are interaction-aware, meaning that they can distinguish between other nodes and subnets;
- time-awareness: the network has knowledge of past events or likely future ones, like in learning-based routing;
- goal-awareness: the HR approach enforces the global goal of high efficiency (e.g., simplifying the search for the shortest path towards a specific destination); goal-awareness is thus implicit in how routing tables are populated and updated;
- meta-self-awareness: a way to implement it is to concurrently apply different routing strategies, choosing the best one at runtime.

4   HR-based Network Exploration

Let us consider a network consisting of N nodes and S subnetworks. We are interested in evaluating the approximate number of forwardings needed by a probe message to propagate through the whole graph. The probe message is generated by a random node, which sends it to one of its neighbors. Then it is forwarded to another neighbor, and so on. A node that receives the probe message is marked as "visited".

The simplest search strategy is the *Random Walk (RW)*, where the next hop is randomly selected among the node's neighbors. Even if the network is recursive and hierarchical, such features are not exploited by RW. As a consequence, RW is quite expensive, in terms of probe message propagations needed to explore the whole network — even if the network topology has good features, such as scale invariance [21].

Another simple strategy is Classical Flooding (CF) (see also Section 2), where every probe message received by a node is forwarded to all the neighbors of the node. To avoid network congestion, probe messages are always associated to a Time To Live (TTL), which is the maximum number of times they can be forwarded. The main advantage of CF is that, if a packet can be delivered, it will actually be (probably multiple times). However, CF can be very costly in terms of wasted bandwidth and also inefficient, as it may originate broadcast storms [6].

Our HR-based network exploration approach takes subnetworks into account and exploits collective self-awareness. Every node is member of a subnetwork $NET s$ ($s \in \{1, .., S\}$) and has an identifier $Node n$ ($n \in \mathbb{N}$) which is unique within that network. The *name* of the generic node is denoted as NET*s*.Node*n*. A node may have neighbors that are members of other subnetworks. For example, the routing table illustrated in Figure 2 allows NET4.Node2 to forward messages i) to other nodes of NET4 that are directly reachable, ii) to NET1 and NET6 through directly reachable nodes that belong to those subnetworks, and iii) to NET9 through NET6. Importantly, the size of the routing table is $O(S)$.

The neighbor to whom the probe is forwarded belongs to the same subnetwork of the sender. If all neighbors of the same subnetwork have been already visited, the probe is forwarded to one neighbor from another subnetwork, excluding the previous hop. If there is only one neighbor belonging to other subnetworks and it is the previous hop, then the neighbor that grants access to the longest route is chosen. The flowchart of the HR-based network exploration approach is shown in Figure 3.

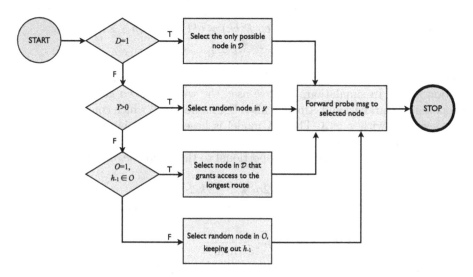

Fig. 3. HR-based network exploration approach. $\mathcal{D}$ is the set of neighbors of the considered peer ($D = |\mathcal{D}|$). $\mathcal{Y}$ is the set of neighbors belonging to the same subnetwork of the considered peer ($Y = |\mathcal{Y}|$), that have not been visited yet. $\mathcal{O}$ is the set of neighbors that belong to other subnetworks ($O = |\mathcal{O}|$). h_{-1} denotes the previous hop.

5   Performance Evaluation

To evaluate the proposed HR-based network exploration algorithm, we adopted the general-purpose discrete event simulation environment called DEUS[2] [22,23], whose purpose is to facilitate the simulation of highly dynamic overlay networks, with several hundred thousands nodes, on a single machine — without the need to simulate also lower network layers (whose effect can be taken into account,

---

[2] DEUS project homepage: https://github.com/dsg-unipr/deus/

in any case, when defining the virtual time scheduling of message propagation events).

To simplify the configuration of the routing tables in simulation, we consider the (sub-optimal) scenario in which every node knows which subnetworks can be reached through its direct neighbors. No further knowledge is necessary, when S is of the same order of magnitude as the mean node degree $\langle k \rangle$ of the network. Instead, for large networks, with $S \gg \langle k \rangle$, further knowledge — provided by neighbors of neighbors (of neighbors etc.) — is necessary to build meaningful collective self-awareness.

We take into account two network topologies, characterized by different statistics for the *node degree*, that is the number of links starting from a node. The node degree of a network is described in terms of probability mass function (PMF) $P(k) = P\{\text{node degree} = k\}$.

The first network topology we consider is *scale-free*, meaning that its PMF decays according to a power law, *i.e.*, a polynomial relationship that exhibits the property of scale invariance (*i.e.*, $P(bk) = b^a P(k), \forall a, b \in \mathbb{R}$), such as:

$$P(k) = ck^{-\tau} \quad \forall k = 0, .., N - 1$$

where $\tau > 1$ (otherwise $P(k)$ would not be normalizable) and c is a normalization factor. In simulation, we have used the widely known generative model proposed by Barabási and Albert (called BA model) [24], which constructs scale-free networks with $\tau \simeq 3$. The BA model is based on two ingredients: growth and preferential attachment. Every node which is added connects to m existing nodes, selected with probability proportional to their node degree. The resulting PMF is

$$P(k) \simeq 2m^2 k^{-3} \quad \forall k > m$$

and the mean node degree is $\langle k \rangle = 2m$. The average path length is

$$\langle l \rangle_{BA} \simeq \frac{\ln N}{\ln \ln N}$$

The second network topology we consider is a purely random one, described by the well-known model defined by Erdös and Rényi (ER model). Networks based on the ER model have N vertices, each connected to an average of $\langle k \rangle = \alpha$ nodes. The presence or absence of a link between two vertices is independent of the presence or absence of any other link, thus each link can be considered to be present with independent probability p. It is trivial to show that

$$p = \frac{\alpha}{N - 1}$$

If nodes are independent, the node degree distribution of the network is binomial:

$$P(k) = \binom{N - 1}{k} p^k (1 - p)^{N-1-k}$$

which, for large values of N, converges to the Poisson distribution

$$P(k) = \frac{\alpha^k e^{-\alpha}}{k!} \qquad \text{with } \alpha = \langle k \rangle = \sigma^2$$

The average path length (*i.e.*, the expected value of the shortest path length between node pairs) is

$$\langle l \rangle_{ER} \simeq \frac{\ln N}{\ln \alpha}$$

Scale-free and ER are the extremes of the range of meaningful network topologies, as they represent the presence of strong hubs and the total lack of hubs, respectively. It has been recently shown that the degree distribution of the Internet is the composition of the contributions of two classes of connections [25]: the provider-customer connections introduce a scale-free distribution, while the peering connections can be modeled using a Weibull distribution. A large contribution to the deviation from the scale-free distribution is given by the Internet Exchange Points (IXPs), that are physical infrastructures which allow ASs to exchange Internet traffic, usually by means of mutual peering agreements, leading to lower costs (and, sometimes, lower latency) than in upstream provider-customer connections. Since IXP introduce a high number of peering relationships, the higher the number of connections identified as crossing IXPs, the larger the deviation of the degree from the scale-free distribution. The structural properties of the Internet AS-level (Autonomous System level) topology graph can be discovered by means of the concept of *community*, which is informally defined as "an unusually densely connected set of ASes" [26]. The task of detecting communities in a graph is very hard for at least two reasons: firstly, because there is no formal definition of a community and, secondly, because most algorithms are computationally demanding and cannot be applied to dense graphs.

Our simulations compared the HR, RW and CF network exploration strategies, with networks consisting of $N = 1000$ nodes, and either $S = 20$ or $S = 100$ subnetworks. With the BA topology, when $m = 5$ and $m = 20$, the mean node degree is $\langle k \rangle = 10$ and $\langle k \rangle = 40$, respectively. To have the same $\langle k \rangle$ values for the ER topology, we set $\alpha = 10$ and $\alpha = 40$. The virtual time VT of the simulation is represented on the x axis. Before $VT = 3000$, the network is grown and configured — *i.e.*, routing tables are filled after $N = 1000$ nodes have been created and connected. Then, at $VT = 3000$ the probe message is generated and forwarded at every virtual time step. Regarding the CF strategy, $TTL = 4$ is necessary to explore the whole network in all the scenarios we consider (which are characterized by $\langle l \rangle_{ER,\alpha=10} \simeq 3$, $\langle l \rangle_{ER,\alpha=40} \simeq 2$ and $\langle l \rangle_{BA} = 3.6$, respectively). Indeed, when the topology is ER with $\alpha = 10$, CF with $TTL = 3$ reaches 69% of the nodes only.

In Figure 4, the fraction of visited nodes (averaged over 25 simulation runs) is reported on the y axis. As the presence of subnetworks does not affect RW and CF, just one curve is plotted for these cases. Conversely, subnetwork awareness plays a fundamental role in HR. From the results we obtained, it is clear that CF allows to explore the whole network much faster than RW and HR. However, CF produces a large burst of probe message forwardings (see num_f in Table 1), to complete the exploration of the network. Instead, HR and RW require fewer messages to achieve the same result (albeit more slowly). In general, HR and RW have to be preferred, with respect to CF. Then, it is also clear that HR outperforms RW. Indeed, for the topologies under consideration, the HR-based

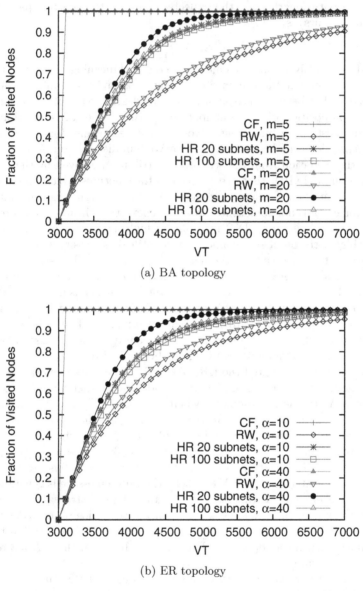

(a) BA topology

(b) ER topology

Fig. 4. Fraction of visited nodes when the network topology is BA or ER.

algorithm requires about 4000 message propagations to visit the whole network. With the same amount of propagations, instead, RW visits only 90% of the network. The difference is more evident when the topology is BA, because using HR the probe message is forced to visit every hub (*i.e.*, one of the most connected nodes) just once, while with RW the probe message visits hubs very frequently.

On the other hand, since neither RW nor HR exploit hubs conveniently, performance is better with the ER topology (where node degree values are more fairly distributed).

Table 1. No-HR vs HR: number of probe message forwardings until the network has been fully explored.

Strategy	Topology	S	num_f
CF	BA, $m = 5$	n.a.	$24 \cdot 10^3$
RW	BA, $m = 5$	n.a.	$> 4 \cdot 10^3$
HR	BA, $m = 5$	20	$4 \cdot 10^3$
HR	BA, $m = 5$	100	$4 \cdot 10^3$
CF	BA, $m = 20$	n.a.	$12 \cdot 10^3$
RW	BA, $m = 20$	n.a.	$> 4 \cdot 10^3$
HR	BA, $m = 20$	20	$3 \cdot 10^3$
HR	BA, $m = 20$	100	$3 \cdot 10^3$
CF	ER, $\alpha = 10$	n.a.	$15 \cdot 10^3$
RW	ER, $\alpha = 10$	n.a.	$> 4 \cdot 10^3$
HR	ER, $\alpha = 10$	20	$3 \cdot 10^3$
HR	ER, $\alpha = 10$	100	$3 \cdot 10^3$
CF	ER, $\alpha = 40$	n.a.	$11 \cdot 10^3$
RW	ER, $\alpha = 40$	n.a.	$> 4 \cdot 10^3$
HR	ER, $\alpha = 40$	20	$2.5 \cdot 10^3$
HR	ER, $\alpha = 40$	100	$3 \cdot 10^3$

6   Conclusions

Collective self-awareness and self-expression, based on the simultaneous application of hierarchy and recursion, make it possible to design efficient and scalable network exploration strategies, with limited extra cost in terms of design complexity.

Other than network exploration, message routing and distributed computing, also distributed sensing, mapping and geo-localization systems may benefit from collective self-awareness and self-expression. The networking and computing research communities have already started studying HR-based strategies, but a lot of work still needs to be done. In our opinion, it will be particularly important to find novel strategies for the efficient maintenance of HR-enabling information.

Consciousness, Self-Awareness and Self-Expression in Psychology and Cognitive Science

Psychology as a scientific discipline started in the second half of the nineteenth century. Almost immediately, the notion of consciousness became central. In 1905, Freud separated the unconscious, the preconscious and the conscious. Since then, several other levels of consciousness have been defined and examined. In 1934, Mead proposed the distinction between focusing attention outward toward the environment (consciousness), and inward toward the self (self-awareness). Self-awareness represents a complex multidimensional phenomenon that comprises various self-domains and corollaries. A very clear and comprehensive survey on this topic has been recently published by Morin [15]. Duval and Wicklund defined self-awareness as *the capacity of becoming the object of one's own attention* [27]. In this state one actively identifies, processes, and stores information about the self. The layered self-awareness model proposed by Neisser [28] is particularly appealing to computer engineers. Indeed, a computational interpretation of that model (illustrated in this paper) has been recently developed by Faniyi *et al.* [14]. The lowest level of self-awareness, denoted as *ecological self*, concerns the reactions to stimuli. The highest level, denoted as *conceptual self*, represents the most advanced form of self-awareness, where the organism is capable of constructing and reasoning about an abstract symbolic representation of itself. Eventually, the conceptual self may allow the organism to become aware of its own self-awareness capabilities (*i.e., meta-self-awareness*). Self-expression, in social psychology literature, is a notion that is closely associated with a multitude of positive concepts, such as freedom, creativity, style, courage, self-assurance. Fundamentally, self-expression is the assertion of one's personal characteristics.

References

1. Bertier, M., Kermarrec, A.-M., Tan, G.: Message-efficient byzantine fault-tolerant broadcast in a multi-hop wireless sensor network. In: IEEE Int'.l Conference on Distributed Computing Systems (ICDCS), Genoa, Italy (2010)
2. Yu, D., Hua, Q.-S., Wang, Y., Yu, J., Lau, F.C.M.: Efficient distributed multiple-message broadcasting in unstructured wireless networks. In: IEEE INFOCOM, Turin, Italy (2013)
3. Khazaei, H., Misic, J., Misic, V.B.: Mobile software agents for wireless network mapping and dynamic routing. In: IEEE Int'.l Conference on Distributed Computing Systems (ICDCS) Workshops, Genoa, Italy (2010)
4. Li, M., Yang, J., An, C., Li, C., Li, F.: IPv6 network topology discovery method based on novel graph mapping algorithms. In: IEEE Symposium on Computers and Communications (ISCC), Split, Croatia (2013)
5. Amoretti, M., Cagnoni, S.: Toward Collective Self-Awareness and Self-Expression in Distributed Systems. IEEE Computer **48**(7), 29–36 (2015)

6. Ni, S., Tseng, Y., Chen, Y., Sheu, J.: The broadcast storm problem in mobile Ad hoc networks. In: ACM MobiCom, Seattle, WA, USA (1999)
7. Garey, M., Johnson, D.: Computers and Intractability: A Guide to the Theory of NP-Completeness. Freeman, San Francisco (1978)
8. Winter, T., et al.: RPL: IPv6 routing protocol for low power and lossy networks. In: RFC 6550, IETF, March 2012
9. Hui, J., Kelsey, R.: Multicast protocol for low power and lossy networks (MPL). In: Work in Progress draft-ietf-roll-trickle-mcast-11, IETF, November 2014
10. La, C.-A., Varga, L.-O., Heusse, M., Duda, A.: Energy-efficient multi-hop broadcasting in low power and lossy networks. In: 17th ACM Int'.l Conference on Modeling, Analysis and Simulation of Wireless and Mobile Systems (MSWiM 2014), Montreal, Canada (2014)
11. Kramer, K.H., Minar, N., Maes, P.: Cooporative mobile agents for dynamic network routing. ACM SIGMOBILE Mobile Computing and Communications Review 3, 12–16 (1999)
12. Houidi, I., Louati, W., Zeghlache, D.: A distributed virtual network mapping algorithm. In: IEEE Int'.l Conference on Communications (ICC), Beijing, China (2008)
13. Shaikh, A., Goyal, M., Greenberg, A., Rajan, R., Ramakrishnan, K.K.: An OSPF Topology Server: Design and Evaluation. IEEE Journal of Selected Areas in Communications 20(4), 746–755 (2002)
14. Faniyi, F., Lewis, P.R., Bahsoon, R., Yao, X.: Architecting self-aware software systems. In: IEEE/IFIP WICSA 2014, pp. 91–94
15. Morin, A.: Self-Awareness Part 1: Definition, Measures, Effects, Functions, and Antecedents. Social and Personality Psychology Compass 5(10), 807–823 (2011)
16. Mitchell, M.: Complex systems: network thinking. Artificial Intelligence 170(18), 1194–1212 (2006)
17. Cabri, G., Capodieci, N., Cesari, L., De Nicola, R., Pugliese, R., Tiezzi, F., Zambonelli, F.: Self-expression and dynamic attribute-based ensembles in SCEL. In: Margaria, T., Steffen, B. (eds.) ISoLA 2014, Part I. LNCS, vol. 8802, pp. 147–163. Springer, Heidelberg (2014)
18. Touch, J., Baldine, I., Dutta, R., Ford, B., Finn, G., Jordan, S., Massey, D., Matta, A., Papadopoulos, C., Reiher, P., Rouskas, G.: A dynamic recursive unified internet design (DRUID). Computer Networks 55(4), 919–935 (2011)
19. Van Meter, R.: Quantum networking and internetworking. In: IEEE Network, July/August 2012
20. Jacobson, V., Smetters, D.K., Thornton, J., Plass, M.F., Briggs, N.H., Braynard, R.L.: Networking named content. In: ACM CoNEXT, Rome, Italy (2009)
21. Amoretti, M.: A Modeling Framework for Unstructured Supernode Networks. IEEE Communications Letters 16(10), 1707–1710 (2012)
22. Amoretti, M., Agosti, M., Zanichelli, F.: DEUS: a discrete event universal simulator. In: 2nd ICST/ACM Int'.l Conference on Simulation Tools and Techniques (SIMUTools 2009), Roma, Italy, March 2009
23. Amoretti, M., Picone, M., Zanichelli, F., Ferrari, G.: Simulating mobile and distributed systems with DEUS and ns-3. In: Int'.l Conference on High Performance Computing and Simulation 2013, Helsinki, Finland, July 2013
24. Barabási, A.-L., Albert, R.: Emergence of Scaling in Random Networks. Science 286(5439), 509–512 (1999)

25. Siganos, G., Faloutsos, M., Krishnamurthy, S., He, Y.: Lord of the links: a framework for discovering missing links in the Internet topology. IEEE/ACM Transactions on Networking **17**(2), 391–404 (2009)
26. Gregori, E., Lenzini, L., Orsini, C.: k-dense Communities in the Internet AS-Level Topology Graph. Computer Networks **57**(1) (2013)
27. Duval, S., Wicklund, A.: A Theory of Objective Self Awareness. Academic Press (1972)
28. Neisser, U.: The roots of self-knowledge: perceiving self, it, and thou. In: Snodgrass, J.G., Thompson, R.L. (eds.) The Self Across Psychology: Self-Recognition, Self-Awareness, and the Self-Concept. Academy of Sciences, New York

Swarm-Based Controller
for Traffic Lights Management

Federico Caselli, Alessio Bonfietti$^{(\boxtimes)}$, and Michela Milano

DISI, University of Bologna, Bologna, Italy
`alessio.bonfietti@unibo.it`

Abstract. This paper presents a Traffic Lights control system, inspired by Swarm intelligence methodologies, in which every intersection controller makes independent decisions to pursue common goals and is able to improve the global traffic performance. The solution is low cost and widely applicable to different urban scenarios. This work is developed within the COLOMBO european project. Control methods are divided into macroscopic and microscopic control levels: the former reacts to macroscopic key figures such as mean congestion length and mean traffic density and acts on the choice of the signal program or the development of the frame signal program; the latter includes changes at short notice based on changes in the traffic flow: they include methods for signal program adaptation and development. The developed system has been widely tested on synthetic benchmarks with promising results.

1 Introduction

Vehicular traffic is among the main plagues of modern cities. The ever increasing number of vehicles, both private and public, sets new challenges in the road network related to traffic optimization. This does not only mean to improve the traffic flow but also aims at reducing pollution and costs of the monitoring infrastructures [1].

Currently, the most common way to sense vehicular traffic is through the use of road deployed sensors. The most common sensor used is the inductive loop, whose installation and maintenance is expensive. To lower installation costs, approaches for traffic management based on V2X vehicular communication technology have started to be investigated. Unfortunately the V2X approaches mainly rely on message exchange between vehicles (V2V), requiring a high penetration rate of equipped vehicles (meaning a high number of V2X enabled vehicles over the total vehicle population) to achieve the desired goals.

One of the objectives of COLOMBO is the design and the development of an innovative, robust[1] and low cost[2] traffic light control system inspired by swarm intelligence methodologies (see [2]). Swarm intelligence is a discipline that studies natural and artificial systems composed of a large number of (typically identical or very similar) individuals called agents, which coordinate with decentralized

---

[1] COLOMBO system can work with very low penetration rates (see Section 5).
[2] Based on V2X technology.

© Springer International Publishing Switzerland 2015
M. Gavanelli et al. (Eds.): AI*IA 2015, LNAI 9336, pp. 17–30, 2015.
DOI: 10.1007/978-3-319-24309-2_2

control and self-organization. Recent research results [3][4] have shown that the principles underlying many natural swarm intelligence systems can be exploited to engineer artificial swarm intelligence systems that show many desirable properties and leads to effective solutions.

Following the principles of Complex Adaptive Systems (CAS), we conjecture that a smart and planned global traffic flow could emerge as the result of local decisions, automatically made by local controllers, executing simple policies in an emergent fashion. Emergent systems are very common in nature and colonies of social insects are one of the most interesting examples for our purposes. Every traffic light controller is a simple agent that controls one or more intersections and operates independently of all other controllers. It relies only on local information coming from the lanes that form the controlled intersection, which are distinguished between incoming and outgoing lanes. This principle is taken from autonomous agents theory, where each agent relies only on local information since there is no central coordination, either by choice or by force.

Following these principles, our system offers unlimited scalability, adaptability to traffic conditions and maximizes road network capabilities, while totally removing at the same time the costs associated to the control center and to the communication infrastructure required in conventional systems.

Our system is based on [5] which is inspired by two academic works: [6] presents local policies able to reach global traffic control through emergence. This is not enough for our purposes, since every policy is thought for a specific traffic density; [7] discusses a mechanism able to choose among different local policies with respect to traffic density, but executes non-reactive policies. Taking into account this work, we developed a control system composed of two levels, with different policies to handle multiple traffic situations in real-time and a high-level policy that selects between them. The system has been extensively tested and compared to a traditional static approach (called *static*) and to a dynamic approach based on inductive loops detections (called *actuated*). Results show that our approach is viable, even in case of low penetration rates.

The rest of the paper is structured as follows: Section 2 presents the related works. Section 3 illustrates the key ideas of the proposed system, which is detailed in Section 4. Then Section 5 presents a wide evaluation of our system. Section 6 concludes pointing out some interesting future development.

2   Related Works

Far from being exhaustive, given the large amount of literature in the field, in the following we present a rapid selection of works which we consider more similar to ours; for a comprehensive survey of existing efforts in the field we refer interested readers to [8].

Urban traffic control systems can be roughly divided into three major categories: centralized, decentralized and fully distributed systems.

- Centralized systems present a unified control center that collects data from the sensors scattered through the city. They have complete knowledge of the controlled network, which is used to create the traffic plan. These systems can also be overridden by traffic experts, if it is necessary. Different solutions basically differ in the evaluation of the control strategy and inherit in some way from the TRANSYT off-line optimization model [9]; for example, SCOOT [10] is a largely deployed centralized solution.

- Decentralized systems present more than one decision-making entity and a master entity that coordinates them. This is the approach adopted by these two systems currently in production: SCATS [11] and UTOPIA [12].
- Fully distributed systems do not have any centralized controller that coordinates or generates traffic plans: every single intersection controller is an independent *agent* that takes its own decision and it is influenced only by its neighborhood. This is an innovative approach without evidence of large scale deployment. Many proposals about distributed solutions exist: they may be agent-based, as in [13], where an agent is in charge of handling the traffic lights in the controlled intersection and performs actions with regard to the local traffic status only, while in other cases the information coming from surrounding agents is also taken into account [14].

Static optimization is not able to adapt to changing needs in traffic. Instead, centralized solutions are able to reach good performances, but are really expensive: they need a unified control center that needs to be connected to every sensor in the network and to every traffic light controller. A system like this one requires high initial installation and maintenance costs. Also, centralized systems do not scale well when used to control big road networks. A partial solution to the scaling problem is given by decentralized system, since they do not require a central controller. Distributed approaches would be simpler and would scale better, but problems may arise for the communication part of these systems. [15] and [16] present this problem in relation to MARL (Multi Agent Reinforcement Learning), which has communication needs that grow exponentially with the number of agents. Communication requires also mutual knowledge, thus reconfiguration of neighboring agents is needed when the topology of the network changes: MARL has no centralized controller but still has high costs due to the communication part.

As previously stated our system is based on [5]. Our controller has enhanced [5] in several key aspects, such as:

- robustness: our system is robust to incomplete traffic information (i.e. different penetration rate), while [5] requires full knowledge.
- dispersion: our system takes into account the non homogeneity of the traffic flows over different lanes of the intersection.
- reliability: the representation of the actual traffic condition is based on the average speed instead of the number of V2X equipped vehicles. In fact, we see that the speed is more robust w.r.t. to changes in the penetration rate.

3   The Concept

Swarm intelligence systems make use of natural metaphors and share a common principle: they present a multitude of simple agents that may be unaware of the system they are part of. Typically, the interactions between the agents are based on alterations of the surrounding environment: this is a form of indirect coordination called *stigmergy* [2].

Our self-organizing system is made up of different independent traffic light controllers. Every controller works in a continuous loop like the one represented in Figure 1. Data about the status of the traffic is acquired by the sensors, translated into *pheromone values* and used as input for the *stimulus functions*. These stimuli are particular functions used to probabilistically determine which policy is the most appropriate to handle the current sensed traffic conditions. The policies are simple rules specifically defined to cope with different traffic conditions. The system also receives feedback from the traffic itself, rewarding or penalizing the choices it takes.

Our proposal abandons the traditional static approach: as implied before, the system decides when it is time to switch to the next phase on the basis of the sensed traffic conditions and not necessarily according to a clock. This makes our system able to react to changes in the traffic density both on the input and on the output lanes of the controlled junction. Communication with the neighboring traffic light controllers is done indirectly through stigmergy exploiting the natural metaphor of the pheromone and without explicit knowledge of the existence of other controllers.

The reason why we chose to forgo a centralized control is that it would be computationally too expensive and difficult to optimize. A centralized system would also need to predict the traffic behavior, which is known to be a hard task since the traffic is a complex system. Moreover, a decentralized system capable of self-organization is simpler to implement and is more reactive to rapidly varying conditions.

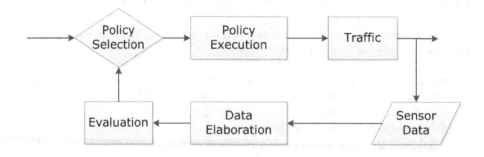

Fig. 1. Traffic lights execution loop.

4 Description of the System

Our system is structured in two levels called *microscopic* and *macroscopic*. The former takes short-term decisions like which lanes should receive green and for how long the green light should be kept while the latter takes more higher-level decisions, like what is the criteria that should be used at low-level. Both judgments are done on the basis of the current traffic conditions.

4.1 Chains

The traditional static execution of a traffic light plan is a continuous loop between phases giving green to a particular set of directions, followed by yellow lights, red lights to all directions and then another green phase to allow traffic transit from a different set of directions. This is done in a static way, which means that green could be kept to lanes even after all waiting vehicles have left or given to lanes with no cars waiting for it. Our idea is that this decision should be taken on-line on the basis of the current traffic conditions. The best moment when this decision can be taken is during the so-called *all-red phase*, which is the phase needed when red is given to all the directions because of safety constraints.

The whole signal sequence can be split in different sub-sequences called *chains*, as shown in Figure 2. The first phase of a chain is called *target phase* and it gives green to a set of lanes identified as *target lanes*. The last phase of a chain, which gives red to all the lanes, is called *commit phase*. When the commit phase is reached, we probabilistically decide if the *microscopic level policy*

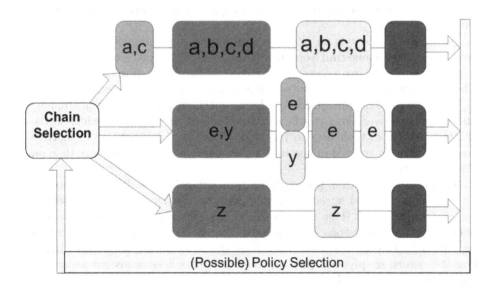

Fig. 2. Chain selection and execution. The letters denote different lanes.

should be changed and, in that case, we proceed by probabilistically selecting it on the basis of the current traffic status. This concludes the selection of the chains. The criteria we use for the selection depends both on the average speed of cars in the incoming lanes in the intersection and on how long these vehicles have been waiting.

Phases are also distinguished between *decisional* and *transient*. The former can have their duration varied between a minimum and a maximum duration time while the latter must be executed for a predetermined amount of time. Transient phases are needed because in some conditions it is not possible to extend a phase. This happens when we have a yellow light phase, whose duration is decided by regulations. The duration of a *decisional* phase is determined by the logic of the currently selected microscopic level policy.

4.2   Macroscopic Level "Swarm" Policy

The goal of the macroscopic level policy, named *Swarm*, is to decide the most appropriate microscopic level policy according to the actual sensed traffic conditions. Since there is some uncertainty in measuring the traffic, e.g. which situations should be interpreted as "high" traffic and which ones should be "low", we rely on the natural metaphor of the *pheromone*.

Pheromone Levels. Measuring the traffic is a hard tasks since there are numerous variables that must be considered, e.g. the number of lanes, the geometry of the intersection and the vehicles' paths. A good measure should also be insensitive to sudden short spikes and be able to react rapidly to more persistent changes. Our idea is to use the natural metaphor of the pheromone to abstract the traffic.

Natural pheromone is an olfactory trail left by some animals. For example, ants leave a pheromone trail from their nest to a source of food along the shortest path between these two points. Pheromone has two interesting characteristics. The first one is that it is additive, so the more ants walk along the same path, the higher the value of the pheromone is. The second characteristic is that pheromone also evaporates over time, this is why ants are able to determine which is the shortest path from their nest to the food: if there are two paths, the shortest one will obviously covered in less time so less pheromone will evaporate than on the other path.

In our implementation, pheromone is proportional to the difference between the maximum allowed speed on a lane and the average speed of the sensed vehicles, limited to 0. This way, the faster the cars, the smaller the values of pheromone, and vice-versa. This approach ignores the number of cars and it is supported by the speed-density relation of the fundamental diagram of traffic flow [17], which simply says that the number of cars does not count as long as the traffic is flowing freely.

We calculate an average value of the pheromone on the *incoming lanes* (i.e. φ_{in}), the ones that enter the controlled intersection, and another average value of the pheromone on the *outgoing lanes* (i.e. φ_{out}), which are the ones that come from the controlled junction.

Note that, averaging the pheromone value, the system cannot be aware of the non homogeneity of traffic density. To avoid this potential drawback we introduce a new pheromone level (i.e. $\varphi_{dsp\ in}$) proportional to the dispersion of the information of the traffic over the incoming lanes. This allows the selection of microscopic policies working better either for homogeneous or non homogeneous traffic conditions. These three pheromone values serve as input for the *stimulus functions*.

Stimulus Function. The stimulus functions, or *stimuli*, are functions associated with the microscopic level policies used to probabilistically determine which policy is more appropriate according to the current traffic conditions. They map their associated policy in a $\varphi_{in} \times \varphi_{out} \times \varphi_{dsp\ in}$ space. This mapping is determined experimentally or via automatic parameter tuning and is used by the Swarm policy to probabilistically select the proper policy given the pheromone on the incoming lanes (φ_{in}), the pheromone on the outgoing lanes (φ_{out}) and the pheromone dispersion over the incoming lanes ($\varphi_{dsp\ in}$).

The idea behind the stimulus function is that the more desirable the associated policy the higher the stimulus should be, given the current pheromone values.

The shape used for the stimulus function is obtained by considering the maximum value of a family of Gaussians. The use of more than one function allows the definition of multiple areas in the pheromone space (i.e. different traffic conditions) where the policy performs best, centering each Gaussian on a different area. As stated before, the parameters defining the shape of the Gaussians are computed off-line via ad-hoc experiments or automatic parameter tuning. The same policy can perform well in different traffic conditions depending on the characteristics of the controlled intersection, so the associated stimulus function may have different parameters for different agents.

The stimulus function must be normalized over its domain. This formulation makes the stimulus function similar to a probability density function, which is important since it expresses the level of specialization of a policy: we want high stimuli in the neighborhood of specialization and a rapid decrease outside it. Figure 3 plots the Gaussians of the stimulus functions for each policy considering only a single function to improve the readability (see Section 4.2 for details).

Fig. 3. Stimulus functions of the microscopic level policies. From left to right: Phase, Platoon, Marching and Congestion.

Policy Selection. The choice of the policy occurs after the evaluation of the stimulus functions in a probabilistic selection fashion. The probability $P(i,j)$ of the i-th agent to choose the j-th policy for its intersection is determined by the following equations:

$$P(i,j) = \frac{T_{\theta,j}(s_{i,j})}{\sum_j T_{\theta,j}(s_{i,j})} \tag{1}$$

$$T_{\theta,j}(s_{i,j}) = \frac{s_{i,j}^2}{s_{i,j}^2 + \theta_{i,j}} \tag{2}$$

Note that the stimulus functions $s_{i,j}$ are not taken into account directly in the calculation of the probability $P(i,j)$. $\theta_{i,j}$ is the sensitivity threshold above which the i-th agent adopts the j-th policy. The sensitivity threshold $\theta_{i,j}$ represents the level of sensitivity of the agent to the adoption of that policy. This threshold is variable in time, decreasing if a policy is selected in a stable way, and increasing in time as a policy is not selected. This is called *reinforcement* because as a policy is selected and found to be working well, it is learned and its probability to be selected again (stable policy) increases. At the same time, even a policy that has a nearly-zero stimulus function will always have a non-zero probability to be selected, because the threshold cannot go above a maximum θ_{max} value, as well as a policy that is stably selected for a long time will not be reinforced too much, as the threshold cannot go below a minimum θ_{min} value. These two values are also bounded: $0 \leq \theta_{min} < \theta_{max} \leq 1$.

The new policy is selected in a probabilistic way: the higher the stimulus function for a policy, the higher the probability that policy will be selected. According to the model given by [18], at each step in simulation, the pheromone levels of the controlled incoming and outgoing lanes are updated and at the end of a commit phase the Swarm selects a new microscopic level policy for execution. The event of selecting another policy is also probabilistic, with probability $0 < p_{change} \leq 1$. It is clear that if $p_{change} = 1$ the process of policy selection occurs every time a commit phase is reached, making the agent highly susceptible to traffic density variations. This probability should be sufficiently low to mitigate the indecision of the agent and to guarantee its reactivity at the same time.

4.3   Microscopic Level Policies

Microscopic level policies take short-term decisions. Usually, they operate on a base sequence of stages and apply variations to this base sequence. The most common decisions taken by an adaptive policy concern the duration of the green phase for different lanes, or including/excluding parts of the whole base sequence.

However, most traffic-dependent policies will always cycle the signal sequence in a pre-fixed order. We have already presented the concept of *chains* in Section 2 and we mentioned the existence of some particular phases called *decisional*. The duration of those phases changes according to the logic of the currently selected microscopic level policy. We are now going to present the implemented microscopic level policies[3]. Most of them make use of a so-called *traffic threshold* as one of the conditions used to end a decisional phase. This is a simple threshold applied to the same indicator we use to determine which chain should be selected.

Phase Policy. This policy will terminate the current chain as soon as another one has reached the traffic threshold, respecting the minimum duration constraint of the decisional phases. In case no other chain wants to be activated, i.e. there is no traffic opposing the current green directions, the current decisional phase is kept indefinitely, regardless of its maximum phase duration.

```
bool canRelease(int elapsed, bool thresholdPassed,
    const MSPhaseDefinition* stage, int vehicleCount){
  if(elapsed >= stage->minDuration)
    return thresholdPassed;
  return false; }
```

This policy is adequate in medium-low traffic situations, where this early termination will not make the traffic lights switch too often.

Platoon Policy. This policy will try to let all the vehicles in the current green directions pass the intersection before releasing the green light. It is similar to *Phase*: as before, a chain is executed in respect of every minimum duration of the decisional phases and until there is another one above the traffic threshold. We also have a third condition: the decisional phase must be executed for its maximum duration time or there must be no other vehicle incoming from the allowed directions.

```
bool canRelease(int elapsed, bool thresholdPassed,
    const MSPhaseDefinition* stage, int vehicleCount){
  if(elapsed >= stage->minDuration)
    if (thresholdPassed)
      return ((vehicleCount == 0) ||
          (elapsed >= stage->maxDuration));
  return false; }
```

This behavior allows the creation of platoons of vehicles, which is positive since it helps to handle the traffic. In intense traffic situations, each decisional phase would be executed for the maximum allowed time, so its definition has a great impact on the performance of the system.

Marching Policy. This policy is adequate when the traffic looks too intense from all directions to take any on-line decision regarding only the incoming lanes. In this case, there are two possible approaches: either use a static duration for decisional stages or consider also the outgoing lanes, not allowing traffic to lanes that are too heavily loaded. This second case may suggest a different, more complex, way to select the chain to execute.

---

[3] These policies were presented in [5].

```
bool canRelease(int elapsed , bool thresholdPassed ,
    const MSPhaseDefinition* stage , int vehicleCount) {
  return (elapsed >= stage−>duration ); }
```

Congestion Policy. This policy is used when the outgoing lanes are saturated and there may be vehicles waiting inside the intersection. In order to avoid grid-locks, all input lanes are inhibited, i.e. the current executing chain is terminated following the pre-defined plan to the commit phase, then no other chain is activated until the congestion has been solved. In doing this, every decisional phase is executed only for their minimum duration time.

As soon as the outgoing lanes are emptying, the pheromone levels will drop allowing Swarm to select another microscopic level policy. This will also enable again the chain selection.

Decay Threshold. With very low penetration rate configurations some of the microscopic level policies may reach a deadlock situation in which there is no chain change. This issue arises whenever the system can not detect any existing vehicles (due to the low penetration rate) on a lane served with red light. The unseen vehicles may wait for green light for a long period of time.

We solved this issue introducing a dynamic traffic threshold, called *decayThreshold*, working as a trigger for the chain selector. The threshold is based upon an exponential decay; it decreases at a rate proportional to its current value following the equation:

$$N(t) = N_0 \exp^{\frac{-t}{\tau}} \tag{3}$$

where $N(t)$ is the quantity at time t and N_0 is the initial quantity, that for our setup was 1. The exponential value $\frac{-t}{\tau}$ is a parameter and should be tuned. This threshold is used in condition that, if satisfied, triggers a chain change.

```
if(random > (1 − decayThreshold))
```

The control is probabilistic. It randomly chooses a number between [0, 1] and it checks the condition. In this way the triggering time is always different, even though it is influenced by the exponential decay value.

5   Evaluation

Several traffic simulators exist in the literature [19]; in COLOMBO we implemented and tested the system inside the microscopic traffic simulator SUMO [20]. The aim of this experimental section is to evaluate the performance of the proposed system, comparing the algorithm with a static and a fully-actuated approach (both implemented in SUMO). We also investigated how our method performs for different penetration rates of equipped vehicles.

The experiments are run on the penetration ratios of 100%, 50%, 25%, 10%, 5%, 2.5% and 1%. These different configurations are obtained not by modifying the number of simulated cars, which would provide unrealistic results, but by assigning to the shadow cars[4] a special SUMO type which allows the simulation of all vehicles but considers only the normal cars when calculating the input information used by the Swarm controller.

We developed a synthetic scenario, composed by 16 traffic lights arranged on 4 by 4 grid, built with the aim to simulate critical traffic conditions. The four central traffic lights are controlled by the Swarm policy while the others use an actuated policy. The structure of each traffic light is based on the German "Richtlinie für Lichtsignalanlagen" (Guidelines for traffic light systems), or RiLSA for short [21]. Simulation have been run using different penetration ratios and measuring how the average waiting time of the vehicles varies (lower values mean better results).

The evaluation scenario is a 4 by 4 grid composed by four horizontal streets and four vertical streets, forming sixteen intersections. The distance between two consecutive neighboring crossing is about 500m. Every street is defined as a single lane, which splits into two lanes near each intersection. The outer traffic lights are used to regulate the traffic flow for the Swarm controlled junctions. This structure gives a good approximation of a real world scenario where the controlled traffic lights are placed in an urban context. The Swarm controllers use the same configuration, however the control of each intersection is independent from the others. The configuration is obtained through off-line automatic parameter tuning executed for every penetration ratio on a simplified road network composed of a single cross-like RiLSA intersection controlled by a Swarm agent with actuated traffic lights at each lane.

Each traffic light controller implements two chains: one that gives green to the south-north and north-south directions and a second one that gives green to the perpendicular directions. Each chain has two green decisional phases: one gives green to the straight, right and left turn directions with by a minimum duration of 10s, a maximum duration of 50s and a default duration of 31s; the other is dedicated to the left turns direction only with a minimum duration of 2s, a maximum duration of 20s and a default duration of 6s. Each chain also has a yellow transient phase of 4s and finally a red commit phase of 4s.

We adopted a traffic generator to create different simulations. Each simulation consists in traffic flows which resembles the traffic patterns as occurring in the real-world by taking into account different realistic daily load curves. The average vehicles per hour of the flows obtained is about 4500v/h.

Figure 4 depicts the performance of the different configurations. For each simulated configuration (x-axis) the figure shows the mean waiting steps (y-axis) obtained by averaging 100 different simulations. Table 1 reports the values outlined in the figure.

---

[4] The vehicles that do not have to be considered on a set penetration ratio. E.g. In a simulation with 20% penetration ratio the shadow vehicles will be the 80%.

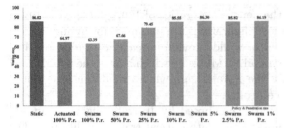

Fig. 4. Average waiting steps for the different configurations.

Table 1. Average waiting steps.

Configuration	Wait steps
Static	86.02
Actuated 100% p.r.	64.97
Swarm 100% p.r.	63.39
Swarm 50% p.r.	67.66
Swarm 25% p.r.	79.45
Swarm 10% p.r.	85.55
Swarm 5% p.r.	86.30
Swarm 2.5% p.r.	85.82
Swarm 1% p.r.	86.19

It is worth mentioning that only our proposal is affected by the low penetration rate of equipped vehicles, since the actuated system relies on inductive loops and the static approach does not sense cars.

At full knowledge our system is slightly better than the actuated approach and it outperforms the static one. Note also that the Swarm algorithm is comparable with the actuated even with 50% penetration rate (which means that our system detects only half of the vehicles).

With very low penetration rates (10% or lower) our system is comparable with the static one. It is finally interesting to highlight that our system performance with very low penetration rates does not degrade with the decrease of the penetration rate.

6   Conclusions

This paper presented a Swarm-based Traffic Lights control system in which every intersection controller makes independent decisions to pursue common goals and is able to improve global traffic performance. This solution is low cost and widely applicable to different urban scenarios. This work is developed within the COLOMBO european project.

The promising results presented in Section 5 show that the proposed approach performance is comparable to more sophisticated systems, like the fully-actuated one (which has full knowledge using the inductive loops detectors). Moreover the experiments showed that the system performance does not degrade depending on the percentage of detectable vehicles.

As future work we plan to (1) adapt the system to include interactions with pedestrians and public transportation and (2) test it on real world scenarios.

References

1. Chen, Y., Richard Yu, F., Zhou, B.: Improving throughput in highway transportation systems by entry control and virtual queue. In: Proceedings of the Third ACM International Symposium on Design and Analysis of Intelligent Vehicular Networks and Applications, DIVANet 2013, pp. 9–14. ACM, New York (2013)

2. Bonabeau, E., Dorigo, M., Theraulaz, G.: Swarm intelligence: from natural to artificial systems, vol. 1. Oxford University Press (1999)
3. Ducatelle, F., Di Caro, G.A., Gambardella, L.M.: Principles and applications of swarm intelligence for adaptive routing in telecommunications networks. Swarm Intelligence 4(3), 173–198 (2010)
4. Ferrante, E., Sun, W., Turgut, A.E., Dorigo, M., Birattari, M., Wenseleers, T.: Self-organized flocking with conflicting goal directions. In: Proceedings of the European Conference on Complex Systems 2012, pp. 607–613. Springer (2013)
5. Slager, G., Milano, M.: Urban traffic control system using self-organization. In: 13th International IEEE Conference on Intelligent Transportation Systems, pp. 255–260 (2010)
6. Gershenson, C.: Self-organizing traffic lights (2004). arXiv preprint nlin/0411066
7. de Oliveira, D., Ferreira Jr, P.R., Bazzan, A.L.C., Klügl, F.: A swarm-based approach for selection of signal plans in Urban scenarios. In: Dorigo, M., Birattari, M., Blum, C., Gambardella, L.M., Mondada, F., Stützle, T. (eds.) ANTS 2004. LNCS, vol. 3172, pp. 416–417. Springer, Heidelberg (2004)
8. Seredynski, M., Arnould, G., Khadraoui, D.: The emerging applications of intelligent vehicular networks for traffic efficiency. In: Proceedings of the Third ACM International Symposium on Design and Analysis of Intelligent Vehicular Networks and Applications, DIVANet 2013, pp. 101–108. ACM, New York (2013)
9. Robertson, D.I.: Transyt: a traffic network study tool (1969)
10. Robertson, D.I., David Bretherton, R.: Optimizing networks of traffic signals in real time: The scoot method. IEEE Transactions on Vehicular Technology 40(1) (1991)
11. Sims, A.G.: The sydney coordinated adaptive traffic system. In: Engineering Foundation Conference on Research Directions in Computer Control of Urban Traffic Systems, 1979, Pacific Grove, California, USA (1979)
12. Peek Traffic. Utopia/spot-technical reference manual. Peek Traffic, Amersfoort,The Netherlands, Tech. Rep (2002)
13. Priemer, C., Friedrich, B.: A decentralized adaptive traffic signal control using v2i communication data. In: 12th International IEEE Conference on Intelligent Transportation Systems, ITSC 2009, pp. 1–6. IEEE (2009)
14. Mizuno, K., Fukui, Y., Nishihara, S.: Urban traffic signal control based on distributed constraint satisfaction. In: Hawaii International Conference on System Sciences, Proceedings of the 41st Annual, pp. 65–65. IEEE (2008)
15. Bazzan, A.L.C.: Opportunities for multiagent systems and multiagent reinforcement learning in traffic control. Autonomous Agents and Multi-Agent Systems 18(3), 342–375 (2009)
16. Busoniu, L., Babuska, R., De Schutter, B.: A comprehensive survey of multiagent reinforcement learning. IEEE Transactions on Systems, Man, and Cybernetics, Part C: Applications and Reviews 38(2), 156–172 (2008)
17. Greenshields, B.D., Channing, W.S., Miller, H.H., et al: A study of traffic capacity. In: Highway research board proceedings, vol. 1935. National Research Council (USA). Highway Research Board (1935)
18. Theraulaz, G., Bonabeau, E., Denuebourg, J.N.: Response threshold reinforcements and division of labour in insect societies. Proceedings of the Royal Society of London. Series B: Biological Sciences 265(1393), 327–332 (1998)

19. Shafiee, K., Lee, J.B., Leung, V.C.M., Chow, G.: Modeling and simulation of vehicular networks. In: Proceedings of the First ACM International Symposium on Design and Analysis of Intelligent Vehicular Networks and Applications, DIVANet 2011, pp. 77–86. ACM, New York (2011)
20. Krajzewicz, D., Erdmann, J., Behrisch, M., Bieker, L.: Recent development and applications of SUMO - Simulation of Urban MObility. International Journal On Advances in Systems and Measurements 5(3&4), 128–138 (2012)
21. FGSV Verlag Forschungsgesellschaft fuer Strassen-und Verkehrswesen. RiLSA - Richtlinien fur Lichtsignalanlagen - Lichtzeichenanlagen fur den Strassenverkehr. FGSV (1999)

Path Relinking for a Constrained Simulation-Optimization Team Scheduling Problem Arising in Hydroinformatics

Maddalena Nonato and Andrea Peano[✉]

Engineering Department, University of Ferrara,
Via Saragat 1, 44121 Ferrara, Italy
{maddalena.nonato,andrea.peano}@unife.it

Abstract. We apply Path Relinking to a real life constrained optimization problem concerning the scheduling of technicians due to activate on site devices located on a water distribution network in case of a contamination event, in order to reduce the amount of consumed contaminated water. Teams travel on the road network when moving from one device to the next, as in the Multiple Traveling Salesperson Problem. The objective, however, is not minimizing travel time but the minimization of consumed contaminated water. This is computed through a computationally demanding simulation given the devices activation times. We propose alternative Path Relinking search strategies exploiting time-based and precedence-based neighborhoods, and evaluate the improvement gained by coupling Path Relinking with state of the art, previously developed, hybrid Genetic Algorithms. Experimental results on a real network are provided to support the efficacy of the methodology.

Keywords: Scheduling · Neighborhood search · Simulation-optimization

1 Introduction

Hydroinformatics is a new, promising, interdisciplinary research field arising at the junction of Hydraulic Engineering and Computer Science, in which complex decision problems related to water management applications are modelled and solved by way of quantitative solution tools developed within well assessed computer science methodological paradigms, such as Constrained Programming on Finite Domain and Mathematical Programming Optimization. Several such examples can be found in the literature which exploit the network based problem structure, taking advantage of solution methodologies already developed for transportation and communication networks, as earlier pointed out by Simonis in a seminal work [34]. Among the most recent contributions, let us mention: the design of the expansion of a Water Distribution System (WDS) combining global search techniques with local search [5], the optimal location on the WDS of water quality sensors in order to early detect water contamination exploiting integer programming based location models [27] and objective function sub-modularity [23], the

© Springer International Publishing Switzerland 2015
M. Gavanelli et al. (Eds.): AI*IA 2015, LNAI 9336, pp. 31–44, 2015.
DOI: 10.1007/978-3-319-24309-2_3

optimal scheduling of devices controlling field irrigation [20] to meet farmers irrigation time demands, the optimal location of isolation valves on the pipes of a WDS to minimize service disruption in case of maintenance operations [7] and [9], and the scheduling of devices activation as a countermeasure to contamination events [10], which is the problem we deal with in this paper.

In all cases, the feasible solutions have to meet complex technological requirements which are modelled by the constraints of the optimization problem, while the objective function describes how the hydraulic system reacts to certain values of the parameters, which are the model variables. Quite often, the system reaction can not be encoded by analytical closed formulas but it is the result of a computational demanding simulation process, which poses a challenge to the development of a solution methodologies able to scale efficiently and tackle real life instances.

In this paper we deal with the last mentioned problem, namely computing the optimal activation time of a set of devices located on the WDS. In case of a contamination event, the devices activation times alter water flow, influence how contaminant spreads in the network, and determine at which concentration contaminant reaches demand nodes where drinking water is consumed by the users. An optimal schedule is a set of activation times which minimizes the volume of consumed contaminated water. A feasible schedule is a set of activation times according to which the teams of technicians, due to manually activate the devices on site, can reach the selected device on time, travelling on the street network when moving from a device to the next. Previous approaches [2], [10], and [11] already improved the state of the art in hydroinformatics [18], where schedules were computed by hand: Genetic Algorithms (GAs) can compute better schedules automatically [28]. Hereby, we build upon previous contributions, and we show how solution approaches for such complicated real life problems can largely benefit from the integration of different search paradigms.

In the rest of the paper, first we introduce the problem and recall the solution strategy based on hybrid GA developed so far, pointing out at some deficiencies. Then, we describe a neighbourhood based search strategy, so called *Path Relinking* (PR) originally proposed by Glover [16], which intensifies the search within a section of the feasible region to which a set of high quality solutions belong. We present how to use PR in our problem, compare two different neighbourhood structures to build the search path connecting two solutions, and asses the efficacy of the approach by experimental results computed on real data for the WDS of a medium size city in Italy, showing how by enhancing our GA with a post optimization PR phase can improve the approach robustness and partially mend the present flaws.

2   Problem Description

WDSs are essential components of our daily life as they bring clean, safe drinking water to customers every day. At the same time, WDSs are among the most vulnerable infrastructures, highly exposed to the risk of contamination by chemical and biological agents, either accidental or intentional. A WDS is a complex arrangement of interconnected pipes, pumps, tanks, hydrants and valves, whose large planimetric extent (a small city network may reach $200km$ and a thousand of pipes and nodes) and sparse topology prevent full surveillance. Therefore monitoring is the only viable alternative. In practice, a set of water quality sensors is located on the WDS to test water safety in real time, looking for the presence of potential contaminants [27]. Their location is strategically determined so that a contamination event is detected as soon as possible, based on a set of contamination scenarios.

Contaminant quickly spreads through the network and population alerting strategies may not entirely ward off users' water consumption. When the network is fully districted, the sector where the alarm is raised can be seamlessly disconnected from the rest of the network, but this is rarely the case. In general, despite of the hazard, water supply can not be completely cut off. The shut down of the entire system would disrupt those security related functions that rely on continuous water supply, such as fire police service or water based cooling systems at large, production intensive, industrial facilities. Therefore, beside population warning procedures, countermeasures devoted to divert the contaminant flow away from high demand concentration sectors must be set up, aiming at mitigating population harm.

An effective way of altering water flow is by activating some of the devices which are part of the system, namely by closing isolation valves and opening hydrants, in order to achieve contaminant isolation, containment, and flushing. In particular, opening hydrants can expel contaminated water, while contaminated pipes can be isolated by closing their isolation valves. Due to the highly non linear functional dependencies that link water flow and the time at which a given device is operated, the global effect of a schedule, i.e., the vector of activation times for the selected devices, can not be decomposed into the sum of the effects of the activation of each individual device, nor the effect of a local change in the schedule can be anticipated. On the contrary, the only way to evaluate the volume of consumed contaminated water due to a schedule is by a computationally intensive simulation. In other words, we are optimizing a black box function and solving a so called simulation-optimization problem [3]. The chosen simulation package is EPANET [32], a discrete event-based simulator which represents the state of the art in Hydraulic Engineering. EPANET is given a schedule, the description of the hydraulic network, the expected water demand, and a contamination scenario, and it yields the volume of contaminated consumed water (we speak of contamination when concentration level is above the danger threshold of $0.3mg/ml$ that causes human death if ingested). Simulation is computationally intensive and it is the bottleneck of any solution approach.

In our case study, each simulation call takes on average $5''$, which poses a limit on the number of function evaluation calls, despite of the fact that the problem is solved offline; this fact influences the choice of the search strategy, as discussed in [11]. Moreover, there is no a priori information on what a good schedule should look like, and common sense inspired criteria such as "the sooner the better" lead to low quality solutions. In conclusion, there is no way to distinguish between a good and a bad schedule without simulation.

So far, it concerns the objective function of our simulation optimization problem. Regarding the solution feasibility, a schedule $t^{\mathcal{F}}$ is feasible provided that there is an assignment of the n devices to the m teams available and, for each team, its devices can be sequenced so that if device j follows device i the difference between the respective activation times in the schedule is equal to τ_{ij}, i.e., the travelling time on the street network from the location of device i to the location of device j. All teams gather at the mobilization point at a given time after the alarm is raised (according to the protocol) and conventionally the departure time is set to 0. This maps the feasible region of our problem into the one of a well known optimization problem, the *open m-Travelling Salesman Problem* [4] (mTSP), providing a graph representation of our problem where the mobilization point is the depot, each device is a client node of the graph and each team visits the assigned devices travelling along a route starting from the depot.

All these features motivated our choice of a Genetic Algorithm (GA) hybridized with a Mixed Integer Linear Programming solver which encapsulates the feasibility constraints within the cross over operators, as described in detail in [11]. In that paper, several computational experiments were carried out to calibrate population size and number of generations for a single GA run. Moreover, we verified the poor quality of the solutions obtained according to heuristic criteria, such as minimum makespan or minimizing the sum of the activation times. Besides, neighbourhood based local search were tested and proved to be not competitive given the limited number of solution evaluations allowed, since the search trajectory remains confined not far from the starting point. On the contrary, the literature confirms that in such cases population based heuristics, which carry on a broader search and are able to explore a wider part of the feasible region, are able to provide better results.

Although we could improve by far and large the best solutions available for our case study, that solution approach has a typical GA flaw, that is, it converges to a local optimum which depends on the starting population. However, the differences among different solutions quality varies depending on the contamination scenario. Since the number of function evaluations is limited, in this study we address the question concerning what is the best way of exploiting the computational resources we are allowed, and if there is a way to take advantage of the knowledge of a set of different, high quality solutions.

In the next section we describe how PR can provide a pattern search that fulfils these expectations.

3   Intensification by Path Relinking

As mentioned, in this application GAs often yield high quality solutions that depend on the initial population: this is due to the existence of several local optima. These different solutions identify a promising subregion of the feasible space, which is worth further inspection, according to some exploration strategy. Classical Local Search transforms a solution gradually: at each step it moves from a solution to an improving one in the current neighbourhood, driven by the objective function. In our case, the search goes from one local optimum to another, by gradually making the current solution more similar to the final one. This search is not guided by the objective function, but rather by a *distance* criterion, and quite often a better solution is found along this search trajectory. This philosophy lies at the heart of an intensification technique named PR [16].

Working on a *reference set* (*rs*) composed of several solutions, PR first selects from *rs* an *initial* reference (*r*) and a *guiding* (often called *target*) (*g*) solution, then it iterates valid moves to transform step by step *r* into *g*. Figure 1 shows a graphical representation of the transformation of *r* into *g*, differing initially on 4 elements; so 3 intermediate solutions are selected, namely r^1, r^2, and r^3. This procedure allows for the exploration of the path between two good solutions, according to the hypothesis that a better one can be found among the feasible solutions in the middle.

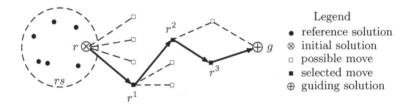

Fig. 1. Graphical representation of Path Relinking

Since PR builds a new solution starting from the features of two elite solutions, it can be also seen as an evolutionary algorithm, in which randomness is substituted by a deterministic search strategy that draws the possible path between two feasible solutions.

The building blocks of a Path Relinking algorithm are:

– the reference set and its construction;
– the reference and the target solutions and their selection;
– the path between two solutions, i.e., the neighbourhood structure.

Several variants and generalizations of PR are possible, which are elegantly discussed in [17], such as truncating the search on a path to resume it on another (either new or existing) path of the same $g - r$ couple, or different policies for choosing the move in the current neighbourhood rather than moving on the best one. In this work we adopt this last classical strategy since we prefer to spend the limited number of solution evaluations to explore the "best" path between different $r - g$ pairs rather than several paths between the same $r - g$ pair.

In our case, the bunch of best populations given by some GA runs provides naturally the dataset the reference set can be built up from, the target g can be easily selected as the best solution in *rs*; and the reference r has to be selected properly through quality and diversity criteria, as Section 3.1 reports.

3.1 Selection of Reference Candidates

As stated before, the reference set can be built up from the final populations of the GAs. In particular, diversity from the target beside quality should be taken into account, since the number of inspected solutions grows with the distance to the target; thus, different metrics can be combined together to filter properly the initial dataset.

The distance between two solutions can be evaluated considering the routes of the teams as well as the activation times. Despite in the former studies the diversity is measured on the graph representation of the routing problems [21, 29–31, 35], in this case the preferred way is to measure the diversity over the time representation. In fact the graph representations of the solution would introduce a huge amount of redundancy [6], this means that the same vector of activation times can be mapped into different trees that may differ a lot wrt the metrics defined for graph representations.

The metrics here proposed for the time representation are the Hamming distance

$$h(g, r) = \sum_{i=1}^{N_{dev}} d_i \tag{1}$$

where g_i (r_i) is the activation time of device i in solution g (r) and $d_i = 1$ if $g_i \neq r_i$ while $d_i = 0$ otherwise, and the euclidean distance

$$e(g, r) = \sqrt{\sum_{i=1}^{N_{dev}} (g_i - r_i)^2} \tag{2}$$

between two vectors of activation times g and r. The former gives a measure about how many elements differ in the vectors, whereas the latter measures how much the vectors differ in terms of activation times. In order to prevent the inclusion of too similar vectors in rs, two thresholds β and γ are defined: given the target g a solution r is included only if $h(g, r) \geq \beta$ and $e(g, r) \geq \gamma$.

As far as the quality of the reference set, a proper metric is to consider only solutions having quality within a certain percentage distance δ from the target's one, i.e., $\frac{q(r) - q(g)}{q(g)} \leq \delta$ holds for any $r \in rs$.

Finally, the choice of β, γ, and δ is really important, even more whenever the number of evaluations is limited. In fact, in this case excluding a promising solution may affect hugely the effectiveness of the approach.

3.2 A Path Relinking Based on Sequences

Path Relinking for routing problems works on symbolic representations of routes, in which any route is expressed by an ordered set of visited customers [21, 30, 35]. For 3 vehicles v_1, v_2, and v_3, and 7 customers, namely $c_1, \ldots, c_7$, a feasible solution assigns a route to each vehicle, e.g., $v_1 = \{c_1, c_2\}$, $v_2 = \{c_3, c_4\}$, $v_3 = \{c_5, c_6, c_7\}$. Equal solutions visit the customers along same routes. To transform a solution into a different one, every customer should be *relocated* into the right position of the right route. In Path Relinking for routing problems, this is done iteratively by relocating one customer at each step.

The same representation can be used in this mTSP variant. Since the routes have no names, the order of the devices within the routes is the valuable information; moreover, in a route any device precedes only one other, thus the order of activation within the routes can be stated by listing the devices' *predecessors*, i.e., a list of tuples "(h_p, h_s)" meaning that the device h_p precedes h_s in the solution. For example, in the solutions in Figure 2, 3 teams work overall on 7 hydraulic devices, namely $1, \ldots, 7$; the *initial* solution r and the guiding g can be represented by the following predecessor lists:

$$P_r = \{(d, 3), (d, 4), (d, 6), (1, 5), (4, 7), (6, 1), (7, 2)\},$$

$$P_g = \{(d, 3), (d, 4), (d, 6), (1, 7), (3, 5), (4, 1), (5, 2)\}.$$

In general, two solutions a and b are equal iff $P_a = P_b$; whereas, whenever two solutions differ, the predecessors of a that are not in b are $P_{a-b} = P_a \setminus P_a \cap P_b$, vice versa for b is $P_{b-a} = P_b \setminus P_a \cap P_b$. Moreover, the cardinality $card(P_{a-b})$ measures the distance of a from b, and vice versa being $card(P_{a-b}) = card(P_{b-a})$. For example, for r and g we have that $P_{r-g} = \{(1, 5), (4, 7), (6, 1), (7, 2)\}$, $P_{g-r} = \{(1, 7), (3, 5), (4, 1), (5, 2)\}$; so, the distance between r and g is 4.

To get closer to b starting from the configuration of a, at least one device $h_s \mid (h_p, h_s) \in P_{b-a}$ should be relocated after its predecessor h_p in b; in this sense, the set P_{b-a} contains the possible moves to transform a into b. For example, $(5, 2) \in P_{g-r}$ means that the device 2 needs to be relocated right after 5, making r more similar and closer to g. This means that the neighbourhood of r with respect to g is the set of solutions $\mathcal{N}(r, g) = \{r_n \mid card(P_{g-r} \setminus P_{g-r_n}) = 1\}$. In other words, the neighbourhood of r with respect to g is the set of solutions r_n obtained by relocating 1 device in r according with P_{g-r}.

At the k-th iteration, PR has to choose which device $h_s \mid (h_s, h_p) \in P_{g-r}^k$ is relocated, in order to move to r^{k+1}. To make this choice it evaluates by EPANET every $r_n^k \in \mathcal{N}^k$, and moves to the one with the lowest volume; this solution becomes the reference of the next step, and it is called r^{k+1}.

Every time a device has been relocated after its new predecessor their link becomes permanent and no further moves can break it. Thus, whenever a device is chosen to be relocated after another one, it carries the following chain of fixed edges along with it; this prevents the current choice to destroy the previous ones. To implement this behaviour the procedure should be enriched with a memory, which stores the previous moves.

Figure 2 shows a possible path from r to g consisting of 4 intermediate steps. Table 1 reports at any step the values of P_{g-r}, the chosen move to r^{k+1}, and the fixed edges, for the example in Figure 2. The first move transforms the initial solution into r^1 by relocating 2 after 5; this edge is now fixed and this move is stored into the memory, represented by a the dashed box. The second move from r^1 to r^2 relocates the chain $5 - 2$ after 3. Then 7 is relocated after 1 achieving

r^3. Finally 1 is relocated together with its fixed successor 7 after 4. The target is finally reached.

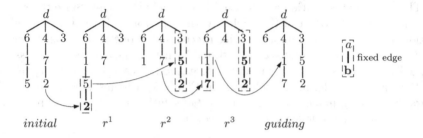

Fig. 2. Feasible References and Target solutions for 7 devices differing on 4 predecessors

Table 1. Iterations of the PR algorithm for the example in Figure 2

k r^i	P_{g-r}^i	move	fixed edges
0 r	$\{(1,7),(3,5),(4,1),(5,2)\}$	$(5,2)$	$\{\}$
1 r^1	$\{(1,7),(3,5),(4,1)\}$	$(3,5)$	$\{(5,2)\}$
2 r^2	$\{(1,7),(4,1),\}$	$(1,7)$	$\{(5,2),(3,5)\}$
3 r^3	$\{(4,1)\}$	$(4,1)$	$\{(5,2),(3,5),(1,7)\}$
4 g	$\{\}$		$\{(5,2),(3,5),(1,7),(4,1)\}$

Recall that in this real application only solutions with 3 routes are considered to be feasible. So we exclude from P^k the moves that vary the number of routes, i.e., moves that either empty a route or add a new route.

This version of PR moves at most N_{dev} times and calls EPANET at most $\frac{N_{dev}(N_{dev}+1)}{2}$ times. Sometimes the procedure visits the same solution twice or more, in such a case the solution would be evaluated by EPANET more times, wasting precious computing resources. For this reason the solving architecture is enriched with a cache, which stores the explored solutions and allows for saving a call to EPANET. It is worth noting that the procedure may explore the entire path between *initial* and *guiding* before the maximum number of EPANET calls expires. In such a case, the procedure selects another reference, and iterates over it. The algorithm continues until it reaches the maximum number of EPANET calls or reference solutions.

From now, we refer to this version as the "routing" PR (PRr).

3.3   The Time Based Variant

Another representation for the mTSP encodes a solution as a vector of activation times [11]. Given the feasible solutions r and g, the indexes of the differing elements is given by $I_{r-g} = \{i \mid r_i \neq g_i\}$. If r equals g then $I_{r-g} = \emptyset$. To transform r into g iteratively, at each step k one element in I_{r-g} should be fixed to its value in g. This decreases by one the distance between r^k and g. Let r^k be the reference vector at the k-th step, $I^k = I_{r^k-g}$, and let $F^k = I_{r-g} \setminus I^k$ be the set of indexes that have been already fixed. The next solution r^{k+1} in the path between r and g is obtained by keeping $r^{k+1}_f = g_f$ for all $f \in F^k$ and fixing the new element $r^{k+1}_i = g_i$ for one $i \in I^k$.

If the remaining elements of r^{k+1} were the same as in r (or r^k) the resulting vector could not correspond to a feasible schedule. So, these elements are chosen by solving a constrained optimisation problem whose constraints depict a mTSP, the elements in $F^k \cup \{i\}$ are fixed, whereas the other (non-fixed) elements are the actual integer variables of the program; the objective is to optimise these variables, so that their values are as close as possible to the ones in r. To do that, the program minimizes the Euclidean distance of the non-fixed elements from r [11], i.e., given $i \in I^k$:

$$dist(r^k, r) = \sum_{j \in I^k \setminus \{i\}} |r^{k+1}_j - r_j| \tag{3}$$

Notice that a feasible vector always exists, being g a feasible solution of the program. The neighbourhood of r^k is then defined as follows:

$$\mathcal{N}^k = \{r^{k+1} \mid card(I^k \setminus I^{k+1}) = 1, \text{minimizes (3)}\}.$$

The procedure explores every solution by varying the index $i \in I^k$, and for each i it calls the optimiser to compute a new feasible vector, finally it evaluates the solution by calling EPANET. The solution in N^k having the lowest volume is selected to be the reference solution for the next step. Figure 3 shows, on a graph with 7 devices, how routes change when an additional element becomes fixed; e.g., in r^6 the activation time "(26)", which was (20) in r^5, has been fixed,

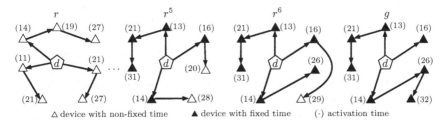

Fig. 3. Feasible routes and activation times for the solution r, r^5, r^6, and g

and the related device is now visited by another route. The minimization of $dist(r^6, r)$ also transforms (28) in r^5 into (29) in r^6, by changing the route of the concerned device; this new activation time is clearly very close to (27) in r.

The constrained optimisation program minimizing (3) can be stated by any declarative paradigm, such as Constraint Programming [8], Constraint Logic Programming [22], Answer Set Programming [15, 24], Mixed Integer Linear Programming [26]; so some suitable solvers are: Gecode [14], ECLiPSe [33], DLV [25], Clasp [12, 13], SCIP [1], Gurobi [19].

Since the number of feasible moves decreases at each iteration of at least one unit, the maximum number of EPANET calls is again $\frac{N_{dev}(N_{dev}+1)}{2}$. Also this PR uses a cache to store the explored solution, so it ends up whenever either no more EPANET calls are available, or rs is empty.

Notice how this technique integrates MILP, hydraulic simulators, and PR into the solving architecture; thus, we will refer to it as the "hybrid" PR (PRh).

3.4   Computational Results

The experiments were performed on the Ferrara's hydraulic network, which supplies drinking water to about 120, 000 inhabitants. 20 contamination scenarios $(A \dots T)$ were tested. Basing on the techniques proposed in [18], 3 teams of technicians were considered to be available to operate on 13 hydraulic devices, namely 7 valves and 6 hydrants. The hydraulic simulator we used was EPANET [32], and takes about 5 seconds to evaluate a schedule of the selected devices on each contamination scenario. Even though EPANET is open-source, the simulation procedures and the network specifications are sensitive data for hydraulic engineers and can not be disclosed.

Genetic Algorithms proposed in [11] were allowed a maximum of 500 EPANET calls, and the population was sized to 20 individuals. These values were calibrated in previous works, and the GAs typically converge within the 500 EPANET calls. As mentioned, we observed that the final solution depends on the initial population as the GA gets stuck on different local optima, so parallel small sized independent GAs explore the search space better than one big sized GA. In this study, Path Relinking is tested to explore the region enclosing such solutions.

The hypothesis tested hereby is whether either PRr or PRh may improve the best solution starting from the final populations of 10 independent GAs; in other words, the reference set was built up from the 10 final populations. The two PRs were compared to an additional independent GA run, to be considered a strengthening run of the same first 10. In this way all the approaches are directly comparable. PRr, PRh, and the additional GA were equipped with 500 EPANET calls each; the total amount of calls is then 5500 for any configuration. To weaken the randomness, the tests were repeated 10 times on each contamination scenario. To disambiguate, these runs are considered to be *global*, wrt the 10 *local* GA runs. So we denote the best of the c-th trial of 10 GAs as s_c^*, while the global optimum is defined as $s^* = \min_c\{s_c^*\}$. The constrained model minimizing (3) was implemented in Mixed Integer Linear Programming and solved by Gurobi; its solving time was negligible.

Table 2 reports the best volumes computed by the different approaches for each scenario. The additional GA (+1 GA) never improves s^*; this is quite

Table 2. The Table reports for each scenario and in this order: the averaged volume of contaminated water in litres (l), the ratio between variance and averaged volume, and the best volume (l) for 10 independent *global* runs of 10 GA; it also reports for 10 independent *global* runs of +1 GA, PRr and PRh: the best volume (l), the number of improvements, the averaged improvement in l; last row reports the average of some of these columns; $min(best)$, $max(impr.\sharp)$, and $max(impr.ave)$ are highlighted in bold.

scen	10 GA			+1 GA			PRr			PRh		
	↓ (sorting key)				impr.			impr.			impr.	
	ave l	$\frac{var}{ave}$	best (s^*) l	best l	$\sharp$	ave l	best l	$\sharp$	ave l	best l	$\sharp$	ave l
A	6,022	0.04	6,000	6,000	0	0	**5,997**	8	9	6,000	7	9
B	7,170	0.10	7,170	7,170	0	0	7,170	1	2	**7,156**	5	14
C	10,868	1.51	10,672	10,672	0	0	**10,569**	3	49	10,623	7	47
D	11,229	1.16	11,021	11,021	0	0	11,021	0	0	**10,993**	7	44
E	12,732	0.21	**12,698**	**12,698**	1	5	**12,698**	1	4	12,698	3	15
F	13,938	0.76	13,793	13,793	1	2	**13,624**	4	69	13,723	7	44
G	15,841	0.22	15,758	15,758	0	0	15,758	4	29	**15,692**	8	57
H	16,991	2.44	16,571	16,571	1	3	**15,708**	7	207	16,351	9	137
I	20,792	7.21	**20,122**	**20,122**	0	0	**20,122**	2	50	**20,122**	5	22
J	22,273	0.39	22,164	22,164	0	0	22,164	2	8	**22,105**	9	85
K	25,138	0.56	**25,043**	**25,043**	0	0	**25,043**	2	21	**25,043**	7	68
L	35,067	1.00	34,662	34,662	0	0	34,662	4	136	**34,536**	7	120
M	36,706	0.52	**36,706**	**36,706**	0	0	**36,706**	1	2	**36,706**	5	103
N	40,121	4.74	39,230	39,230	1	21	39,230	4	121	**39,128**	10	215
O	42,019	1.68	**41,595**	**41,595**	0	0	**41,595**	0	0	**41,595**	6	79
P	44,470	0.34	44,286	44,286	1	10	44,286	0	0	**44,188**	2	13
Q	46,452	1.11	46,175	46,175	1	2	46,175	0	0	**46,144**	8	137
R	52,531	1.47	52,210	52,210	1	15	52,210	3	57	**52,205**	5	77
S	77,397	0.16	77,232	77,232	0	0	77,232	2	21	**76,999**	6	123
T	144,622	0.07	144,409	144,409	1	8	144,409	2	24	**144,350**	8	82
ave					0	3		3	38		7	76

expected because the additional GA follows the same exploration pattern as any other GA. Anyway, for 8 scenarios, for one c of 10 the additional GA improves s_c^*. On average PRr improves s_c^* 3 times out of 10, while PRh does it 7 times. For 5 scenarios out of 20 neither PRr nor PRh were able to improve s^*; in 4 scenarios (A,C,F, and H), PRr outperforms PRh in terms of global best (s^*), and only for one of these scenarios PRh was not able to improve s^*. On the contrary, in 11 scenarios PRh improves s^* whereas PRr doesn't. Only in scenario A PRr improves s_c^* more times than PRh. Moreover, PRh on average improves the s_c^* twice as many times as PRr (see last row), decreasing the volume of contaminated water than double the PRr (76 vs 38).

Notice that, the higher is the averaged volume the higher is the outperforming rate of PRh wrt +1 GA and PRr. In fact, from the scenario M onwards,

PRh outperforms the others in terms of global best (s^*), averaged number of improvements and averaged improvement in volume.

The variance of s_c^*, whose normalization over the average is given by $\frac{var}{ave}$ in Table 2, is not correlated to number of improvements the PRs may achieve. In fact, PR is able to improve s^* even for scenarios whose variance is low; this happens mostly when distant local optima have similar quality. Also, since PR's exploration capability grows with the distance between r and g, PR should be always coupled to strengthen parallel Genetic Algorithms, even when low variance would suggest that no further improvement is possible.

Finally, despite of the fact that PRh is generally better performing than PRr, there is not a real dominance (A,C,F,H), which suggests to integrate both techniques in future works.

4   Conclusions

Genetic Algorithms are used to optimise the scheduling of operations in case of contamination events in Water Distribution Systems [11]; the final populations may contain distant solutions both in terms of similarity and quality. A local search paradigm can improve the solutions by exploiting the knowledge about these local optima.

Two Path Relinking (PR) variants have been developed and tested for a real life hydraulic network, namely the Ferrara's one, to optimise the scheduling of 3 technicians teams over a set of 13 among valves and hydrants, with the aim of reacting to contamination events and minimizing the volume of contaminated water consumed by the users. 20 contamination scenarios were simulated and tested. The hydraulic simulator EPANET was used to compute the volume given a scheduling of the devices; since EPANET takes about 5 seconds to evaluate each solution, we tackled with a computationally intensive consuming simulation-optimisation problem.

A PR was developed to optimise the *routes* of the teams and was named PRr. The other was developed to design directly the activation times of the devices; it was named PRh, from *hybrid*, because it exploits solvers for constrained optimisation programs to compute feasible times; a Mixed Integer Linear Programming implementation was used in this specific case. The two PRs proved to be very effective in improving solutions' quality starting from the final populations of parallel Genetic Algorithms. In the future, more sophisticated PR variants will be tested, e.g., truncated PR, greedy randomized adaptive PR, and others.

Acknowledgments. We thank Stefano Alvisi and Marco Franchini for assistance with the hydraulic simulator and the instance they developed, and for the fruitful discussions about the hydraulic engineering.

References

1. Achterberg, T.: SCIP: Solving constraint integer programs. Mathematical Programming Computation **1**(1), 1–41 (2009). http://mpc.zib.de/index.php/MPC/article/view/4

2. Alvisi, S., Franchini, M., Gavanelli, M., Nonato, M.: Near-optimal scheduling of device activation in water distribution systems to reduce the impact of a contamination event. Journal of Hydroinformatics **14**(2), 345–365 (2012)
3. April, J., Glover, F., Kelly, J.P., Laguna, M.: Simulation-based optimization: practical introduction to simulation optimization. In: Proceedings of the 35th Conference on Winter Simulation: Driving Innovation, WSC 2003, pp. 71–78. Winter Simulation Conference (2003)
4. Bektas, T.: The multiple traveling salesman problem: an overview of formulations and solution procedures. Omega **34**(3), 209–219 (2006)
5. Bent, R., Coffrin, C., Judi, D., McPherson, T., van Hentenryck, P.: Water distribution expansion planning with decomposition. In: 14th Water Distribution Systems Analysis Conference, WDSA 2012, 24–27 September 2012 in Adelaide, South Australia, p. 305. Engineers Australia (2012)
6. Carter, A.E., Ragsdale, C.T.: A new approach to solving the multiple traveling salesperson problem using genetic algorithms. European Journal of Operational Research **175**(1), 246–257 (2006)
7. Cattafi, M., Gavanelli, M., Nonato, M., Alvisi, S., Franchini, M.: Optimal placement of valves in a water distribution network with CLP(FD). Theory and Practice of Logic Programming **11**(4–5), 731–747 (2011)
8. Frühwirth, T., Abdennadher, S.: Essentials of Constraint Programming. Springer (2003)
9. Gavanelli, M., Nonato, M., Peano, A.: An ASP approach for the valves positioning optimization in a water distribution system. Journal of Logic and Computation (2013, in press). doi:10.1093/logcom/ext065
10. Gavanelli, M., Nonato, M., Peano, A., Alvisi, S., Franchini, M.: Genetic algorithms for scheduling devices operation in a water distribution system in response to contamination events. In: Hao, J.-K., Middendorf, M. (eds.) EvoCOP 2012. LNCS, vol. 7245, pp. 124–135. Springer, Heidelberg (2012)
11. Gavanelli, M., Nonato, M., Peano, A., Alvisi, S., Franchini, M.: Scheduling countermeasures to contamination events by genetic algorithms. AI Communications **28**(2), 259–282 (2015)
12. Gebser, M., Kaufmann, B., Kaminski, R., Ostrowski, M., Schaub, T., Schneider, M.: Potassco: The Potsdam Answer Set Solving Collection. AI Communications **24**(2), 107–124 (2011)
13. Gebser, M., Kaufmann, B., Schaub, T.: Conflict-driven answer set solving: From theory to practice. Artificial Intelligence **187–188**, 52–89 (2012)
14. Gecode Team. Gecode: Generic constraint development environment (2006). http://www.gecode.org
15. Gelfond, M.: Answer sets. In: van Harmelen, F., Lifschitz, V., Porter, B. (eds.) Handbook of Knowledge Representation, chap. 7, pp. 285–316. Elsevier Science (2008)
16. Glover, F., Laguna, M., Martí, R.: Fundamentals of scatter search and path relinking. Control and Cybernetics **39**, 653–684 (2000)
17. Gonçalves, J.F., de Magalhães Mendes, J.J., Resende, M.G.C.: A hybrid genetic algorithm for the job shop scheduling problem. European Journal of Operational Research **167**(1), 77–95 (2005)
18. Guidorzi, M., Franchini, M., Alvisi, S.: A multi-objective approach for detecting and responding to accidental and intentional contamination events in water distribution systems. Urban Water **6**(2), 115–135 (2009)
19. Gurobi Optimization, Inc., Gurobi optimizer reference manual (2014). http://www.gurobi.com

20. Haq, Z.U., Anwar, A.A.: Irrigation scheduling with genetic algorithms. Journal of Irrigation and Drainage Engineering **136**(10), 704–714 (2010)
21. Ho, S., Gendreau, M.: Path relinking for the vehicle routing problem. Journal of Heuristics **12**(1–2), 55–72 (2006)
22. Jaffar, J., Maher, M.J.: Constraint logic programming: A survey. Journal of Logic Programmig **19**(20), 503–581 (1994)
23. Krause, A., Leskovec, J., Guestrin, C., VanBriesen, J., Faloutsos, C.: Efficient sensor placement optimization for securing large water distribution networks. Journal of Water Resources Planning and Management **134**(6), 516–526 (2008)
24. Leone, N.: Logic programming and nonmonotonic reasoning: from theory to systems and applications. In: Baral, C., Brewka, G., Schlipf, J. (eds.) LPNMR 2007. LNCS (LNAI), vol. 4483, pp. 1–1. Springer, Heidelberg (2007)
25. Leone, N., Pfeifer, G., Faber, W., Eiter, T., Gottlob, G., Perri, S., Scarcello, F.: The DLV system for knowledge representation and reasoning. ACM Transactions on Computational Logic (TOCL) **7**(3), 499–562 (2006)
26. Martin, R.: Large Scale Linear and Integer Optimization: A Unified Approach. Springer, US (1999)
27. Murray, R., Hart, W., Phillips, C., Berry, J., Boman, E., Carr, R., Riesen, L.A., Watson, J.-P., Haxton, T., Herrmann, J., Janke, R., Gray, G., Taxon, T., Uber, J., Morley, K.: US environmental protection agency uses operations research to reduce contamination risks in drinking water. Interfaces **39**(1), 57–68 (2009)
28. Peano, A.: Solving Real-Life Hydroinformatics Problems with Operations Research and Artificial Intelligence. PhD thesis, University of Ferrara (2015)
29. Prins, C., Prodhon, C., Calvo, R.: Solving the capacitated location-routing problem by a grasp complemented by a learning process and a path relinking. 4OR **4**(3), 221–238 (2006)
30. Rahimi-Vahed, A., Crainic, T., Gendreau, M., Rei, W.: A path relinking algorithm for a multi-depot periodic vehicle routing problem. Journal of Heuristics **19**(3), 497–524 (2013)
31. Reghioui, M., Prins, C., Labadi, N.: GRASP with path relinking for the capacitated arc routing problem with time windows. In: Giacobini, M. (ed.) EvoWorkshops 2007. LNCS, vol. 4448, pp. 722–731. Springer, Heidelberg (2007)
32. Rossman, L.A.: EPANET 2 users manual. National Risk Management Research Laboratory, Office of research and development, U.S. Environmental Protection Agency, USA (2000)
33. Schimpf, J., Shen, K.: Ecl^{i}ps^e - from LP to CLP. TPLP **12**(1–2), 127–156 (2012)
34. Simonis, H.: Constraint applications in networks. Handbook of constraint programming **2**, 875–903 (2006)
35. Sörensen, K., Schittekat, P.: Statistical analysis of distance-based path relinking for the capacitated vehicle routing problem. Computers & Operations Research **40**(12), 3197–3205 (2013)

Dynamical Properties of Artificially Evolved Boolean Network Robots

Andrea Roli[1]($\boxtimes$), Marco Villani[2], Roberto Serra[2], Stefano Benedettini[1],
Carlo Pinciroli[3], and Mauro Birattari[4]

[1] Department of Computer Science and Engineering,
Alma Mater Studiorum Università di Bologna, Bologna, Italy
andrea.roli@unibo.it
[2] Department of Physics, Informatics and Mathematics,
Università di Modena e Reggio Emilia & European Centre for Living Technology,
Venice, Italy
[3] MIST, École Polytechnique de Montreal, Montreal, Canada
[4] IRIDIA-CoDE, Université libre de Bruxelles, Brussel, Belgium

Abstract. In this work we investigate the dynamical properties of the
Boolean networks (BN) that control a robot performing a composite
task. Initially, the robot must perform phototaxis, i.e. move towards a
light source located in the environment; upon perceiving a sharp sound,
the robot must switch to antiphototaxis, i.e. move away from the light
source. The network controlling the robot is subject to an adaptive walk
and the process is subdivided in two sequential phases: in the first phase,
the learning feedback is an evaluation of the robot's performance in
achieving only phototaxis; in the second phase, the learning feedback
is composed of a performance measure accounting for both phototaxis
and antiphototaxis. In this way, it is possible to study the properties of
the evolution of the robot when its behaviour is adapted to a new operational requirement. We analyse the trajectories followed by the BNs in
the state space and find that the best performing BNs (i.e. those able
to maintaining the previous learned behaviour while adapting to the
new task) are characterised by generalisation capabilities and the emergence of simple behaviours that are dynamically combined to attain the
global task. In addition, we also observe a further remarkable property:
the complexity of the best performing BNs increases during evolution.
This result may provide useful indications for improving the automatic
design of robot controllers and it may also help shed light on the relation
and interplay among robustness, evolvability and complexity in evolving
systems.

1 Introduction

Genetic regulatory networks (GRNs) model the interaction and dynamics
among genes. From an engineering and computer science perspective, GRNs are
extremely interesting because they are capable of producing complex behaviours,
notwithstanding the compactness of their description. Cellular systems are also

© Springer International Publishing Switzerland 2015
M. Gavanelli et al. (Eds.): AI*IA 2015, LNAI 9336, pp. 45–57, 2015.
DOI: 10.1007/978-3-319-24309-2_4

both robust and adaptive, i.e. they can maintain their basic functions in spite of damages and noise, and they are able to adapt to new environmental conditions. Such a complex behaviour can be interpreted from an artificial system design's viewpoint, suggesting the possibility of achieving robust and adaptive behaviours in agents, robots, and group of robots, by exploiting the properties of GRN models.

Among the most studied models for GRNs, are Boolean networks (BNs), first introduced by Kauffman [11]. A BN is a discrete-time discrete-state dynamical system whose state is a N-tuple in $\{0,1\}^N$, $(x_1, \ldots, x_N)$. The state is updated according to the composition of N Boolean functions $f_i(x_{i_1}, \ldots, x_{i_{K_i}})$, where K_i is the number of inputs of node i, which is associated to Boolean variable x_i. Each function f_i governs the update of variable x_i and depends upon the values of variables $x_{i_1}, \ldots, x_{i_{K_i}}$. Most works on BNs deal with so-called *autonomous* networks, i.e. systems that evolve in time without input from the external—at most, only the initial state may be extrenally imposed. Usually, BNs are subject to a deterministic, synchronous and parallel node update, even if other update schemes are possible [28]. In the synchronous and deterministic update scheme, every state has a unique successor and the trajectory is composed of a transient and a state cycle (possibly a fixed point, i.e. a cycle of length 1).

BNs have received considerable attention in the community of complex system science. Works in complex systems biology show that BNs provide powerful model for cellular dynamics [26,29], cellular differentiation [25,31] and interactions among cells and environment [24]. A specific dynamical regime at the boundaries between order and chaos, called the *critical* regime, is of notable interest. Critical networks enjoy important properties, such as the capability of optimally balancing evolvability and robustness [1] and maximising the average mutual information among nodes [21]. Hence the conjecture that living cells, and living systems in general, are critical [17].

In recent works, it has been shown that such kind of BNs can be used to control robots [6,22,23]. In this case, the BN evolution in time also depends on the values of some "input" nodes which are set depending on the robot's sensor readings. The BN is trained by means of a learning algorithm that manipulates the Boolean functions. The algorithm employs as learning feedback a measure of the performance of the BN-controlled robot (in the following, BN-robot) on the task to perform. The effectiveness of this approach was demonstrated through experiments on both simulated and real robots.

In this contribution, we outline some results on the analysis of the BN-robot's dynamics along the learning process. We analyse the trajectories followed by the BN-robot in the space of BN states and compute significant features, such as state number and frequency of state occurrence in sample trajectories. In addition, we compute the number of *fixed points*, i.e. BN states repeated as long as the BN inputs do not change. The number of fixed points is an indicator of the generalisation capabilities of the system, as they represent simple functional building blocks of the type `while <condition> do <action>`, which compose the overall system dynamics. Moreover, we estimate the statistical complexity of the system by means of a complexity measure called the *LMC complexity* [15]. The dynamics of a complex system is neither totally disordered (as an ideal gas at equilibrium), nor perfectly ordered (as a crystal); therefore we expect that a measure of the distance of a system from these two conditions should have very high values when the system exhibits complex behaviours. While we are of course aware of the fact that there is no general agreement on an all-encompassing definition of a measure of complexity, LMC seems particularly interesting in this case, as it will be discussed in Section 3.

We found that the successful performing BN-robots, which show the capability of attaining the learned behaviours also in spite of noise and perturbations (*robustness*) while adapting to new tasks to perform (*evolvability*), are characterised by both number of fixed points and LMC complexity higher than those of unsuccessful ones. These preliminary results may provide useful indications for improving the automatic design of robot controllers and may help shed light on the relation and interplay among robustness, evolvability and complexity in evolving systems.

The structure of the paper is as follows. After a summary of the experimental setting in Section 2, we discuss the main results of the analysis of the dynamics of the BNs controlling the robot in Section 3 and we conclude with a discussion and an outlook to future work in Section 4.

2   Experimental Setting

In this experiment, we control an *e-puck* robot [16] by means of a BN. The values of a set of network nodes (BN input nodes) are imposed by the robot's sensor readings, and the values of another set of nodes (BN output nodes) are observed and used to encode the signals for maneuvering the robot's actuators. The BN controlling the robot is subject to synchronous and parallel update. As described in the following, the Boolean functions are set by a search process, whilst the topology of networks is set at random.[1] The sensors consists of four light sensors and one sound sensor, while the actuators correspond to right and left wheel speed controllers. The Boolean values of the output nodes are sent to wheel actuators after a preprocessing consisting in a moving average, so as to feed the motors with signals in the range [0,1]. The robot is placed in a random position and with random orientation in a squared arena, with one light source in a corner. The BN-robot must accomplish the following task: initially, it must perform phototaxis, that is, move towards the light source; upon perceiving a sharp sound, the BN-robot must switch to antiphototaxis, that is, move away from the light source.[2] The robot is trained in simulation by means of an *adaptive walk*: the process starts from a randomly generated BN, it iteratively mutates its functions and keeps only the changes that either improve the BN-robot's performance or do not decrease it. Mutation is implemented by

---

[1] The choice for a random topology is not a limitation, as discussed in [22].

[2] A video of a typical run of a best performing BN-robot is available at https://www.youtube.com/watch?v=6ZF9Ijpwkd8.

randomly choosing a node and an entry in its Boolean function truth table and flipping it. The algorithm is therefore a stochastic descent in the space of Boolean functions. [3] Advanced search strategies can of course be devised so as to attain a higher performance; nevertheless, this subject is beyond the scope of this paper.

The BN-robot is trained in two sequential phases. In the first phase, the learning feedback is an evaluation of the robot's performance in achieving only phototaxis. In the second phase, the learning feedback is composed of a performance measure accounting for both phototaxis and antiphototaxis. In this way, we can study the properties of the evolution of the BN-robot when its behaviour has to be adapted to a new operational requirement. We define the performance of a BN-robot as a function of an error $E \in [0, 1]$. The smaller is the error, the better is the robot performance. The error function is given by a weighted sum of phototaxis and antiphototaxis errors: at each time step $t \in \{1, \ldots, T\}$, the robot is rewarded if it is moving in the correct direction with respect to the light. Let t_c be the time instant at which the clap is performed. The error function E is defined as follows:

$$E = \alpha \left(1 - \frac{\sum_{i=1}^{t_c} s_i}{t_c} \right) + (1 - \alpha) \left(1 - \frac{\sum_{i=t_c+1}^{T} s_i}{T - t_c} \right),$$

where:

$$\forall i \in \{1, \ldots, t_c\}, \ s_i = \begin{cases} 1 & \text{if the robot goes towards to the light at step } i \\ 0 & \text{otherwise} \end{cases}$$

$$\forall i \in \{t_c+1, \ldots, T\}, \ s_i = \begin{cases} 1 & \text{if the robot moves away from the light at step } i \\ 0 & \text{otherwise} \end{cases}$$

In the first phase of the training is $\alpha = 1$, whilst in the second phase α is set to 0.5 so as to take into account both phototaxis and antiphototaxis.

One hundred independent runs of the entire training process were executed,[4] starting from 100 initial BNs generated at random (with 20 nodes, 3 inputs per node and no self-connections).

During the training process BN-robots are subject to random perturbations, so as to train them also for operating in noisy environments. Along the training process we tested the BN-robot and collected statistics on the BN states traversed.

The experiments in simulation have been run by means of the open source simulator ARGoS [19].

3   Analysis of BN Dynamics

A significant fraction of the training experiments—about 30%—leads to a successful BN-robot, i.e. a robot able to perform both phototaxis and antiphototaxis

---

[3] Details can be found in [22].

[4] The experiments reported in [22] were re-run so as to have a greater number of replicas.

and to switch between the first and the second behaviour when it perceives a sharp sound signal. The unsuccessful BN-robots are either able to perform photo-taxis only or not even that task. In the successful cases, the phototaxis capability acquired by the BN-robot in the first training phase is maintained while also the antiphototaxis behaviour is learned. Whence these systems have the possibility of successfully balancing robustness and evolvability.[5]

A question may rise at to what extent topology affects the results: after visual inspection of a sample of the BNs we discover that topology has an impact only in pathological cases, such as complete disconnection of all sensors or actuators. Notably, one of the best performing BNs has a topology in which the South light sensor is disconnected, which means that the network was anyway able to integrate this piece of information.

We studied the properties of the BN trajectories as they control the robot during its actions. The BN that controls a robot is coupled with the environment, as some of its nodes are forced to values imposed by the sensors and some of its outputs control the robot actuators (the wheels in this case). As a consequence, the network itself is *embodied* and its dynamics must be studied in the operational setting in which the robot is functioning, characterised by a specific sensors–actuators loop mediated by the environment. Therefore, we studied the dynamics of such BNs by means of the properties of their trajectories in the state space collected during robot runs. More precisely, for each BN-robot we run the robot starting from 1000 different initial conditions and recorded the sequence of BN states the network traverses during the run. This collection of state sequences is then merged into a graph whose vertices are the network states traversed by the BN and the edges the observed transitions between two states (see a typical trajectory graph in Figure 1). Moreover, the frequency of occurrence of states in the trajectories is recorded. This information is used to compute several features of the BN dynamics, which will be detailed in the following. The statistics that will be shown are computed by subdividing the BN-robots into three classes: BN-robots able to attain correctly the task (*both* class, about 30/100), BN-robots able to perform phototaxis only (*pt* class, about 60/100) and totally failing robots (*none* class, about 10/100).

3.1   Number of States

The number of unique states in the collection of trajectories—i.e. the number of different states in the set collecting all the states in the sampled trajectories— is an indicator of the size of the state space the BN dynamics occupies, as it represents the portion of state space actually explored by the BN. The smaller this size, the greater the generalisation capability of the network. Indeed, a large number of unique states denotes BN trajectories that do not overlap, which in turn means that the network has simply learned collections of successful examples. Conversely, a small number of unique states denotes trajectories that share

---

[5] We use the terms robustness and evolvability with the same meaning as in the work by Aldana et al. [1]

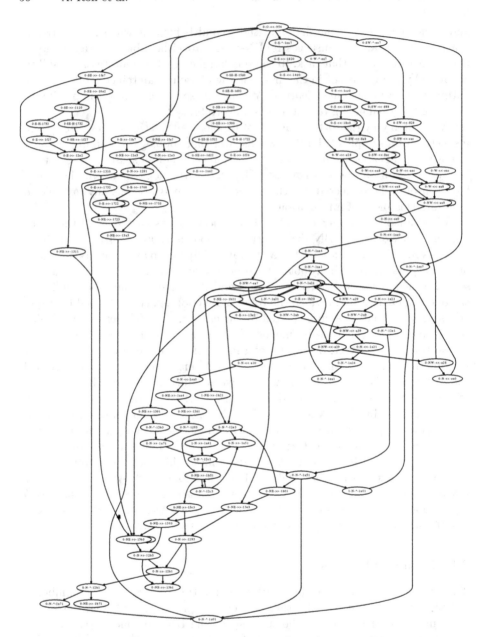

Fig. 1. Typical trajectory graph of a BN-robot. Transitions between nodes occur either for internal network state update or caused by input change. Node labels—not relevant for this context—denote the encoded binary state of the network.

a large fraction of transitions, which is a property of a system that was able to generate a compact model of the world.

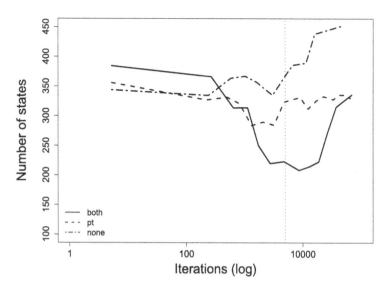

Fig. 2. Average number of different states in the trajectory collections as a function of learning algorithm's iteration. Averages are taken across three different behaviour classes: *both* class ↔ successful BN-robots, *pt* class ↔ BN-robots able to perform phototaxis only, *none* class ↔ failing BN-robots.

In Figure 2 the average number of states in the trajectory collection for each class of robots is plotted along the training phase. The dashed vertical line denotes iteration 5000 at which the objective function was changed so as to include also the evaluation on the antiphototaxis behaviour. We observe that the successful BN-robots are characterised by a decreasing number of unique states up to iteration 5000, when the BN is forced to accomplish a more complex behaviour and the number of states starts to increase, meaning that the training process is still acting so as to adapt the BN-robot to achieve the compound task. BN-robots able to perform phototaxis only show a similar but far less marked pattern, whilst—as expected—worse BN-robots show no tendency to generalise.

3.2   Number of Fixed Points

Some states in BN-robot trajectories are repeated until a change occurs in the input. With slight abuse of term, we call these states *fixed points*. These states represent simple functional building blocks of the type `while <condition> do <action>` (e.g. "turn right until the light input changes") which are combined to achieve a global behaviour. The emergence of fixed points reveals that the BN is able to extract regularities in the environment and to classify them.

The curves in Figure 3 show that the average number of fixed points in the successful BN-robots increases with training and it consistently increases when the more complex task has to be learned. Instead, the BN-robots of the other

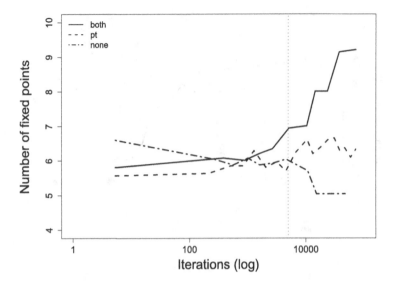

Fig. 3. Average number of fixed points as a function of learning algorithm's iteration. Averages are taken across three different behaviour classes: *both* class ↔ successful BN-robots, *pt* class ↔ BN-robots able to perform phototaxis only, *none* class ↔ failing BN-robots.

two classes maintain approximately the same number of fixed points along the training.

3.3   Statistical Complexity

An analysis of the trajectories of a system may also be focused to capture a further notable dynamical property, which is usually called *statistical complexity* [3,7,8,15,20,27]. This quantity is aimed at estimating to what extent a system works at the edge of order and disorder, i.e. in critical regime. Critical regimes may provide an optimal trade-off between reliability and flexibility, i.e. they make the system able to react consistently with the inputs and, at the same time, capable to provide a sufficiently large number of possible outcomes. This conjecture has been introduced with the expression "computation at the edge of chaos" [4,13,18] and it is supported by results on different computational models such as ϵ-machines [30], cellular automata [9], and neural networks of different kinds [2,12,14].

A system that does not change in time (i.e. in the ordered regime), as well as a system characterised by random behaviour (i.e. in the disordered regime) should be evaluated with low complexity. High complexity is expected to characterise systems in the critical regime accomplishing non trivial tasks. Several measures have been proposed [20] to account for *statistical complexity* (SC), i.e., the algorithmic complexity of a program that reproduces the statistical proper-

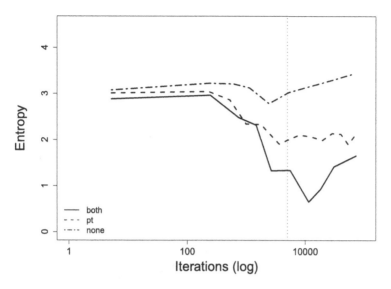

Fig. 4. Entropy of the BN controller as a function of the learning algorithm's iteration. Averages are taken across three different behaviour classes: *both* class ↔ successful BN-robots, *pt* class ↔ BN-robots able to perform phototaxis only, *none* class ↔ failing BN-robots.

ties of a system. In this light, the SC of both a constant and a random sequence is low.

Among various measures of SC, we have chosen a simple yet effective one, which is called LMC complexity, by the name of its inventors [15]. The idea is rather simple: if we want the SC of a system to be high in intermediate regions between order and disorder, we can define it as the product of a measure that increases with disorder and another which decreases with it. The first measure is the Shannon *entropy*, computed over the frequency of the states traversed by the BN-robot. If the BN-robot traverses states $x \in X$ with probabilities $P(x)$ estimated by means of their frequencies, the entropy is defined as:

$$H(X) = - \sum_{x \in X} P(x) \, log P(x)$$

In the definition of $H(X)$ we assume $0 \, log 0 = 0$.

The second measure contributing to SC is *disequilibrium*:

$$D(X) = \sum_{x \in X} \left(P(x) - \frac{1}{|X|} \right)^2$$

The disequilibrium estimates the extent to which a system exhibits patterns far from equidistribution. For example, if the trajectory of a system is composed of only few of the possible states (e.g., a short cyclic attractor), then it has a high disequilibrium.

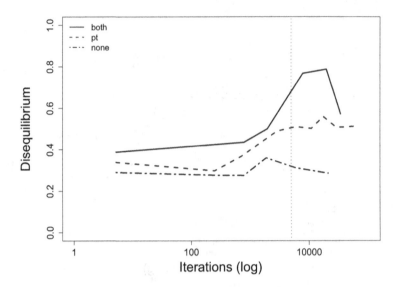

Fig. 5. Disequilibrium of the BN controller as a function of the learning algorithm's iteration. Averages are taken across three different behaviour classes: *both* class ↔ successful BN-robots, *pt* class ↔ BN-robots able to perform phototaxis only, *none* class ↔ failing BN-robots.

Finally, the LMC complexity is defined as:

$$C(X) = H(X) \cdot D(X)$$

A high entropy means that the sequences of states in the BN trajectories are highly diversified. Conversely, a high disequilibrium among the states characterises trajectories mostly composed of the repetition of few states. It is conjectured that a complex system operates in a dynamical regime such that a balance between these two quantities is achieved [15].

It is quite informing to separately observe the three measures, namely entropy, disequilibrium and complexity. In Figure 4 the entropy of BN controllers is shown along the adaptive process. As in previous graphs, the average value for the three performance classes is plotted. Notably, the entropy of well performing BN-robots decreases up to the fitness function change, providing evidence that the adaptive process is successfully achieving generalisation of the task. At iteration 5000, when the fitness function is change so as to include also antiphototaxis, the entropy starts to increase as the BN is adapting to the new task to be accomplished. The reason for this increase has to be ascribed to the adaptive process which does not seem to be completed for all the best performing BNs at the 10000th iteration. The entropy of BN robots that do not perform the complete task shows instead a different behaviour, as it just slightly decreases in the case of BN-robots performing phototaxis only, while it even increases for the worst performing BN-robots.

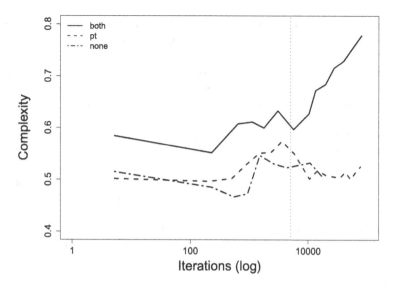

Fig. 6. Complexity of the BN controller as a function of the learning algorithm's itera-tion. Averages are taken across three different behaviour classes: *both* class ↔ successful BN-robots, *pt* class ↔ BN-robots able to perform phototaxis only, *none* class ↔ failing BN-robots.

Disequilibrium shows a complementary behaviour with respect to entropy, as illustrated by Figure 5. Finally, as shown in Figure 6, the complexity of the successful BN-robots increases steadily during the training process, whilst it is almost constant for the unsuccessful ones. This result supports the conjecture that complexity characterises systems that perform non-trivial tasks [4,15]. Nev-ertheless, this point deserves further investigations, especially to be compared with previous work on similar subjects [5,10] in which the relation between fit-ness and complexity is addressed.

4   Conclusion and Future Work

The main finding of the analysis of the trajectories of BN-robots is that the networks that optimally balance robustness and evolvability are characterised by generalisation capability and high statistical complexity of their trajectories. Even if preliminary, these results suggest that also artificial systems that has to cope with changing environments may have an advantage in enjoying the same properties. In the settings in which this hypothesis turned out to hold, additional information for both training and analysing these systems would be available. In particular, the evaluation of features such as fixed points and complexity may be profitably incorporated into the objective function of the adaptive process, with the goal of guiding it towards high performing networks.

Conversely, experiments on simple artificial systems provide a controlled envi-ronment for studying general properties of living systems. The use of BNs and their trajectories make it possible to link results in digital worlds with biological ones, as these models have been proven to capture relevant biological phenomena.

The results presented in this paper concern preliminary experiments on the subject, which may be further investigated in several directions. The robustness of the results against changes in the search strategy and input and output encoding should be assessed.

In the next future, we plan to investigate the relation between complexity measures and performance of BN-robots in noisy and varying environments. First of all, this is expected to provide guidelines for the automatic design of truly adaptive robotic systems; furthermore, we aim at contributing elucidate the elusive interplay among complexity, robustness and evolvability.

Acknowledgments. We thank the anonymous referees who carefully read the paper and provided pertinent and valuable suggestions for preparing this final version.

References

1. Aldana, M., Balleza, E., Kauffman, S., Resendiz, O.: Robustness and evolvability in genetic regulatory networks. Journal of Theoretical Biology **245**, 433–448 (2007)
2. Bertschinger, N., Natschläger, T.: Real-time computation at the edge of chaos in recurrent neural networks. Neural Computation **16**, 1413–1436 (2004)
3. Crutchfield, J.: The calculi of emergence: Computation, dynamics, and induction. Physica D **75**, 11–54 (1994)
4. Crutchfield, J., Young, K.: Computation at the onset of chaos. In: Complexity, Entropy, and Physics of Information. Addison Wesley (1990)
5. Edlund, J., Chaumont, N., Hintze, A., Koch, C., Tononi, G., Adami, C.: Integrated information increases with fitness in the evolution of animats. PLOS Computational Biology **7**(10), e1002236:1–e1002236:13 (2011)
6. Garattoni, L., Roli, A., Amaducci, M., Pinciroli, C., Birattari, M.: Boolean network robotics as an intermediate step in the synthesis of finite state machines for robot control. In: Liò, P., Miglino, O., Nicosia, G., Nolfi, S., Pavone, M. (eds.) Advances in Artificial Life, ECAL 2013, pp. 372–378. The MIT Press (2013)
7. Gell-Mann, M., Lloyd, S.: Information measures, effective complexity, and total information. Complexity **2**(1), 44–52 (1996)
8. Grassberger, P.: Randomness, information, and complexity, August 2012. arXiv:1208.3459
9. Hordijk, W.: The EvCA project: A brief history. Complexity **18**, 15–19 (2013)
10. Joshi, N., Tononi, G., Koch, C.: The minimal complexity of adapting agents increases with fitness. PLOS Computational Biology **9**(7), e1003111:1–e1003111:10 (2013)
11. Kauffman, S.: The Origins of Order: Self-Organization and Selection in Evolution. Oxford University Press, UK (1993)
12. Kinouchi, O., Copelli, M.: Optimal dynamical range of excitable networks at criticality. Nature Physics **2**, 348–351 (2006)
13. Langton, C.: Computation at the edge of chaos: Phase transitions and emergent computation. Physica D **42**, 12–37 (1990)

14. Legenstein, R., Maass, W.: Edge of chaos and prediction of computational performance for neural circuit models. Neural Networks **20**, 323–334 (2007)
15. Lopez-Ruiz, R., Mancini, H., Calbet, X.: A statistical measure of complexity. Physics Letters A **209**, 321–326 (1995)
16. Mondada, F., Bonani, M., Raemy, X., Pugh, J., Cianci, C., Klaptocz, A., Magnenat, S., Zufferey, J.C., Floreano, D., Martinoli, A.: The e-puck, a robot designed for education in engineering. In: Gonçalves, P., Torres, P., Alves, C. (eds.) Proceedings of the 9th Conference on Autonomous Robot Systems and Competitions, vol. 1, pp. 59–65 (2009)
17. Nykter, M., Price, N., Aldana, M., Ramsey, S., Kauffman, S., Hood, L., Yli-Harja, O., Shmulevich, I.: Gene expression dynamics in the macrophage exhibit criticality. In: Proceedings of the National Academy of Sciences, USA, vol. 105, pp. 1897–1900 (2008)
18. Packard, N.: Adaptation toward the edge of chaos. In: Dynamic Patterns in Complex Systems, pp. 293–301 (1988)
19. Pinciroli, C., Trianni, V., O'Grady, R., Pini, G., Brutschy, A., Brambilla, M., Mathews, N., Ferrante, E., Di Caro, G., Ducatelle, F., Birattari, M., Gambardella, L., Dorigo, M.: ARGoS: a modular, multi-engine simulator for heterogeneous swarm robotics. Swarm Intelligence **6**(4), 271–295 (2012)
20. Prokopenko, M., Boschetti, F., Ryan, A.: An information-theoretic primer on complexity, self-organization, and emergence. Complexity **15**(1), 11–28 (2008)
21. Ribeiro, A., Kauffman, S., Lloyd-Price, J., Samuelsson, B., Socolar, J.: Mutual information in random Boolean models of regulatory networks. Physical Review E **77**, 011901:1–011901:10 (2008)
22. Roli, A., Manfroni, M., Pinciroli, C., Birattari, M.: On the design of boolean network robots. In: Di Chio, C., Cagnoni, S., Cotta, C., Ebner, M., Ekárt, A., Esparcia-Alcázar, A.I., Merelo, J.J., Neri, F., Preuss, M., Richter, H., Togelius, J., Yannakakis, G.N. (eds.) EvoApplications 2011, Part I. LNCS, vol. 6624, pp. 43–52. Springer, Heidelberg (2011)
23. Roli, A., Villani, M., Serra, R., Garattoni, L., Pinciroli, C., Birattari, M.: Identification of dynamical structures in artificial brains: an analysis of boolean network controlled robots. In: Baldoni, M., Baroglio, C., Boella, G., Micalizio, R. (eds.) AI*IA 2013. LNCS, vol. 8249, pp. 324–335. Springer, Heidelberg (2013)
24. Serra, R., Villani, M.: Modelling bacterial degradation of organic compounds with genetic networks. Journal of Theoretical Biology **189**(1), 107–119 (1997)
25. Serra, R., Villani, M., Barbieri, A., Kauffman, S., Colacci, A.: On the dynamics of random Boolean networks subject to noise: Attractors, ergodic sets and cell types. Journal of Theoretical Biology **265**(2), 185–193 (2010)
26. Serra, R., Villani, M., Semeria, A.: Genetic network models and statistical properties of gene expression data in knock-out experiments. Journal of Theoretical Biology **227**, 149–157 (2004)
27. Shalizi, C.: Methods and techniques of complex systems science: An overview, March 2006. arXiv:nlin/0307015
28. Shmulevich, I., Dougherty, E.: Probabilistic Boolean Networks: The Modeling and Control of Gene Regulatory Networks. SIAM, Philadelphia (2009)
29. Shmulevich, I., Kauffman, S., Aldana, M.: Eukaryotic cells are dynamically ordered or critical but not chaotic. PNAS **102**, 13439–13444 (2005)
30. Strogatz, S.: Nonlinear dynamics and chaos. Perseus Books Publishing (1994)
31. Villani, M., Serra, R.: On the dynamical properties of a model of cell differentiation. EURASIP Journal on Bioinformatics and Systems Biology **4**, 1–8 (2013)

Adaptive Tactical Decisions in Pedestrian Simulation: A Hybrid Agent Approach

Luca Crociani, Andrea Piazzoni, Giuseppe Vizzari$^{(\boxtimes)}$, and Stefania Bandini

CSAI - Complex Systems & Artificial Intelligence Research Center, University
of Milano-Bicocca, Viale Sarca 336/14, 20126 Milano, Italy
{luca.crociani,giuseppe.vizzari,stefania.bandini}@disco.unimib.it,
andrea.piazzoni@campus.unimib.it

Abstract. Tactical level decisions in pedestrian simulation are related
to the choice of a route to follow in an environment comprising several
rooms connected by gateways. Agents are supposed to be aware of the
environmental structure, but they should also be aware of the level of
congestion, at least for the gateways that are immediately in sight. This
paper presents the tactical level component of a hybrid agent architecture
in which these decisions are enacted at the operational level by mean of a
floor-field based model, in a discrete simulation approach. The described
model allows the agent taking decisions based on a static a-priori knowledge of the environment and dynamic perceivable information on the
current level of crowdedness of visible path alternatives.

Keywords: Pedestrian simulation · Tactical level · Hybrid agents

1 Introduction

Simulation is one of the most successful areas of application of agent-based
approaches: models and techniques employed by researchers in different disciplines are not necessarily in line with the most current results in the computer
science and engineering (see, e.g., [2]), and yet the area still presents interesting
opportunities for agent research and computer science in general. Pedestrians
and crowds simulation is an example of this situation: both the automated analysis and the synthesis of pedestrian and crowd behavior, as well as attempts to
integrate these complementary and activities [13], present open challenges and
potential developments in a smart environment perspective [11].

Modeling human decision making activities and actions is an extremely challenging goal, even if we only consider choices about walking behavior: different
types of decisions are taken at different levels of abstraction: [12][1] provides a well-
known scheme to model the pedestrian dynamics, describing 3 levels of behavior:
(i) **Strategic level**, managing abstract plans and final objectives motivating the
overall decision to move (e.g. "I am going to the University today to follow my

---

[1] A similar classification can be found in vehicular traffic modeling from [10].

© Springer International Publishing Switzerland 2015
M. Gavanelli et al. (Eds.): AI*IA 2015, LNAI 9336, pp. 58–71, 2015.
DOI: 10.1007/978-3-319-24309-2_5

courses and meet my friend Paul"); (ii) **Tactical level**, constructing sequences of activities to achieve the defined objectives (e.g. "I'll take the 7:15 AM train from station X, get off at Y and then walk to the Department, then ... "); (iii) **Operational level**, physically executing the defined plans (i.e. creating a precise walking trajectory, such as a sequence of occupied cells and related simulation turn in a discrete simulation).

Most of the literature has been focused on the reproduction of the physics of the system, so on the lowest level: this is partly due to the fact that data on the fundamental diagram achieved with different set of experiments and in different environment settings (see, e.g[15]) supports a robust validation of the models. Relevant recent works, such as [7] and [14], start exploring the implications of tactical level decisions during evacuation. In particular, [7] modifies the floor-field Cellular Automata approach for considering pedestrian choices not based on the shortest distance criterion but considering the impact of congestion on travel time. [14] explores the implications of four strategies for the route choice management, given by the combination of applying the shortest or quickest path, with a local (i.e., minimize time to vacate the room) or global (i.e., minimize overall travel time) strategy. The global shortest path is calculated with the well-known Floyd-Warshall algorithm, implying computational times that can become an issue by having a large number of nodes or by considering special features in the simulated population (i.e. portion of the path where the cost differs from an agent to another). The work in this paper will propose an alternative and efficient approach to find a global path, where each agent will be able to consider additional costs in sub-paths without adding particular weight to the computation.

We must emphasize the fact that the measure of success and validity of a model is not the *optimality* with respect to some cost function, as in robotics, but the *plausibility*, the similarity of results to data acquired by means of observations or experiments. Putting together *tactical* and *operational* level decisions in a comprehensive framework, preserving and extending the validity that, thanks to recent extensive observations and analyses (see, e.g., [4]), can be achieved at the operational level, represents an urgent and significant open challenge.

The following Sect. will present the adaptive tactical level part of the model whereas its experimental application in benchmark scenarios showing the adequacy in providing adaptiveness to the contextual situation will be given in Section 3. Conclusions and future developments will end the paper.

2   A Model for Tactical Level of Pedestrians

The model described in this paper provides an approach to deal with tactical choices of agents in pedestrian simulation systems. For sake of space, the description of the operational level components of the model is omitted and it can be found in [1].

2.1   A Cognitive Representation of the Environment for Static Tactical Choices

The framework that enables agents performing choices on their plans implies a graph-like, topological, representation of the walkable space, whose construction is defined in [6] and only briefly reported in this section. This model allows agents to perform a static path planning, since dynamical information such as congestion is not considered in the graph. These additional elements will be considered in the extension that is presented in the next section and represent the innovative part of this paper.

The environment abstraction identifies *regions* (e.g. a room) as node of the labeled graph and *openings* (e.g. a door) as edges. This form of *cognitive map* is computed starting from the information of the simulation scenario, provided by the user and necessarily containing: (i) the description of the walkable space, that is, the size of the simulated environment and the positions of obstacles and walls; (ii) the position of final destinations (i.e. exits) and intermediate targets (e.g. a ticket machine); (iii) borders of the logical regions of the environment that, together with the obstacles, will define them. Approaches to automatically configure a graph representation of the space, without any additional information by the user, have been already proposed in the literature (e.g. [9]), but they are not leading to a cognitively logical description, i.e., a *topological* map.

The cognitive map is defined as a graph $\mathcal{CM} = (\mathcal{V}, \mathcal{E})$ generated with a procedure included to the floor field diffusion, starting from the statements that each user-defined opening generates a floor field from its cells and spread only in the regions that it connects, and that each region has a flag indicating its properties among its cells. The floor fields diffusion procedure iteratively adds to $\mathcal{CM}$ the couple of nodes found in the diffusion (preventing duplicates) and respectively labeled with the region and edge identifiers. Each *final destination*, different from the normal openings since it is located in only one region, will compose an edge linking the region to a special node describing the external *universe*. Intermediate targets will be mapped as attributes of their region's node.

To allow the calculation of the *paths tree*, that will be described in the following section, functions $Op(\rho)$ and $Dist(\omega_1, \omega_2)$ are introduced describing respectively: the set of openings accessible from the region ρ and the distance between two openings linking the same arbitrary region. Since, in general, an opening is associated to a set of cells associated, the value of the floor field in the center cell of ω_1, ω_2 will be used for the computation of the distance among them.

2.2   Modeling Adaptive Tactical Decisions with A Paths Tree

To enhance the route choice and enable dynamical, adaptive, decisions of the agents in a efficient way, a new data structure has been introduced, containing information about the cost of *plausible* paths towards the exit from each region of the scenario.

The well-known Floyd-Warshall algorithm, in fact, can solve the problem but it introduces issues in computational time: the introduction of dynamical elements in the paths cost computation (i.e. congested paths) implies a re-computation of the cost matrix underlying the algorithm every step. More in details, the penalty of a congested path is a subjective element for the agents, since they are walking with different desired velocities, thus the calculation cost increases also with the number of agents.

The approach proposed here implies an off-line calculation of the data-structure that we called *paths tree*, but is computationally efficient during the simulation and provides to the agents direct information about the travel times describing each path.

The Paths Tree. We define the *Paths Tree* as a tree data-structure containing the set of plausible paths towards a destination, that will be its root.

A path is defined as a finite sequence of openings $X \to Y \to \ldots \to Z$ where the last element represents the final destination. It is easy to understand that not every sequence of openings represents a path that is walkable by an agent.

First, a walkable path must be a sequence of consecutive oriented openings in the physical space: an opening E connects two regions R_1 and R_2, can be formally defined as $E = R_1, R_2$; (R_1, E, R_2) and (R_2, E, R_1) are the oriented representations of E. Consecutive openings E_1 and E_2 are such that (R_i, E_1, R_j) and (R_j, E_2, R_k).

In addition to this constraint, a valid walkable path must lead to a *universe* region (i.e. towards a final target). In particular, an agent will consider only valid paths towards its goal, starting from the region where the agent is located.

An important element in the definition of the adopted approach is the expected travel time associated to given path p:

Definition 1. *Let p a path, $T(p)$ is the function which return the expected travel time from the first opening to the destination.*

$$T(p) = \sum_{i \in [1, |p|-1]} \frac{Dist(opening_i, opening_{i+1})}{speed} \tag{1}$$

We consider that a plausible path must be loop-free: by assuming the aim to minimize the time to reach the destination, a plan passing through a certain opening more than once would be not plausible. This will not imply that an agent cannot go through a certain opening more than once during the simulation, and that this could actually happen only with a change of the agent plan, due for instance to an unexpected congestion perceived in a point of the planned path.

Should only convex regions be present in the simulated space, we could easily achieve the set of plausible paths by extending previous constraint and consider not plausible a path passing twice in the same region. However, since the definition of region describes also rooms, concave regions must be considered. Some paths may, thus, imply to pass through another region and then return to the first one to reduce the length of the path.

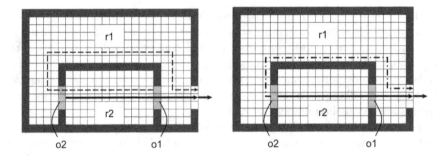

Fig. 1. In the left, a concave region can imply the plausibility of a path crossing it twice, but its identification is not elementary: only the path represented by the continuous line is plausible. All the correct paths for this environment are shown on the right. Inside $r2$ the choice between the two openings is determined by the level of congestion on $o1$.

As we can see by the Figure 1, on the left, both paths start from r_1, go through r_2, and then return to r_1. However, only the path represented by the continuous line is plausible. To support the definition of the constraint that identifies the correct paths, the concept of *sub-path* and a minimality rule must be defined.

Definition 2 (Sub-path). *Let p a path, a sub-path p' of p is a sub-sequence of oriented openings denoted as $p' \subset P$ which respects the order of appearance for the openings in p, but the orientation of openings in p' can differ from the orientation in p. p' must be a valid path.*

The reason of the orientation change can be explained with the example in Fig. 1 in the right: given the path $p = (r1, o2, r2) \rightarrow (r2, o1, r1) \rightarrow end$, the path $p' = (r2, o2, r1) \rightarrow end$ is a valid path and is considered as a sub-path of p, with a different orientation of $o2$. In addition, given the path $p_1 = (r_2, o_2, r_1) \rightarrow end$, the path $p_2 = (r_1, o_2, r_2) \rightarrow (r_2, o_1, r_1) \rightarrow end$ is as well a minimal path if and only if the travel time of p_2 is less than p_1. It is easy to understand that this situation can emerge only if r_1 is concave. As we can see, the starting region of the two paths is different, but the key element of the rule is the position of the opening o_2. If this rule is verified in the center position of the opening o_2, this path will be a considerable path by the agents surrounding $o2$ in $r1$.

In Figure 1, on the right, the correct paths for this example environment are shown. An agent located in $r2$ can reach $r1$ and then the destination D using both of the opening considering the congestion in the environment at the time of planning. An agent located in $r1$ can go directly to the exit or chose the path $o2 \rightarrow o1 \rightarrow D$, according to its starting position.

Definition 3 (Minimal Path). *p is a minimal path if and only if it is a valid path and $\forall p' \subset p : S(p') = S(p) \land D(p') = D(p) \implies T(p') > T(p)$*

The verification of this rule is a sufficient condition for the opening loop constraint and it solves the problem on the region loop constraint independently from the configuration of the environment (i.e. convex or concave regions).

Given this constraint on path minimality, which we consider an indication of plausibility, the complete set of minimal paths towards a destination can be built. It must be noted that an arbitrary path, through the notion of sub-paths, represents a set of paths itself: an agent, in fact, could select a sub-path of a larger minimal path. So a minimal representation of the set is a tree-like structure defined as:

Definition 4 (Paths Tree). *Given a set of minimal paths towards a destination, the Paths-Tree is a tree where the root represents the final destination and a branch from every node to the root describes a minimal path, crossing a set of openings (other nodes) and region (edges). Each node is associated to an attribute describing the expected travel time to the destination.*

An Algorithm to Compute the Paths Tree. The proposed algorithm constructs the Paths Tree recursively, starting from a path containing only the destination and adding nodes if and only if the generated path respects the definition of minimality.

Formally the Paths Tree is defined as $PT = (N, E)$ where N is the set of nodes and E the set of edges. Each node $n \in N$ is defined as a triple (id, o, τ) where $id \in \mathbb{N}$ is the id of the node, $o \in \mathcal{O}$ is the name of the opening and $\tau \in \mathbb{R}^+$ is the expected travel time for the path described by the branch. Each edge $e \in E$ is defined as a triple (p, c, r) where $p \in \mathcal{O}$ is the id of the parent, $c \in \mathcal{O}$ is the id of the child and $r \in \mathcal{R}$ is the region connecting the child node to its parent. To allow a fast access to the nodes describing a path that can be undertaken from a certain region, we added a structure called M that maps each region r in the list of edges $e : (p, c, r) \in E$ (for every c).

Given a destination $D = (r_x, universe)$, the paths tree computation is defined with the following procedures.

Algorithm 1. Paths tree computation

1: add $(0, D, 0)$ to N
2: add 0 to $M[r_x]$
3: $\forall s \in \mathcal{O}$ ShortestPath$[s] \leftarrow \infty$
4: ExpandRegion$(0, D, 0, R_x, ShortestPath)$

With the first line, the set N of nodes is initialized with the destination of all paths in the tree, marking it with the id 0 and expected travel time 0. In the third row the set of *ShortestPath* is initialized. This will be used to track, for each branch, the expected travel time for the shortest sub-path, given a start opening s. *ExpandRegion* is the core element of the algorithm, describing the

---

Algorithm 2. ExpandRegion

Require: input parameters $(parentId, parentName,$
$\quad parentTime, RegionToExpand, ShortestPath)$
1: $expandList \leftarrow \emptyset$
2: $opList = Op(RegionToExpand) \setminus parentName$
3: **for** $o \in opList$ **do**
4: $\quad \tau = parentTime + \frac{D(o,parentName)}{speed}$
5: $\quad$ **if** $CheckMinimality(ShortestPath, o, \tau) ==$ True **then**
6: $\quad\quad$ add (id, o, τ) to N
7: $\quad\quad$ add $(parentId, id, r)$ to E
8: $\quad\quad$ $ShortestPath[o] \leftarrow \tau$
9: $\quad\quad$ $nextRegion = o \setminus r$
10: $\quad\quad$ add id to $M[nextRegion]$
11: $\quad\quad$ add $(id, o, \tau, nextRegion)$ to $expandList$
12: $\quad$ **end if**
13: **end for**
14: **for** $el \in expandList$ **do**
15: $\quad$ $ExpandRegion(el, ShortestPath)$
16: **end for**

---

recursive function which adds new nodes and verifies the condition of minimality. The procedure is described by Alg. 2.

In line 2 a list of openings candidates is computed, containing possible extensions of the path represented by $parentId$. Selecting all the openings present in this region (except for the one labeled as $parentName$) will ensure that all paths eventually created respect the validity constraint.

At this point, the minimality constraint 3 has to be verified for each candidate, by means of the function $CheckMinimality$ explained by Alg. 3. Since this test requires the expected travel time of the new path, this has to be computed before. A failure in this test means that the examined path – created by adding a child to the node $parentId$ – will not be minimal. Otherwise, the opening can be added and the extension procedure can recursively continue.

In line 6, id is a new and unique value to identify the node, which represents a path starting from the opening o and with expected travel time τ; line 7 is the creation of the edge from the parent to the new node. In line 8, $ShortestPath[o]$ is updated with the new discovered value τ. in line 9 the opening is examined as a couple of region, selecting the one not considered now. In fact, the element $nextRegion$ represents the region where is possible to undertake the new path. In line 10 the id of the starting opening is added to $M[nextRegion]$, i.e., the list of the paths which can be undertaken from $nextRegion$. In line 11 the node with his parameter is added to the list of the next expansions, which take place in line 13-14. This passage has to be done to ensure the correct update of $ShortestPath$.

To understand how the constraint of minimality is verified, two basic concepts of the procedure need to be clarified. Firstly, the tree describes a set of paths towards a unique destination, therefore given an arbitrary node n, the path described by the parent of n is a subpath with a different starting node and

Algorithm 3. CheckMinimality

Require: input parameter $(ShortestPath, o, \tau)$
1: **if** $ShortestPath[o] > \tau$ **then**
2:   **return** True
3: **else**
4:   **return** False
5: **end if**

---

leading to the same destination. Furthermore, the expansion procedure implies that once reached a node of depth l, all the nodes of its path having depth $l - k$, $k > 0$ have been already expanded with all child nodes generating other minimal paths.

Note that the variable $ShortestPath$ is particularly important since, given p the current path in evaluation, it describes the minimum expected travel time to reach the destination (i.e. the root of the tree) from each opening already evaluated in previous expansions of the branch. Thus, if τ is less than $ShortestPath[o]$, the minimality constraint 3 is respected.

Congestion Evaluation. The explained approach of the paths tree provides information on travel times implied by each path towards a destination. By only using this information, the choice of the agents would be still static, essentially describing the shortest path. This could lead to an increase of the experienced travel times, since congestion may emerge without being considered.

For the evaluation of congestion, we provide an approach that estimates, for each agent, the additional time deriving by passing through a jam. The calculation considers two main aspects: the size of the possibly arisen congestion around an opening; the average speed of the agents inside the congested area. Since the measurement of the average speed depends on the underlying model that describes the physical space and movement of the agents, we will just clarify that the speed is estimated through the adoption of an additional grid counting the recent *blocks* (i.e. when agents maintain positions at the end of the step although desired to move) in the surrounding area of each opening. The average number of blocks influences the probability to move into the area per step, thus the speed of the agents. For the size of the area, our approach is to define a minimum radius of the area and (i) increase it when the average speed becomes too low or (ii) reduce it when it returns normal.

As we can see in Figure 2, the presence of an obstacle in the room is well managed by using floor field while defining the area for a given radius. If a many agents try to go through the same opening at the same time, a congestion will arise, reducing the average speed and increasing the size of the monitored area until included agents are no more involved in blocks.

During this measurement the average speed value for each radius is stored. Values for sizes smaller than the size of the area will be used by the agents inside it, as will be explained in the next section. Two function are introduced

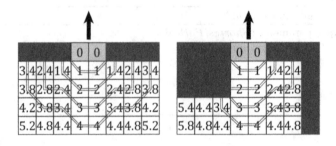

Fig. 2. Examples of surroundings of different sizes, for two configurations of the environment.

for the calculation: $size(o)$: return the size of the congestion around the opening, $averageSpeed(o, s)$: return the average speed estimated in the area of size s around the opening o.

Agents Dynamic Path Choice. At this point we have defined which information an agent will use to make its decision: (i) the Paths Tree, computed before the simulation, will be used as a list of possible path choice; (ii) the position of the agent, will be used to adjust the expected travel time considering the distance between the agent and the first opening of a path: $d(a, o)$; (iii) the information about congestion around each opening, computed during the simulation, will be used to estimate the delay introduced by each opening in the path.

The agent, who knows in which region R_x he is located, can access the Paths Tree using the structure $M[R_x]$. The structure returns a list of nodes, each representing the starting opening for each path. At this point the agent can compute the expected travel time to reach each starting opening and add it to the travel time τ of the path.

To consider congestion, the agent has to estimate the delay introduced by each opening in a path, by firstly obtaining the size of the jammed area.

$$size_a(o) = \begin{cases} size(o) & \text{if } d(a, o) \geq i(x) \\ d(a, o) & \text{otherwise} \end{cases} \tag{2}$$

At this point, the agent can suppose that for the length of the area it will travel at the average speed around the opening.

$$delay(o) = max(size_a(o)\big(\frac{1}{averageSpeed(o)} - \frac{1}{speed_a}\big), 0) \tag{3}$$

If the agent is slower than the average speed around an opening, the delay will be lower than 0. In this case it is assumed that the delay is 0, implying that the congestion will not increase his speed.

At this point the agent can estimate the delay introduced by all openings.

$$pathDelay(p) = \sum_{o \in p} delay(o) \tag{4}$$

We can consider that agents only have access to delay information about openings that are present in the area it is located into, whereas the delay is considered zero (in an optimistic hypothesis) in openings that are far from its perception.

$$Time(p) = \tau_p + \frac{d(a, S(p))}{speed_a} + pathDelay_a(p) \tag{5}$$

Where:

- τ_p : the expected travel time of the path p
- $\frac{d(a,S(p))}{speed_a}$: the expected time to reach $S(p)$ from the position of the agent
- $pathDelay_a(p)$: the estimation of the delay introduced by each opening in the path, based on the memory of the agent (which may or may not be updated for each opening).

3   Applications with an Experimental Scenario

In order to show the potential and the possibility to fine tune the proposed approach, the evacuation in a hypothetical scenario has been simulated with a consistent incoming flow of people. A graphical representation of the environment and flow configuration is depicted in Fig. 3(a): it illustrates a sample situation in which two flows of pedestrians enter an area with six exits, distributed among 3 equal rooms, at a rate of 10 pedestrians per second. An important peculiarity is the slightly asymmetrical configuration of the environment, that causes shorter distances towards the three southern exits. This is reflected by the illustrated paths tree in Fig. 3(b) where, to give an example, the paths starting from o4 and o5 and leading out through o2 take a little more time than the ones going out by using o7. This variation significantly affected the results of the simulations, here shown with cumulative mean density maps [5][2] in Fig. 4.

In particular, the results of two simulations in which different approaches have been implemented for the dynamic estimation of the path traveling times by the agents are shown. In the first approach, shown in the top row, all the agents perceived the same *congestion time* for the openings that they can detect during the simulation (i.e. the travel time corrected considering the path delay discussed in the previous section). In the second approach, instead, a random error of ±10% has been added to the overall calculation of the traveling time

---

[2] These heat maps describe the mean local density value in each cell. It is calculated in a time window of 50 steps where, at each step, only values of occupied cells are collected.

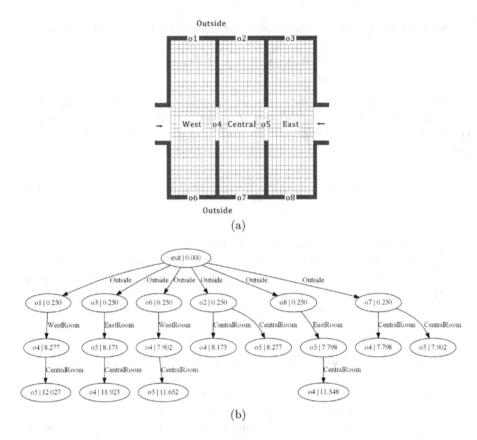

Fig. 3. The experimental scenario (a) and the associated paths tree (b).

Time(p) in order to consider the fact that pedestrians do not have an *exact* estimation of distances and delays caused by perceived congestion, in a more commonsense spatial reasoning framework [3].

By comparing the results it is possible to notice that, counter–intuitively, the insertion of the random perturbation caused an optimization of the flows in this overcrowded scenario. In the firsts 100 steps of the simulations, the dynamics for the two approaches is similar and described by the missed usage of the central room, since the distance between the northern and southern exits is quite small. The less precise calculation causes the agents to start using the central room and associated exits earlier than in the precise delay estimation case, in particular, around 130th step vs 150th step in the first scenario, generating lower level of densities and, thus, higher outgoing flow rates. Moreover, this error balances the attractiveness of middle southern and northern exits that are more evenly adopted than in the precise calculation approach (as shown in Fig. 4 (b) and (e)), leading not only to a more efficient but especially more plausible space utilization.

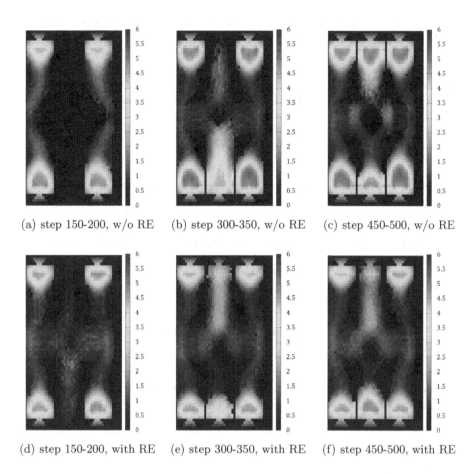

(a) step 150-200, w/o RE   (b) step 300-350, w/o RE   (c) step 450-500, w/o RE

(d) step 150-200, with RE   (e) step 300-350, with RE   (f) step 450-500, with RE

Fig. 4. The test scenario respectively without and with a random perturbation of the agent estimated travel time.

4   Conclusions

The paper has presented a hybrid agent architecture for modeling tactical level decisions in pedestrian simulations. The agents make decisions based on a static a-priori knowledge of the environment and dynamic perceivable information on the current level of congestion of visible path alternatives. The model was experimented in a sample scenario showing the adequacy in providing adaptiveness to the contextual situation while preserving a plausible overall pedestrian dynamic: congestion is detected and, when possible, longer trajectories are adopted granting overall shorter travel times. The actual validity of this approach must still be proven, both in evacuations and other kinds of situations: this represents an open challenge, since there are no comprehensive data sets on human tactical level decisions and automatic acquisition of this kind of data from video cameras is still a challenging task [8].

Acknowledgments. This work was partly supported by the ALIAS project ("Higher education and internationalization for the Ageing Society"), funded by Fondazione CARIPLO.

References

1. Bandini, S., Crociani, L., Vizzari, G.: Heterogeneous speed profiles in discrete models for pedestrian simulation. In: Proceedings of the 93rd Transportation Research Board Annual Meeting (2014). http://arxiv.org/abs/1401.8132
2. Bandini, S., Manzoni, S., Vizzari, G.: Agent based modeling and simulation: An informatics perspective. Journal of Artificial Societies and Social Simulation **12**(4), 4 (2009)
3. Bandini, S., Mosca, A., Palmonari, M.: Common-sense spatial reasoning for information correlation in pervasive computing. Applied Artificial Intelligence **21**(4&5), 405–425 (2007). doi:10.1080/08839510701252676
4. Boltes, M., Seyfried, A.: Collecting pedestrian trajectories. Neurocomputing **100**, 127–133 (2013). http://www.sciencedirect.com/science/article/pii/S0925231212003189
5. Castle, C.J.E., Waterson, N.P., Pellissier, E., Bail, S.: A comparison of grid-based and continuous space pedestrian modelling software: analysis of two uk train stations. In: Peacock, R.D., Kuligowski, E.D., Averill, J.D. (eds.) Pedestrian and Evacuation Dynamics, pp. 433–446. Springer, US (2011). doi:10.1007/978-1-4419-9725-8_39
6. Crociani, L., Invernizzi, A., Vizzari, G.: A hybrid agent architecture for enabling tactical level decisions in floor field approaches. Transportation Research Procedia **2**, 618–623 (2014)
7. Guo, R.Y., Huang, H.J.: Route choice in pedestrian evacuation: formulated using a potential field. Journal of Statistical Mechanics: Theory and Experiment **2011**(04), P04018 (2011)
8. Khan, S.D., Vizzari, G., Bandini, S.: Identifying sources and sinks and detecting dominant motion patterns in crowds. Transportation Research Procedia **2**, 195–200 (2014). http://www.sciencedirect.com/science/article/pii/S2352146514000660, the Conference on Pedestrian and Evacuation Dynamics 2014 (PED 2014), 22-24 October 2014, Delft, The Netherlands
9. Kretz, T., Bönisch, C., Vortisch, P.: Comparison of various methods for the calculation of the distance potential field. In: Klingsch, W.W.F., Rogsch, C., Schadschneider, A., Schreckenberg, M. (eds.) Pedestrian and Evacuation Dynamics 2008, pp. 335–346. Springer, Heidelberg (2010). doi:10.1007/978-3-642-04504-2_29
10. Michon, J.A.: A critical view of driver behavior models: what do we know, what should we do? In: Evans, L., Schwing, R.C. (eds.) Human Behavior and Traffic Safety, pp. 485–524. Springer, US (1985). doi:10.1007/978-1-4613-2173-6_19
11. Pianini, D., Viroli, M., Zambonelli, F., Ferscha, A.: HPC from a self-organisation perspective: the case of crowd steering at the urban scale. In: International Conference on High Performance Computing and Simulation, HPCS 2014, Bologna, Italy, 21–25 July, 2014, pp. 460–467. IEEE (2014). doi:10.1109/HPCSim..6903721

12. Schadschneider, A., Klingsch, W., Klüpfel, H., Kretz, T., Rogsch, C., Seyfried, A.: Evacuation dynamics: empirical results, modeling and applications. In: Meyers, R.A. (ed.) Encyclopedia of Complexity and Systems Science, pp. 3142–3176. Springer (2009)
13. Vizzari, G., Bandini, S.: Studying pedestrian and crowd dynamics through integrated analysis and synthesis. IEEE Intelligent Systems **28**(5), 56–60 (2013). doi:10.1109/MIS.2013.135
14. Wagoum, A.U.K., Seyfried, A., Holl, S.: Modelling dynamic route choice of pedestrians to assess the criticality of building evacuation. Advances in Complex Systems **15**(07), 15 (2012)
15. Zhang, J., Klingsch, W., Schadschneider, A., Seyfried, A.: Transitions in pedestrian fundamental diagrams of straight corridors and t-junctions. Journal of Statistical Mechanics: Theory and Experiment **2011**(06), P06004 (2011). http://stacks.iop.org/1742-5468/2011/i=06/a=P06004

Computer Vision

Using Stochastic Optimization to Improve the Detection of Small Checkerboards

Hamid Hassannejad, Guido Matrella, Monica Mordonini,
and Stefano Cagnoni$^{(\boxtimes)}$

Dipartimento di Ingegneria dell'Informazione,
Università degli Studi di Parma, Parma, Italy
{hamid.hassannejad,guido.matrella,monica.mordonini,
stefano.cagnoni}@unipr.it

Abstract. The popularity of mobile devices has fostered the emergence of plenty of new services, most of which rely on the use of their cameras. Among these, diet monitoring based on computer vision can be of particular interest. However, estimation of the amount of food portrayed in an image requires a size reference. A small checkerboard is a simple pattern which can be effectively used to that end. Unfortunately, most existing off-the-shelf checkerboard detection algorithms have problems detecting small patterns since they are used in tasks such as camera calibration, which require that the pattern cover most of the image area. This work presents a stochastic model-based approach, which relies on Differential Evolution (DE), to detecting small checkerboards. In the method we propose the checkerboard pattern is first roughly located within the image using DE. Then, the region detected in the first step is cropped in order to meet the requirements of off-the-shelf algorithms for checkerboard detection and let them work at their best. Experimental results show that, doing so, it is possible to achieve not only a significant increase of detection accuracy but also a relevant reduction of processing time.

Keywords: Checkerboard detection · Model-based object detection · Differential evolution · Size reference

1 Introduction

Apps for mobile devices that record and analyze food intake are becoming more and more popular. However, they often rely on a mostly manual procedure which compels users to record meals, enter data about the food type and, above all, personally estimate the food amount. This last task can be particularly troublesome, error-prone, and user-dependent, while being, at the same time, the most critical for meeting the apps' goals. Therefore, many efforts are being made to make the procedure faster, easier and more precise by automating it. In particular, many researchers are trying to use food images, possibly taken by mobile devices, to automatically extract information [3,4,17]. The main tasks that are being automated are the recognition of food type and the estimation of its amount from one or more images.

© Springer International Publishing Switzerland 2015
M. Gavanelli et al. (Eds.): AI*IA 2015, LNAI 9336, pp. 75–86, 2015.
DOI: 10.1007/978-3-319-24309-2_6

Food volume estimation is the most direct approach to calculating calories or nutrients of food intake, once the food type is known. Volume can be estimated from images using different image analysis procedures, but only up to a scale factor, if the picture is not taken in controlled conditions (i.e., at a certain distance, from a certain angle, etc.). Therefore, in many studies, an object of known size is used as reference. In [1,7], for instance, the user is required to put her/his finger besides the dish. Later, the finger is detected and used as reference. In this approach, the variation of finger pose and of the environmental conditions in which the picture is taken introduces new problems. In [5], a specific pattern of known size printed on a card is used as reference, showing how such a standardized reference can improve the accuracy of volume estimation. Many studies have followed this idea using a checkerboard as reference [8–10,16,17]. In some of these studies, like [8–10], the checkerboard is also used as reference for colors.

The regular structure of checkerboards and the existence of effective algorithms to detect them are some of the good reasons for choosing such patterns over other options. Nevertheless, off-the-shelf checkerboard detection algorithms are usually designed to be a step in camera-calibration or pose-detection algorithms and are usually tuned for specific situations like: flat checkerboards, a large single checkerboard, etc. In fact, checkerboards which are used as size references usually consist of few squares and occupy a relatively small portion of the image, since they need to appear aside the main object which is to be detected and measured. This situation makes it difficult for 'standard' checkerboard detectors to be as effective as within the settings for which they have been originally designed. Thus, in other applications, different algorithms, or modified versions of the most popular ones, are needed. For example, [11] introduces a method to detect checkerboards in blurred or distorted images and [13] a method to detect checkerboards on non-planar surfaces.

Two popular algorithms for checkerboard detection are provided by OpenCV and Matlab. The OpenCV algorithm, which is available through the *findChessboardCorners()* function, is based on using a graph of connected quads, while the Matlab algorithm, which is available in the Computer Vision System Toolbox through the *detectCheckerboardPoints* function, analyses selected corner patterns to detect a checkerboard. Both algorithms have problems in detecting small checkerboards. Figure 1 shows three examples in which both algorithms fail. In the case of food intake monitoring, missing the checkerboard in an image would cost the user the trouble of taking another image of the food and, even so, there would not be any guarantee that, under the same conditions, the algorithm would work with the new image.

In this paper we introduce a pre-processing step which improves the effectiveness of the checkerboard detectors provided by Matlab and OpenCV (hereafter, referred as *basic methods*). The main idea is to roughly locate the image region where the checkerboard is located and then accurately detect the pattern by running one of the two basic methods on such a region, cropped to allow the

Fig. 1. Above: typical images used for camera calibration. Below: Examples of images where checkerboards are used as size reference for food volume estimation, on which both OpenCV and Matlab functions for detecting checkerboards fail to detect the pattern.

algorithm to operate in conditions which are closer to the ones for which it has been designed.

To locate the checkerboard, we use the same approach as [14] and [15]. Following that method, a hypothesis of checkerboard location can be evaluated by rigidly transforming a model of the checkerboard and then projecting it onto the image according to a perspective transform. The likelihood of the hypothesis is evaluated using a similarity measure between the projection of the model onto the image and the actual content of the corresponding image region. This procedure allows one to turn object detection into an optimization problem, in which the parameters to be tuned are the coefficients of the rigid transformation and of the projection. Using such a similarity measure as target function, a meta-heuristic is then used to generate hypotheses, until similarity reaches a predefined value, which means that the checkerboard has been located. In this work, Differential Evolution (DE) [12] is employed for optimization. One of the advantages over other model-based approaches is that this approach does not need any preliminary preprocessing of the image (like color segmentation) or projection of many points [15].

2 Locating Objects in an Image

The procedure we adopted estimates the pose of an object based on a 3D model and can be utilized with any projection system and any general object model. In this case the model is obtained by sampling the different parts of the object to be detected, obtaining a set of 3D points which describe its shape. Then, once an object pose has been hypothesized, the points are projected onto the image plane according to a transformation which maps points in the camera reference frame onto the image, and are then matched to the actual image content. The likelihood of the detection is evaluated using a similarity measure based on pixel intensity.

This generally-applicable object detection algorithm therefore includes the following steps:

1. Consider a set of key points, of known coordinates with respect to a general model of the object to detect.
2. Hypothesize a pose for the object, then apply the corresponding rototranslation transform to the points which represent it and project them onto the image.
3. Verify that the pixel intensities of the points in the set match those which can be observed in the image region where they have been projected, to assess the presence of the object.

In our case, the model of the object (a checkerboard) consists of 73 key points (Figure 2), corresponding to the center, edges and corners of each square.

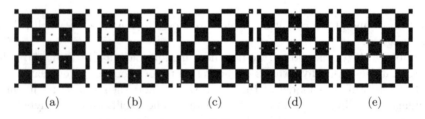

(a)   (b)   (c)   (d)   (e)

Fig. 2. Key points are evaluated in five steps to compute the similarity degree between the projected model and the image.

After locating the checkerboard in the image with a degree of confidence above a pre-set threshold which the similarity function must reach, the region thus identified is cropped and the basic methods to detect checkerboard corners are applied to the resulting image.

3   Differential Evolution

Problems which involve global optimization over continuous spaces are ubiquitous in real life. Any design problem can be virtually considered an optimization problem, which is often multi-objective and constrained. In general, the task is to optimize certain properties of a system by pertinently choosing some parameters which describe its behaviour and are represented, for convenience, as a vector.

Among the many available metaheuristics for solving continuous optimization problems, Differential Evolution [12] is a relatively simple evolutionary method. It iteratively tries to improve the solution to a problem with respect to a given measure of quality, using principles which may be referred back to both swarm intelligence and evolutionary computation methods. In DE, first, a random set of solutions (population) is generated. Then, in turn, the solutions are evaluated by the fitness function and new solutions are generated for the next "generation". Such new individuals are created by combining members of the current population. Every individual acts as a parent vector corresponding to which a donor vector is created. In the basic version of DE, the donor vector V_i for the ith parent (X_i) is generated by combining three random and distinct individuals X_{r1}, X_{r2} and X_{r3}, according to the following rule (mutation of difference vectors):

$$V_i = X_{r1} + F(X_{r2} - X_{r3})$$

where F (scale factor) is a parameter that strongly influences DE's performances and typically lies in the interval $[0.4, 1]$. Recently, several other mutation strategies have been designed for DE, experimenting with different base vectors and different numbers of vectors to generate perturbations. For example, the method explained above is called $DE/rand/1$, which means that the first element of the donor vector equation X_{r1} is randomly chosen and only one difference vector (in our case $X_{r2} - X_{r3}$) is added. After mutation, every parent-donor pair generates a child (U_i), called trial vector, by means of a crossover operation.

$$U_{i,j} = \begin{cases} V_{i,j} & \text{if } (rand_{i,j} \leq C_r \text{ or } j = j_{rand}) \\ X_{i,j} & \text{otherwise} \end{cases}$$

As described in the above equation, the jth component/dimension of the ith donor vector is obtained by means of uniform (binomial) crossover, where $rand_{i,j}$ is a random number drawn from a uniform distribution in the range $[0, 1]$, C_r is the crossover rate, and j_{rand} is a randomly selected dimension. The newly-generated individual U_i is evaluated by comparing its fitness to its parent's. The best survives and will be part of the next generation.

There are several reasons for researchers to consider DE as an attractive optimization tool: DE is simple and straightforward to implement, and exhibits much better performance in comparison with several other methods on problems onto which many real-world problems can be mapped. Moreover, the number of control parameters in DE is very small, and its space complexity is low [2].

One of the main characteristics of DE, common to most meta-heuristics, is its stochastic nature. If it fails to find the checkerboard in a run, it is always possible for it to succeed in another run. A restart strategy in case of failure is allowed, on the one hand, by DE's simple formulation and, on the other hand, by its intrinsically parallel nature, which makes it possible to implement the algorithm very efficiently. This ability to explore the solution space effectively and globally, even away from the region where, a priori, the solution is most likely to be found, is something that is missing in the basic methods and offers the DE-based method higher performance in the presence of exceptions or outliers; this is obviously compensated, in principle, by lower repeatability and accuracy upon success. However, as shown by the results we report in Section 5, within this problem, the advantages strongly overcome such drawbacks.

4   Fitness Function

Let $f : \Re^n \to \Re$ be the fitness function (cost function) which must be maximized (minimized). The function takes a candidate solution in the form of a vector as argument and produces a real number as output. The goal of global optimization algorithms like DE is to find a solution s such that $f(s) >= f(c)$ ($f(s) <= f(c)$ for a minimization problem) for all possible values of c in the search-space, i.e., s is the global maximum (minimum).

In this work the input argument of the fitness function is a 6-parameter vector:

$$V = [x, y, z, \alpha, \beta, \gamma]$$

These parameters represent the pose of the object with respect to the camera. The first three parameters represent a translation in the 3D coordinate system while the other three represent rotations around the coordinate system axes. These parameters are the coefficients of the transformation which matches the world coordinate system to the camera coordinate system (see Figure 3).

Fig. 3. Camera coordinate system (green,left) and world coordinate system (blue, right). The input vector of the fitness functions represents the transformation (translation and rotation) which matches the world coordinate system to the camera coordinate system.

The fitness function calculates the degree of similarity between the reprojected model and the image. To do so, 73 key points are used to describe the checkerboard model, as shown in figure 2. The similarity degree is calculated in five steps, in each of which a different subset of points is matched to the content of the image region onto which they are projected, generating a score which is proportional to the matching degree. The following step is taken only if at least a certain degree of similarity has been obtained in the previous one, otherwise the hypothesized pose is rejected and a low fitness value is returned.

In fact, during these procedure a checkerboard model is grown gradually and compared with the re-projected points step by step. Algorithm 1 presents the pseudo-code of the procedure. The thresholds are defined based on the importance of the step and they have been tuned empirically. A perfect match will produce a score of 54, however any score higher than 45 can be considered to represent an acceptable match for the purposes of this work.

Algorithm 1. Fitness Function

---

function FITNESSFUNCTION(PoseVector)

   Calculate Rotation (R) & Translation (T) matrices from PoseVector

   $Score \leftarrow Score + FirstLevelCenterCheck([R,\ T])$
   if Score > 6 **then**
      $Score \leftarrow Score + SecondLevelCenterCheck([R,\ T])$
   end if
   if Score > 20 & the center is black **then**
      $Score \leftarrow Score + PoseCheck([R,\ T])$
   end if
   if Score > 23 **then**
      $Score \leftarrow Score + EdgesCheck([R,\ T])$
      $Score \leftarrow Score + verticesCheck([R,\ T])$
   end if
return Score

end function

---

5   Experimental Results

We used an implementation of DE with binomial crossover and a population of 60 individuals, with scale factor F and crossover rate set to 0.6 and 0.8, respectively. Each run of DE lasted for 1000 generations on every image, unless a fitness value greater than 45 was obtained before. If lower fitness value had been returned, DE was restarted from scratch with a new population. For every image, DE was allowed to run up to six times before the detection was aborted.

The algorithm has been tested over four sets of images, for a total of 451 images. The images in each set are taken by a different mobile device and in different environments, as shown by figures 1 and 4. Motorola Moto-G, Samsung Galaxy Note 1, and Samsung Galaxy S3 smartphones were used to take the pictures. In the case of Samsung Galaxy S3, since the images at standard resolution produced worse results using both basic detection algorithms, a scaled version of the images was tested as well. The fourth set therefore contains the images in the third set scaled by a factor of 0.5. Figure 4 shows some examples of the located checkerboards.

Fig. 4. The same images of Figure 1, where checkerboards have been correctly located (image regions marked on the card).

Table 1. Results of the DE-based checkerboard locating algorithm.

Camera	N. of images	Detections (%)	Detections in the first run(%)	DE repetitions (avg.)
Motorola MotoG	179	176 (98.32%)	143 (79.99%)	1.34
Samsung Galaxy Note 1	12	12 (100%)	11 (91.67%)	1.25
Samsung Galaxy S3	130	128 (98.46%)	123 (94.61%)	1.2
Samsung Galaxy S3 scaled (0.5)	130	127 (97.70%)	125 (96.15%)	1.13
Total	451	443 (98.22%)	402 (89.14%)	1.23

A PC was used to perform the tests. It was equipped with an *Intel Core i7 @2.80 GHz* CPU, *6 GB* RAM, and *Windows 7 64-bit*.

Table 1 reports the results of the algorithm in locating checkerboards. In more than 98% of cases the checkerboard was correctly located. In 89% of cases it was found within the first DE run; on average, DE needed to be repeated 1.23 times for each image before the match was found.

After the DE-based algorithm had located the checkerboard, the basic methods were applied. Table 2 and Table 3 show the results of a comparison between the new approach and the basic methods alone. As can be observed, there was a noticeable improvement in both speed and performance of the algorithms when they were applied to the images pre-processed by DE. In fact, the harder the problem for them, the longer the basic algorithms run. In the case of OpenCV, the introduction of the DE-based method reduced the total processing time (including pre-processing) by more than 20%, whereas the number of correctly detected checkerboards increased by more than 41%. Using the Matlab algorithm, the introduction of the DE-based method reduced the total processing time (including pre-processing) up to 38% while the number of detected checker-

Table 2. Results of checkerboard corner detection using the OpenCV algorithm.

Camera	Number of Images	Method	Accuracy (%)	Time (ms)
Motorola MotoG	179	OpenCV algorithm	118 (65.92%)	3368393
		with Preprocessing	133 (74.30%)	2652066
		Improvement	12.71%	21.3%
Samsung Galaxy Note 1	12	OpenCV algorithm	4 (33.33%)	258969
		with Preprocessing	7 (58.33%)	206385
		Improvement	75%	20.31%
Samsung Galaxy S3	130	OpenCV algorithm	24 (18.46%)	16276198
		with Preprocessing	51 (39.23%)	12904769
		Improvement	112.5%	20.61%
Samsung Galaxy S3 scaled (0.5)	130	OpenCV algorithm	94 (72.30%)	909017
		with Preprocessing	101 (77.69%)	744299
		Improvement	7.44%	18.12%

Table 3. Results of checkerboard corner detection using the Matlab algorithm.

Camera	Number of Images	Method	Accuracy (%)	Time (ms)
		Matlab algorithm	159 (88.82%)	1267700
Motorola MotoG	179	with Preprocessing	171 (95.53%)	303183
		Improvement	7.54%	76.09%
		Matlab algorithm	11 (91.67%)	25412
Samsung Galaxy Note 2	12	with Preprocessing	12 (100%)	33209
		Improvement	9.09%	-30.68%
		Matlab algorithm	91 (70%)	626590
Samsung Galaxy S3	130	with Preprocessing	105 (80.77%)	614923
		Improvement	15.38%	1.87%
Samsung Galaxy S3		Matlab algorithm	121 (93.08%)	195250
scaled (0.5)	130	with Preprocessing	129 (99.23%))	137695
		Improvement	6.61%	29.48%

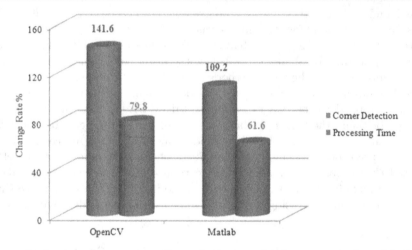

Fig. 5. Reduction of processing time and increase of detection rate after adding the DE-based pre-processing step.

boards increased by about 10%. Image set 2 was the only one where an increase in processing time was measured. This anomaly could be due to the high success rate of the basic methods alone with such pictures and to the small number of pictures in such a set, which makes the performance estimation less reliable. Figure 5 shows a summary of the results.

6   Conclusions and Future Work

The algorithm we propose can improve automatic detection of small checkerboards in images by adding a pre-processing step to existing checkerboard detection algorithms. In the pre-processing step the region where the checkerboard is expected to be located is determined by employing DE to optimize a projection transform by which a model of the checkerboard to detect is matched to the image content.

Experimental results showed that this approach significantly improves the performances of the algorithms provided by Matlab and OpenCV, as regards both processing time and successful detection rate of small checkerboards.

A small checkerboard, printed on a PVC card, is a proper measurement reference for many applications in which such a pattern is expected to occupy only a small region in the image. Therefore, this algorithm could be a valuable asset in automatic size-estimation applications and, in particular, for food amount estimation.

Even if we chose off-the-shelf algorithms as references for performance evaluation, the DE-based procedure we propose can improve the performance of even more sophisticated checkerboard detection algorithms which, as most such algorithms do, expect the checkerboard to be either the only object in the scene or to be of larger or, at least, pre-set size. In fact, the main effect of DE-based pre-processing is to make such algorithms substantially scale-invariant.

In rather sporadic cases, a symmetric checkerboard as the one used in this work may cause inconsistent pose detections. The procedure we described is applicable to any pattern, so an asymmetric checkerboard could be used in applications in which the accuracy of DE-based object location is more critical. Finally, taking into consideration the intrinsic parallel nature of DE, parallelizing the algorithm on GPU or any other massively parallel architecture using platforms like CUDA or OpenCL could be considered to further reduce processing time [6].

Acknowledgments. This work is partially funded by EU within the "Helicopter" Project (a H2020 Ambient Assisted Living Joint Project). The authors would like to thank, in particular, the project partner METEDA s.r.l.

References

1. Almaghrabi, R., Villalobos, G., Pouladzadeh, P., Shirmohammadi, S.: A novel method for measuring nutrition intake based on food image. In: 2012 IEEE International Instrumentation and Measurement Technology Conference (I2MTC), pp. 366–370, May 2012
2. Das, S., Suganthan, P.N.: Differential evolution: a survey of the state-of-the-art. IEEE Transactions on Evolutionary Computation **15**(1), 4–31 (2011)
3. Kitamura, K., Yamasaki, T., Aizawa, K.: Foodlog: capture, analysis and retrieval of personal food images via web. In: Proceedings of the ACM Multimedia 2009 Workshop on Multimedia for Cooking and Eating Activities, pp. 23–30. ACM (2009)
4. Kong, F., Tan, J.: Dietcam: Automatic dietary assessment with mobile camera phones. Pervasive and Mobile Computing **8**(1), 147–163 (2012)
5. Martin, C.K., Kaya, S., Gunturk, B.K.: Quantification of food intake using food image analysis. In: Annual International Conference of the IEEE Engineering in Medicine and Biology Society, EMBC 2009, pp. 6869–6872. IEEE (2009)

6. Nashed, Y.S., Ugolotti, R., Mesejo, P., Cagnoni, S.: Libcudaoptimize: an open source library of GPU-based metaheuristics. In: Proceedings of the 14th Annual Conference Companion on Genetic and Evolutionary Computation, pp. 117–124. ACM (2012)
7. Pouladzadeh, P., Villalobos, G., Almaghrabi, R., Shirmohammadi, S.: A novel SVM based food recognition method for calorie measurement applications. In: ICME Workshops, pp. 495–498 (2012)
8. Puri, M., Zhu, Z., Yu, Q., Divakaran, A., Sawhney, H.: Recognition and volume estimation of food intake using a mobile device. In: 2009 Workshop on Applications of Computer Vision (WACV), pp. 1–8. IEEE (2009)
9. Rahman, M.H., Li, Q., Pickering, M., Frater, M., Kerr, D., Bouchey, C., Delp, E.: Food volume estimation in a mobile phone based dietary assessment system. In: 2012 Eighth International Conference on Signal Image Technology and Internet Based Systems (SITIS), pp. 988–995. IEEE (2012)
10. Rahmana, M.H., Pickering, M.R., Kerr, D., Boushey, C.J., Delp, E.J.: A new texture feature for improved food recognition accuracy in a mobile phone based dietary assessment system. In: 2012 IEEE International Conference on Multimedia and Expo Workshops (ICMEW), pp. 418–423. IEEE (2012)
11. Rufli, M., Scaramuzza, D., Siegwart, R.: Automatic detection of checkerboards on blurred and distorted images. In: IEEE/RSJ International Conference on Intelligent Robots and Systems, IROS 2008, pp. 3121–3126. IEEE (2008)
12. Storn, R., Price, K.: Differential evolution-a simple and efficient adaptive scheme for global optimization over continuous spaces, vol. 3. ICSI, Berkeley (1995)
13. Sun, W., Yang, X., Xiao, S., Hu, W.: Robust checkerboard recognition for efficient nonplanar geometry registration in projector-camera systems. In: Proceedings of the 5th ACM/IEEE International Workshop on Projector camera systems, p. 2. ACM (2008)
14. Ugolotti, R., Nashed, Y.S.G., Cagnoni, S.: Real-time GPU based road sign detection and classification. In: Coello, C.A.C., Cutello, V., Deb, K., Forrest, S., Nicosia, G., Pavone, M. (eds.) PPSN 2012, Part I. LNCS, vol. 7491, pp. 153–162. Springer, Heidelberg (2012)
15. Ugolotti, R., Nashed, Y.S., Mesejo, P., Ivekovič, Š., Mussi, L., Cagnoni, S.: Particle swarm optimization and differential evolution for model-based object detection. Applied Soft Computing 13(6), 3092–3105 (2013)
16. Weiss, R., Stumbo, P.J., Divakaran, A.: Automatic food documentation and volume computation using digital imaging and electronic transmission. Journal of the American Dietetic Association 110(1), 42–44 (2010)
17. Zhu, F., Bosch, M., Woo, I., Kim, S., Boushey, C.J., Ebert, D.S., Delp, E.J.: The use of mobile devices in aiding dietary assessment and evaluation. IEEE Journal of Selected Topics in Signal Processing 4(4), 756–766 (2010)

Multi Agent Systems

Empowering Agent Coordination
with Social Engagement

Matteo Baldoni[✉], Cristina Baroglio, Federico Capuzzimati,
and Roberto Micalizio

Dipartimento di Informatica, Università degli Studi di Torino,
c.so Svizzera 185, 10149 Torino, Italy
{matteo.baldoni,cristina.baroglio,federico.capuzzimati,
roberto.micalizio}@unito.it

Abstract. Agent coordination based on Activity Theory postulates that
agents control their own behavior from the outside by using and creating artifacts through which they interact. Based on this conception, we
envisage social engagements as first-class resources that agents exploit
in their deliberative cycle (as well as beliefs, goals, intentions), and propose to realize them as artifacts that agents create and manipulate along
the interaction, and that drive the interaction itself. Consequently, agents
will base their reasoning on their social engagement, instead of relying on
event occurrence alone. Placing social engagement at the center of coordination promotes agent decoupling and also the decoupling of the agent
specifications from the specification of their coordination. The paper also
discusses JaCaMo+, a framework that implements this proposal.

Keywords: Social engagement · Commitments · Agents and artifacts ·
Agent programming · Implementation

1 Introduction

We propose an agent programming approach that is inspired by, and extends,
the *environment programming* methodology proposed in [18]. In this methodology, the environment is seen as a programmable part of the (multiagent) system.
The methodology takes advantage of the **A&A** meta-model [17,23] that, having
its roots in the Activity Theory [13], extends the agent paradigm with the *artifact* primitive abstraction. An artifact is a computational, programmable system
resource, that can be manipulated by agents.

In this context, programming a Multiagent System (MAS) consists of two
main activities: (1) programming the agents as autonomous software entities
designed to accomplish user-defined goals; and (2) programming the environment
(i.e., the artifacts) that provide agents with those functionalities they can exploit
while performing their own activities. In other words, the environment where the
agents operate is thought of as a set of artifacts, where each artifact is a resource,
or tool, that is made available to the agents. Indeed, the framework is much more
powerful as it allows agents not only to access and use artifacts, but also to create
new artifacts, to adapt them, and even to link two (or more) artifacts.

© Springer International Publishing Switzerland 2015
M. Gavanelli et al. (Eds.): AI*IA 2015, LNAI 9336, pp. 89–101, 2015.
DOI: 10.1007/978-3-319-24309-2_7

An artifact provides the agents using it with a set of *operations* and a set of *observable properties*. Operations are computational processes that are executed inside the artifact itself, and that can be triggered by agents or other artifacts. Observable properties are state variables that are observable by all those agents using, or *focusing* on, the artifact. Of course, observable properties can change over time as a result of the operations occurred inside the artifact. It must be noticed that, although artifacts are substantially a coordination mechanism, direct communication between agents is still possible in the framework proposed by Ricci et al. [18]

The *Coordination by Social Engagement* (CoSE) programming approach proposed in this paper is a characterization of the A&A meta-model that aims at further simplifying the design and programming of MASs. First of all, in CoSE agents are not allowed to communicate directly, not even via message exchange; agent interaction can occur only by way of artifacts (as postulated by Activity Theory [13]). In addition, CoSE better characterizes the content of artifact observable properties. These properties are not only (shared) state variables, but are also, and more importantly, *social relationships*. Namely, structures that explicitly represent the dependencies existing between any two agents that interact through a same artifact.

In this paper we show how the programming of interacting agents can be systematically approached by relying on the explicit representation of the agents' *social engagement*. The basic idea is that when agents can directly handle social relationships as *resources*, the coding phase can be organized in a precise sequence of steps. Specifically, we focus on social relationships that can be captured as *social commitments* [19]. The advantages are both on the *software engineering* perspective (decoupling of code), and on the *modeling* perspective (agents may consider truly social dependencies, and thus other agents, in their deliberative cycle).

2   Coordination via Social Engagement

We consider social relationships that can be represented as *social commitments* [19]. A commitment $C(x, y, s, u)$ captures that agent x (debtor) commits to agent y (creditor) to bring about the consequent condition u when the antecedent condition s holds. Antecedent and consequent conditions are conjunctions or disjunctions of events and commitments. Commitments have a *life cycle*. We adopt the commitments life cycle proposed in [21]. Briefly, a commitment is *Null* right before being created; *Active* when it is created. Active has two substates: *Conditional* (as long as the antecedent condition did not occur), and *Detached* (when the antecedent condition occurred). In the latter case, the debtor is now engaged in the consequent condition of the commitment. An Active commitment can become: *Pending* if suspended; *Satisfied*, if the engagement is accomplished; *Expired*, if it will not be necessary to accomplish the consequent condition; *Terminated* if the commitment is canceled when Conditional or released when Active; and finally, *Violated* when its antecedent has been satisfied, but its consequent will be forever false, or it is canceled when Detached (the debtor will be considered liable for the violation). As usual, commitments are manipulated by the commitment operations *create* (an agent creates a commitment toward someone), *cancel* (a debtor withdraws an own commitment), *release* (an agent withdraws a commitment of which it is the creditor), *assign* (a new creditor is specified by the previous one), *delegate* (a new debtor is specified by the previous one), *discharge* (consequent condition u holds). Since debtors are expected to behave so as to satisfy their engagements, commitments create social expectations on the agents' behaviors [9]. Moreover, since we implement commitments as observable properties of an artifact, they can be used by agents in their practical reasoning together with beliefs, intentions, and goals for taking into account other agents and the conditions the latter committed to have achieved.

Programming a MAS in our setting requires, as in the environment programming methodology, to program both *agents* and *artifacts*. CoSE provides a *commitment-driven methodology* for programming agents.

Artifacts. We consider artifacts that include commitments among their observable properties; whenever this happens, we say that the artifact has a *social state*. Thus, an artifact *Art* is formally represented as a tuple $\langle S, O, R, \rho \rangle$, where:

1. S is a social state, namely, a set of state variables and social relationships represented as commitments;
2. O is a set of artifact operations made available to the agents focusing on the artifact; each operation *op* in O has a social meaning: The execution of *op* creates a new social relationship (i.e., creates a new commitment), or it makes an existing social relationship evolve (i.e., the state of a commitment changes). In other words, whenever an artifact operation is executed, the artifact social state changes not only because some state variables change their values, but also because some social relationships evolve (e.g., new commitments are created, or existing commitments are detached/satisfied). Along the line discussed in [2], we say that the type τ of an operation *op*, $op : \tau$, is given by the set $\{c_1, \ldots, c_n\}$ of commitments that are created by the execution of *op*.
3. R is a set of role names exposed by the artifact: an agent which intends to use the artifact must enact one the roles it exposes.
4. $\rho : R \to 2^O$ is a function mapping role names to subset of operations in O; for each $r \in R$, $\rho(r)$ denotes the subset of operations that an agent playing role r can legally perform on the artifact.

Since operations are typed, and since roles map to operations, we can associate each role $r \in Art.R$ with a type. Formally, a role $r \in Art.R$ has a type τ, $r : \tau$, such that

$$\tau = \bigcup_{op_i \in \rho(r) | op_i : \tau_i} \tau_i,$$

namely, the type of a role is the union of all the types of the operations that are associated with that role.

Agents. For the purposes of this paper, an agent Ag can be abstracted as a triple $\langle \Sigma, B, G \rangle$; where Σ is the agent internal state, inspectable by Ag only, and B is a set $\{b_1, \ldots, b_m\}$ of *behaviors*, each of which enables Ag to perform a given activity. In other terms, each behavior represents a piece of software that the designer foresees in the agent specification phase, and, then, the programmer actually implements in the development phase. Also behaviors can be associated with types. According to [2], a behavior b has type τ, denoted as $b : \tau$, where τ is, as before, a set of commitments $\{c_1, c_2, \ldots, c_n\}$. A behavior b of type τ is capable of satisfying the commitments in the type. [2] also pointed out how the usage of typed behaviors enables a dynamic type-checking of agents, guaranteeing that an agent can only enact roles for which it can satisfy all the involved commitments. Finally, G is a set of goals assigned to the agent.

CoSE Methodology. Programming an agent Ag, thus, comes down to implementing the set of its behaviors. To this aim, the CoSE methodology suggests a programmer the following steps. Let $\mathcal{E}nv$ be a programming environment consisting of a set $\{Art_1, \ldots, Art_n\}$ of artifacts, let G be a set $\{g_1, \ldots, g_k\}$ of goals assigned to Ag, and let B be the initial (possibly empty) set of behaviors of Ag. For each goal $g_i \in G$:

1. If g_i can be obtained by Ag without the need of interacting with other agents, then, program the behavior b for achieving g_i and add b to B;
2. Otherwise, Ag needs to interact with other agents. To this aim, select a suitable artifact Art_i in $\mathcal{E}nv$. This choice is made by relying on the artifact roles and operations. Specifically, the following matches are considered:
   (a) $\exists op \in Art_i.O$ *whose social meaning is* $create(\mathsf{C}(x, y, g_i, p))$. Intuitively, an agent playing the role x *offers* to have the consequent condition p achieved if some other agent, playing role y, will have g_i achieved. Thus, if Ag will play the role x, by performing op it will become the debtor of a conditional commitment, created in the social state $Art_i.S$. If another agent, playing role y, will bring about the antecedent g_i (i.e., the goal Ag is interested in), Ag will be engaged in bringing about the consequent.
   (b) $\mathsf{C}(Ag', y, q, g_i) \in Art_i.S$ *and* $\exists op \in Art_i.O$, *with social meaning* $detach(\mathsf{C}(Ag', y, q, g_i))$. Intuitively, the social state $Art_i.S$ already contains an offer by agent Ag', and if Ag plays role y, then, by performing op, it will accept that offer. Thus, by achieving the antecedent condition q, Ag will bind the other agent to bring about g_i.
3. Once a suitable artifact Art_i has been selected, the programmer has to identify the role(s) in $Art_i.R$ agent Ag could play during its execution. Such a step is partially based on the set of operations that have been previously recognized as useful for achieving g_i. Note, however, that function ρ just maps roles to subsets of operations, but it does not induce a partition on $Art_i.O$; that is, given any two roles $r1$ and $r2$ in R, $\rho(r1) \cap \rho(r2)$ is not necessarily empty. Therefore, once the programmer has identified the operations Ag needs for achieving g_i, the programmer has to select, among all roles in $Art_i.R$ enabling such operations, a role r that better than others fits her/his needs. Let *Roles* be such a set of selected role(s) in Art_i.

4. For each role $r \in Roles$, let r be of type $\tau=\{c_1,\ldots,c_n\}$. Then, for each commitment $c_j \in \tau$, program a behavior b agent Ag assumes whenever a state change occurs on c_j, and add b in B.

Following these steps, the programmer will implement an agent incrementally, by considering one goal at a time, and by focusing on subproblems (i.e., behaviors), that either directly or via interaction will be programmed to obtain the goal at hand.

CoSE enjoys the following properties:

- *Agent-to-Agent Decoupling*: Since agents can only interact through artifacts, the separation between agents is even more neat than in the direct message exchange case. The exchange of messages, in fact, assumes that the two agents (the sender and the receiver) share a common language. In CoSE it is only required that agents are capable of using artifact operations. The advantage is to promote agent openness and heterogeneity.
- *Agent-Logic-to-Coordination-Logic Decoupling*: Programming an agent just requires the designer to consider two main aspects: (1) the domain-dependent process for reaching the goal the agent is devised for, and (2) the behavior of the agent as a response to changes in the social states of the artifacts the agent is focusing on. The coordination logic is no longer part of the agent; rather, the coordination logic is only implemented within the artifact. Among the main advantages, *code verification*, i.e. the interaction logic is programmed just in one precise portion of the system and verified once only, and *Code Maintainability*, i.e. the interaction logic is not spread across the agents, changes to the interaction logic just involve the artifact, while agents do not need to change.

3   JaCaMo+: Programming Coordination with Social Relationships

JaCaMo+ builds on the seminal work [1] and on *JaCaMo* [6], a platform integrating Jason (as an agent programming language), CArtAgO (as a realization of the A&A meta-model), and Moise (as a support to the realization of organizations). A MAS realized in JaCaMo is a Moise agent organization, which involves a set of Jason agents, all working in CArtAgO environments. CArtAgO environments can be designed and programmed as a dynamic set of shared artifacts, possibly distributed among various nodes of a network, that are collected into workspaces. By *focusing* on an artifact, an agent registers to be notified of events that are generated inside the artifact; e.g., when other agents execute some action.

Jason [7] implements in Java, and extends, the agent programming language AgentSpeak(L). Jason agents are characterized by a BDI architecture: Each of them has (1) its belief base, which is a set of ground (first-order) atomic formulas; and (2) its set of plans (plan library). It is possible to specify two types of goals:

achievement goals (atomic formulas prefixed by the '!' operator) and *test goals* (prefixed by '?'). Agents can reason on their beliefs/goals and react to events, amounting either to belief changes (occurred by sensing their environment) or to goal changes. Each plan has a triggering event (an event that causes its activation), which can either be the addition or the deletion of some belief or goal. The syntax is inherently declarative. In JaCaMo, the beliefs of Jason agents can also change due to operations performed by some agent of the MAS on the CArtAgO environment, whose consequences are automatically propagated.

JaCaMo+ extends JaCaMo along different directions. In particular, JaCaMo+ reifies the social relationships (commitments) as resources that are available to the interacting agents. This was obtained by enriching CArtAgO artifacts with an explicit representation of commitments and of commitment-based interaction protocols. In this way, JaCaMo+ seamlessly integrates Jason BDI agents with social commitments. The resulting class of artifacts reifies the execution of commitment-based protocols, including the social state, and enables Jason agents to be notified about the social events, and to perform practical reasoning about *social expectations* thanks to commitments: Agents expect that the commitment debtors behave so as to satisfy the corresponding consequent conditions, and use such information to decide about their own behavior and goals.

A *protocol artifact* is a JaCaMo+ artifact that implements a commitment-based protocol, structured into a set of roles, which can manipulate the protocol social state. By enacting a role, an agent receives "social powers", whose execution has public social consequences, expressed in terms of commitments. A JaCaMo+ agent, focusing on a protocol artifact, has access to the social state of the artifact. The implementation, actually, maps the social state onto a portion of the belief base each such agent has: any change occurred in the artifact's social state is instantaneously propagated to the belief bases of all the focusing agents. Agents are, thereby, constantly aligned with the social state.

An agent playing a role can only execute the protocol actions that are associated with such a, otherwise the artifact raises an exception that is notified to the violator. When a protocol action is executed, the social state is updated accordingly by adding new commitments, or by modifying the state of existing commitments. The artifact is responsible for maintaining the social state up-to-date, following action execution and the commitment life cycle.

JaCaMo+ allows specifying Jason plans, whose triggering events involve social relationships; i.e., commitments. In JaCaMo+, a commitment is represented as a term $cc(debtor, creditor, antecedent, consequent, status)$ where *debtor* and *creditor* identify the involved agents, while *antecedent* and *consequent* are the commitment conditions. As a difference with standard commitment notation, we explicitly represent a commitment state by means of the *Status* parameter. Commitments can be used inside a *plan context* or *body*. Differently than beliefs, commitment assertion/deletion can only occur through the artifact, as a consequence of a change of the social state. For example, this is the case that deals with commitment addition:

$$+cc(debtor, creditor, antecedent, consequent, status) : \langle context \rangle \leftarrow \langle body \rangle.$$

The plan is triggered when a commitment that unifies with the one specified in the plan head appears in the social state. The syntax is the standard for Jason plans. *Debtor* and *creditor* are to be substituted by the proper role names. The plan may be aimed at achieving a change of the commitment status (e.g., the debtor will try to bring about the consequent and satisfy the commitment), or at allowing the agent to do something as a reaction (e.g., collecting information). Similar schemata can be defined to tackle commitment deletion and the addition (deletion) of social facts. JaCaMo+ allows using commitments also in contexts and plans as test goals $?cc(\dots)$, or achievement goals $!cc(\dots)$. Addition or deletion of such goals can, as well, be managed by plans. For example:

$$+!cc(debtor, creditor, antecedent, consequent, status) : \langle context \rangle \leftarrow \langle body \rangle.$$

The plan is triggered when the agent creates an achievement goal concerning a commitment. Consequently, the agent will act upon the artifact so as to create the desired social relationship. After the execution of the plan, the commitment $cc(debtor, creditor, antecedent, consequent, status)$ will hold in the social state, and will be projected onto the belief bases of each agent focusing on the artifact.

4   CoSE Methodology in Action

We explain the impact of our proposals by comparing the JaCaMo implementation [18] of Dijkstra's Dining Philosophers, with an implementation obtained via the CoSE methodology in JaCaMo+. The problem involves two roles and, thus, two kinds of agents: *waiter*, which is in charge of initializing the artifact, and *philosopher*. (We omit *waiter* because trivial and not interactive.)

In the JaCaMo implementation, coordination is obtained by relying on a CArtAgO artifact. Goal start initializes information about the philosopher's forks, and starts the main loop (*thinking* and then *eating*).

```
1  !start .
2  +!start  <-  .my_name(Me);
3             in("philo_init",Me,Left,Right);
4             +my_left_fork(Left); +my_right_fork(Right);
5             !living.
6  +!living  <-  !thinking;
7             !eating;
8             !!living.
9  +!eating  <-  !acquireRes; !eat; !releaseRes.
10 +!acquireRes:  my_left_fork(Left) & my_right_fork(Right)
11    <-  in("ticket");
12        in("fork",Left); in("fork",Right).
13 +!releaseRes:  my_left_fork(Left) & my_right_fork(Right)
14    <-  out("fork",Left); out("fork",Right);
15        out("ticket").
16 +!thinking  <-  .my_name(Me); println(Me," thinking").
17 +!eat  <-  .my_name(Me); println(Me," eating").
```

Listing 1.1. The philosopher in JaCaMo [18].

Eating requires using the artifact for gaining forks and also a ticket that is used for avoiding deadlocks. Each agent implements the coordination policy in its plans, through the artifact operations *in* and *out*, as it directly manages the acquisition and release of forks (and tickets). So, even though Agent-to-Agent decoupling is achieved, there is *no clear separation of concerns* between the agent programming and the artifact programming for what concerns the coordination logic; consequently, there is a tight coupling between the agents and the artifacts that allow their interaction. This hinders the specification of an agent programming methodology independent of that of the artifacts.

A first improvement to this solution (still in JaCaMo) could be placing the coordination logic inside the artifact; this could be achieved by: (1) moving the calls of *in* and *out* inside new higher level operations (in the following, *askForks* and *returnForks* respectively); and (2) introducing an observable property *availableForks* that is notified in the agent's belief base when forks are available:

```
1 +!eating:  my_left_fork(Left) & my_right_fork(Right)
2       <- askForks(Left, Right).
3 +availableForks(Left, Right) <- !eat; returnForks(Left, Right).
```

This second solution, though improved, is still not completely satisfactory. The relation between *askForks* and *availableForks* (the latter is a consequence of the former), that is fundamental to the programmer, is *hidden* inside the artifact. The agent invokes *askForks* (a service, at all respects) and, when forks are available, the artifact tells the agent through the observable property *availableForks*. Observable properties are *signals*, the agent is programmed to react to signals. Indeed, the plan corresponding to *availableForks* is activated as a reaction to the creation of the corresponding belief: The relation, that ties the plan to the event that activates it, is *causal*, but since it is not expressed explicitly, the agent is not enabled to perform any kind of reasoning. Also in this second solution, thus, such a relation depends on the coordination logic that is contained in the artifact. Once again, the lack of separation of concerns is troublesome for the specification of an agent programming methodology independent of artifacts.

```
1 /* Initial goals */
2 !counter(0).
3 /* Plans */
4 !start.
5 +!start: true
6       <- focusWhenAvailable("philoArtifact");
7          enact("philosopher").
8 +enacted(Id,"philosopher",Role_Id)
9       <- +enactment_id(Role_Id);
10         .my_name(Me);
11         in("philo_init",Me,Left,Right);
12         +my_left_fork(Left);
13         +my_right_fork(Right);
14         !!living.
15 +!living: counter(C)
16       <- !thinking;
17          !eating.
18 +!eating: my_left_fork(Left) & my_right_fork(Right) & counter(C)
19       <- .my_name(Me); ?enactment_id(Role_Id);
20          askForks(Left, Right, C).
21 +cc(My_Role_Id, "philosopher", available(Left,Right,C),
22       returnForks(Left,Right,C),"DETACHED")
23    :  enactment_id(My_Role_Id) & my_left_fork(Left) &
24       my_right_fork(Right) & counter(C)
25    <-  !eat(Left, Right, C);
26        returnForks(Left, Right, C).
27 +cc(My_Role_Id, "philosopher", available(Left,Right,C),
28       returnForks(Left,Right,C), "SATISFIED")
```

```
29      :    enactment_id(My_Role_Id) & my_left_fork(Left)
30      <-   ?counter(C); -+counter(C+1); !living.
31 +!eat(Left, Right, C): my_left_fork(Left) & my_right_fork(Right)
32           & available(Left, Right, C) & counter(C)
33      <-   .my_name(Me); ?enactment_id(Role_Id).
34           println(Me, " ", Role_Id, " eating").
35 +!thinking: counter(C)
36      <-   .my_name(Me); ?enactment_id(Role_Id);
37           println(Me, " ", Role_Id, " thinking, time ",C).
```

Listing 1.2. The philosopher agent program in JaCaMo+.

In JaCaMo+, instead (Listing 1.2), coordination relies on social engagement. The agent program is built by exploiting the CoSE methodology. As in the JaCaMo solution, we suppose our agent has a *!living* main cycle (ln. 15) that alternates the goals *!thinking* and *!eating*. Coordination is needed just for eating: to this aim, forks must be available. Hence, we consider an environment that contains at least one artifact which satisfies one of the cases of step (2) in the methodology. Suppose case (a) is satisfied by an artifact exposing a role *philosopher*, which is empowered with an operation *askForks*, that will let the agent on stand-by until forks are available and, then, will create a commitment to return the assigned forks. The agent who executes the operation is the debtor of such a commitment, any other philosopher is the creditor. The antecedent condition is that forks are available and the consequent is that forks will be returned. Note that fork assignment is decided by way of a coordination policy that is implemented in the artifact.

Now, we need to program the agent behavior in occurrence of the state changes of such a commitment; indeed, only state changes that are meaningful to the aims of the agent are to be tackled. In our case, only *Detached* and *Satisfied* are meaningful. When the commitment is detached, the agent eats and then executes *returnForks*, thus satisfying its commitment. When the commitment is satisfied, the agent can re-start its main cycle (*!living*).

In this case, *askForks* is *not a mere service* by the artifact; it creates a social engagement, whose debtor is the requesting philosopher, and the creditor is the whole class of *philosophers*. This is done by the commitment *C(My_Role_Id, "philosopher", available(Left,Right,C), returnForks(Left,Right,C))*. The agent is requested to include one or more behaviors for managing such a commitment and, in particular, for managing the case in which it is *Detached*. This is possible because *askForks* and the *event +cc(My_Role_Id , "philosopher" , available(Left,Right,C), returnForks(Left,Right,C), "DETACHED")*[1] are *tied by the social meaning* of the operation in an explicit way, and this information is available to the programmer, who does not need to know the coordination logic that is implemented inside the artifact. Knowing the social meanings of artifact operations is sufficient for coordinating with others correctly. The connection between the event "commitment detached" and the associated plan is not only causal, but rather the plan has the aim of satisfying the consequent condition of the commitment (*returnForks*). Once again, there is not need of knowing or using logics that are internal to the artifact. Thus, we achieve not only

---

[1] Meaning that a belief of type *cc*(...) was added to the agent's belief base.

Agent-to-Agent decoupling, but also a real Agent-Logic-to-Coordination-Logic Decoupling. Social meanings are the key element that enables the definition of an agent programming methodology.

5   Discussion and Conclusions

CoSE extends environments by realizing Engeström's *activity systems*, rather than mere artifacts. Citing [13, page31], activity systems extend the classical triadic model (subject, object, and mediating artifact) in that the outcome is no longer momentary (situational), but consists of *new, objectified meanings* and *relatively lasting patterns of interaction*. In our case, objectified meanings are supplied by reified commitments, and all interactions are driven by such meanings instead than by the events (signs or signals), that are "physically" executed by the agents. Reified commitments also specify expected behaviors in terms of *what* is to be achieved rather than *how*.

The introduction of a commitment-based shared semantics of events allows for the design of software that meets many *software engineering principles*. First of all, *abstraction*: a failure to separate *behavior* (i.e., what) from *implementation* (i.e., how) is a common cause of unnecessary coupling among the components in a system. The what-level is tackled, in our proposal, by working at the level of commitments, and this is the only information that matters for carrying on an interaction/coordination. An agent who takes on a commitment assumes the responsibility that something will occur. Now then, who will make it occur, and which steps will bring to the outcome, are left to the how-level of single agent programming. Such steps may depend on the context, accommodating emerging opportunities, or managing specific difficulties. In other words, it is generally not necessary to impose any strict causal chain in signal generation.

Indeed, CoSE facilitates a *separation of concerns* between agent programming and the programming of agent coordination: As social relationships abstract the actual events upon which agents should coordinate, programmers can consider the programming of agents and the programming of coordination artifacts as two distinct problems. Programmers can define an agent's behavior on the sole basis of the *semantics* of social relationships, rather than on low-level events. Focusing on artifacts, programmers focuses on how the occurrence of an event changes the states of a set of social relations. Since social relationships are more abstract than events, the decoupling brings along further beneficial properties like *modularity* and *reuse*.

The *generality* principle calls for the development of software that is free from unnatural restrictions and limitations. This is precisely what is achieved by an approach that focuses on what rather than on how and that relies on a declarative, rather than on a procedural, representation. The *incremental development* principle advocates the incremental realization of software; e.g., one case at a time. Our proposal meets this requirement in that agent software can be developed by tackling one commitment at a time, or even one commitment state change at a time. A carefully planned incremental development process can also simplify the management of changes in requirements.

Concerning agent-based design, many proposals are found in the literature. Briefly, SODA [15] is an agent-oriented methodology for the analysis and design of agent-based systems, adopting a layering principle and a tabular representation. It focuses on inter-agent issues, like the engineering of societies and environment for MAS, and relies on a meta-model that includes both agents and artifacts. However, SODA does not foresee social engagements nor it provides an agent programming methodology. GAIA [24] is a methodology for developing a MAS as an organization, not for programming agents. The 2CL Methodology [5] is an extension of [12]. It supports the design of commitment-based business protocols that include temporal constraints, and allows the verification of properties. As such, it could be used for helping the realization of artifacts, although the methodology is general and not oriented to this specific abstraction.

Social engagements are at the basis also of organization-oriented programming, of which JaCaMo is a prominent example. Recently, [25] proposed a JaCaMo extension that introduces Interaction Components to encode–in an automaton-like shape–protocols, where transitions are associated with (undirected) *obligations*. Such protocols provide *guidelines* of how organizational goals should be achieved. However, organization-driven guidelines are but a kind of interaction. We claim that when guidelines are missing, interaction should be supported based on the fundamental notions of goal and engagement. So, our proposal complements [25], and more in general organizational and normative approaches [10,11,14,16], in this respect. Social commitments [19], differently from obligations, are taken by agents as a result of internal deliberative processes, and can be directly manipulated by the agents, In addition, [22] shows how goals and commitments are strongly interrelated. Commitments are, thus, evidence of the agents' capacity to take responsibilities autonomously. Citing Singh [20], an agent would become a debtor of a commitment based on the agent's own communications: either by directly saying something or having another agent communicate something in conjunction with a prior communication of the debtor. That is, there is a *causal path* from the establishment of a commitment to prior communications by the debtor of that commitment. Such causal relationships are at the heart of the CoSE methodology. By contrast, obligations can result from a deliberative process which is outside the agent; this is the case of the interaction component in [25]. For a detailed discussion of the differences between obligations and commitments see [3].

Consequently, our proposal differs deeply also from proposals like [8], which also account for a social dimension of the MAS. That work, for instance, presents the SOPL language, that allows including, in each of the agents' programs, states of affairs that the agent tolerates (even though they are not explicit goals of its own), and rules to reason about the other agents' mental states (social conditions). Social conditions comprise possible evolutions depending on how other agents behave, they can vary from agent to agent, and are used by the agent in the process of deciding how to act. Nevertheless, mental states are private to the agents and the absence of a semantics of actions based on mutual agreement makes speculations about the others' behavior fragile, because expectations do not base upon explicit engagements. From a software engineering perspective, then, since evolutions of interactions are encoded in the very agent programs, the proposal does not support the decoupling of agents from the interaction logic.

Future work will concern tackling the formal notions of *social context* and of *enactment of a protocol in a social context* introduced in [4], as well as further exploring the use of typing systems, along the lines of [2].

Acknowledgments. This work was partially supported by the *Accountable Trustworthy Organizations and Systems (AThOS)* project, funded by Università degli Studi di Torino and Compagnia di San Paolo (CSP 2014).

References

1. Baldoni, M., Baroglio, C., Capuzzimati, F.: A Commitment-based Infrastructure for Programming Socio-Technical Systems. ACM Transactions on Internet Technology, Special Issue on Foundations of Social Computing **14**(4), 23:1–23:23 (2014)
2. Baldoni, M., Baroglio, C., Capuzzimati, F.: Typing multi-agent systems via commitments. In: Dalpiaz, F., Dix, J., van Riemsdijk, M.B. (eds.) EMAS 2014. LNCS, vol. 8758, pp. 388–405. Springer, Heidelberg (2014)
3. Baldoni, M., Baroglio, C., Capuzzimati, F., Micalizio, R.: Leveraging commitments and goals in agent interaction. In: Ancona, D., Maratea, M., Mascardi, V. (eds.) Proc. of XXX Italian Conference on Computational Logic, CILC (2015)
4. Baldoni, M., Baroglio, C., Chopra, A.K., Singh, M.P.: Composing and verifying commitment-based multiagent protocols. In: Wooldridge, M., Yang, Q. (eds.) Proc. of 24th International Joint Conference on Artificial Intelligence, IJCAI 2015, Buenos Aires, Argentina, July 25–31, 2015
5. Baldoni, M., Baroglio, C., Marengo, E., Patti, V., Capuzzimati, F.: Engineering commitment-based business protocols with the 2CL methodology. JAAMAS **28**(4), 519–557 (2014)
6. Boissier, O., Bordini, R.H., Hbner, J.F., Ricci, A., Santi, A.: Multi-agent oriented programming with JaCaMo. Science of Computer Programming **78**(6), 747–761 (2013)
7. Bordini, R.H., Fred Hübner, J., Wooldridge, M.: Programming Multi-Agent Systems in AgentSpeak Using Jason. John Wiley & Sons (2007)
8. Buccafurri, F., Caminiti, G.: Logic programming with social features. Theory and Practice of Logic Programming (TPLP) **8**(5–6), 643–690 (2008)
9. Conte, R., Castelfranchi, C., Dignum, F.P.M.: Autonomous norm acceptance. In: Papadimitriou, C., Singh, M.P., Müller, J.P. (eds.) ATAL 1998. LNCS (LNAI), vol. 1555, pp. 99–112. Springer, Heidelberg (1999)
10. Criado, N., Argente, E., Noriega, P., Botti,: Reasoning about norms under uncertainty in dynamic environments. International Journal of Approximate Reasoning (2014)
11. Dastani, M., Grossi, D., Meyer, J.-J.C., Tinnemeier, N.A.M.: Normative multi-agent programs and their logics. In: Normative Multi-Agent Systems, 15.03. - 20.03.2009. Dagstuhl Seminar Proceedings. Schloss Dagstuhl - Leibniz-Zentrum für Informatik, Germany, vol. 09121 (2009)
12. Desai, N., Chopra, A.K., Singh, M.P.: Amoeba: A methodology for modeling and evolving cross-organizational business processes. ACM Trans. Softw. Eng. Methodol. **19**(2) (2009)

13. Engeström, Y., Miettinen, R., Punamäki, R.-L. (eds.): Perspectives on Activity Theory. Cambridge University Press, Cambridge (1999)
14. Meneguzzi, F., Luck, M.: Norm-based behaviour modification in BDI agents. In: AAMAS, vol. 1, pp. 177–184. IFAAMAS (2009)
15. Molesini, A., Omicini, A., Denti, E., Ricci, A.: SODA: a roadmap to artefacts. In: Dikenelli, O., Gleizes, M.-P., Ricci, A. (eds.) ESAW 2005. LNCS (LNAI), vol. 3963, pp. 49–62. Springer, Heidelberg (2006)
16. Okouya, D., Fornara, N., Colombetti, M.: An infrastructure for the design and development of open interaction systems. In: Winikoff, M. (ed.) EMAS 2013. LNCS, vol. 8245, pp. 215–234. Springer, Heidelberg (2013)
17. Omicini, A., Ricci, A., Viroli, M.: Artifacts in the A&A meta-model for multi-agent systems. JAAMAS **17**(3), 432–456 (2008)
18. Ricci, A., Piunti, M., Viroli, M.: Environment programming in multi-agent systems: an artifact-based perspective. Autonomous Agents and Multi-Agent Systems **23**(2), 158–192 (2011)
19. Singh, M.P.: An ontology for commitments in multiagent systems. Artif. Intell. Law **7**(1), 97–113 (1999)
20. Singh, M.P.: Commitments in multiagent systems some controversies, some prospects. In: Paglieri, F., Tummolini, L., Falcone, R., Miceli, M. (eds.) The Goals of Cognition. Essays in Honor of Cristiano Castelfranchi, vol. 31, pp. 601–626. College Publications, London (2011)
21. Telang, P.R., Singh, M.P., Yorke-Smith, N.: Relating goal and commitment semantics. In: Dennis, L., Boissier, O., Bordini, R.H. (eds.) ProMAS 2011. LNCS, vol. 7217, pp. 22–37. Springer, Heidelberg (2012)
22. Telang, P.R., Yorke-Smith, N., Singh, M.P.: Relating goal and commitment semantics. In: Dennis, L., Boissier, O., Bordini, R.H. (eds.) Programming Multi-Agent Systems. LNCS, vol. 7212, pp. 22–37. Springer, Heidelberg (2012)
23. Weyns, D., Omicini, A., Odell, J.: Environment as a first class abstraction in multiagent systems. JAAMAS **14**(1), 5–30 (2007)
24. Zambonelli, F., Jennings, N.R., Wooldridge, M.: Developing multiagent systems: The Gaia methodology. ACM Trans. Softw. Eng. Methodol. **12**(3), 317–370 (2003)
25. Zatelli, M.R., Hübner, J.F.: The interaction as an integration component for the JaCaMo platform. In: Dalpiaz, F., Dix, J., van Riemsdijk, M.B. (eds.) EMAS 2014. LNCS, vol. 8758, pp. 431–450. Springer, Heidelberg (2014)

Anticipatory Coordination in Socio-Technical Knowledge-Intensive Environments: Behavioural Implicit Communication in $\mathcal{MoK}$

Stefano Mariani$^{(\boxtimes)}$ and Andrea Omicini

DISI, Alma Mater Studiorum–Università di Bologna, Bologna, Italy
{s.mariani,andrea.omicini}@unibo.it

Abstract. Some of the most peculiar traits of socio-technical KIE (knowledge-intensive environments) – such as unpredictability of agents' behaviour, ever-growing amount of information to manage, fast-paced production/consumption – tangle coordination of information, by affecting, e.g., reachability by knowledge prosumers and manageability by the IT infrastructure. Here, we propose a novel approach to coordination in KIE, by extending the $\mathcal{MoK}$ model for knowledge self-organisation with key concepts from the cognitive theory of BIC (behavioural implicit communication).

1 Introduction

Socio-technical systems (STS) arise when cognitive and social interaction are mediated by information technology, rather than by the natural world alone [18]: in other words, any system in which the infrastructure enabling and constraining interaction is technological, but the evolution of the system is driven by social and cognitive interactions, is a STS. By definition, STS are heavily interaction-centred, so they need proper *coordination* mechanisms at the infrastructure level to harness the intricacies of run-time dependencies between the agents (either software or human) participating the system [8]. However, designing effective coordination is made complex by, at least, two aspects of STS:

unpredictability — By definition, STS have "humans-in-the-loop", and, whereas software behaviour is programmable and predictable, human's one is not. Accordingly, the coordination infrastructure may only draw the boundaries within which user behaviour can stretch, by defining the set of admissible actions and interactions at users' disposal.

scale — STS are typically physically-distributed open systems, often large-scale ones, connecting an ever-increasing number of people, devices, data. Hence, the coordination infrastructure of STS should exploit decentralised coordination mechanisms to be able to scale in/out upon need.

In addition, STS are often deployed within *knowledge-intensive environments* (KIE), that is, workplaces in which sustainability of the organisation long-term goals is influenced by (if not even dependant on) the evolution of knowledge

© Springer International Publishing Switzerland 2015
M. Gavanelli et al. (Eds.): AI*IA 2015, LNAI 9336, pp. 102–115, 2015.
DOI: 10.1007/978-3-319-24309-2_8

embodied within the organisation itself [1]. The fact that knowledge is an *organised* combination of data, procedures, and operations, continuously interacting and evolving driven by human users' practice and (learnt) experience [1], motivates why, usually, KIE are computationally supported by STS. Therefore, KIE, too, call for suitable coordination mechanisms, whose development is far from trivial, mostly due to the following key aspects of KIE:

size — KIE store a massive amount of raw data (knowledge-intensive in space), aggregated information, reification of procedures and best-practices, and the like. The coordination infrastructure should then minimise the overhead of additional information needed for coordination-related functional and non-functional requirements, by relying as much as possible on the information already in the KIE.

pace — Likewise, data within KIE is produced and consumed at a fast pace (knowledge-intensive in time): when the system features a huge number of users, an ever-increasing computational load is inevitably charged on the underlying coordination infrastructure. Hence, coordination mechanisms adopted to organise information should be as simple and efficient as possible.

In order to tackle the issues above, coordination models and technologies draw inspiration from *distributed collective intelligence* phenomena in natural systems, looking for self-organising and adaptive coordination mechanisms—as witnessed, e.g., by [9,11,15,16,19]. Similarly, in this paper we focus on the "social layer" of STS, looking for novel coordination approaches inspired by the latest cognitive and social sciences research results. In particular, we take as a reference the $\mathcal{M}$olecules of $\mathcal{K}$nowledge ($\mathcal{MoK}$) coordination model for knowledge self-organisation in KIE [11], and extend it toward the notion of *anticipatory coordination* – as an efficient form of *collective intelligence* arising by emergence from a number of distributed non-intelligent agents –, according to the theory of *behavioural implicit communication* (BIC) [3].

Accordingly, the remainder of the paper is structured as follows: Section 2 summarises BIC and recaps the key features of $\mathcal{MoK}$; Section 3 presents the main contribution of the paper, that is, the BIC-oriented extension of $\mathcal{MoK}$ supporting anticipatory coordination; Section 4 reports on an early validation of the model; finally, Section 5 provides for concluding remarks and further works.

2   Background

2.1   Behavioural Implicit Communication

Behavioural implicit communication (BIC) is a form of implicit interaction with no specialised signal conveying the message, since the message is the practical behaviour itself [3]. This presupposes advanced *observation* capabilities: participants should be able to observe others' actions, as well as to *mind-read* the intentions behind them. Mind-reading enables the process of *signification*, that is, the ability to ascribe goals and intentions to actions and their effects (*traces*), or, in other words, meanings to signs. In turn, signification enables *anticipatory coordination*, that is, the ability to foresee possible interferences/opportunities so as to adapt accordingly, or, at least, to plan suitable coordinated actions [2].

The crucial point of BIC is that it applies to human beings, to both cognitive and non-cognitive agents, and to computational environments as well [17]. This paves the way towards the notion of *smart environments*, that is, pro-active, intelligent working environments able to autonomously and spontaneously adapt their behaviour according to users' interactions [3]—which is, not by chance, the very notion of anticipatory coordination. Also, smart environments enable BIC based on the observation of traces of actions, too. *Trace-based communication* is related to the notion of *stigmergy*, introduced in the biological study of social insects [6] to explain the coordination of termites building their nest without exchanging messages—another form of distributed collective intelligence. Adopting the perspective taken in [3], stigmergy is communication via environment modifications which are not specialised signals: so, stigmergy can be interpreted as a special form of BIC, where the addressee does not directly perceive the behaviour, but just other post-hoc traces and outcomes of it.

In [14], an abstract model for smart environments, supporting BIC in the context of multi-agent systems (MAS), defines two types of environment:

c-env — A *common environment*, where agents can observe only the state of the environment, not the actions of their peers. A trace is modelled as a part of the environment, instead of as a product of other agents. *c-env* enables agents to modify environment state while keeping track of such changes.

s-env — A *shared environment*, as an enhanced *c-env* enabling different forms of observability of actions, and awareness of this observability—by the agents, and by the environment itself as well.

Accordingly, three fundamental features are required for a computational environment to fully support BIC-based coordination, closely related to observation, mind-reading, and signification abilities [14]: *(i)* observability of (human / software) agent actions, and of their traces as well, should be an intrinsic property of the environment; *(ii)* agents and the environment should be able to understand actions and their traces, possibly inferring intentions and goals motivating them—regardless of whether they are intelligent enough to perform true reasoning, or merely programmed to react properly; *(iii)* agents and the environment should also be able to understand the effects of their activity on other agents, so as to exploit the opportunity to obtain a desired reaction.

Section 3 describes how such requirements can be met in the specific case of a *MoK*-coordinated socio-technical KIE, and how *MoK compartments* [11] can be extended to support the notions of *c-env* and *s-env*.

2.2   The *Molecules of Knowledge* Model

Molecules of Knowledge (*MoK*) is a coordination model promoting *self-organisation* of information [11]. Drawing inspiration from *biochemical tuple spaces* [15] and *stigmergic coordination* [12], $\mathcal{MoK}$ pursues two main goals: *(i)* self-aggregation of information into more complex heaps, possibly reifying useful knowledge previously hidden; *(ii)* diffusion of information toward the interested agents, that is, those agents needing it to achieve their goals. The $\mathcal{MoK}$ model is built around the following abstractions:

seeds — The sources of information. Seeds continuously and spontaneously inject atoms (data chunks) into compartments (tuple-based repositories).

compartments — The repositories of information. Compartments are the computational and topological abstraction of $\mathcal{MoK}$, *(i)* defining the notions of *locality* and *neighborhood*, *(ii)* responsible for storing atoms, molecules and enzymes, and *(iii)* in charge of reactions scheduling and execution.

catalysts — The information *prosumers* (consumer + producer). Catalysts are the agents willing to exploit information living within the $\mathcal{MoK}$ system for their own purposes. As a side effect of their activity, catalysts influence the way in which information spontaneously aggregate and diffuse within compartments – in one word, evolves – driven by $\mathcal{MoK}$ reactions.

atoms — The atomic unit of information. Continuously injected into compartments by seeds, atoms are subject to $\mathcal{MoK}$ reactions and agents actions.

molecules — The composite unit of information. Molecules are the reification of similarities between atoms, spontaneously tied together by $\mathcal{MoK}$ reactions.

enzymes — The reification of catalysts' actions. Enzymes are automatically produced by the compartment within which the action is being done, then exploited by $\mathcal{MoK}$ reactions to influence information evolution.

reactions — The "laws of nature" driving information evolution. Reactions are the *coordination laws* dictating how information evolves, and how catalysts may influence such process. $\mathcal{MoK}$ features five reactions[1]:
 - *injection* extracts atoms from seeds and puts them into compartments
 - *aggregation* ties together semantically related atoms into molecules, or molecules into other molecules
 - *diffusion* moves atoms and molecules between neighboring compartments
 - *decay* destroys atoms and molecules as time passes by
 - *reinforcement* consumes enzymes to increase *concentration* of atoms and molecules (relevance w.r.t. others in the same compartment)

A $\mathcal{MoK}$-coordinated system is thus a network of $\mathcal{MoK}$ compartments (tuple-space-like information repositories), in which $\mathcal{MoK}$ seeds (sources of information) continuously and spontaneously inject $\mathcal{MoK}$ atoms (information pieces). $\mathcal{MoK}$ atoms may then aggregate (into molecules, more complex information chunks), diffuse, being reinforced, decay. Such autonomous and decentralised processes are driven by $\mathcal{MoK}$ reactions (coordination laws) and influenced by $\mathcal{MoK}$ enzymes (reification of user actions), transparently released by $\mathcal{MoK}$ catalysts (users, either human or software agents) while performing their activities. $\mathcal{MoK}$ reactions are scheduled by $\mathcal{MoK}$ compartments according to Gillespie's *chemical dynamics simulation* algorithm [5], so as to promote chemical-inspired self-organisation based on *locality*, *situatedness*, and *stochasticity* [11].

---

[1] In [11] reactions were four; injection was added in [10].

3   Towards Anticipatory Coordination

In this section, the $\mathcal{M}$olecules of $\mathcal{K}$nowledge model is extended toward anticipatory coordination, by borrowing BIC concepts [3]. In particular, Subsection 3.1 extends the notion of compartment according to the definition of smart environments provided in [14], while Subsection 3.2 extends the definition of enzymes and introduces the trace abstraction into the $\mathcal{M}o\mathcal{K}$ model. In addition, Subsection 3.3 proposes a set of actions that catalysts may use to interact with a $\mathcal{M}o\mathcal{K}$ system, along with their impact on enzymes and traces generation.

3.1 $\mathcal{M}o\mathcal{K}$ Compartments as Shared Smart Environments

$\mathcal{M}o\mathcal{K}$ compartments may play the role of smart environments, since they model locality in $\mathcal{M}o\mathcal{K}$, and also constitute $\mathcal{M}o\mathcal{K}$ computational environment. *Neighbourhoods* are another topological abstraction, defined in $\mathcal{M}o\mathcal{K}$ as the set of compartments connected by some infrastructural relationship—in the simplest case, physical spatial proximity or direct ("1-hop") network reachability. Since neighbourhoods, too, define a notion of locality and computational environment – being $\mathcal{M}o\mathcal{K}$ diffusion reaction explicitly bound to neighbouring compartments – they can be regarded as smart environments, too. Recursively, the characterisation of smart environment can be extended to the network of compartments—therefore, to the whole $\mathcal{M}o\mathcal{K}$-coordinated system.

According to [11], the only sort of smart environment enabled in $\mathcal{M}o\mathcal{K}$ is *c-env*, mapped upon a compartment, because: *(i)* a $n : 1$ relationship is assumed between compartments and catalysts—no sharing of working environments is supported; *(ii)* enzymes are visible only to $\mathcal{M}o\mathcal{K}$ reactions; *(iii)* enzymes cannot diffuse, thus neighbourhoods cannot perceive them [11]. So, $\mathcal{M}o\mathcal{K}$ does not support *s-env* since there is no observable action reification in shared environments. Also, support to *c-env* is limited to compartments – not neighbourhoods – since enzymes cannot diffuse. Hence, an extension of the notions of compartment and enzyme is needed to enable *s-env* and improve support to *c-env*:

- each compartment no longer belongs to a single catalyst
- enzymes are: *(i)* *diversified* to resemble the epistemic nature of the action they reify; *(ii)* made observable to users sharing the compartment they live in; *(iii)* no longer consumed by reinforcement reaction, but now subject to decay; *(iv)* now generating *traces* through a *deposit* reaction
- traces are introduced as the $\mathcal{M}o\mathcal{K}$ abstraction resembling (side) effects of actions; as such, traces are: *(i)* different in kind, according to their father enzyme; *(ii)* observable only by $\mathcal{M}o\mathcal{K}$ reactions; *(iii)* subject to diffusion, decay and to an enzyme-dependant *perturbation reaction*—novel in $\mathcal{M}o\mathcal{K}$

This enables full support to the notions of *s-env* and *c-env* in $\mathcal{M}o\mathcal{K}$, and makes it possible to match the three requirements for anticipatory coordination mentioned in Subsection 2.1.

Now, *compartments* represent *s-env*, as the shared working environment where catalysts' actions are made observable to others, and to the environment itself. Also, neighbourhoods represent *c-env*, where action traces may diffuse, becoming part of the environment as they participate $\mathcal{MoK}$ reactions. Observability is now an intrinsic environment property, since compartments enable observability by design. Also, actions may be observed either directly or via their traces (their effects), making it easier to infer goals, as well as to understand how actions affect peers—in particular, when epistemic actions are concerned.

3.2 Enzymes and Traces as BIC Enablers

In [11], traces, along with both perturbation and deposit reactions, are missing, whereas enzymes and reinforcement reaction are formalised as follows:

$$\texttt{enzyme}(mol)_c \qquad \texttt{enzyme}(mol') + mol_c \xrightarrow{r_{reinf}} mol_{c+1}$$

where subscript c denotes concentration, and mol, mol' are supposed to match according to some matching criteria—e.g., LINDA matching [4] or OWL subsumption [7]. Traces, perturbation reaction, and deposit reaction are defined below, while enzymes and reinforcement are re-defined accordingly:

$$\texttt{enzyme}(species,\ s,\ mol)_c$$
$$\texttt{enzyme}(species,\ s,\ mol') + mol_c \xrightarrow{r_{reinf}} \texttt{enzyme}(species,\ s,\ mol') + mol_{c+s}$$
$$\texttt{trace}(enzyme,\ p,\ mol)_c$$
$$\texttt{trace}(enzyme,\ p,\ mol') + mol_c \xrightarrow{r_{pert}} .\texttt{exec}\,(p,\ trace,\ mol)$$
$$enzyme \xrightarrow{r_{dep}} enzyme + \texttt{trace}(enzyme,\ p[species],\ mol)$$

where *species* defines the epistemic nature of the action, s is the strength of reinforcement, p is the perturbation the trace wants to perform, and .*exec* starts execution of perturbation p—notice, p is implicitly defined by *species*, as highlighted by notation $p[species]$. Also, decay reaction is extended to enzymes and traces, whereas diffusion to traces solely. Thus, in the extended version of $\mathcal{MoK}$:

- enzymes belong to a certain *species*, reflecting the epistemic nature of actions, and determine the perturbation action performed by generated traces; enzymes also provide a bounded feedback (strength s)
- reinforcement reaction no longer consumes enzymes, which now decay
- traces belong to enzymes—defining (through species) perturbation action p
- perturbation reaction consumes a trace and the related molecule, then triggers execution of the perturbation action
- deposit reaction generates traces from enzymes, without consuming them

The role played by enzymes and traces in anticipatory coordination is then fundamental: they are the abstractions supporting observation of catalysts' actions by both other users and by the environment. In addition, reinforcement and perturbation reactions are the mechanisms enabling mind-reading and signification on the environment side. Reinforcement is meant to influence relevance of the information users manipulate during their workflows, according to the nature and frequency of their actions, so as to better support them in pursuing their goals. Enzymes cannot participate in diffusion reaction because the actions they reify are *situated*, that is, happen at a precise time as well as in a precise space (the compartment). Mind-reading and signification are supported by assuming that users manipulating a given corpus of information are interested in that information more than other. Perturbation is meant to influence relevance, location, content, namely any domain-specific trait of information, in response to users' actions and according to their nature (enzymes' species), with the goal of easing and optimising users' workflows.

Thus, traces are free to wander in the network of $\mathcal{MoK}$ compartments looking for a chance to apply their perturbation action, actually enabling the environment not only to perceive users' action traces, but also to exploit them for the profit of the coordination process—promoting the distributed collective intelligence leading to anticipatory coordination. Mind-reading and signification are supported by assuming that every user action may be interpreted by the environment without the need to directly estimate users' intentions and goals, but inferring them from the characteristics of the business domain within which the $\mathcal{MoK}$-coordinated socio-technical KIE is deployed.

3.3   Tacit Messages to Steer Anticipatory Coordination

Based on a survey of heterogeneous socio-technical KIE – such as Facebook, Twitter, Mendeley and Storify – we devised the most common actions provided to users: here we discuss the BIC *tacit messages* they could convey, and the kind of *perturbation actions* that could be designed accordingly.

Tacit messages are proposed in [3] to describe the kind of messages a practical action (and its traces) may implicitly send to the observers:

1. *presence* — "Agent X is here". Since an action (trace) is observable in shared compartments (neighbourhoods), any agent therein becomes aware of X existence and location—likewise for the environment.
2. *intention* — "X plans to do action b". If the agents' workflow determines that action b follows action a, peers (as well as the environment) observing X doing a may assume X next intention to be "do b". Accordingly, the environment may decide to undertake anticipatory coordination actions easing/hindering action b—e.g. because action b is computationally expensive.
3. *ability* — "X is able to do $a_{i=1,...,n}$". Assuming actions $a_{i=1,...,n} \in A$ have similar pre-conditions, agents (and the environment) observing X doing a_i may infer X is also able to do $a_{j \neq i}$. Accordingly, the environment may further (no longer) support such pre-conditions, enabling (prohibiting) actions $\in A$.
4. *opportunities* — "$[e_1, \ldots, e_n]$ is the set of pre-conditions for doing a". Agents observing X doing a may infer that $[e_1, \ldots, e_n]$ hold, thus, they may take the opportunity to do a immediately. The environment in turn, making similar observations, may act as seen for tacit message 3.
5. *accomplishment* — "X achieved S". If S is the state of affairs reachable as a consequence of doing action a, agents observing X doing a may infer that X is now in state S. Since the environment too can make a similar inference, it may anticipate X next intentions from, e.g., its estimated state S.

6. *goal* — "X has goal g". By observing X doing action a, peers of X may infer X's goal to be g, e.g. because action A is part of a workflow aimed at achieving g—likewise for the environment. Accordingly, the environment may act similarly to what seen for tacit message 2.

7. *result* — "Result r is available". If peer agents know that action a brings result r, whenever agent X does a they can expect result r to be soon available—in case action a completes successfully. The environment in turn, may start planning coordination actions involving result r, e.g., synchronisation of parallel activities for agents waiting for r.

Since agents can undertake the above described inferences, *MoK* compartments actually act as BIC-based enablers of distributed collective intelligence phenomena—e.g., anticipatory coordination emerging due to agent interaction.

The above categorisation is general enough to suit several different application domains and practical actions. In the case of socio-technical KIE, we identified a set of fairly-common actions, in spite of the diversity in scope of the software platforms—e.g. Facebook vs. Mendeley:

- *quote/share* — re-publishing or mentioning someone else's information can convey, e.g., tacit messages 1, 3, 5. If X shares Y's information through action a, every other agent observing a becomes aware of existence and location of both X and Y (1). The fact that X is sharing information I from source S lets X's peers infer X can manipulate S (3). If X shared I with Z, Z may infer, e.g., that X expects Z to somehow use it (5).
- *like/favourite* — marking as relevant a piece of information can convey, e.g., tacit messages 1, 4. If the socio-technical platform lets X be aware of Y marking information I as relevant, X may infer that Y exists (1). If Y marks as relevant I belonging to X, X may infer that Y is interested in her work, perhaps seeking for collaborations (4).
- *follow* — subscribing for updates regarding some piece of information or some user can convey tacit messages 2, 4. Since X manifested interest in Y's work through subscription, Y may infer X intention to use it somehow (2). Accordingly, Y may infer the opportunity for, e.g., collaboration (4).
- *search* — performing a search query to retrieve information can convey, e.g., tacit messages 1, 2, 4—notice however, which assumptions to make about a search action heavily depends on which search criteria are supported. If X search query is observable by peer agents, they can infer X existence and location (1). Also, they can infer X goal to acquire knowledge related to its search query (2). Finally, along the same line, they can take the chance to provide matching information (4).

Accordingly, perturbation actions may range from sending discovery messages informing agents about the presence and location of another (1), to establishing privileged communication channels so as to ease collaborations (4); from undertaking coordination actions enabling/forbidding some interaction protocol (2, 3, 6), to proactively notifying users about availability of novel information (4, 7).

4   Experiment

In the following we simulate a citizen journalism scenario, where users share a $\mathcal{MoK}$-coordinated IT platform for retrieving and publishing news stories. Users have personal/shared devices (smartphones, tablets, pcs, workstations) running the $\mathcal{MoK}$ middleware, which they use to search the IT platform for relevant information. Searches can spread up to a logical neighbourhood of the searched compartment – for a number of reasons: limiting bandwidth consumption, boosting security, optimising information location, etc. –, including those of colleagues interested in stories belonging to similar topics. User searches leave traces that the $\mathcal{MoK}$ middleware exploits to attract similar information, actually enacting anticipatory coordination.

Fig. 1, 2a-2b demonstrate how the emergent collective intelligence phenomena enabling anticipatory coordination is effectively supported by suitable BIC-inspired abstractions and mechanisms. The coordination infrastructure does not know in advance the effectiveness of its coordination activities in supporting users' workflows: it can only try to react to users' activities at its best, according to its own interpretation of users' goals. This is exactly what anticipatory coordination is: the infrastructure tries to foresee the user coordination needs even before users do, with the aim of satisfying them at best.

Fig. 1a shows the initial configuration: information molecules (coloured dots) are randomly scattered throughout the grid (black squares)—light-blue little squares represent links between compartments, allowing diffusion. Fig. 1b highlights two compartments in which enzymes (coloured flags) have just been released, thus traces begin to spawn and diffuse (coloured arrows): green enzymes in the bottom-left one, cyan enzymes in the top-right one[2]. Fig. 1c demonstrates that the expected clusters appear: red molecules (brought by green traces' perturbation action) have the highest concentration in the bottom-left (highlighted) compartment, likewise magenta molecules (brought by cyan traces) in the top-right one. Fig. 1d-1f demonstrate that clusters are transient: they last as long as users' action effects (enzymes and traces) last. In fact, besides new clusters appearing (magenta molecules, top-left and yellow molecules, bottom-right), the previous ones either disappear (magenta cluster, top-right) or are replaced (orange cluster, bottom-left). This *adaptiveness* feature is confirmed by Fig. 2a-2b, plotting the oscillatory trend of clustered ("still") molecules and traces. Also, Fig. 1d-1f highlight other desirable features of $\mathcal{MoK}$, stemming from its biochemical inspiration and BIC, respectively: *locality* and *situatedness* (of both computations and interactions). In fact, as neighbouring compartments can influence each other through diffusion, they can also act independently by, e.g., aggregating different molecules.

As a last note, we remark how the extended $\mathcal{MoK}$ model deals with the typical issues of socio-technical KIE highlighted in Section 1. In terms of unpredictability,

---

[2] Colours represent semantic differences for different matches: red molecules match green enzymes/traces, orange molecules match lime enzymes/traces, yellow molecules match turquoise enzymes/traces, magenta molecules match cyan enzymes/traces, pink molecules match sky blue enzymes/traces.

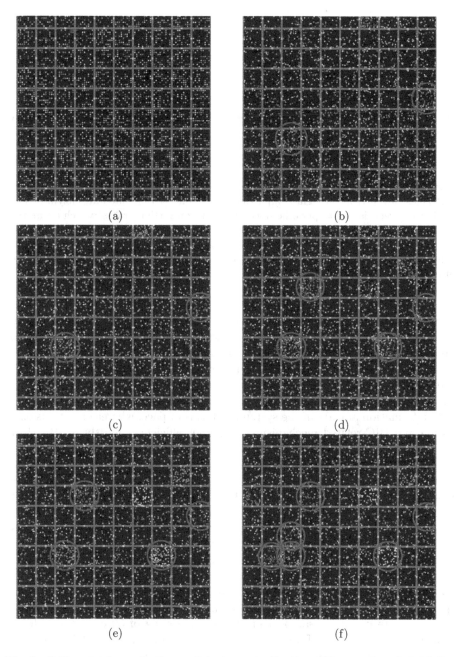

Fig. 1. *Self-organising, adaptive anticipatory coordination.* Whereas data is initially randomly scattered across workspaces (a), as soon as users interact (b) clusters appear by emergence thanks to *BIC-driven self-organisation* (c, d). Whenever new actions are performed by catalysts, the $\mathcal{MoK}$ infrastructure *adaptively* re-organises the spatial configuration of molecules (e) so as to better tackle the new coordination needs (f).

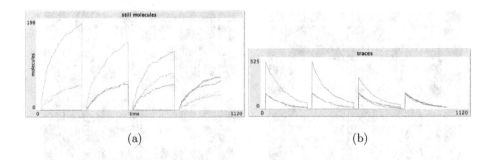

(a)   (b)

Fig. 2. On the left, concentration of still molecules over time. Still molecules represent molecules currently in the right compartment—the one storing matching enzymes. The oscillatory trend is due to periodic injection of enzymes (thus traces) which clears the "still" state of molecules. The different colours correspond to the different molecules. On the right, concentration of traces over time. Traces move molecules to the right compartment—the one storing matching enzymes. The oscillatory trend is due to decay of traces over time. The different colours correspond to the different traces.

$\mathcal{MoK}$ anticipates user coordination needs, not based on future behaviour prediction, but rather on present actions and its mind-reading and signification abilities. In terms of scale, $\mathcal{MoK}$ reactions act only locally, thus self-organisation exploits local information only. In terms of size, $\mathcal{MoK}$ decay helps mitigating the issue by destroying information and meta-data as time passes; furthermore, the overhead brought by $\mathcal{MoK}$ BIC-based extension is minimal, since it exploits only information already in the system. In terms of pace, whereas reactions execution and BIC-related mechanisms are rather efficient – mostly due to their local nature – there is a fundamental issue still to be addressed in $\mathcal{MoK}$: the semantic similarity measure ($\mathcal{F}_{\mathcal{MoK}}$ in [11]). On the one hand, an accurate measure likely leads to more meaningful clusters; on the other hand, it often requires more expensive computations. Thus, a tradeoff is needed—our efforts for further developments of $\mathcal{MoK}$ are also devoted to investigate this issue.

Technical details of the experiment are as follows[3]: 100 $\mathcal{MoK}$ compartments are networked in a grid (4 neighbours per compartment, except border)—see Fig. 1; 2500 molecules, split in 5 non-overlapping semantic categories (representing matching with different enzymes), are uniformly sampled then randomly scattered in the grid—statistically, 500 molecules per category; 250 enzymes, split in the same categories, are generated in 5 random compartments; enzymes' categories are uniformly sampled in batches consisting of 50 enzymes each, so that generated enzymes of a given category are always multiple of 50; enzymes are generated *periodically* (every 250 time steps) and subject to decay; 2 traces

---

[3] Simulation tool used is NetLogo 5.0.5, available from http://ccl.northwestern.edu/netlogo/. Videos of the simulations are available on YouTube (https://youtu.be/8ibkXdukTfk). Source code of the simulations are to be released as a NetLogo model, available from http://ccl.northwestern.edu/netlogo/models/community/.

per enzyme are generated, coherently with enzymes' category and according to the same time interval; traces too are subject to decay, although at a lower rate w.r.t. enzymes—due to their different purpose: representing *long-term effects* of actions for the former, reifying *situated* actions for the latter.

The simulations proceed as follows: molecules randomly diffuse among neighbouring compartments; enzymes reify a search action which successfully collects a set of molecules from the local compartment; enzymes stand still in the compartment where the action took place until decay, generating traces; traces, representing tacit message 2, randomly diffuse among neighbouring compartments until either (i) decay or (ii) find a matching molecule to apply their perturbation action to; the perturbation action makes the involved molecule diffuse toward the compartment where the trace's father enzyme belong.

5   Conclusion and Further Work

In this paper we propose a novel approach to coordination in socio-technical KIE. In particular, we extend the $\mathcal{M}$olecules of $\mathcal{K}$nowledge model [11] to support the notion of *anticipatory coordination* [3]. To this end, concepts from the cognitive theory of BIC are brought within the $\mathcal{MoK}$ model both by extending existing abstractions – compartments and enzymes – and by introducing new abstractions and mechanisms—traces, deposit reaction and perturbation action. To evaluate our proposal, we simulate how to obtain intelligent spatial distribution of information with $\mathcal{MoK}$, based solely on user interaction, as an example of distributed collective intelligence—in particular, anticipatory coordination.

Although our experiment focusses on one specific pattern of anticipatory coordination, we believe that the results achieved are more than encouraging, thus deserve further investigation. In particular, simulations of other $\mathcal{MoK}$ behaviours – e.g., re-arrange the network of compartments so as to reflect the current collaborations among catalysts – are actually in progress, and will help further validating both the extended $\mathcal{MoK}$ model, and the practice of applying BIC theory to coordination in socio-technical KIE.

Furthermore, our efforts are currently devoted to fully implement and run a $\mathcal{MoK}$ coordinated system on a large-scale scenario—e.g. the one here simulated. In fact, although a prototype implementation of $\mathcal{MoK}$ exists, such a large-scale deployment has not been achieved, yet. As far as implementation is concerned, special care will be paid to the semantic similarity measure. In our experience, ontology-based semantic matching is rather unfeasible, except for basic relationships only, e.g., subsumption alone. On the contrary, purely syntactical matching has too low expressiveness. Viable tradeoffs may be usage of wildcards, e.g. as in Java *regular expressions*[4], or of synonymy relationships only (hyperonymy for "is-a" relationships, meronymy fort "part-of", etc.), e.g., as done in [13] using WordNet[5].

---

[4] http://docs.oracle.com/javase/tutorial/essential/regex/
[5] http://wordnet.princeton.edu

References

1. Bhatt, G.D.: Knowledge management in organizations: Examining the interaction between technologies, techniques, and people. Journal of Knowledge Management **5**(1), 68–75 (2001)
2. Castelfranchi, C.: Modelling social action for AI agents. Artificial Intelligence **103**(1–2), 157–182 (1998)
3. Castelfranchi, C., Pezzullo, G., Tummolini, L.: Behavioral implicit communication (BIC): Communicating with smart environments via our practical behavior and its traces. International Journal of Ambient Computing and Intelligence **2**(1), 1–12 (2010)
4. Gelernter, D.: Generative communication in Linda. ACM Transactions on Programming Languages and Systems **7**(1), 80–112 (1985)
5. Gillespie, D.T.: Exact stochastic simulation of coupled chemical reactions. The Journal of Physical Chemistry **81**(25), 2340–2361 (1977)
6. Grassé, P.P.: La reconstruction du nid et les coordinations interindividuelles chez Bellicositermes natalensis et Cubitermes sp. la théorie de la stigmergie: Essai d'interprétation du comportement des termites constructeurs. Insectes Sociaux **6**(1), 41–80 (1959)
7. Horrocks, I.: OWL: a description logic based ontology language. In: Gabbrielli, M., Gupta, G. (eds.) ICLP 2005. LNCS, vol. 3668, pp. 1–4. Springer, Heidelberg (2005)
8. Malone, T.W., Crowston, K.: The interdisciplinary study of coordination. ACM Computing Surveys **26**(1), 87–119 (1994)
9. Mamei, M., Zambonelli, F.: Programming pervasive and mobile computing applications: The TOTA approach. ACM Transactions on Software Engineering and Methodology (TOSEM) **18**(4), July 2009
10. Mariani, S.: Parameter engineering vs. parameter tuning: the case of biochemical coordination in MoK. In: Baldoni, M., Baroglio, C., Bergenti, F., Garro, A. (eds.) CEUR Workshop Proceedings of the From Objects to Agents, vol. 1099, pp. 16–23. Sun SITE Central Europe, RWTH Aachen University, Turin, December 2–3, 2013
11. Mariani, S., Omicini, A.: Molecules of knowledge: self-organisation in knowledge-intensive environments. In: Fortino, G., Badica, C., Malgeri, M., Unland, R. (eds.) IDC 2012. SCI, vol. 446, pp. 17–22. Springer, Heidelberg (2012)
12. Van Dyke Parunak, H.: A survey of environments and mechanisms for human-human stigmergy. In: Weyns, D., Van Dyke Parunak, H., Michel, F. (eds.) E4MAS 2005. LNCS (LNAI), vol. 3830, pp. 163–186. Springer, Heidelberg (2006)
13. Pianini, D., Virruso, S., Menezes, R., Omicini, A., Viroli, M.: Self organization in coordination systems using a WordNet-based ontology. In: Gupta, I., Hassas, S., Jerome, R. (eds.) 4th IEEE International Conference on Self-Adaptive and Self-Organizing Systems (SASO 2010), pp. 114–123. IEEE CS, Budapest (2010)
14. Tummolini, L., Castelfranchi, C., Ricci, A., Viroli, M., Omicini, A.: "Exhibitionists" and "Voyeurs" do it better: a shared environment for flexible coordination with tacit messages. In: Weyns, D., Van Dyke Parunak, H., Michel, F. (eds.) E4MAS 2004. LNCS (LNAI), vol. 3374, pp. 215–231. Springer, Heidelberg (2005)
15. Viroli, M., Casadei, M.: Biochemical tuple spaces for self-organising coordination. In: Field, J., Vasconcelos, V.T. (eds.) COORDINATION 2009. LNCS, vol. 5521, pp. 143–162. Springer, Heidelberg (2009)

16. Viroli, M., Pianini, D., Beal, J.: Linda in space-time: an adaptive coordination model for mobile ad-hoc environments. In: Sirjani, M. (ed.) COORDINATION 2012. LNCS, vol. 7274, pp. 212–229. Springer, Heidelberg (2012)
17. Weyns, D., Omicini, A., Odell, J.J.: Environment as a first-class abstraction in multi-agent systems. Autonomous Agents and Multi-Agent Systems **14**(1), 5–30 (2007)
18. Whitworth, B.: Socio-technical systems. Encyclopedia of human computer interaction, 533–541 (2006)
19. Zambonelli, F., Castelli, G., Ferrari, L., Mamei, M., Rosi, A., Di Marzo, G., Risoldi, M., Tchao, A.E., Dobson, S., Stevenson, G., Ye, Y., Nardini, E., Omicini, A., Montagna, S., Viroli, M., Ferscha, A., Maschek, S., Wally, B.: Self-aware pervasive service ecosystems. Procedia Computer Science **7**, 197–199 (2011)

A Kinetic Study of Opinion Dynamics in Multi-agent Systems

Stefania Monica$^{(\boxtimes)}$ and Federico Bergenti

Dipartimento di Matematica e Informatica, Università degli Studi di Parma,
Parco Area delle Scienze 53/A, 43124 Parma, Italy
{stefania.monica,federico.bergenti}@unipr.it

Abstract. In this paper we rephrase the problem of opinion formation from a physical viewpoint. We consider a multi-agent system where each agent is associated with an opinion and interacts with any other agent. Interpreting the agents as the molecules of a gas, we model the opinion evolution according to a kinetic model based on the analysis of interactions among agents. From a microscopic description of each interaction between pairs of agents, we derive the stationary profiles under given assumption. Results show that, depending on the average opinion and on the model parameters, different profiles can be found, with different properties. Each stationary profile is characterized by the presence of one or two maxima.

1 Introduction

This paper describes a model for opinion formation in multi-agent systems. Many kinds of approaches have been investigated in the literature to study opinion evolution among agents based, e.g., on graph theory [1], cellular automata [2], and thermodynamics [3]. Recently, social interactions in multi-agent systems have been described according to microscopic models based on kinetic theory. Typically, kinetic theory is used to derive macroscopic properties of gases by analyzing the details of the collisions of the molecules [4]. Similarly, from the description of the details of each interaction between pairs of agents, the global opinion can be described from a macroscopic point of view [5].

The research interest related to the application of kinetic approaches to the description of multi-agent systems gave birth to new disciplines, namely *econophysics* and *sociophysics* [6]. Econophysics deals with the description of the evolution of market economy and wealth distribution in a society [7]. Sociophysics, instead, aims at characterizing the evolution of social features, such as opinion, in a society [8]. These disciplines are based on the fact that the formalism that describes the interactions between molecules in a gas can be adapted to describe the effects of interactions between agents. In particular, the kinetic framework can be used to outline agent-based cooperation models, such as that in [9], to study large scale systems, such as those in [10], and to model wireless sensor networks (see, e.g., [11]).

© Springer International Publishing Switzerland 2015
M. Gavanelli et al. (Eds.): AI*IA 2015, LNAI 9336, pp. 116–127, 2015.
DOI: 10.1007/978-3-319-24309-2_9

In this paper we focus on studying the opinion evolution in a multi-agent system on the basis of a kinetic formulation. Under the assumption that each agent is associated with an opinion $v \in I \subseteq \mathbb{R}$, we investigate how the opinion of the considered system changes on the basis of given rules that describe the effects of single interactions. The model that we consider is introduced in [12]. In particular, we assume that each agent can interact with any other agent in the system and that the opinion of each changes due to two different reasons, namely compromise and diffusion [13]. Compromise is related to the fact that, as a consequence of an interaction, an agent can be persuaded to change its opinion in favour of that of the interacting agent. This phenomenon is modeled as a deterministic process. Diffusion is instead modeled as a random process and it is related to the fact that agents can change their opinions autonomously.

The paper is organized as follows. Section 2 describes the considered kinetic model from an analytical viewpoint. Section 3 derives explicit formulas for the stationary profiles in a specific case. Section 4 shows simulation results for different values of the parameters of the model. Section 5 concludes the paper.

2   Kinetic Model of Opinion Formation

Sociophysics is based on the idea that the same laws that describe the interactions among molecules can be generalized to describe the effects of social interactions among agents. As a matter of fact, while molecules are typically associated with their velocities, agents can be associated with an attribute which represents one of its characteristics that can be, for instance, its opinion. In the following, we associate to each agent a parameter v defined in the interval $I = [-1, 1]$. According to such an assumption, ± 1 represent extremal opinions, while 0 correspond to the middle point of the interval of interest I.

Kinetic theory relies on the definition of a function $f(v, t)$ which represents the density of opinion v at time t and which is defined for each opinion $v \in I$ and for each time $t \geq 0$. Since $f(v, t)$ is a density function, the following equality needs to hold

$$\int_I f(v, t) \mathrm{d}v = 1. \tag{1}$$

In order to describe the opinion evolution using a kinetic approach, we assume that the function $f(v, t)$ evolves on the basis of the Boltzmann equation. In particular, we consider the following formulation of the Boltzmann equation

$$\frac{\partial f}{\partial t} = \mathcal{Q}(f, f)(v, t) \tag{2}$$

where $\frac{\partial f}{\partial t}$ represents the temporal evolution of the distribution function and $\mathcal{Q}$ is the *collisional operator* which takes into account the effects of interactions. In order to derive an explicit formula for the collisional operator $\mathcal{Q}$, the details of the binary interactions need to be described. In the considered model, the post-interaction opinions of two interacting agents are obtained by adding to their pre-interaction opinions a contribution related to compromise and a contribution related to diffusion, according to the following formula

$$\begin{cases} v' = v + \gamma C(|v|)(w - v) + \eta_* D(|v|) \\ w' = w + \gamma C(|w|)(v - w) + \eta D(|w|). \end{cases} \tag{3}$$

where the pair (v', w') denotes the post-interaction opinions of the two agents, whose pre-interaction opinions were (v, w). In (3) the second terms on the right hand side of the two equations are related to compromise, according to the parameter γ, which is defined in $(0, \frac{1}{2})$, and the function $C(\cdot)$; the third terms are related to diffusion, through the random variables η and η_* and the function $D(\cdot)$. The functions $C(\cdot)$ and $D(\cdot)$, which describe the impact of compromise and diffusion, respectively, depend on the absolute value of the opinion, namely they are symmetrical with respect to the middle point of I. Moreover, we assume that both functions are not increasing with respect to the absolute value of the opinion, coherently with the fact that, typically, extremal opinions are more difficult to change. Finally, we assume that

$$0 \le C(|v|), D(|v|) \le 1 \qquad \forall v \in I.$$

From (3), since both γ and $C(\cdot)$ are positive, the contribution of compromise is positive each time an agent interacts with another agent whose opinion value is greater while it is negative otherwise. Hence, the idea of compromise is respected, since the difference between the opinions of the two agents is reduced after the considered interaction if the diffusion term is neglected.

The contribution of diffusion, instead, can be either positive or negative depending on the value of the random variables η and η_*. In the following, we assume that such random variables have the same statistics. In particular, we assume that their average value is 0 and their variance is σ^2, namely

$$\int \eta \vartheta(\eta) \mathrm{d}\eta = \int \eta_* \vartheta(\eta_*) \mathrm{d}\eta_* = 0$$
$$\int \eta^2 \vartheta(\eta) \mathrm{d}\eta = \int \eta_*^2 \vartheta(\eta_*) \mathrm{d}\eta_* = \sigma^2. \tag{4}$$

where $\vartheta(\cdot)$ is the probability density function. In order to take into account the effects of diffusion we need to define the *transition rate*

$$W(v, w, v', w') = \vartheta(\eta)\vartheta(\eta_*)\chi_I(v')\chi_I(w') \tag{5}$$

where χ_I is the indicator function relative to the set I (equal to 1 if its argument belong to I, and to 0 otherwise) and it is meant to make sure that the post-interaction opinions are in I.

Under these assumptions, the explicit expression of the collisional operator Q defined in (2) can be finally written as

$$Q(f, f) = \int_{\mathbb{B}^2} \int_I \left[{}'W \frac{1}{J} f({}'v) f({}'w) - W f(v) f(w) \right] \mathrm{d}w \mathrm{d}\eta \mathrm{d}\eta_*$$

where $\mathbb{B}$ is the support of ϑ, $'v$ and $'w$ are the pre-interaction variables which lead to v and w, respectively, $'W$ is the transition rate relative to the 4−uple $('v,'w,v,w)$ and J is the Jacobian of the transformation of $('v,'w)$ in (v,w) [12].

Instead of solving (2) we consider its weak form. In functional analysis, the weak form of a differential equation is obtained by multiplying both sides of the considered equation by a test function $\phi(v)$, namely a smooth function with compact support, and then integrating the obtained equation with respect to v. The weak form of the Boltzmann equation can be derived from (2) and, using a proper change of variable in the integral, it can be written as

$$\frac{\mathrm{d}}{\mathrm{d}t} \int_I f(w,t)\phi(v)\mathrm{d}v = \int_{\mathbb{B}^2} \int_{I^2} W f(v)f(w)(\phi(v') - \phi(v))\mathrm{d}v\mathrm{d}w\mathrm{d}\eta\mathrm{d}\eta_* \tag{6}$$

If we consider $\phi(v) = 1$ in (6) then the following equation is obtained

$$\frac{\mathrm{d}}{\mathrm{d}t} \int_I f(v,t)\mathrm{d}v = 0 \tag{7}$$

which says that the number of agents is constant. This property is analogous to mass conservation of the molecules in a gas.

Considering $\phi(v) = v$ as a test function in (6) and recalling (3) we obtain

$$\begin{aligned}\frac{\mathrm{d}}{\mathrm{d}t} \int_I f(w,t)v\mathrm{d}v = \gamma \int_{\mathbb{B}^2} \int_{I^2} W f(v)f(w)C(|v|)(w - v)\mathrm{d}v\mathrm{d}w\mathrm{d}\eta\mathrm{d}\eta_* \\ + \int_{\mathbb{B}^2} \int_{I^2} W f(v)f(w)\eta D(|v|)\mathrm{d}v\mathrm{d}w\mathrm{d}\eta\mathrm{d}\eta_*\end{aligned} \tag{8}$$

Defining the average value of the opinion at time t as

$$u(t) = \int_I f(w,t)v\,\mathrm{d}v \tag{9}$$

the left hand side of (8) corresponds to the derivative $\dot{u}(t)$ of the average opinion. The first integral in the right hand side of (8) can be written as

$$\gamma \int_I f(v)C(|v|)\mathrm{d}v \int_I vf(v)\mathrm{d}v - \gamma \int_I f(v)C(|v|)v\mathrm{d}v. \tag{10}$$

The second integral in (8) is 0 because the average value of ϑ is 0, according to (4). Therefore, from (8) and (10) it can be obtained that the variation of the average opinion u can be written as

$$\dot{u}(t) = \gamma \int_I f(v)C(|v|)\mathrm{d}v \int_I vf(v)\mathrm{d}v - \gamma \int_I f(v)C(|v|)v\mathrm{d}v. \tag{11}$$

Observe that if C is constant then (10) is 0 for symmetry and (11) becomes

$$\dot{u}(t) = 0 \tag{12}$$

i.e., the average opinion is conserved, namely $u(t) = u(0)$. This property corresponds to the conservation of momentum.

We are interested in studying the behaviour of the distribution function $f(v, t)$ for large values of the time t and to derive, eventually, stationary profiles. In order to simplify notation we first define a new temporal variable τ

$$\tau = \gamma t \tag{13}$$

where γ is the coefficient related to compromise which appear in (3). Assuming that $\gamma \simeq 0$, namely that each interaction causes small opinion exchange,

$$g(v, \tau) = f(v, t) \tag{14}$$

describes the asymptotic behaviour of $f(v, t)$. The weak form of a Fokker-Planck equation can be derived by substituting $f(v, t)$ with $g(v, \tau)$ in (6) and using a Taylor series expansion of $\phi(v)$ around v in (6) [12]:

$$\frac{dg}{d\tau} = \frac{\lambda}{2} \frac{\partial^2}{\partial v^2} (D(|v|)^2 g) + \frac{\partial}{\partial v} ((v - u)g) \tag{15}$$

where

$$\lambda = \sigma^2 / \gamma. \tag{16}$$

We are now interested in studying stationary solutions g_∞ of (15), which satisfy

$$\frac{dg_\infty}{d\tau} = 0. \tag{17}$$

In next section we analyze these solutions for different values of λ.

3   Stationary Behaviour of Opinion Distribution

In this section we derive some stationary profiles for the opinion density g. Such profiles are defined as solutions of (17) and, therefore, they depend on parameters u and λ and on the choice of the diffusion function.

In the remaining of the paper, we assume that the compromise function $C(|v|)$ is constant and equal to 1. As observed in Section 2, this choice leads to a constant value of the average opinion, which is denoted as u in the following. We consider the following distribution function

$$D(|v|) = 1 - v^2 \tag{18}$$

which is a non increasing function of $|v|$, as discussed at the beginning of the previous section. According to this assumption, the effects of the interactions between pairs of agents described in (3) are

$$\begin{cases} v' = v + \gamma(w - v) + \eta(1 - v^2) \\ w' = w + \gamma(v - w) + \eta_*(1 - w^2) \end{cases} \tag{19}$$

In order to guarantee that the post-interaction opinions still belong to the interval of interest I we need to define the support $\mathbb{B}$ of the distribution function $\vartheta(\cdot)$ of η and η_*. Considering the first equation in (19), we can conclude that

$$|v'| \leq (1-\gamma)|v| + \gamma + |\eta|(1-v^2)$$

from which it can be derived that if $|\eta| \leq M = \frac{1-\gamma}{1+|v|}$ then $|v'| \leq 1$. Analogous considerations hold for $|w'|$ when taking into account the second equation of (19). Since the minimum value of M is obtained in correspondence of the maximum values of $|v'|$ and γ, namely when $|v| = 1$ and $\gamma \simeq 1/2$, then it can be concluded that if $|\eta| \leq \frac{1}{4}$, then $|v'| \leq 1$ independently of the pre-interaction opinion v. The same holds for $|w'|$, therefore from now on we assume that $\mathbb{B} = (-1/4, 1/4)$. We are now interested in finding the stationary solutions, namely the functions which satisfy (17). From (15) the stationary solutions satisfy

$$\frac{\lambda}{2}\frac{\partial}{\partial v}((1-v^2)^2 g) + (v-u)g = C \tag{20}$$

where u is the average opinion (which is constant) and C is a constant. Observe that the constant C must be 0. As a matter of fact, by integrating (20) one obtains

$$\frac{\lambda}{2}\int_{-v_1}^{v_2}\frac{\partial}{\partial v}((1-v^2)^2 g) + \int_{-v_1}^{v_2}(v-u)g = C(v_2 + v_1). \tag{21}$$

From the previous equation, if $v_1 \to 1$ and $v_2 \to 1$ then the first integral is 0 for symmetry and the second integral can be written as

$$\int_I vg dv - u\int_I g dv = u - u = 0.$$

It can then be concluded from (21) that $C = 0$.

Using classical analysis in (20), one obtains

$$\frac{g'}{g} = \frac{4v}{1-v^2} + \frac{2(u-v)}{\lambda(1-v^2)^2}. \tag{22}$$

The left hand side of the previous equation is the derivative of $\log g$. Integrating the right hand side of (22) leads to an explicit expression of $\log g$, and, therefore, of g. The first added on the right hand side of (22) can be written as

$$\frac{d}{dv}(-2\log(1-v^2)) \tag{23}$$

Concerning the remaining terms in (22), first observe that

$$\frac{2u}{\lambda}\frac{1}{(1-v^2)^2} = \frac{d}{dv}\left(\frac{u}{2\lambda}\log\left(\frac{1+v}{1-v}\right) + \frac{uv}{\lambda(1-v^2)}\right). \tag{24}$$

Moreover one can calculate

$$-\frac{2v}{\lambda(1-v^2)^2} = \frac{1}{2\lambda}\left(\frac{1}{(1+v)^2} - \frac{1}{(1-v)^2}\right) = -\frac{1}{2\lambda}\frac{d}{dv}\left(\frac{1}{1+v} + \frac{1}{1-v}\right)$$
$$= -\frac{1}{\lambda}\frac{d}{dv}\frac{1}{1-v^2}. \tag{25}$$

Finally, using (23), (24), and (25), equation (22) can be written as

$$\frac{d}{dv} \log g(v) = \frac{d}{dv}\left[-2\log(1-v^2) + \frac{u}{2\lambda}\log\left(\frac{1+v}{1-v}\right) + \frac{uv-1}{\lambda(1-v^2)} \right]$$

and, therefore,

$$\log g(v) = \log(1-v^2)^{-2} + \log\left(\frac{1+v}{1-v}\right)^{\frac{u}{2\lambda}} + \frac{uv-1}{\lambda(1-v^2)} + \alpha_{u,\lambda}. \qquad (26)$$

where $\alpha_{u,\lambda}$ is a constant depending on the average opinion u and on the value of λ. Taking the exponential of (26) the following expression for the stationary solution is derived

$$g_\infty(v) = c_{u,\lambda}(1+v)^{-2+\frac{u}{2\lambda}}(1-v)^{-2-\frac{u}{2\lambda}} \exp\left(\frac{uv-1}{\lambda(1-v^2)}\right) \qquad (27)$$

where $c_{u,\lambda}$ must be determined in order to satisfy

$$\int_I g_\infty(w) = 1. \qquad (28)$$

Observe that if $u = 0$, then $g_\infty(v)$ is an even function.

In order to see if the stationary profile is characterized by maxima and/or minima, we now aim at studying the derivative of g_∞. From (22) the derivative of g_∞ can be written as

$$g'_\infty(v) = g_\infty(v)\left(\frac{4\lambda v(1-v^2) + 2(u-v)}{\lambda(1-v^2)^2}\right) \qquad (29)$$

and, therefore,

$$g'_\infty(v) = 0 \quad \Longleftrightarrow \quad g_\infty(v) = 0 \quad \vee \quad 4\lambda v^3 + (2-4\lambda)v - 2u = 0. \qquad (30)$$

From (27), $g_\infty(v) = 0$ if and only if $v = \pm 1$, namely in the extremes of the considered interval I. Hence, we are interested in finding the solutions of the second condition in (30), namely the solutions of

$$v^3 + \left(\frac{1}{2\lambda} - 1\right)v - \frac{u}{2\lambda} = 0. \qquad (31)$$

Observe that equation (31) is a polynomial equation of degree 3 and therefore it always admits at least one real solution.

If $u = 0$, namely if the average opinion is the middle point of I, (31) becomes

$$v^3 + \left(\frac{1}{2\lambda} - 1\right)v = 0 \qquad (32)$$

and in this case the solutions are

$$v_1 = 0 \qquad v_{2,3} = \pm\sqrt{1 - \frac{1}{2\lambda}} \qquad (33)$$

Observe that if $\lambda \leq \frac{1}{2}$ the only real root of (32) is $v_1 = 0$ and its multiplicity is 1 if $\lambda < \frac{1}{2}$ while it is 3 if $\lambda = \frac{1}{2}$. In these cases, v_1 is a maximum point. If $\lambda > \frac{1}{2}$, instead, equation (32) admits three real roots. In this last case, v_1 is a point of minimum while v_2 and v_3 are points of maximum.

If $u \neq 0$ the solution of (31) requires the use of Cardano's formula for the solution of polynomial equations of degree 3, according to which

$$v_1 = \sqrt[3]{-\frac{q}{2} + \sqrt{\Delta}} + \sqrt[3]{-\frac{q}{2} - \sqrt{\Delta}} \tag{34}$$

is a real root of (31), where

$$\Delta = \frac{q^2}{4} + \frac{p^3}{27} \qquad p = \left(\frac{1}{2\lambda} - 1\right) \qquad q = -\frac{u}{2\lambda} \tag{35}$$

If $\Delta \leq 0$ then equation (31) has three real roots, which, besides v_1, are

$$v_{2,3} = -\frac{v_1}{2} \pm \frac{1}{2}\sqrt{-4p - 3v_1^2}. \tag{36}$$

Hence, if $\Delta < 0$, the stationary profile g_∞ has three singular points. If $\Delta = 0$, then from (34) it can be concluded that v_1 has the following simplified expression

$$v_1 = 2\sqrt[3]{-\frac{q}{2}}.$$

Substituting this result in (36) one obtains that $-4p - 3v_1^2 = 0$ and, therefore, $v_2 = v_3 = -v_1/2$, namely equation (31) has three real roots, two of which are coincident. In the case with $\Delta = 0$ the singular points of g_∞ are two and one of them is also an inflection point. Finally, if $\Delta > 0$ then equation (31) has v_1 as the only real root, hence g_∞ has only one singular point.

From (35), the value of Δ can be expressed as a function of λ and u as

$$\Delta = \frac{27u^2\lambda + 2(1 - 2\lambda)^3}{432\lambda^3} \tag{37}$$

Since, from (16), λ is defined as the ratio between two positive quantities, one can conclude that

$$\Delta < 0 \iff u^2 < M(\lambda) = \frac{2}{\lambda}\left(\frac{1-2\lambda}{3}\right)^3. \tag{38}$$

Since $u \in I$, then $0 \leq u^2 \leq 1$ and, therefore, if $M(\lambda) \geq 1$ the inequality on the right hand side of (38) is satisfied for all the values of u, while if $M(\lambda) < 0$ the previous inequality is never satisfied. It can be shown that

$$M(\lambda) < 0 \iff \lambda < \frac{1}{2} \qquad M(\lambda) \geq 1 \iff \lambda \geq 2. \tag{39}$$

Hence, the following considerations hold:

- if $0 < \lambda < \frac{1}{2}$ then the condition $u^2 < M(\lambda)$ is never satisfied and, therefore, $\Delta > 0$ and the stationary profile g_∞ has only one singular point
- if $\lambda \geq 2$ then the condition $u^2 < M(\lambda)$ is satisfied for all the values of the average opinion u and, therefore, $\Delta < 0$ and the stationary profile g_∞ has three singular points
- if $\frac{1}{2} \leq \lambda < 2$, the number of stationary points of g_∞ depends on the value of the average opinion u.

4 Numerical Results

In this section, various stationary profiles for different values of u and λ are shown. We start by considering $u = 0$ so that the average opinion corresponds to the middle point of I. In this case, the stationary profile g_∞ is an even function.

In Fig. 1, the stationary profiles $g_\infty(v)$ are shown for various values of λ, namely $\lambda = 1/4$ (blue line), $\lambda = 1/2$ (red line), $\lambda = 1$ (green line), and $\lambda = 3$ (black line). Fig. 1 shows that if $\lambda = 1/4$, then $g_\infty(v)$ has only one maximum (corresponding to $u = 0$), in agreement with (33). If $\lambda = 1/2$, then $v = 0$ is the only stationary point of $g_\infty(v)$, but in this case the multiplicity of $v = 0$ as a solution of (32) is 3. Observe that the value of the maximum is smaller compared to that relative to $\lambda = 1/4$. If $\lambda > 1/2$, according to (33), the function $g_\infty(v)$ admits three stationary points. In particular, Fig. 1 shows that if $\lambda = 1$ there is a minimum in correspondence of $v = 0$ and two maxima in $v = \pm 1/\sqrt{2}$. In this case, the value of the two maxima is similar to that of the maximum obtained with $\lambda = 1/2$. Fig. 1 also shows the stationary profile $g_\infty(v)$ when $\lambda = 3$. In this case, the two points of maximum are closer to the extremes of the interval I where the opinion is defined and the values of the maxima are approximately the double of those relative to $\lambda = 1$. The value of the minimum

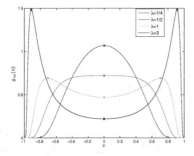
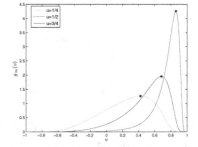

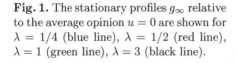

Fig. 1. The stationary profiles g_∞ relative to the average opinion $u = 0$ are shown for $\lambda = 1/4$ (blue line), $\lambda = 1/2$ (red line), $\lambda = 1$ (green line), $\lambda = 3$ (black line).

Fig. 2. The stationary profiles g_∞ relative to the value $\lambda = 1/4$ are shown for $u = 1/4$ (yellow line), $u = 1/2$ (green line), $u = 3/4$ (violet line).

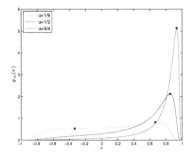
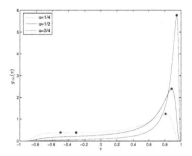

Fig. 3. The stationary profiles g_∞ relative to the value $\lambda = 3/4$ are shown for $u = 1/9$ (yellow line), $u = 1/2$ (green line), $u = 3/4$ (violet line).

Fig. 4. The stationary profiles g_∞ relative to the value $\lambda = 1$ are shown for $u = 1/4$ (yellow line), $u = 1/2$ (green line), $u = 3/4$ (violet line).

corresponding to 0, instead, is nearly halved with respect to the previous case. Fig. 1 shows that, if $u = 0$, small values of λ, corresponding to $\sigma^2 \leq 1/2\gamma$, namely to small contributes of diffusion in (3), lead to stationary profiles where opinions are near the middle of I. At the opposite, an increase of the value of λ corresponds to stationary profiles where the agents are divided into two groups. As λ increases, the two points of maximum get closer to the extremes of I and the corresponding value of the maxima increases, showing that if the contribute of diffusion is greater than that of compromise extremal opinions tend to prevail.

From now on, we consider values of u different from 0. For symmetry reasons, we only focus on positive values of u. First, we set $\lambda = 1/4$. According to (39) and (38), in this case $\Delta < 0$ regardless of the value of the average opinion and, therefore, the stationary profile $g_\infty(v)$ always has one stationary point, namely a maximum point. Fig. 2 shows the stationary profiles for $u = 1/4$ (yellow line), $u = 1/2$ (green line), and $u = 3/4$ (violet line). The maxima are marked with a black asterisk. From Fig. 2 it can be observed that as the average opinion increases the value of the corresponding maximum also increases, in agreement with the idea that if the average opinion gets closer to 1 (namely, to one of the extremes of the interval I) the opinions of all agents tend to be more concentrated near the value of u.

We now set $\lambda = 3/4$. According to (39) and (38), in this case: $\Delta < 0$ if $|u| < 1/9$; $\Delta = 0$ if $|u| = 1/9$; $\Delta > 0$ if $|u| > 1/9$. Fig. 3 shows the stationary profiles for $u = 1/9$ (yellow line), $u = 1/2$ (green line), $u = 3/4$ (violet line), and the stationary points are marked with a black asterisk. As expected, if $u = 1/9$ the function $g_\infty(v)$ has two stationary point, namely a point of maximum in v_1 and an inflection point in $v_2 = v_3 = -v_1/2$. Greater values of u, instead, lead to a unique stationary point, namely a point of maximum.

In Fig. 4 the stationary profiles for $\lambda = 1$ and for the average opinions $u = 1/4$ (yellow line), $u = 1/2$ (green line), $u = 3/4$ (violet line) are shown. If $\lambda = 1$ then: $\Delta < 0$ if $|u| < \sqrt{2/27}$; $\Delta = 0$ if $|u| = \sqrt{2/27}$; $\Delta > 0$ if $|u| > \sqrt{2/27}$. Therefore,

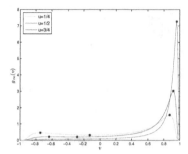   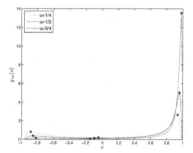

Fig. 5. The stationary profiles g_∞ relative to the value $\lambda = 3/2$ are shown for $u = 1/4$ (yellow line), $u = 1/2$ (green line), $u = 3/4$ (violet line).

Fig. 6. The stationary profiles g_∞ relative to the value $\lambda = 3$ are shown for $u = 1/4$ (yellow line), $u = 1/2$ (green line), $u = 3/4$ (violet line).

the function $g_\infty(v)$ has three stationary points if $u = 1/4$, while it has only a stationary point if $u = 1/2$ and $u = 3/4$.

Fig. 5 shows the stationary profiles $g_\infty(v)$ for $\lambda = 3/2$. In this case: $\Delta < 0$ if $|u| < \sqrt{24}/9$; $\Delta = 0$ if $|u| = \sqrt{24}/9$; $\Delta > 0$ if $|u| > \sqrt{24}/9$. We consider the same values of u as in the previous case and, since $\sqrt{24}/9 \simeq 0.61$, it is expected that if $u = 1/4$ and $u = 1/2$ the function $g_\infty(v)$ has three stationary points while if $u = 3/4$ the stationary profile only admits a point of maximum. These results are confirmed in Fig. 5 where $g_\infty(v)$ is shown for $u = 1/4$ (yellow line), $u = 1/2$ (green line), and $u = 3/4$ (violet line).

Finally, Fig. 6 shows the stationary profiles $g_\infty(v)$ for $\lambda = 3$ and $u = 1/4$ (yellow line), $u = 1/2$ (green line), $u = 3/4$ (violet line). According to (39) and (38), in this case $\Delta < 0$ for all the possible values of the average opinion u, and, therefore, $g_\infty(v)$ always has three stationary points.

5   Conclusions

In this paper the temporal evolution of opinion in a multi-agent system is investigated through a kinetic approach. More precisely, we studied the asymptotic behaviour of the opinion distribution on the basis of a model inspired from the molecules interactions in a gas. Assuming that the opinion of each agent can change because of two reasons, namely compromise and diffusion, stationary profiles with different characteristics can be derived as the parameters of the model change. For a particular choice of the compromise function and of the diffusion function, we showed that the asymptotic distribution is characterized by one, two, or three stationary points, depending on the average opinion and on the parameters of the model.

Further analysis on this subject, which also involves simulation results, is currently under investigation. In particular, we are interested in adopting the kinetic framework in scenarios that could use general-purpose industrial strength technology (see, e.g., [14,15]) and in modeling wireless sensor networks for localization purposes (see, e.g., [16,17]).

References

1. Tsang, A., Larson, K.: Opinion dynamics of skeptical agents. In: Proceedings of 13th International Conference on Autonomous Agents and Multiagent Systems (AAMAS 2014), Paris, France, May 2014
2. Monica, S., Bergenti, F.: A stochastic model of self-stabilizing cellular automata for consensus formation. In: Proceedings of 15th Workshop "Dagli Oggetti agli Agenti" (WOA 2014), Catania, Italy, September 2014
3. Schweitzer, F., Holyst, J.: Modelling collective opinion formation by means of active brownian particles. European Physical Journal B **15** (2000)
4. Groppi, M., Monica, S., Spiga, G.: A kinetic ellipsoidal BGK model for a binary gas mixture. EPL: Europhysics Letter **96**, December 2011
5. Weidlich, W.: Sociodynamics: A Systematic Approach to Mathematical Modelling in the Social Sciences. Harwood Academic Publisher, Amsterdam (2000)
6. Chakraborti, B.K., Chakrabarti, A., Chatterjee, A.: Econophysics and Sociophysics: Trends and Perspectives. Wiley, Berlin (2006)
7. Slanina, F.: Inelastically scattering particles and wealth distribution in an open economy. Physical Review E **69**, 46–102 (2004)
8. Sznajd-Weron, K., Sznajd, J.: Opinion evolution in closed community. International Journal of Modern Physics C **11**, 1157–1166 (2000)
9. Bergenti, F., Poggi, A., Somacher, M.: A collaborative platform for fixed and mobile networks. Communications of the ACM **45**(11), 39–44 (2002)
10. Bergenti, F., Caire, G., Gotta, D.: Large-scale network and service management with WANTS. In: Industrial Agents: Emerging Applications of Software Agents in Industry, pp. 231–246. Elsevier, 2015
11. Monica, S., Ferrari, G.: Accurate indoor localization with UWB wireless sensor networks. In: Proceedings of the 23rd IEEE International Conference on Enabling Technologies: Infrastructure for Collaborative Enterprises (WETICE 2014), Parma, Italy, pp. 287–289, June 2014
12. Toscani, G.: Kinetic models of opinion formation. Communications in Mathematical Sciences **4**, 481–496 (2006)
13. Ben-Naim, E.: Opinion dynamics: Rise and fall of political parties. Europhysics Letters **69**, 671–677 (2005)
14. Bergenti, F., Caire, G., Gotta, D.: Agents on the move: JADE for android devices. In: Proceedings of 15th Workshop "Dagli Oggetti agli Agenti" (WOA 2014), Catania, Italy, September 2014
15. Bergenti, F., Caire, G., Gotta, D.: Agent-based social gaming with AMUSE. In: Procs. 5th Int'l Conf. Ambient Systems, Networks and Technologies (ANT: 2014) and 4th Int'l Conf. Sustainable Energy Information Technology (SEIT 2014), ser. Procedia Computer Science, pp. 914–919. Elsevier (2014)
16. Monica, S., Ferrari, G.: An experimental model for UWB distance measurements and its application to localization problems. In: Proceedings of the IEEE International Conference on Ultra Wide Band (ICUWB 2014), Paris, France, pp. 297–302, September 2014
17. Monica, S., Ferrari, G.: Swarm intelligent approaches to auto-localization of nodes in static UWB networks. Applied Soft Computing **25**, 426–434 (2014)

Cooperating with Trusted Parties Would Make Life Easier

Pasquale Caianiello, Stefania Costantini, Giovanni De Gasperis$^{(\boxtimes)}$,
and Subhasis Thakur

Dipartimento di Ingegneria e Scienze Dell'Informazione e Matematica,
Universitá Degli Studi Dell'Aquila, Via Vetoio 1, 67100 L'Aquila, Italy
{pasquale.caianiello,stefania.costantini,giovanni.degasperis,
subhasis.thakur}@univaq.it

Abstract. We experimentally analyze the performance of a heterogeneous population of agents playing the Iterated Prisoner's Dilemma with a possible prior commitment ad a posterior punishment for defection. We argue that the presence of agents with a probabilistic strategy that depends on trust and reputation enforces a better performance of typically cooperative agents.

1 Introduction

This paper deals with cooperation enforcement in the Prisoner's Dilemma (PD) game. Though the PD is a non-cooperative game, in several application fields, ranging from biology to social sciences to multi-agent systems (MAS), it has become the leading paradigm to model and discuss cooperative behavior. In one-shot Prisoner's Dilemma, two players simultaneously decide to either cooperate (C) or defect (D). If both play C, they get more than if both play D, otherwise in case one defects and the other cooperates, then the defector gets the highest payoff, while the cooperator gets the lowest. Consequently, rational choice would imply that it is safer for each player to defect, even though both would get a better payoff in case of cooperation. The situation were both players do not cooperate is the Nash equilibrium of the PD game. Several approaches for promoting cooperation have been proposed, some introducing for instance voluntary rather than compulsory participation [1] with punishment for non cooperating agents, some others introducing prior commitments and possibly posterior punishment for non cooperation [2]. The main aspect to consider is that both a-priori negotiation for reaching a commitment and a-posteriori administration of punishment have costs.

As a matter of fact, negotiation and punishment are complementary in the way they try to induce cooperation: prior commitments function better with "compliant" agents, and punishments with "free riders" which pursue their momentarily best interest. Commitment definition and formation has been extensively studied (cf. [3] and the references therein) and both mechanisms have been considered in the field of software agents and multi agent systems

© Springer International Publishing Switzerland 2015
M. Gavanelli et al. (Eds.): AI*IA 2015, LNAI 9336, pp. 128–135, 2015.
DOI: 10.1007/978-3-319-24309-2_10

(MAS) where commitment are usefully employed in many fields that include inter-agent communication. Prior commitment, though costly, may be applied on a probabilistic basis [4]. Recent studies (see, e.g. [5]) discuss the conditions when a strategy that combines the two mechanisms is better than either strategy by itself in a MAS. It has been advocated in more general terms [6] that the tendency to making prior agreements rather than just requiring a posteriori compensations emerges from a variety of examples in biological and social contexts, thus suggesting that this behavior could have been shaped by natural selection, and, therefore, "good agreements make good friends" [6].

In this paper, we cope with the Iterated Prisoner's Dilemma, where players engage in Prisoner's Dilemma repeatedly and change their strategy according to a shared indicator built upon previous actions of all involved agents. We assume the implicit existence of a game manager that provides payments for game payouts and collects punishment fines. In such a research setting, both a-priori commitments and a-posteriori punishment for defeating commitments have been used in the existing literature in support of "apology" (see [7,8]). We propose to use forms of public trust evaluation as indicators for implementing strategies eventually leading to more successful course of actions. In particular, we argue that long term gain would result from effective commitment reached with trusted parties, where trust evaluation evolves dynamically with game repetitions. We provide results of computational simulations showing that the adoption of trust evaluation (cf. [9,10] and the reference therein for a discussion about the notion of trust and of trust-update mechanisms) enforces a higher final gain for agents playing the iterated PD.

1.1 Game Theory Basic Notions

Game theory is a study of strategic decision making involving cooperative and non-cooperative agents. This paper falls into the field of non-cooperative game theory as we study how a group selfish agents make decisions that maximize their respective utility. Nash equilibrium is a way to model the equilibrium of such decision making process. In this paper agents decisions are based on the evaluations of the opponent, by means of trust and reputation. While in cooperative game theory we study how a group of agents can decide on the rules of cooperation using their respective share of the utility gained from such cooperation, we argue that the evaluation of trust and reputation may be a viable way to promote cooperation as concepts from both cooperative and non-cooperative game theory may be used together in the study of multi-agent decision making.

2 Model and Methods

As we are addressing the question of commitment in the Prisoner Dilemma, we consider the scenario where the game is played in several iterated rounds among a non homogeneous population of different agents. In each round two players are selected at random with uniform probability to play a one-shot game.

Before making their choice (C or D), either player may simultaneously propose a commitment to cooperation which the other player may accept or deny. If both players propose, then commitment is established; if only one proposes, then commitment is established only if the other accepts; finally if either does not accept or neither proposes then no commitment is established. Commitment proposal has a cost ϵ that is shared in equal parts if the commitment is established. On the other end, if no commitment is established, the entire cost is charged only to the proposing player, if there is one.

After the commitment proposal/accepting preliminary move, the players make their simultaneous choice C or D, they get their resulting payoff and, if either agent plays D in a round when a commitment was established, it will have to pay a penalty δ to the deceived opponent. In Fig. 1 we show the payoff matrix where $T > R > P > S$.

	C	D
C	R,R	S,T
D	T,S	P,P

Fig. 1. Payoff matrix

We present simulation results by computing agents cumulative wealth obtained as a net outcome from rounds payoff, commitment, and penalty payments.

2.1 Players Typologies and Profiling

The two-moves version of the prisoner dilemma game allows the definition of typical agents with deterministic behaviors. We consider a population of several different playing agents of two major classes as described in Fig. 2. Agents in the first class behave according to a strategy that does not change over time, the names associated to their behavior are already established in related literature, [5]. We describe them with minor differences and some new entries, in particular the BASTARD agent who always tries to establish a commitment but afterwards always deceives, and the SCHIZO agent, who always tries to establish a commitment but afterwards behaves inconsistently by playing D when the commitment is established and plays C when there is no commitment.

Among the all theoretically possible agent behaviors, the only missing is the one that never proposes or accepts and then plays C anyway. In fact although it seems that there are 2^4 possible deterministic behaviors , it should be noted if an agent proposes, then according to the game rules it will always accept. So no player can always propose and never accept.

In the bottom lines of Fig. 2 we describe a class of agents whose strategy is probabilistic an depends on the global profiling of their opponent represented as trustworthiness θ measuring the agent's disposition to comply to commitment, and reputation ρ, measuring the agent's willingness to play C role. Such profiling is globally updated at each round with the simple reinforcement rule $x(t) := x(t-1) + \Delta x$ that increases (or decreases) their trustworthiness and reputation, x is θ or ρ, by a fraction Δ of what they miss to get to the maximum (or minimum):

$$\Delta\theta = \begin{cases} +\alpha(1 - \theta) \ if \ commit \ and \ play \ C \\ -\alpha\theta \ if \ commit \ and \ play \ D \\ 0 \ if \ no \ commit \end{cases}$$

$$\Delta\rho = \begin{cases} +\alpha(1 - \rho) \ if \ play \ C \\ -\alpha\rho \ if \ play \ D \end{cases}$$

where $0 < \alpha < 1$ and drives the rate of change of θ and ρ during subsequent rounds.

The class of agents with probabilistic behavior that we consider include: RANDOM who in any game and with an opponent just flips a coin to decide what to do, and the others, but DIPLOMAT, play C with probability that is equal to the opponent reputation ρ if no commitment is established, and modulate their moves in establishing commitment with a probabilistic choice depending on the opponent profiling. In particular TRUSCoop who decides whether to establish a commitment with a probability θ, always plays C if a commitment is established; TRUST who proposes a commitment with probability θ, never accepts commitments and plays C with probability θ if a commitment is established;

name	propose	accept	coop on commit	coop on no commit
C	always	always	always	always
D	never	never	*irrelevant*	never
COMP	always	always	always	never
FAKE	never	always	never	never
FREE	never	always	always	never
BASTARD	always	always	never	never
SCHIZO	always	always	never	always
RANDOM	P=1/2	P=1/2	P=1/2	P=1/2
TRUST CooP	P=θ	P=θ	always	P=ρ
TRUST	P=θ	always	P=θ	P=ρ
REP Only	never	never	P=ρ	P=ρ
DIPLOMAT	always	always	P=$\theta*\rho$	P=ρ

Fig. 2. The probability P of playing C for agents in the simulation, yellow are pure deterministic agents, orange are probabilistic.

REPonly who never commits, and DIPLOMAT who always tries to establish a commitment and, when it is established than plays D with a probability that is equal to the product of θ and ρ, and otherwise plays D when the commitment is not established.

We stress the fact that in the present work we deliberately kept "adaptive" agents strategies simply depend on the global profiling of agents trustworthiness and reputation, in order to gain a preliminary insight on possible outcomes where trust and reputation are involved in agents decisions. Obviously more complex adaptive strategies taking into account different aspects of agents behavior may be conceived of, and will be the object of sequel work.

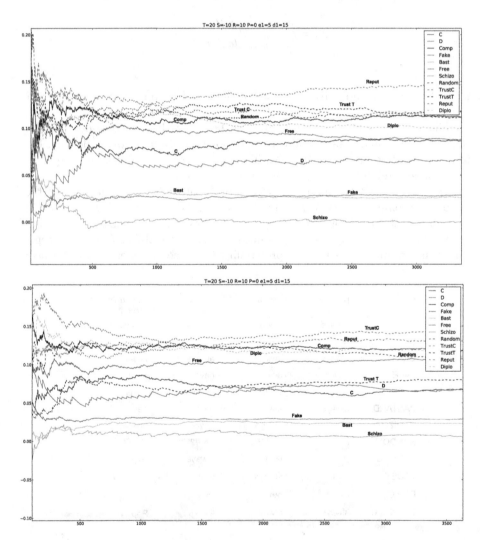

Fig. 3. Trend of agents relative wealth in two different simulations with same parameters values.

3   Simulation Results

We have performed a number of experiments that we show and discuss below. The outcome is that the introduction of trust and reputation increases the level of cooperation while decreasing the cost for both single agents and overall MAS. Each simulation is initialized with a random population of 1000 agents chosen with uniform probability among the 12 agent described in Fig 2. Each simulation is run for 10000 rounds where two players are chosen at random uniformly in the population.

3.1   Agents Performance

In Fig. 3 we plot the relative agents wealth in two different simulations, with the same choice of ϵ and δ and other parameters. We noticed that different runs of the simulation with same parameters do lead to somewhat different results in agents relative performance. That, in fact, it's due to the complexity of the game that we are simulating and testifies for a dependency of the final result on the initial random choices. Chance and luck do play a role in the Iterated Prisoner Dilemma.

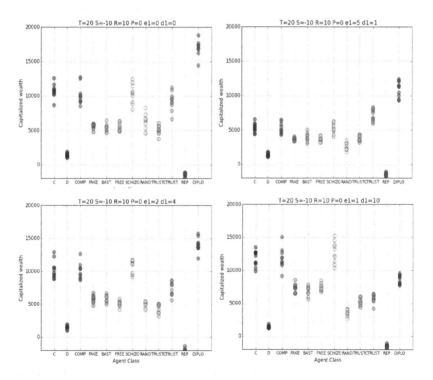

Fig. 4. Simulation results for different choice values of the commitment cost ϵ (e1) and penalty δ (d1).

To overcome this problem in order to arrive at more definite relative evaluation of agents performance we decided to get a better idea by looking at ten different simulations and look at the cluster of final results that the agents obtained. In Fig. 4 we report about the final wealth obtained when running ten simulations of the game with different values of ϵ and δ, as obtained in the simulations and presented in order of increasing values of the ratio δ/ϵ. As we see there is a definite best performance of the DIPLOMAT, except for high values of δ when the DIPLOMAT performance suffers for its probabilistic choice, and consistently cooperative agents as C and COMP perform better. Notice that SCHIZO does have an appeal in almost any situations.

3.2   Agents Wealth vs Commitment Cost and Punishment

In order to appreciate the possible influence of the chosen values of ϵ and δ, in Fig. 5 we plot the average wealth obtained by all the agents of the same type against a few chosen values of ϵ and δ plotted for increasing δ/ϵ

We argue that a combination of the commitment cost plus violator's punishment with a trust mechanism, where involved agents are keener to pay commitment cost if dealing with trusted agents, should result in better agents performance. Moreover, the fact that the population includes different types of agents, many of which behave according to a probabilistic strategy depending on trust and reputation, also modifies relative performance of agents with a fixed behavior.

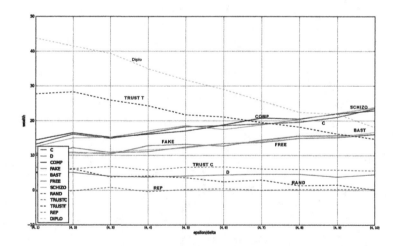

Fig. 5. Agents average wealth for choice values of ϵ (e1) and δ (d1) plotted for increasing (ϵ, δ)

4   Concluding Remarks

In this paper, we have advocated trust evaluation to promote cooperation in the Iterated Prisoner's Dilemma with prior commitment, and performed simulations with a mixed population of deterministic and probabilistic agents whose move depends on a simple profiling of opponents trustworthiness and reputation. The experimental results suggest that some probabilistic agents relying on trustworthiness and reputation perform consistently better, though their performance decreases for high values of the ratio of punishment over commitment costs in favor of more consistent typically cooperative agents. We can argue that, on the one end trust and reputation provide a solid playground for agents to achieve a better payoff in cooperation games, on the other a reasonable penalty for defecting commitments could promote cooperation.

References

1. Hauert, C., Szabó, G.: Prisoner's dilemma and public goods games in different geometries: Compulsory versus voluntary interactions. Complexity 8(4), 31–38 (2003)
2. Pereira, L.M., Santos, F.C., et al.: The emergence of commitments and cooperation. In: Proceedings of the 11th International Conference on Autonomous Agents and Multiagent Systems, vol. 1, pp. 559–566. International Foundation for Autonomous Agents and Multiagent Systems (2012)
3. Singh, M.P.: Commitments in multiagent systems: some history, some confusions, some controversies, some prospects. In: Paglieri, F., Tummolini, L., Falcone, R., Miceli, M. (eds.) The Goals of Cognition: Essays in Honor of Cristiano Castelfranchi, pp. 601–626. College Publications, London (2012)
4. Chen, X., Szolnoki, A., Perc, M.: Probabilistic sharing solves the problem of costly punishment. New Journal of Physics 16(8) (2014)
5. Han, T.A., Lenaerts, T.: The efficient interaction of costly punishment and commitment. In: Proceedings of the 2015 International Conference on Autonomous Agents and Multiagent Systems, AAMAS 2015, Istanbul, Turkey, May 4–8, 2015, pp. 1657–1658 (2015)
6. Han, T.A., Pereira, L.M., Santos, F.C., LenaertsJ, T.: Good agreements make good friends. Scientififc Reports 3 (2013). (Online, Open-Access)
7. Han, T.A., Pereira, L.M., Santos, F.C., LenaertsJ, T.: Why is it so hard to say sorry? evolution of apology with commitments in the iterated prisoner's dilemma. In: Rossi, F. (ed.) IJCAI 2013, Proceedings of the 23rd International Joint Conference on Artificial Intelligence, Beijing, China, August 3–9, 2013, IJCAI/AAAI (2013)
8. Okamoto, K., Matsumura, S.: The evolution of punishment and apology: an iterated prisoner's dilemma model. Evolutionary Ecology 14(8), 703–720 (2000)
9. Castelfranchi, C., Falcone, R.: Trust theory: A socio-cognitive and computational model, vol. 18. John Wiley & Sons (2010)
10. Wang, Y., Singh, M.P.: Formal trust model for multiagent systems. In: Veloso, M.M. (ed.) IJCAI 2007, Proceedings of the 20th International Joint Conference on Artificial Intelligence, pp. 1551–1556 (2007)

Agent Based Simulation of Incentive Mechanisms on Photovoltaic Adoption

Valerio Iachini, Andrea Borghesi$^{(\boxtimes)}$, and Michela Milano

DISI, University of Bologna, Bologna, Italy
valerio.iachini@gmail.com, {andrea.borghesi3,michela.milano}@unibo.it

Abstract. Sustainable energy policies are becoming of paramount importance for our future, shaping the environment around us, underpinning economic growth, and increasingly affecting the geopolitical considerations of governments world-wide. Renewable energy diffusion and energy efficiency measures are key for obtaining a transition toward low carbon energy systems.

A number of policy instruments have been devised to foster such a transition: feed-in-tariffs, tax exemptions, fiscal incentives, grants. The impact of such schemes on the actual adoption of renewable energy sources is affected by a number of economic and social factors.

In this paper, we propose a novel approach to model the diffusion of residential PV systems and assess the impact of incentives. We model the diffusion's environment using an agent-based model and we study the emergent, global behaviour emerging from the interactions among the agents. While economic factors are easily modelled, social ones are much more difficult to extract and assess. For this reason, in the model we have inserted a large number of social parameters that have been automatically tuned on the basis of past data. The Emilia-Romagna region of Italy has been used as a case study for our approach.

1 Introduction

Energy policies affect and are affected by a number of interconnected social, economical and environmental aspects. The transition toward a sustainable and low-carbon economy should be fostered by governments worldwide. Energy efficiency measures and renewable energy sources are two key enablers for such a transition. For this reason a number of policy instruments have been implemented to push stakeholders toward virtuous energy-aware behaviour: feed-in tarifs, tax exemption, investment grants, fiscal incentives. Stakeholders involved in energy policies have conflicting interests that should be taken into account to understand and forecast the impact of energy policy instruments.

We propose here the design and assessment of a predictive model that takes into account both economic and social drivers pushing stakeholders (households in particular) toward the adoption of photovoltaic plants. A large portion of the total installed PV power, in fact, comes from photovoltaic panels installed by private citizens and enterprises. For this reason, policy makers cannot directly

© Springer International Publishing Switzerland 2015
M. Gavanelli et al. (Eds.): AI*IA 2015, LNAI 9336, pp. 136–148, 2015.
DOI: 10.1007/978-3-319-24309-2_11

decide on the total photovoltaic power installed, but they have to foster the PV power generation through indirect means, usually in the form of incentives to the PV energy (i.e. feed-in-tariffs for the electricity generated by PV systems).

We propose an agent-based model[5,9,17] (ABM) that simulates the behavior of single households and government entities (*micro*-level) in order to evaluate and explain emergent phenomena (*macro*-level). We tackled the challenge of reproducing the household's decisions to install a PV system for their houses. We take into account both economic aspects (return on investment, income, interest on loans), territorial aspects (position, roof width, population distribution) and social aspects (imitation, network effect on knowledge diffusion). Since it is very difficult to a priori calibrate these parameters, we employ automatic parameter tuning techniques to tune these social aspects to meet an emerging behaviour that is taken from past data. Past data concern policy instruments present in the past and photovoltaic adoption[1].

2   Related Works

Many scholars have tried to model the diffusion of innovations. Their works had evidenced that the diffusion of innovation is a social process. Many proposed models are ABMs, where the agents are connected to form a *small-world* network. The information exchanged between entities in the network influences the diffusion of the innovation. In this direction [1] implemented a threshold model based on the so called "bandwagon effect". In this model the increase of the number of adopters generates new information about the innovation, which in turn produces high pressure on people who have not yet adopted the innovation. An important factor in this process is the estimation of the profitability of the innovation made by potential adopters. Since potential adopters may be unsure about the correctness of their profitability assessment, other people who have already adopted the innovation could influence their decision. The authors of [1] express the bandwagon assessment of innovation of a potential adopter as a function that involves the evaluation of profitability of the new technology and the amount of information received regarding the innovation, weighted by the amount of "trust" placed on such information.

Another approach used to predict the diffusion of an innovation has been proposed by [4]. In this model the price and the performance of the innovation influence people decisions. The potential adopter knows the price of innovation but he ignores its performance. The performance is based on the perception that the potential adopter has of the innovation. Over time, potential adopters receive information about the performance by word-of-mouth from other adopters and consequently the uncertainty about the innovation potential is reduced. The diffusion of residential PV systems could be modeled using the previously described models. The innovation is the PV technology and potential adopters are the households: models estimate the benefits deriving from the adoption of a PV system and a household decides whether to install or not a PV system.

---

[1] Data are available (in Italian) on http://www.gse.it

The authors of [18] proposed a two level threshold agent-based model that is specifically aimed at estimating the diffusion of PV systems. In this model agents represent households that choose whether or not to install a PV system. The low level component simulates electric consumption for each agent and provides the payback time of the investment. The high level component models the behaviour of agents toward PV adoption. Four factors affect the decision: payback period, household income, neighbourhood and advertisement. The adoption of a PV system by a household is determined by its "desire level", computed as the linear combination of the four factors. If the desire level of the household exceeds the threshold the household installs a PV system. [10] has proposed an ABM inspired by the work of [18]: the household's decision is again a linear combination of four factors but in the latter work these factors are weighted differently according to the social class of the household. Moreover in the [10] model agents are connected to form a small-world network in such a way that those who are in the same social class are more likely to be linked together.

Another factor that we may consider is the geographical location of buildings. [12] proposed a model that uses a geographic information system (GIS) along with an ABM to study the diffusion of PV systems. Including the real topology of the area under consideration allows to analyse the effects of solar exposure and population density on diffusion of PV systems. Agents who have a similar opinion on technology could influence each other.

3   The Simulated Model

Our model simulates the behaviour of households in presence of different incentive mechanisms, to predict the diffusion of PV systems. We focused mainly on families living in the Emilia-Romagna region (and especially considering the 2007-2013 period), but the process described below is valid for any region or country. In [3] [2] we proposed a preliminary agent-based model to simulate the impact of national and regional incentives on the installation of PV panels in the Emilia-Romagna region. The model we discuss in this paper largely improves the previous version, especially in the social interactions among the agents; our work was partially inspired by related works cited in Section 2, in particular by [10]. The simulation model was developed in Netlogo [15].

We define two kinds of agents: the households and the region. The households make the decision whether or not to install a PV system. Each household is described by a set of attributes: age class, education level, income, family size, consumption, roof area, budget, geographical coordinates and social class. We use these attributes to define the household behaviour and to build the social network. The region agent regulates the regional incentive; it defines the type of incentive offered, the amount of available budget and who is eligible to obtain the incentive. The regional incentives are provided by the region on top on the national ones given by the Italian government during the considered timespan, i.e. a feed-in tariff (see [3] for a detailed survey on the incentives).

The simulator is structured into two phases: 1) the configuration phase and 2) the evolution phase. In the *configuration phase*, the simulator sets up the virtual environment and creates the agent-based model. Firstly, it places the agents on the virtual environment recreating the actual population density. WE have used data on the buildings of the Emilia-Romagna region to obtain the positions and roof areas of houses. The simulator uses this information to assign a building and a roof to each household. Then the simulator builds the social network taking into account the physical distance between the household and the proximity between their attributes. The social network determines how the information about PV systems is exchanged.

In the *evolution phase* the simulator runs the model, recreating the behaviour of agents and updating the virtual environment. A simulation consists of a sequence of steps with a time frame of six months. The PV adoption by each household/agent is affected by the payback period of the investment, environmental sensitivity, the household's income and the communications with other households. We express these factor using a combination of four functions (which we refer to as *utility* function) to determine the "desire level" of a household. If the desire level exceeds a threshold the household installs the PV system.

3.1 Configuration Phase

In the configuration phase the simulator initializes the virtual world creating the starting conditions. The simulator loads the dataset containing the household's descriptions and places the families in the virtual world following the actual density distribution. The geographical coordinates and roof areas are obtained by associating each family to a building: buildings are sorted by their roof size and families with the highest income and the largest number of members are assigned to the biggest ones.

We acquired the buildings by analysing the Ersi shape-files provided by Emilia-Romagna region[2]. Those shape-files contain a polygon for each building detected by the region. Since our model requires only the positions of buildings and the areas of the roofs, it was necessary to process these files to extract the relevant information. We used QGIS [11], a free and open source Geographic Information System (GIS), to manipulate these shape-files and calculate the position and size of the houses.

Each household is described by a vector of attributes: age class, education level, income, family size, energy consumption and social class. The distribution of each attribute is obtained by analysing the Survey on Household Income and Wealth (SHIW) provided by Bank of Italy[3]. In addition each household establishes a budget for purchasing a PV system. The key idea is that to different household income classes correspond different spending powers. If a family has an income around the mean, the family will expect to pay the average PV system price. Conversely, if the family income is lower or higher than the average, the family will aim to spend less or more for a PV system.

---

[2] http://dati.emilia-romagna.it
[3] https://www.bancaditalia.it/statistiche/indcamp/bilfait/

To assign an income to each family we used a linear regression model based on a set of explanatory variables extracted from the data provided in SHIW, i.e. the number of earners and members of the family, age class, etc.

The Social Network. During the configuration phase the simulator initializes also the social network: for each family, a list of friends is provided. Since the families are geographically distributed on the region we use the extended version of the rank-based model proposed by [8] to get the small-world properties. In such a model the probability that a link between node u and node v exists is proportional to a ranking function which depends both on the geographical proximity of the nodes (physical neighbours) and on the attribute proximity of the nodes (how the nodes are similar w.r.t. their attributes). After we build a network using the extended rank-based method, we randomize it to add long-range links. These links drastically reduce the average path length because they connect distant parts of the network. The randomization process takes every edge and rewires it with an empirically obtained probability p.

Social Classes. An innovation has a very high price at the beginning due to the high costs of production. However the price decreases over time because of technological improvements, especially those in the production phase, that make manufacturing more efficient. Indeed, many technologies follow an S-shape curve that relates the investments made by the company with the performance of the technology [13][14]. In the first stage the performance improvement is slow because the technology still needs to be fully understood. Afterwards, as researchers and producers obtain a better knowledge of the technology, the improvement accelerates. However, when the technology reaches its natural limit of performance, the improvements tend to slow down.

Similarly the diffusion of innovations follows an S-curve. In the initial stage, the adoption is slow because the technology is poorly understood. When the knowledge about the technology has spread, the innovation enters the mass market and the rate of adoption increases. Finally the adoption rate decreases when the market has been fully saturated. [13] identifies five categories of different adopters: 1) innovators, 2) early adopters, 3) early majority, 4) late majority and 5) laggards. In Figure 1 the five different classes of adopters considered in order to model the technology adoption rate are shown (bell-shaped curve); in the Figure we also report the S-shaped curve which represents the innovation diffusion. During the configuration phase we adopt the model of [13] to group the households into social classes. Households that belong to the same social class have similar characteristics and behaviours. Since the utility function models the households' behaviours, the social class is reflected in different sets of weights which combine the four factors.

We consider each family as a point in three-dimensional space: age class, education level and income. We use the K-means clustering technique to subdivide these points in five social classes. The K-means is a prototype-based technique that attempts to find a user-specified number of clusters (k). Each cluster C_i with $i = 1, .., k$ is represented by its prototype c_i, defined as the centroid of the

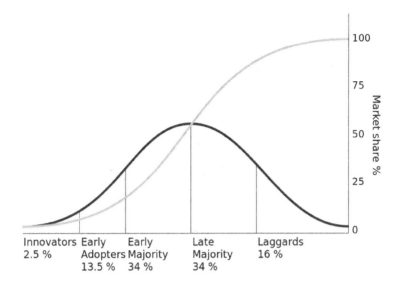

Fig. 1. Adopters Classes, source: http://en.wikipedia.org/wiki/Diffusion_of_innovations

group of points. The K-means algorithm attempts to minimise the total intra-cluster variance, repositioning the centroid at every step until the centroids do not change. It starts with a random set of centroids $c_1, .., c_k$ and then assigns each point x to the nearest centroid. After that, it calculates the new centroids by averaging the points in a cluster. The results of the clustering are shown in Figure 2. The five different colors in the figure represent the five different clusters identified on the base of the three parameters considered (age class, AGE, education level, EDL, and income, INC).

It is difficult to evaluate the goodness of a clustering because we do not know the class labels to be used as a reference. When the ground truth is unknown unsupervised techniques can be used to evaluate the clustering; these methods measure the goodness of a clustering structure without using external information. A common unsupervised method is the silhouette coefficient that relates the cohesion of a cluster with the separation between clusters [16]. The silhouette coefficient is defined for each sample i and it is composed of two scores: 1) the cohesion $a(i)$, the mean distance between the sample i and all other points in the same cluster; 2) the separation $b(i)$, the mean distance between the sample i and all other points in the next nearest cluster. The value of the silhouette coefficient can vary between -1 and 1. If the value is negative, the sample i is closer to the objects of another cluster than other objects of its cluster. Samples with a large $s(i)$ (almost 1) are very well clustered. An overall measure of the goodness of a cluster can be obtained by computing the average silhouette coefficient of all

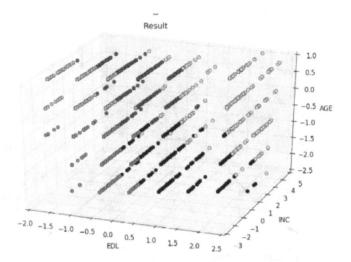

Fig. 2. Clustering Results

samples. Using K-means clustering we get a silhouette coefficient of 0.35, which is not an optimal value, but it is good enough for our purposes.

3.2   Evolution Phase

In the evolution phase the simulator recreates the behaviour of households for a period from the first half of 2007 to the second half of 2036; actually the PV systems are installed only during the 2007-2013 period but we have also to take into account the long lifetime of PV panels and the incentive durations. In each semester (from 2007 to 2013) households proceed to evaluate the adoption of PV system; in the remaining part of the simulation (from 2014 onwards) no new panels are installed. As previously mentioned the desire level for adoption of a PV system is estimated through the utility function described in the configuration phase that every agent computes according to its own characteristics. we remind that the utility function takes into account the income of the family, the payback period, the environmental benefits and the relationships with other families. These factors are weighted differently depending on the social group of the family. The weights for each group are determined by calibrating the model on real data over the 2007-2013 period.

The estimation of the ROE considers costs and gains for a 20 years period, which is the estimated lifetime of a PV system. We calculate the cash flow for each year as the difference between total earnings and total expenditure related to the PV. The expenses that are taken into consideration are the cost of the system, the maintenance costs and the loan interests. The sources of income are the electricity bill savings due to the self-consumption and sales to the grid operator. The national and regional incentives affect the gains and the expenses in different ways: for example earnings are related to the Italian national feed-in tariff and expenditures are influenced by the initial cost (modified by the investment grants offered by the regions) and loan interests (a target of several incentive mechanisms). A household solves the optimization problem to find the size of the system that provides the highest ROE.

We assume that families get advice from PV installers and they become well-informed about their options. Thus households install the PV system that maximise their reward in terms of production and saving.

The Utility Function. The core of our model is the utility function, or desire level, which is responsible for the agent decision to invest in a PV panel or not. The function is defined by the following equation:

$$U(v) = w_{pp}(cls_v)u_{pp}(v) + w_{budget}(cls_v)u_{budget}(v)$$
$$+ w_{env}(cls_v)u_{env}(v) + w_{com}(cls_v)u_{com}(v)$$

where cls_v is the class of agent v. The equation is a linear combination of four factors: the household payback time ($u_{pp}(v)$), the household budget ($u_{budget}(v)$), the neighbourhood influence ($u_{com}(v)$) and the environmental benefit of investing in a PV system ($u_{env}(v)$). These four factors are multiplied by four different weights $w_{pp}(cls_v)$, $w_{budget}(cls_v)$, $w_{com}(cls_v)$ and $w_{env}(cls_v)$, which depend on the class of the agent. The proper calibration of these parameters is a crucial aspect and it will be discussed in Section 4.

The partial utility $u_{pp}(v)$ estimates the expected payback period pp of an agent v. As the function value range is between 0 and 1, we map the actual payback period which could range between zero and twenty years to the range [0,1]; we subtract the $min(pp)$ considered, namely one year, and then divide the value obtained by $max(pp) - min(pp)$, where $max(pp)$ is 21 years, because 20 years is the expected useful life for PV systems. According to [10] we compute $u_{pp}(v)$ as:

$$u_{pp}(v) = \frac{21 - pp(v)}{20}$$

where $pp(v)$ is the payback period for the initial investment. The payback period requires the net present value (NPV) of the PV system: when the NPV value turns from negative to positive a household recovers from its initial investment. The NPV computation is based on the yearly cash flows. The regional and national incentives act on the payback utility factor because they reduce the payback period.

The household budget $u_{budget}(v)$ is given by:

$$u_{budget}(v) = 1 / \frac{v_{equity}}{e^{v_{budget}}}$$

where v_{equity} is the initial investment obtained by subtracting any incentives that affect the PV panel installation price; v_{budget} is the budget available to the agent.

The $u_{env}(v)$ captures the sensitivity toward the environmental benefits related to the adoption of a PV system. It is calculated as the oil saved - clearly correlated to the amount of CO_2 produced - thanks to the PV panel. We use the conversion factor from MWh of energy to TOE (Tonne of Oil Equivalent) provided by the Italian Regulatory Authority for Electricity and Gas[4]. The ecological benefits are expressed as:

$$u_{env}(v) = \frac{1}{e^{oil_{notConsumed} - oil_{consumed}}}$$

Finally the partial utility $u_{com}(v)$ describes how the social interaction affects the agents behaviour. The neighbourhood of an agent is defined by the nodes it shares a link with. The communication factor is calculated as follows:

$$u_{com}(v) = \frac{1}{1 + e^{\frac{1}{2}L_{v,tot}L_{v,adopter}}}$$

with $L_{v,tot}$ the total number of links of agent v and $L_{v,adopter}$ the number of links shared with adopters.

4   Model Calibration

A critical aspect of our simulator is the correct calibration of the model parameters, since they have a great influence on the final outcome; a commonly known weak point of agent-based models is exactly the difficulty to find good parameters. The solution we chose to employ is to devise an automated fine-tuning process which allowed us to test numerous combinations of parameters and select those which provide better results. Our goal is to obtain a simulator able to reproduce the impact that real incentive strategies (along with economic and social aspects) have had on the PV adoption in the ER region; hence we use the real data from the period 2007-2013 to calibrate our model.Real data are taken from the GSE website that records every PV plant along with its power and geographical position. In practice we want to obtain a good fit between the observed trend of PV power installation and the simulated one.

The parameters we need to fine-tune are the weights of the utility function: $w_{pp}(cls_v)$, $w_{budget}(cls_v)$, $w_{com}(cls_v)$ and $w_{env}(cls_v)$. We chose to employ a Genetic Algorithm (GA) [6] to find the configuration of utility function weights which better fit the Emilia-Romagna PV power installation curve. In our model the parameters are correlated, i.e. if we increase the communication factor the remaining ones are necessarily affected, since the sum of all the weights is always equal to one (linear combination). Moreover since the choice is influenced also by the social interaction with the neighbours there is also a dependency among the weights of the different social classes: if we change the weights for the utility function of a node this may change the number of installed PV panels in the neighbourhood; this in turn would produce consequences on the whole network. To summarize, every slight change in the weights of each agent could have

---

[4] A TOE is defined as the amount of energy released by burning one tonne of oil, or 0.187 TOE for each MWh produced.

extremely great impacts on the final outcome and in such circumstances genetic algorithms have been proved to be very effective.

We defined a new set of parameters, $a_i, b_i, c_i \quad \forall i = 0, .., N - 1$, where N is the number of clusters. Each parameter is a real number in the range $[0,1]$; the utility function weights are computed as a linear combination of these new parameters. The following equations define such relations:

$$w_{budget}(cls) = a_{cls}$$
$$w_{pp}(cls) = (1 - w_{budget}(cls))b_{cls}$$
$$w_{env}(cls) = (1 - (w_{budget}(cls) + w_{pp}(cls)))c_{cls}$$
$$w_{com}(cls) = 1 - (w_{budget}(cls) + w_{pp}(cls) + w_{env}(cls))$$

The genetic algorithm starts generating a random initial population of parameters configurations (also called "individuals"). Then it proceeds to evaluate the entire population by running the model. After the evaluation phase, the GA selects the next generation of individuals. We use a strategy called *tournament selection* [7], which selects k individuals from the actual population using n tournaments of j individuals. Each tournament is composed of j random individuals and the individual with the highest fitness is selected for the next population.

Before the next evaluation the GA applies crossover and mutation on the offspring. The crossover randomly selects two individuals and generates one or more children from them. We use one-point crossover where a single crossover point on both parents' configuration is selected. The crossover method selects a random value from the two parents and then produces a new configuration by swapping the values beyond the crossover point. Finally the mutation randomly selects one individual and alters one or more values. The evolution process is repeated for 400 times.

4.1   Results

In Figure 3 we show the results of the fine-tuned simulator. We used a model composed by 2000 agents; each simulations requires around 20 seconds with a 2.40GHz Intel Pentium DualCore CPU e2220 with a 2GB RAM. The genetic algorithm uses a population of 50 individuals and the overall time required to calibrate the machine is around 40 hours.

The Figure shows the observed photovoltaic power installation trend in the Emilia-Romagna Region in the considered timespan (solid line) and the trend obtained through our fine-tuned simulator (dotted line). The year is displayed in the x-axis while the y-axis tells the yearly PV power growth in percentage. The figure clearly reveals that the installed PV power predicted by the agent-based model correctly follows the real trend. Both curves indicates that after an initial slow diffusion phase in the first years (2007-2009) there is a peak in the power capacity in 2011 - possibly due to the combination of high level of national incentives and more widespread knowledge of the technology (see [2] for more insight on the correlation between national incentives and PV diffusion in

ER). The summed square of residuals (SSE) of our forecast w.r.t. the real data
is equal to 8.56 and the R-squared is 0.984, a very good value. There is still a not
entirely negligible gap between the simulated results and the real data but we
are currently working to refine the model; we are confident that better results
could also be achieved through a more accurate fine-tuning of the model.

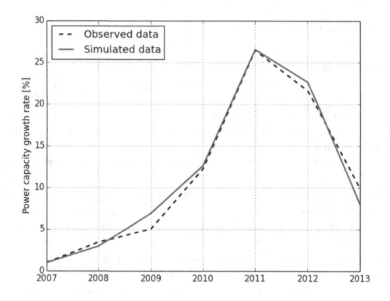

Fig. 3. Model Calibration Results

5 Conclusions

In this paper we proposed an agent-based model to simulate the diffusion of
photovoltaic systems with the goal to assist policy makers in their decisions.
With this model it is also possible to study the impact that different incentive
mechanisms can have on the overall amount of PV power installed. Due to the
difficult nature of the problem and especially since a key factor is given by the
extremely complex nature of social interactions, an agent-based model was a
natural way to cope with this problem. The main advantage of such type of
model is the possibility to define the behaviour of each agent at a micro-level
and observe the emergent trends at macro-level.

The agents in our model are the households which decide whether install a
PV panel on their roof or not. The decision is taken w.r.t. to four factors: 1)
economic sustainability, 2) economic return, 3) social interaction and 4) environmental benefits. How much these four factors influence the final decision (their
weights) is the subject of another problem we considered in this paper. To understand which were the best weights we used a Genetic Algorithm, employing as

a training set the data of the photovoltaic power installed in Emilia-Romagna region in the period 2007-2013.

The fine-tuning of the parameters provided us with a model able to correctly predicts the trend of the installed PV power within an acceptable margin of error. It is nevertheless possible to achieve even better results and this is a research direction we are currently exploring. Along with the model refinements other future research include the development of different parameters calibration techniques and experiments with others data sets. Future works will also try to scale-up the number of agents in the model to get closer to the real number of households in the Emilia-Romagna region (a few millions). We also plan to study more in detail the impact of the incentives on the overall PV installed power, integrating these results within a complete framework capable to aid policy makers in each phase of the decision process.

Acknowledgment. The research leading to these results has received funding from the European Union Seventh Framework Programme (FP7/2007-2013) under grant agreement n. 288147.

References

1. Abrahamson, E., Rosenkopf, L.: Social network effects on the extent of innovation diffusion: A computer simulation. Organization Science **8**(3), 289–309 (1997)
2. Borghesi, A., Milano, M.: Multi-agent simulator of incentive influence on PV adoption. In: 2014 International Conference on Renewable Energy Research and Application (ICRERA), pp. 556–560, October 2014
3. Borghesi, A., Milano, M., Gavanelli, M., Woods, T.: Simulation of incentive mechanisms for renewable energy policies. In: ECMS2013: Proceedings of the European Conference on Modeling and Simulation (2013)
4. Chatterjee, R.A., Eliashberg, J.: The innovation diffusion process in a heterogeneous population: A micromodeling approach. Management Science **36**(9), 1057–1079 (1990)
5. Gilbert, N.: Computational Social Science. SAGE (2010)
6. Goldberg, D.E.: Genetic Algorithms in Search, Optimization and Machine Learning, 1st edn. Addison-Wesley Longman Publishing Co., Inc., Boston (1989)
7. Goldberg, D.E., Deb, K.: A comparative analysis of selection schemes used in genetic algorithms. Foundations of genetic algorithms **1**, 69–93 (1991)
8. Liben-Nowell, D., Novak, J., Kumar, R., Raghavan, P., Tomkins, A.: Geographic routing in social networks. Proceedings of the National Academy of Sciences of the United States of America **102**(33), 11623–11628 (2005)
9. Matthews, R., Gilbert, N., Roach, A., Polhill, G., Gotts, N.: Agent-based land-use models: a review of applications. Landscape Ecology **22**(10) (2007)
10. Palmer, J., Sorda, G., Madlener, R.: Modeling the diffusion of residential photovoltaic systems in Italy: An agent-based simulation (2013)
11. QGIS Development Team. QGIS Geographic Information System. Open Source Geospatial Foundation (2009)
12. Robinson, S.A., Stringer, M., Rai, V., Tondon, A.: GIS-integrated agent-based model of residential solar PV diffusion. In: 32nd USAEE/IAEE North American Conference, pp. 28–31 (2013)

13. Rogers, E.M.: Diffusion of preventive innovations. Addictive Behaviors **27**(6), 989–993 (2002)
14. Schilling, M.A., Izzo, F.: Gestione dell'innovazione. Collana di istruzione scientifica. Serie di discipline aziendali. McGraw-Hill Education (2013)
15. Sklar, E.: NetLogo, a multi-agent simulation environment. Artificial Life **13**(3), 303–311 (2011)
16. Tan, P.-N., Steinbach, M., Kumar, V., et al.: Introduction to data mining, vol. 1. Pearson Addison Wesley, Boston (2006)
17. Troitzsch, K.G., Mueller, U., Gilbert, G.N., Doran, J.: Social science microsimulation. J. Artificial Societies and Social Simulation **2**(1) (1999)
18. Zhao, J., Mazhari, E., Celik, N., Son, Y.-J.: Hybrid agent-based simulation for policy evaluation of solar power generation systems. Simulation Modelling Practice and Theory **19**(10), 2189–2205 (2011)

Knowledge Representation
and Reasoning

Feature-Based Modelling and Information Systems for Engineering

Emilio M. Sanfilippo[1,2][✉] and Stefano Borgo[1]

[1] Laboratory for Applied Ontology (LOA-ISTC),
National Council of Research (CNR), via Alla Cascata 56/C, Povo,
38123 Trento, Italy
sanfilippo@loa.istc.cnr.it
[2] Ph.D. School in ICT, University of Trento, Trento, Italy

Abstract. We use methods based on ontology engineering to individuate the shortcomings of feature-based modelling approaches in product lifecycle data management, and propose an alternative view.

Our aim is to contribute to the development of information systems for the integrated management of product lifecycle knowledge. In particular, we are looking for suitable approaches to model the variety of engineering features as used in intensive knowledge-based product development tasks, in particular dealing with manufacturing and engineering design.

Keywords: Feature · Manufacturing · Ontology engineering · Design

1 Introduction

Product development is a knowledge intensive task in which several teams interact at different times and from distributed geographic places by using heterogeneous computer modelling systems [1]. In order to be machine-processable and cognitively transparent to software agents and to the variety of stakeholders, product knowledge has to be represented in computational languages, with formal semantics, and driven by experts' conceptualisations.

Traditional computer-based technologies for product data modelling, like Computer-Aided Design (CAD) systems, as well as conceptual and data models for engineering are mainly focused on geometric specifications of product knowledge. Nowadays, however, experts need to represent and share qualitative knowledge about the product at hand, that is, knowledge concerning the engineering intents, like functional and material knowledge as well as constraints on machining tools, product management and costs [2]. The quest to add qualitative knowledge into quantitative product models led in the 1970s to the development of *feature-based* product modelling approaches and technologies [3].

Much of the research work in this area has been focused on the development of algorithms for the automatic detection of features in design models to allow the integration of CADs with downstream applications like Computer-Aided

© Springer International Publishing Switzerland 2015
M. Gavanelli et al. (Eds.): AI*IA 2015, LNAI 9336, pp. 151–163, 2015.
DOI: 10.1007/978-3-319-24309-2_12

Manufacturing (CAM) and Computer-Aided Process Planning (CAPP) systems. This has stimulated the development of Artificial Intelligence-based methods, among which knowledge-based expert systems for the automatic generation of manufacturing process plans from a set of input constraints [4].

Despite the amount of work, the use of feature-based technologies is hampered by the lack of a robust methodology for feature representation. Ontology engineering approaches are being actively exploited for product development purposes but even in this case the lack of a shared framework has lead to a number of disconnected and application-based ontologies that deal with feature-based applications in very different ways. Today's engineering ontologies concentrate on formal representations of the concepts for specific application requirements without attempting a deep characterisation of their meaning according to experts' conceptualisations, i.e., giving up to cross-community interoperability.

We aim to fill this gap. The development of information systems is a complex engineering process and the task we are concerned with, namely the formalisation of a broadly applicable knowledge base framework for CAD/CAM integrated systems, requires to systematically analyse the concepts at stake, and that of feature foremost, before moving into application concerns.

The paper is organised as follows. In Section 2 we provide a quick overview of feature concepts as used today in product modelling. The state of art of feature models in engineering is given in Section 3. The problems in existing feature-based modelling approaches are discussed in Section 4. The ontological analysis and formal representation of the notion of feature are described in Section 5 and Section 6, respectively; Section 7 adds an example.

2   Features for Product Modelling

Feature-based systems have represented an evolution of computer-based geometric modelling approaches since the 1970s, and are nowadays the prevalent approach for computer-aided product development. These systems provide support for product data modelling behind the specification of geometric constraints by managing product lifecycle information required during the different stages of product development [5]. In particular, features are used to represent and reason over multiple quantitative and qualitative aspects of product lifecycle, spanning from geometry to e.g. functional information, manufacturability constraints, production costs and material tolerances. Feature-based approaches have stimulated the development of expert systems for engineering design and manufacturing purposes, as well as concurrent and collaborative modelling environments for different product development tasks [6].

Historically, much of the work in the area of feature-based modelling focused on the so-called *geometric features*, namely shapes recurrently used in engineering projects like counter bore, slot, chamfer and rib. This focus broadened over the years leading to the introduction of qualitative feature information typically based on specific requirements, e.g., non-geometrical information needed for design applications, manufacturing process planning or mechanical stress analysis [3]. As a consequence of this variety, feature-based models and terminologies tend to be driven by application concerns, that is, the information attached to the identified feature is tuned to either the product lifecycle phases at stake, or to the application domain in which their use is considered [7].

Consider, for instance, the manufacturing and the engineering design domains. In manufacturing one of the main application concerns of the feature-based approaches is the creation of process plans according to design specifications [8]. In CAPP applications, the design model of the part under consideration is analysed to find the most appropriate solution for its manufacturing. In these systems, machining method, tool access direction, workpiece set up constraints, among other information, is attached to geometric features, giving rise to the so-called *manufacturing feature* [4,9]. These are understood as portion of material to be removed (subtractive feature), or added (additive features) to obtain the desired final geometry. For instance, a hole feature is the volume removed by a drilling cutter [9]; if the cutter penetrates the material frame, resulting in a set of circularly connected inner boundaries, the feature is a *through hole*; if the cutter does not penetrate the frame leaving a base face, the feature is a *blind hole* [10]. In the case of engineering design, among other feature types, the so-called *functional* features are particularly used to merge information on a geometrical shape with details concerning its purpose(s) and expected behaviour(s) within a certain product [11]. A pocket, for example, is a functional feature when it has the function to allow a certain assembly constraint to hold.

Other research communities broaden the meaning of feature in other directions, for instance, aiming to merge shape information with product's characteristics and sub-assemblies. Groover [12, p.634] defines product features as "the characteristics of a product that result from design". Similarly, Brown [11] considers features as things like product's colour, mass, portions of surfaces, etc.

In the area of civil engineering, Nepal et al. [13] take features to be "meaningful real world entities to which one can associate construction-specific information" [13, p.13]. Along the same lines, in mechanical engineering, Anjum and colleagues [14] consider physical items like metal components (e.g. screws) as features for assembly purposes.

3   Features in Engineering Models

Several initiatives focus on the development of feature specifications (data modelling standards, computational ontologies, taxonomies) for disparate applications within the product lifecycle information modelling.

The ISO standard *Automation systems and integration–Product data representation and exchange*, commonly known as STEP (ISO10303) [15] is considered the most relevant effort towards the standardisation of product data across the entire product life-cycle. Within STEP, AP224 is an application protocol dedicated to feature-based product modelling. It specifies recurrent shapes used in manufacturing scenarios. At the core of the AP224 is the concept of manufacturing feature, meant as volume of material to be removed and that results from machining. STEP provides a classification of several feature types, which are employed in various research projects and modelling systems [16].

Ma and colleagues [6] proposed to look at features as general modelling elements resulting from the aggregation of geometric and non-geometric parameters. Their purpose is to provide a layout for feature data specifications in the form of a schema specifying the type of the data to be included for feature representation. The key advantage of their approach is to provide a general and adaptable method for feature data specification, by which the commonalities and differences between different representations can be checked while remaining independent from specific application domains.

Different research communities have proposed to use computational ontologies for feature-based product knowledge representation and data sharing between CAD systems, to facilitate the integration of CADs with downstream applications like CAM and CAPP systems, as well as to provide formal tools for feature recognition and manufacturing verification. For example, the *Core Product Model* (CPM) ontology represent an engineered product as the aggregation of form, function and feature, where the latter is meant as "a subset of the form of an object that has some function assigned to it" [2]. The CPM is reused across different research projects. Dartigues et al. [17] extend it to the integration of CAD/CAPP systems. Their *Feature Ontology* is formalised in KIF.

The *Common Design-Feature Ontology* (CDFO) is an OWL ontology for feature-based CAD models exchange [18]. Feature classes are extracted from CAD systems like Catia V5, Pro/Engineering, SolidWorks and classified into a taxonomy.

The *Manufacturing Core Ontology* (MCCO) was presented by researchers at Loughborough University [14] as a common semantic foundation for modelling and sharing manufacturing knowledge. The concept of feature, meant as "a distinctive attribute or aspect of something" plays a key role within MCCO, because manufacturing operations and tools information is attached to the part to be manufactured with respect to its geometric features. The ontology is specified in Common Logic.

Kim et al. [1] proposed a classification and OWL/SWRL formalisation of assembly features to automatically reason over product knowledge, to reuse assembly models and to facilitate data sharing across applications. The classification is enriched with classes about manufacturing processes, products and materials, among others, so that it can be used to foster CAD/CAM integration.

Recently, Wang and Yu [10] proposed a feature ontology split in two modules, the STEP Box and the Feature Box. The former consists of a partial OWL formalization of ISO10303-AP203. The latter is a feature library that describes features as combinations of the STEP Box elements using OWL axioms and SWRL rules. The authors show how their system is able to automatically recognise a number of STEP features in design models.

4   Bottlenecks of Feature-Based Modelling Approaches

Despite the amount of work in the engineering community on the formal representation of features, as witnessed in the previous two sections, the use of feature-based approaches and systems is hampered by the lack of a shared and systematic understanding of what counts as a feature, and the diversity of methodologies that this situation led to. Overall, we can say that today features are taken to be macro modelling elements with little machine-processable knowledge attached to them. This is probably explained by the early success obtained by the formal representation of form features and, in contrast, the puzzling heterogeneity of non-morphological information. The lack of a unifying framework for the new types of information makes integrated features much harder to model and manage. Without a solid system for non-morphological features, relevant information for engineering purposes cannot be shared or even modelled, limiting the development of CAD/CAM integrated systems [4].

Additionally, research communities have pointed out from the initial development of the feature-based approaches a contrast between the application nature of the feature-based proposals [3] and the guiding idea that feature models should serve as means to reliably share and integrate information spanning all the production phases, thus independently from application needs [18]. This situation has led to modelling approaches that treat features as aggregations of geometric and non-geometric parameters [6] without addressing the basic issue of what features are supposed to be. As a result, if we assume that geometric elements and features are different things, due to the kind of knowledge they carry, it is unclear how to separate them. Assuming they are similar as one would think working in manufacturing applications, it remains unclear why certain geometric configurations correspond to one feature and others to several [4,5].

From the ontological perspective, these issues point to lack of understanding of the entity one is modelling. This concerns the identity and unity criteria that guide the notion of feature: it is neither clear what a feature is (identity), nor how a feature can be considered as a whole entity (unity).

Let us consider the following case. The block in Fig. 1 can be considered from the conceptual design perspective as a single functional feature, because a functional meaning can be attached to the whole geometry of the piece, which is e.g. functional for assembly purposes. From the detailed design perspective, one can consider the geometric feature formed by A and B and bisected by rib C as a single slot feature, whereas from the machining viewpoint one can consider two different features, namely A and B. This happens because for machining two different operations may be required for the realisation of A and B, while these point to a single morphological element from the design perspective. Additionally, one might want to specify the materials used for the piece, as e.g. A, B and C realised on wood.

Imagine now to have four models of the block: i) the conceptual design, ii) the detailed geometry, the iii) the manufacturing and the iv) material models. Which methodology and criteria should support their integration? We can rely on a formal representation of the geometry of the features, e.g., by following the approach proposed by [4]. The geometry would constitute a basic layer upon which further application- and domain-driven formalisations can be added to enrich the integrated feature-based model. However, the formal representation of geometrical entities does not tell us whether A and B constitute a single feature:

unity criteria of physical entities are quite complex and cannot (and should not) be derived from the choice of a geometrical formalism. Additionally, geometrical formalism by itself is not suitable to manage qualitative knowledge. For instance, it cannot support how to attach functional specifications to feature geometry and topology, or the integration of product morphology to its raw materials.

A mathematical approach to product-related knowledge is well-suited for the development of algorithmic procedures for feature extraction from CAD models, but does not suffice to embed qualitative expert's knowledge into models. From this perspective, what is needed is a qualitative representation of the elements used in a engineering system for product modelling, that is, a formal treatment for engineering concepts, feature above all.

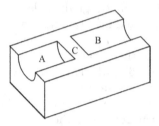

Fig. 1. An example of feature adapted from [5]

Previous work about the application of ontology engineering for feature-based systems has led to the release of multiple ontologies. Nevertheless, research efforts have focused either on application requirements, or on the logical representation of the modelling elements at hand. Little attention has been given to the issue of understanding what features are, how they can be distinguished from pure geometric entities, how they can be enriched with qualitative knowledge and how to characterise feature notions in a way that is stable and re-usable across communities and applications.

5   Classifying Features: An Ontological Viewpoint

From the analysis of the literature two contrasting notions of feature emerge:

F1-feature: Feature as the modelling component of product modelling systems that supplement quantitative geometric models with qualitative engineering knowledge. In this view, a feature is a set of information entities added to a product model for reasoning about the device under design.

F2-feature: Feature as an element of a physical product like a characteristic (e.g. a quality on a par with color and weight), a physical component (a wall of a building), or a geometric configuration (a hole, step, chamfer, etc.). In this view a feature, with its qualifying properties, is related to a physical product by means of specific relationships, depending on the feature types (cf. Sect.6).

These two views have been co-existing and exploited in the literature for at least 20 years. In the former case, a feature exists only within the context of a model: a hole is seen as a helpful, yet abstract, notion that allows a CAD model to convey a variety of useful information about a concavity in the product like why it is needed and how it is obtained. Salomons and colleagues [7] had this view in mind when they stated that a feature is "a carrier of product information that may aid design or communication between design and manufacturing, or between other engineering tasks". In the same years, Shah and Mantyla [3] pointed to the second view claiming that features are "the generic shapes or characteristics of a product with which engineers can associate certain attributes and knowledge useful for reasoning about that product". Here a feature is a fully fledged entity of the physical world: a hole is seen as an actual part of the product.

These two perspectives are strictly related: F2-features are the result of manufacturing activities, the very activities that are set with the goal to realise the features in the sense of F1, i.e., the modelling elements. At the same time the F1- and F2-features cannot be confused: a CAD model may specify that a hole feature of the designed part has a diameter of 0,5 cm with a tolerance value of 0,1mm. Yet, each realisation of the CAD model will have a hole which, while compliant with the specification, has its own specific diameter within the tolerance range. Analogously, a feature is present in a physical product [3] only in the sense of F2, as it would make no sense to claim that a computer-based modelling element is a constituent of a material product.

From this perspective, a F1-feature is an "information aggregate" that satisfies some unity condition for an application purpose, in the sense that various information models can be aggregated to count as a single whole element. For instance, the geometry, functional and manufacturing models of the example in Fig.1, while being three different information models, can also be taken to represent a unique modelling feature in, e.g., a CAD/CAM integrated system. A F1-feature is therefore a *whole* element that exists only within a (computer-based) model, and is part of a larger element, typically the product represented in the model.

At a closer look, a F2-feature may be considered not a feature *per se*. Rather, one could claim, it is a feature only *within* an engineering context. Imagine, for example, an engineer performing a quality test to verify whether a hole on a block of wood is within the prescribed tolerance limits. In a weak reading of F2-features, the hole is seen as a feature during the test activity since it has to be checked against some given specification. Yet, the hole as such, i.e., outside this activity, is not a feature. This view suggests that, according to the terminology in [19], F2-features are *anti-rigid* entities, i.e., *being a F2-feature* is a property that an entity has only within some engineering concern or activity: the particular slot A of Fig. 1 may stop to be a feature once the product is complete while remaining the very same slot.

6   Formal Representation

In order to formalise the readings of the F1- and F2-feature notions introduced in Sect. 5, we now adopt the DOLCE foundational ontology as presented in [20] (DOLCE-CORE).

Foundational ontologies are formal theories for the specification of general, upper-level notions, like object, quality, region, which are common to different modelling scenarios. Differently from domain- and application-driven ontologies, which are focused on specific modelling tasks, a foundational ontology has a large scope and can be highly reusable for different purposes. Its notions are based on the philosophical theories of Formal Ontology, which guarantee solid conceptual bases to its categories. Furthermore, since a foundational ontology is mainly aimed at providing a semantic transparent conceptual framework, it requires the use of a rich axiomatisation; therefore, expressive formal languages are preferred over computational and tractable ones. There is nowadays a spread consensus among the scientific community about the impossibility of a unique foundational ontology for all modelling scenarios, since different research communities do not often share the same ontological commitments. It is rather favoured the development of a library of foundational ontologies, including formal mappings among the different modules to facilitate their comparison.

DOLCE has been explicitly designed with a cognitive-bias aimed at capturing the ontological categories underlying natural language and common-sense thinking. It has been employed in various knowledge representation tasks, from social roles and organisations, to business process modelling, engineering design and manufacturing scenarios. Its conceptual framework is limited to *particulars*, entities that, differently from *properties*, exist in time and cannot instantiate themselves. Examples of particulars are Maradona, the Pisa tower and the authors of this paper. Particulars in DOLCE-CORE include *object*, *quality* and *concept*, which will be shortly introduced. The DOLCE-CORE axioms are indicated by **DLn** where n is the axiom number in [20]; we write **DLn^*** for the axioms of DOLCE-CORE which are only informally given in [20].

In DOLCE-CORE an object (O) has primarily a spatial quality (SQ) identifying its location (DL1*); $I(y,x)$ is red as "y inheres in x", and refers to the inherence relationship holding between a quality and its bearer. A quality (Q), among which SQ, existentially depends on its bearer (DL22), namely it cannot exist without it. Intensional properties are introduced in the domain of quantification as (reified) concepts (C) and classification (CF) is used as a sort of (possibly intensional) instance-of relation between a concept and the entities satisfying the properties it describes. $CF(x,y,t)$ holds if y, at the time t in which it is present (PRE), satisfies the property x (DL18). Then, only concepts can classify other entities (D17). Concepts are classified in DOLCE-CORE by a finite number of disjoint spaces, called SP_i, (DL2*), whose structure we do not discuss here.

DL1* $O(x) \rightarrow \exists y\,(SQ(y) \wedge I(y,x))$
DL22 $Q(x) \rightarrow \exists y(I(x,y))$
DL18 $CF(x,y,t) \rightarrow PRE(y,t)$

DL17 $CF(x, y, t) \rightarrow C(x)$
DL2* $C(x) \leftrightarrow \bigvee_{i \in \{1, \dots n\}} SP_i(x)$

For the purposes of this work, we concentrate on the DOLCE-CORE concepts (C) that refer to the "content" of engineering models. In this sense, we distinguish between *what* is described by e.g. a CAD model, i.e. the set of properties that the corresponding physical products have to satisfy (to be considered of a certain type), from the support (a CAD file, or a piece of paper) in which these properties are represented (by means of a graphical or verbal language). We call the latter *representational artefact* (RA): it has the function of *representing* various concepts specified in modelling languages. For instance, by looking at Fig.1 we need to distinguish: (i) its content, i.e., a geometric form with a number of feature; (ii) the content's specification in a graphical language, namely the drawing; (iii) the representational artefact, i.e., the specific page when this article is printed, or the video screen when Fig.1 is digitally visualised.[1] Clearly, one and the same concept can be represented in different representational artefacts.

Formally, we introduce RA specialising the object class O (A1). The relationship of representation RPT holds between a representational artefact in RA and a concept C at a certain time T (A2). A representational artefact implies the co-existence of the represented concept (A3). An instance of RA may represent more than one concept (A4) but in this case there must exist a concept of which all these are parts (A5).[2] Informally, this says that a concept can be complex, e.g., the concept of a car includes the information entities about its components (frame, engine, seats and so on).

A1 $RA(x) \rightarrow O(x)$
A2 $RPT(x, y, t) \rightarrow RA(x) \wedge C(y) \wedge T(t)$
A3 $RA(x) \wedge PRE(x, t) \rightarrow \exists y\, RPT(x, y, t)$
A4 $RPT(x, y, t) \wedge P(z, y) \rightarrow RPT(x, z, t)$
A5 $RA(x) \rightarrow \exists wt(RPT(x, w, t) \wedge \forall zt(RPT(x, z, t) \wedge P(z, w)))$

In the DOLCE-CORE framework concepts have a static nature as they are invariant across time. In design, however, it seems reasonable to allow concepts to evolve. For instance, the concept of a product under design might change over time due to customers' requirements or to the designer's activity. This can be modelled by adding a temporal parameter to CF: $CF(x, t, y, t')$ holds if entity y, as it is at time t', satisfies x, as it is at time t (A6). We thus adopt (A6) as a replacement of axiom (DL18). Additionally, we want to talk about relationships holding among concepts themselves: $CH(x, y, t)$ says that concept x, as it is at time t, is characterised by concept y (A7). By (A8), we have that if concept x classifies entity y and z characterises x, then y is also classified by z. For instance, if a plank concept is characterised by the concept *being rectangular*, the instances of the plank have to be instances of *being rectangular*.

---

[1] In another view, which we do not exploit here, the physical support is the ink on the paper.
[2] We assume that a "reading" of the RA is (explicitly or implicitly) fixed.

A6 $CF(x,t,y,t') \to C(x) \land PRE(x,t) \land PRE(y,t')$
A7 $CH(x,y,t) \to C(x) \land C(y)$
A8 $CF(x,t,y,t') \land CH(x,z,t) \to CF(z,t,y,t')$

We can now introduce the class of feature modelling elements, the F1-features, indicated by FC (feature concept), as a specialisation of C. In particular, a feature x implies the existence of a concept y that x characterises (A9). The features as product element, the F2-features, form the class PF (physical feature). Here we concentrate on the strong reading of F2-feature described at the end of Section 5. That is, we assume that a F2-feature is a feature *per se* independently of specific engineering concerns and activities.

FC serves to classify the members of PF. So a feature concept can only classify physical features (A10), while a physical feature can be an object, a quality or a DOLCE-feature (A11). Recall that DOLCE-features are physical entities constantly dependent on other objects, like edges, bumps and holes, see (A12) where we write DP for the dependence relation. Note that we now have three distinct notions of feature at play: F1-features (FC), F2-features (PF) and DOLCE-features (F). The first two are engineering-based notions, the third is ontological.

A9 $FC(x) \to \exists yt\, CH(y,x,t)$
A10 $CF(x,t,y,t') \land FC(x) \to PF(y)$
A11 $PF(x) \to O(x) \lor Q(x) \lor F(x)$
A12 $F(x) \land PRE(x,t) \to \exists y(O(y) \land PRE(y,t) \land DP(x,y))$

Regarding PF features, we need to distinguish three cases. Let x be a physical feature, then: If x is an object, then there is an object y, not a PF, of which x is proper part (A13); if x is a quality, then it inheres in an object x (A14); if x is a DOLCE-feature, then there is an object, which is not a PF, upon which x depends (A15). (Clearly, there are important interrelations among these cases but we do not exploit them here.)

A13 $PF(x) \land O(x) \to \exists y\, (O(y) \land PP(x,y) \land \neg PF(y))$
A14 $PF(x) \land Q(x) \to \exists y\, (O(y) \land I(x,y))$
A15 $PF(x) \land F(x) \to \exists y\, (O(y) \land DP(x,y) \land \neg PF(y))$

As noted in the analysis of the literature, F1-features (FC) can be associated to domain information depending on the modelling lifecycle phase, or to application scenarios. We provide the formal representation of some application-driven FCs, but the same modelling methodology can be used for others.

Form features (FC_{Fr}) are defined as the elements in FC whose instances have proper parts which satisfy a unity criterion (U), namely they constitute a whole entity (D2). Material features (FC_{Mt}) are the elements in FC characterised by some material concept, called C_{Mt} (D3). Similarly, functional feature (FC_{Ft}) are feature characterised by some functional concept (C_{Ft}) (D4).

D2 $FC_{Fr}(x) \triangleq FC(x) \land \forall ytt'\, (CF(x,t,y,t') \to$
$$\exists zv(CF(z,t,v,t') \land P(v,y,t') \land U(v,t')))$$

D3 $FC_{Mt}(x) \triangleq FC(x) \land \exists yt(C_{Mt}(y) \land CH(x,y,t))$
D4 $FC_{Ft}(x) \triangleq FC(x) \land \exists yt(C_{Ft}(y) \land CH(x,y,t))$

The formal representation of manufacturing features is about different onto-logical entities as in this case one has to consider the manufacturing process required for the feature realisation, possibly together with its sub-processes and the required machining tools. Therefore, we need to talk about a manufactur-ing plan, that is, a manufacturing concept (C_{Mf}) classifying a manufacturing process (E). In this case, the classification holds between C_{Mf} and E relatively to the time of E itself. Also, we have that a physical feature (typically present at the end of E) depends on an object (O) which participates "passively" in the process (PC_p). Informally, this amounts to say that O is the workpiece, i.e., it undergoes the manufacturing process. Finally, other objects participate "actively" in E, e.g., the manufacturing resources employed during the process. Given these qualifications on the complexity of predicate C_{Mf}, (D5) gives the general definition for manufacturing features.

D5 $FC_{Mf}(x) \triangleq FC(x) \land \exists yt(C_{Mf}(y) \land CH(x,y,t))$

7   Ontology-Based Feature Modelling: An Example

The ontology-based modelling approach introduced in the previous section is now applied to the formal representation of the features in Fig.1. As noted in Sect. 4, current approaches presented in the literature do not provide sufficent support for the integration of multiple qualitative knowledge aspects.

We formalise four different perspectives on the product features, namely the form, the functional, the material and the manufacturing perspectives. Let f be the F1-feature of the product concept *cob* in Fig. 1, then f classifies the F2-feature pf of any instance of *cob* and pf has three parts: the F2-slot feature on the left (pf_1), the F2-slot feature on the right (pf_2), both classified by the same F1-slot feature (f_s), and the F2-rib feature (pf_3) classified by the F1-rib feature (f_r). See formula ($f1$). We thus have pf as the complex F2-feature relative to the geometric information of A, B and C in Fig.1.

Since f is also a F1-functional feature, it is characterised by the functionality concept *cft* (f2). Similarly, f is characterised by material concept *cmt* (f3) while the manufacturing perspective is given in (f4). Since f is characterised by *cft*, *cmt* and *cmf*, we obtain that its corresponding pf satisfies the functionality, the material and the manufacturing concepts (T1).

f1 $FC_{Fr}(f) \land CH(cob, f, t) \land CF(f, t, pf, t') \land CF(f_s, t, pf_1, t') \land CF(f_s, t, pf_2, t') \land$
$CF(f_r, t, pf_3, t') \land pf = pf_1 + pf_2 + pf_3$
f2 $FC_{Ft}(f) \land C_{Ft}(cft) \land CH(f, cft, t)$
f3 $FC_{Mt}(f) \land C_{Mt}(cmt) \land CH(f, cmt, t)$
f4 $FC_{Mf}(f) \land C_{Mf}(cmf) \land CH(f, cmf, t)$

T1 From f1, f2, f3, f4 and A8:

$$CF(f,t,pf,t') \wedge CH(f,cft,t) \wedge CH(f,cmt,t) \wedge CH(f,cmf,t) \rightarrow$$
$$CF(cft,t,pf,t') \wedge CF(cmt,t,pf,t') \wedge CF(cmf,t,pf,t')$$

We have just showed the general modelling approach by which qualitative knowledge relevant to Fig. 1 can be specified by means of our theory. A more detailed formalisation requires to specialise further the relationships across the types of features and the ontological entities. For example, the overall functionality of pf may be subdivided across its physical feature parts and, similarly, the internal structure of the event relative to the manufacturing feature can be used to clarify how the F2-feature is realised.

8 Conclusion

The development of knowledge-based system for product-lifecycle management is a challenging task, as it requires the formal representation of detailed engineering knowledge, as well as the integration of various qualitative knowledge aspects. As we stressed in the paper, no stable, nor well-founded approach is currently available for this purpose.

We presented an ontological analysis of feature-based product modelling notions that is aimed at supporting both product knowledge specification and qualitative knowledge integration. We concentrated on the classification of feature notions by distinguishing between modelling elements and real-world entities, and by investigating their dependencies upon other ontological and engineering notions. In one case, features are meant to embed qualitative knowledge into product models, while in another they are actual entities on a par with the associated physical products. From this distinction, we argued that feature types should be distinguished at the modelling element level. In particular, we discussed engineering features as objects, as qualities and as DOLCE-features although it is still unclear whether these categories are exhaustive. In the end, we showed an approach to formalise and integrate various features qualitative models following our analysis and provided an example related to design and manufacturing.

Acknowledgments. This work was partially funded by the VISCOSO project financed by the Autonomous Province of Trento through the "Team 2011" funding programme, and the FourByThree project funded by the European Horizon 2020 program (grant agreement 637095).

References

1. Kim, K.-Y., Manley, D.G., Yang, H.: Ontology-based assembly design and information sharing for collaborative product development. Computer-Aided Design **38**, 1233–1250 (2006)
2. Fenves, S., Foufou, S., Bock, C., Sriram, R.D.: CPM: A core model for product data. Journal of Computing and Information Science in Engineering **8** (2008)

3. Shah, J., Mantyla, M.: Parametric and Feature Based CAD/CAM. Concepts, Techniques, Applications. John Wiley and Sons (1995)
4. Zhou, X., Qiu, Y., Hua, G., Wang, H., Ruan, X.: A feasible approach to the integration of CAD and CAPP. Computer-Aided Design 39, 324–338 (2007)
5. Mantyla, M., Nau, D., Shah, J.: Challenges in feature-based manufacturing research. Communications of the ACM 39(2), 77–85 (1996)
6. Ma, Y.S., Chen, G., Thimm, G.: Paradigm shift: unified and associative feature-based concurrent and collaborative engineering. Journal of Intelligent Manufacuring 19, 625–641 (2008)
7. Salomons, O.W., Houten, F., Kals, H.J.J.: Review of research in feature-based design. Journal of Manufacturing Systems 12(2), 113–132 (1993)
8. Amaitik, S.M., Kilic, S.E.: STEP-based feature modeller for computer-aided process planning. International Journal of Production Research 43(15), 3087–3101 (2005)
9. Han, J.H., Pratt, M., Regli, W.C.: Manufacturing feature recognition from solid models: A status report. IEEE Transactions on Robotics and Automation 16(6), 782–796 (2000)
10. Wang, Q., Yu, X.: Ontology based automatic feature recognition framework. Computers in Industry 65, 1041–1052 (2014)
11. Brown, D.C.: Functional, behavioral and structural features. In: ASME 2003 International Design Engineering Technical Conferences and Computers and Information in Engineering Conference. American Society of Mechanical Engineers (2003)
12. Groover, M.P.: Automation, Production Systems, and Computer-integrated Manufacturing. Prentice Hall Press (2007)
13. Nepal, M.P., Staub-French, S., Pottinger, R., Zhang, J.: Ontology-based feature modeling for construction information extraction from a building information model. Journal of Computing in Civil Engineering 27(5), 555–569 (2013)
14. Anjum, N.A., Harding, J.A., Young, R.I.M., Case, K.: Manufacturability verification through feature-based ontological product models. Proceedings of the Institution of Mechanical Engineers, Part B: Journal of Engineering Manufacture, 1086–1098 (2012)
15. ISO, Industrial Automation Systems and Integration - Product Data Representation and Exchange. Part 1: Overview and fundamental principles, iso 10303–1:1994(e) ed. (1994)
16. Amaitik, S.M., Kilic, S.E.: An intelligent process planning system for prismatic parts using STEP features. International Journal of Advanced Manufacturing Technology 31, 978–993 (2007)
17. Dartigues, C., Ghodous, P., Grüninger, M., Pallez, D., Sriram, R.: CAD/CAPP integration using feature ontology. Concurrent Engineering 12(2), 237–249 (2007)
18. Abdul-Ghafour, S., Ghodous, P., Shariat, B., Perna, E., Khosrowshahi, F.: Semantic interoperability of knowledge in feature-based CAD models. Computer-Aided Design 56, 45–57 (2014)
19. Guarino, N., Welty, C.: An overview of ontoclean. In: Staab, S., Studer, R. (eds.) Handbook on Ontologies, pp. 201–220. Springer-Verlag, Berlin, Heidelberg (2009)
20. Borgo, S., Masolo, C.: Foundational choices in DOLCE. In: Staab, S., Studer, R. (eds.) Handbook on Ontologies. Springer Verlag, Berlin, Heidelberg (2009)

A Multi-engine Theorem Prover
for a Description Logic of Typicality

Laura Giordano[1], Valentina Gliozzi[2], Nicola Olivetti[3],
Gian Luca Pozzato[2(⊠)], and Luca Violanti[4]

[1] DISIT - Universitá Piemonte Orientale - Alessandria, Alessandria, Italy
laura.giordano@unipmn.it
[2] Dipartimento Informatica - Universitá di Torino, Torino, Italy
{valentina.gliozzi,gianluca.pozzato}@unito.it
[3] Aix Marseille Université - ENSAM, Université de Toulon, LSIS UMR 7296,
Toulon, France
nicola.olivetti@univ-amu.fr
[4] NCR Edinburgh - United Kingdom, Edinburgh, UK
luca.violanti@gmail.com

Abstract. We describe DysToPic, a theorem prover for the preferential Description Logic $\mathcal{ALC} + \mathbf{T}_{min}$. This is a nonmonotonic extension of standard $\mathcal{ALC}$ based on a typicality operator $\mathbf{T}$, which enjoys a preferential semantics. DysToPic is a multi-engine Prolog implementation of a labelled, two-phase tableaux calculus for $\mathcal{ALC} + \mathbf{T}_{min}$ whose basic idea is that of performing these two phases by different machines. The performances of DysToPic are promising, and significantly better than the ones of its predecessor PreDeLo 1.0 recently introduced.

1 Introduction

Recently, a large amount of work has been done in order to extend the basic formalism of Description Logics (for short, DLs) with nonmonotonic reasoning features [1,3–5,7–9,16,17,19,21,22]; the purpose of these extensions is that of allowing reasoning about *prototypical properties* of individuals or classes of individuals. The most well known semantics for nonmonotonic reasoning have been used to the purpose, from default logic [1], to Circumscription [3], to Lifschitz's nonmonotonic logic MKNF [7,21], to preferential reasoning [4,9,16], to rational closure [5,6].

In this work we focus on the simple but powerful nonmonotonic extension of DLs proposed in [10,15,16]. In this approach "typical" or "normal" properties can be directly specified by means of a "typicality" operator $\mathbf{T}$ enriching the underlying DL; the idea is that, given a concept C, the operator $\mathbf{T}$ singles out the typical instances of C. In this formalism, one can express properties holding for all the elements belonging to the extension of C with standard inclusions $C \sqsubseteq D$, as well as properties holding only for the "most normal" elements of C with inclusions of the form $\mathbf{T}(C) \sqsubseteq D$. The typicality operator $\mathbf{T}$ is essentially characterized by the core properties of nonmonotonic reasoning axiomatized by

© Springer International Publishing Switzerland 2015
M. Gavanelli et al. (Eds.): AI*IA 2015, LNAI 9336, pp. 164–178, 2015.
DOI: 10.1007/978-3-319-24309-2_13

either *preferential logic* [18] or *rational logic* [20]. In these logics one can consistently express defeasible inclusions and exceptions such as "normally, newborns have a high level of hematocrit, whereas typical newborns who are affected by neonatal anemia have a low hematocrit":

$\mathbf{T}(Newborn) \sqsubseteq HighHematocrit$
$\mathbf{T}(Newborn \sqcap \exists HasNeonatalDisease.Anemia) \sqsubseteq LowHematocrit$
$HighHematocrit \sqcap LowHematocrit \sqsubseteq \bot.$

In order to perform useful inferences, in [16] we have introduced a nonmonotonic extension of $\mathcal{ALC}$ plus $\mathbf{T}$ based on a minimal model semantics. Intuitively, the idea is to restrict our consideration to models that maximize typical instances of a concept: more in detail, we introduce a preference relation among $\mathcal{ALC}$ plus $\mathbf{T}$ models, then we define a *minimal entailment* restricted to models that are minimal with respect to such preference relation. The resulting logic, called $\mathcal{ALC}+\mathbf{T}_{min}$, supports typicality assumptions, so that if one knows that Giuseppe is a newborn, one can nonmonotonically assume that he is also a *typical* newborn and therefore that he has a high level of hematocrit. As an example, for a TBox specified by the inclusions above, in $\mathcal{ALC} + \mathbf{T}_{min}$ we can infer that:

1. TBox $\models_{\mathcal{ALC}+\mathbf{T}_{min}} \mathbf{T}(Newborn \sqcap Bald) \sqsubseteq HighHematocrit$
2. TBox $\cup \{Newborn(Lino)\} \models_{\mathcal{ALC}+\mathbf{T}_{min}} HighHematocrit(Lino)$
3. TBox $\cup \{Newborn(Lino), \exists HasNeonatalDisease.Anemia(Lino)\} \models_{\mathcal{ALC}+\mathbf{T}_{min}}$
   $LowHematocrit(Lino)$
4. TBox $\cup \{Newborn(Lino), Bald(Lino)\} \models_{\mathcal{ALC}+\mathbf{T}_{min}} HighHematocrit(Lino)$
5. TBox $\cup \{\exists HasBrother. Newborn(Luciano)\} \models_{\mathcal{ALC}+\mathbf{T}_{min}}$
   $HasBrother. HighHematocrit(Luciano)$

In 1 and 4, it can be seen that $\mathcal{ALC} + \mathbf{T}_{min}$ captures a form of *irrelevance*: being a bald newborn is irrelevant with respect to the level of hematocrit, therefore the logic allows to conclude a general property $\mathbf{T}(Newborn \sqcap Bald) \sqsubseteq HighHematocrit$, as well as the fact that the newborn Lino has a high level of hematocrit also in case we further know that he is bald. In 3, it can be seen that $\mathcal{ALC} + \mathbf{T}_{min}$, in case of conflict, allows to give preference to more specific information: Lino is a newborn, but he is affected by neonatal anemia, therefore the logic allows to conclude that he has a low level of hematocrit. Minimal consequence applies also to individuals not explicitly named in the ABox as well, without any ad-hoc mechanism, as shown in 5, where defeasible inferences are applied to the newborn brother of Luciano.

In this work we focus on theorem proving for nonmonotonic extensions of DLs. We introduce DysToPic, a theorem prover for $\mathcal{ALC} + \mathbf{T}_{min}$. DysToPic implements the labelled tableaux calculus for this logic introduced in [16] performing a two-phase computation: in the first phase, candidate models falsifying a given query are generated (complete open branches); in the second phase the minimality of candidate models is checked by means of an auxiliary tableau construction. DysToPic is a multi-engine theorem prover, whose basic idea is that the two phases of the calculus are performed by different machines: a "master" machine M, called the *employer*, executes the first phase of the tableaux calculus, whereas other computers are used to perform the second phase on open branches detected by M. When M finds an open branch, it invokes the second phase on the calculus on a different "slave" machine, called *worker*, S_1, while M goes on performing the first phase on other branches, rather than waiting for the result of S_1. When another open branch is detected, then another machine S_2 is involved in the procedure in order to perform the second phase of the calculus on that branch. In this way, the second phase is performed simultaneously on different branches, leading to a significant increase of the performance.

Labelled tableaux calculi are implemented in Prolog, following the line of the predecessor PreDeLo 1.0, introduced in [14]: DysToPic is inspired by the methodology introduced by the system lean$T^A P$ [2], even if it does not fit its style in a rigorous manner. The basic idea is that each axiom or rule of the tableaux calculus is implemented by a Prolog clause of the program: the resulting code is therefore simple and compact.

In general, the literature contains very few proof methods for nonmonotonic extensions of DLs. We provide some experimental results to show that the performances of DysToPic are promising, in particular comparing them to the ones of PreDeLo 1.0. DysToPic is available for free download at:

http://www.di.unito.it/~pozzato/theoremprovers.html

2   The Logic $\mathcal{ALC} + \mathbf{T}_{min}$

The logic $\mathcal{ALC} + \mathbf{T}_{min}$ is obtained by adding to $\mathcal{ALC}$ the typicality operator $\mathbf{T}$ [10,15]. The intuitive idea is that $\mathbf{T}(C)$ selects the *typical* instances of a concept C. We can therefore distinguish between the properties that hold for all instances of concept C ($C \sqsubseteq D$), and those that only hold for the normal or typical instances of C ($\mathbf{T}(C) \sqsubseteq D$).

The language $\mathcal{L}$ is defined by distinguishing *concepts* and *extended concepts*. Given an alphabet $\mathcal{C}$ of concept names, $\mathcal{R}$ of role names, and $\mathcal{O}$ of individual constants, $A \in \mathcal{C}$ and $\top$ are *concepts* of $\mathcal{L}$; if $C, D \in \mathcal{L}$ and $R \in \mathcal{R}$, then $C \sqcap D, C \sqcup D, \neg C, \forall R.C, \exists R.C$ are *concepts* of $\mathcal{L}$. If C is a concept, then C and $\mathbf{T}(C)$ are *extended concepts*, and all the boolean combinations of extended concepts are extended concepts of $\mathcal{L}$. A KB is a pair (TBox,ABox). TBox contains inclusion relations (subsumptions) $C \sqsubseteq D$, where C is an extended concept of the form either C' or $\mathbf{T}(C')$, and $D \in \mathcal{L}$ is a concept. ABox contains expressions of the form $C(a)$ and $R(a,b)$, where $C \in \mathcal{L}$ is an extended concept, $R \in \mathcal{R}$, and $a, b \in \mathcal{O}$.

In order to provide a semantics to the operator $\mathbf{T}$, we extend the definition of a model used in the "standard" Description logic $\mathcal{ALC}$. The idea is that the operator $\mathbf{T}$ is characterized by a set of postulates that are essentially a reformulation of the Kraus, Lehmann and Magidor's axioms of *preferential logic* $\mathbf{P}$ [18]. Intuitively, the assertion $\mathbf{T}(C) \sqsubseteq D$ corresponds to the conditional assertion $C \mathrel{|\!\sim} D$ of $\mathbf{P}$. $\mathbf{T}$ has therefore all the "core" properties of nonmonotonic reasoning as it is axiomatized by $\mathbf{P}$. The idea is that there is a global preference relation among individuals, in the sense that $x < y$ means that x is "more normal" than y, and that the typical members of a concept C are the minimal elements of C with respect to this relation. In this framework, an element $x \in \Delta$ is a *typical instance* of some concept C if $x \in C^I$ and there is no element in C^I *more typical* than x. The typicality preference relation is partial.

Definition 1. *Given an irreflexive and transitive relation $<$ over Δ and $S \subseteq \Delta$, we define $Min_<(S) = \{x : x \in S$ and $\nexists y \in S$ s.t. $y < x\}$. We say that $<$ is well-founded if and only if, for all $S \subseteq \Delta$, for all $x \in S$, either $x \in Min_<(S)$ or $\exists y \in Min_<(S)$ such that $y < x$.*

Definition 2. *A model of $\mathcal{ALC} + \mathbf{T}_{min}$ is any structure $\langle \Delta, <, I \rangle$, where: Δ is the domain; I is the extension function that maps each extended concept C to $C^I \subseteq \Delta$, and each role R to a $R^I \subseteq \Delta \times \Delta$; $<$ is an irreflexive, transitive and well-founded (Definition 1) relation over Δ. I is defined in the usual way (as for $\mathcal{ALC}$) and, in addition, $(\mathbf{T}(C))^I = Min_<(C^I)$.*

Given a model $\mathcal{M}$ of Definition 2, I can be extended so that it assigns to each individual a of $\mathcal{O}$ a distinct element a^I of the domain Δ (unique name assumption). We say that $\mathcal{M}$ satisfies an inclusion $C \sqsubseteq D$ if $C^I \subseteq D^I$, and that $\mathcal{M}$ satisfies $C(a)$ if $a^I \in C^I$ and $R(a, b)$ if $(a^I, b^I) \in R^I$. Moreover, $\mathcal{M}$ satisfies TBox if it satisfies all its inclusions, and $\mathcal{M}$ satisfies ABox if it satisfies all its formulas. $\mathcal{M}$ satisfies a KB (TBox,ABox), if it satisfies both TBox and ABox.

The semantics of the typicality operator can be specified by modal logic. The interpretation of $\mathbf{T}$ can be split into two parts: for any x of the domain Δ, $x \in (\mathbf{T}(C))^I$ just in case (i) $x \in C^I$, and (ii) there is no $y \in C^I$ such that $y < x$. Condition (ii) can be represented by means of an additional modality $\square$, whose semantics is given by the preference relation $<$ interpreted as an accessibility relation. The interpretation of $\square$ in $\mathcal{M}$ is as follows: $(\square C)^I = \{x \in \Delta \mid$ for every $y \in \Delta$, if $y < x$ then $y \in C^I\}$. We immediately get that $x \in (\mathbf{T}(C))^I$ if and only if $x \in (C \sqcap \square \neg C)^I$.

Even if the typicality operator $\mathbf{T}$ itself is nonmonotonic (i.e. $\mathbf{T}(C) \sqsubseteq E$ does not imply $\mathbf{T}(C \sqcap D) \sqsubseteq E$), what is inferred from a KB can still be inferred from any KB' with KB $\subseteq$ KB'. In order to perform *nonmonotonic* inferences, in [16] we have strengthened the above semantics by restricting entailment to a class of minimal (or preferred) models. Intuitively, the idea is to restrict our consideration to models that *minimize the non-typical instances of a concept.*

Given a KB, we consider a finite set $\mathcal{L}_\mathbf{T}$ of concepts: these are the concepts whose non-typical instances we want to minimize. We assume that the set $\mathcal{L}_\mathbf{T}$ contains at least all concepts C such that $\mathbf{T}(C)$ occurs in the KB or in the query F, where a *query* F is either an assertion $C(a)$ or an inclusion relation $C \sqsubseteq D$. As we have just said, $x \in C^I$ is typical for C if $x \in (\square \neg C)^I$. Minimizing the non-typical instances of C therefore means to minimize the objects falsifying $\square \neg C$ for $C \in \mathcal{L}_\mathbf{T}$. Hence, for a model $\mathcal{M} = \langle \Delta, <, I \rangle$, we define $\mathcal{M}_{\mathcal{L}_\mathbf{T}}^{\square^-} = \{(x, \neg \square \neg C) \mid x \notin (\square \neg C)^I$, with $x \in \Delta, C \in \mathcal{L}_\mathbf{T}\}$.

Definition 3 (Preferred and Minimal Models). *Given a model $\mathcal{M} = \langle \Delta, <, I \rangle$ of a knowledge base KB, and a model $\mathcal{M}' = \langle \Delta', <', I' \rangle$ of KB, we say* that $\mathcal{M}$ is preferred to $\mathcal{M}'$ w.r.t. $\mathcal{L}_\mathbf{T}$, and we write $\mathcal{M} <_{\mathcal{L}_\mathbf{T}} \mathcal{M}'$, if (i) $\Delta = \Delta'$, (ii) $\mathcal{M}_{\mathcal{L}_\mathbf{T}}^{\square^-} \subset \mathcal{M}'^{\square^-}_{\mathcal{L}_\mathbf{T}}$, (iii) $a^I = a^{I'}$ for all $a \in \mathcal{O}$. $\mathcal{M}$ is a minimal model for KB (w.r.t. $\mathcal{L}_\mathbf{T}$) if it is a model of KB and there is no other model $\mathcal{M}'$ of KB such that $\mathcal{M}' <_{\mathcal{L}_\mathbf{T}} \mathcal{M}$.

Definition 4 (Minimal Entailment in $\mathcal{ALC}+\mathbf{T}_{min}$). *A query F is minimally entailed in $\mathcal{ALC} + \mathbf{T}_{min}$ by KB with respect to $\mathcal{L}_\mathbf{T}$ if F is satisfied in all models of KB that are minimal with respect to $\mathcal{L}_\mathbf{T}$. We write* KB $\models_{\mathcal{ALC}+\mathbf{T}_{min}} F$.

As an example, consider the TBox of the Introduction. We have that TBox $\cup \{Kid(daniel)\} \models_{\mathcal{ALC}+\mathbf{T}_{min}} ChocolateEater(daniel)$, since $daniel^I \in (Kid \sqcap \square\neg Kid)^I$ for all minimal models $\mathcal{M} = \langle \Delta <, I \rangle$ of the TBox. In contrast, by the nonmonotonic character of minimal entailment, we have TBox $\cup \{Kid(daniel),$ $\exists HasIntolerance.Lactose(daniel)\} \models_{\mathcal{ALC}+\mathbf{T}_{min}} \neg ChocolateEater(daniel)$.

3   A Tableau Calculus for $\mathcal{ALC} + \mathbf{T}_{min}$

In this section we recall the tableau calculus $\mathcal{TAB}_{min}^{\mathcal{ALC}+\mathbf{T}}$ for deciding whether a query F is minimally entailed from a KB in $\mathcal{ALC} + \mathbf{T}_{min}$ introduced in [16]. The calculus performs a two-phase computation: in the first phase, a tableau calculus, called $\mathcal{TAB}_{PH1}^{\mathcal{ALC}+\mathbf{T}}$, simply verifies whether KB $\cup \{\neg F\}$ is satisfiable in a model of Definition 2, building candidate models; in the second phase another tableau calculus, called $\mathcal{TAB}_{PH2}^{\mathcal{ALC}+\mathbf{T}}$, checks whether the candidate models found in the first phase are *minimal* models of KB, i.e. for each open branch of the first phase, $\mathcal{TAB}_{PH2}^{\mathcal{ALC}+\mathbf{T}}$ tries to build a model of KB which is preferred to the candidate model w.r.t. Definition 3. The whole procedure is formally defined at the end of this section (Definition 5).

$\mathcal{TAB}_{min}^{\mathcal{ALC}+\mathbf{T}}$ tries to build an open branch representing a minimal model satisfying KB $\cup \{\neg F\}$, where $\neg F$ is the negation of the query F and is defined as follows: if $F = C(a)$, then $\neg F = (\neg C)(a)$; if $F = C \sqsubseteq D$, then $\neg F = (C \sqcap \neg D)(x)$, where x does not occur in KB. $\mathcal{TAB}_{min}^{\mathcal{ALC}+\mathbf{T}}$ makes use of labels, denoted with $x, y, z, \ldots$, representing individuals either named in the ABox or implicitly expressed by existential restrictions. These labels occur in *constraints*, that can have the form $x \xrightarrow{R} y$ or $y < x$ or $x : C$, where x, y are labels, R is a role and C is a concept of $\mathcal{ALC} + \mathbf{T}_{min}$ or has the form $\square\neg D$ or $\neg\square\neg D$.

3.1   The Tableaux Calculus $\mathcal{TAB}_{PH1}^{\mathcal{ALC}+\mathbf{T}}$

A tableau of $\mathcal{TAB}_{PH1}^{\mathcal{ALC}+\mathbf{T}}$ is a tree whose nodes are pairs $\langle S \mid U \rangle$. S is a set of constraints, whereas U contains formulas of the form $C \sqsubseteq D^L$, representing inclusion relations $C \sqsubseteq D$ of the TBox. L is a list of labels, used in order to ensure the termination of the tableau calculus. A branch is a sequence of nodes $\langle S_1 \mid U_1 \rangle, \langle S_2 \mid U_2 \rangle, \ldots, \langle S_n \mid U_n \rangle \ldots$, where each node $\langle S_i \mid U_i \rangle$ is obtained from its immediate predecessor $\langle S_{i-1} \mid U_{i-1} \rangle$ by applying a rule of $\mathcal{TAB}_{PH1}^{\mathcal{ALC}+\mathbf{T}}$,

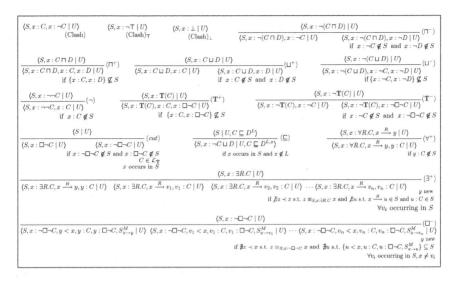

Fig. 1. The calculus $\mathcal{TAB}_{PH1}^{ALC+\mathbf{T}}$. $x \equiv_S y$ denotes that x and y label the same concepts in S. We define $S_{x \to y}^M = \{y : \neg C, y : \Box \neg C \mid x : \Box \neg C \in S\}$.

having $\langle S_{i-1} \mid U_{i-1} \rangle$ as the premise and $\langle S_i \mid U_i \rangle$ as one of its conclusions. A branch is closed if one of its nodes is an instance of a (Clash) axiom, otherwise it is open. A tableau is closed if all its branches are closed.

The rules of $\mathcal{TAB}_{PH1}^{ALC+\mathbf{T}}$ are presented in Fig. 1. Rules $(\exists^+)$ and $(\Box^-)$ are called *dynamic* since they can introduce a new variable in their conclusions. The other rules are called *static*. We do not need any extra rule for the positive occurrences of $\Box$, since these are taken into account by the computation of $S_{x \to y}^M$ of $(\Box^-)$. The (cut) rule ensures that, given any concept $C \in \mathcal{L}_\mathbf{T}$, an open branch built by $\mathcal{TAB}_{PH1}^{ALC+\mathbf{T}}$ contains either $x : \Box \neg C$ or $x : \neg \Box \neg C$ for each label x: this is needed in order to allow $\mathcal{TAB}_{PH2}^{ALC+\mathbf{T}}$ to check the minimality of the model corresponding to the open branch. As mentioned above, given a node $\langle S \mid U \rangle$, each formula $C \sqsubseteq D$ in U is equipped with the list L of labels to which the rule $(\sqsubseteq)$ has already been applied. This avoids multiple applications of such rule to the same subsumption by using the same label.

In order to check the satisfiability of a KB, we build its *corresponding constraint system* $\langle S \mid U \rangle$, and we check its satisfiability. Given KB=(TBox,ABox), its *corresponding constraint system* $\langle S \mid U \rangle$ is defined as follows: $S = \{a : C \mid C(a) \in \text{ABox}\} \cup \{a \xrightarrow{R} b \mid R(a,b) \in \text{ABox}\}$; $U = \{C \sqsubseteq D^\emptyset \mid C \sqsubseteq D \in \text{TBox}\}$. KB is satisfiable if and only if its corresponding constraint system $\langle S \mid U \rangle$ is satisfiable. In order to verify the satisfiability of KB $\cup \{\neg F\}$, we use $\mathcal{TAB}_{PH1}^{ALC+\mathbf{T}}$ to check the satisfiability of the constraint system $\langle S \mid U \rangle$ obtained by adding the constraint corresponding to $\neg F$ to S', where $\langle S' \mid U \rangle$ is the corresponding constraint system of KB. To this purpose, the rules of the calculus $\mathcal{TAB}_{PH1}^{ALC+\mathbf{T}}$ are applied until either a contradiction is generated (*clash*) or a model satisfying $\langle S \mid U \rangle$ can be obtained.

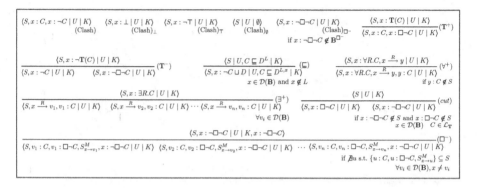

Fig. 2. The calculus $\mathcal{TAB}_{PH2}^{\mathcal{ALC}+\mathbf{T}}$. To save space, we omit rules for $\sqcup$, $\sqcap$, $\neg$.

The rules of $\mathcal{TAB}_{PH1}^{\mathcal{ALC}+\mathbf{T}}$ are applied with the following *standard strategy*: 1. apply a rule to a label x only if no rule is applicable to a label y such that $y \prec x$ (where $y \prec x$ says that label x has been introduced in the tableaux later than y); 2. apply dynamic rules only if no static rule is applicable.

Theorem 1. *Given $\mathcal{L_T}$, $KB \models_{\mathcal{ALC}+\mathbf{T}_{min}} F$ if and only if there is no open branch $\mathbf{B}$ in the tableau built by $\mathcal{TAB}_{PH1}^{\mathcal{ALC}+\mathbf{T}}$ for the constraint system corresponding to $KB \cup \{\neg F\}$ such that the model represented by $\mathbf{B}$ is a minimal model of KB.*

Thanks to the side conditions on the application of the rules and the blocking machinery adopted by the dynamic ones, in [16] it has been shown that any tableau generated by $\mathcal{TAB}_{PH1}^{\mathcal{ALC}+\mathbf{T}}$ for $\langle S \mid U \rangle$ is finite.

3.2 The Tableaux Calculus $\mathcal{TAB}_{PH2}^{\mathcal{ALC}+\mathbf{T}}$

Let us now introduce the calculus $\mathcal{TAB}_{PH2}^{\mathcal{ALC}+\mathbf{T}}$ which checks whether each open branch $\mathbf{B}$ built by $\mathcal{TAB}_{PH1}^{\mathcal{ALC}+\mathbf{T}}$ represents a minimal model of the KB.

Given an open branch $\mathbf{B}$ of a tableau built from $\mathcal{TAB}_{PH1}^{\mathcal{ALC}+\mathbf{T}}$, let $\mathcal{D}(\mathbf{B})$ be the set of labels occurring in $\mathbf{B}$. Moreover, let $\mathbf{B}^{\square^-}$ be the set of formulas $x : \neg\square\neg C$ occurring in $\mathbf{B}$, that is to say $\mathbf{B}^{\square^-} = \{x : \neg\square\neg C \mid x : \neg\square\neg C$ occurs in $\mathbf{B}\}$.

A tableau of $\mathcal{TAB}_{PH2}^{\mathcal{ALC}+\mathbf{T}}$ is a tree whose nodes are tuples of the form $\langle S \mid U \mid K \rangle$, where S and U are defined as in $\mathcal{TAB}_{PH1}^{\mathcal{ALC}+\mathbf{T}}$, whereas K contains formulas of the form $x : \neg\square\neg C$, with $C \in \mathcal{L_T}$. The basic idea of $\mathcal{TAB}_{PH2}^{\mathcal{ALC}+\mathbf{T}}$ is as follows. Given an open branch $\mathbf{B}$ built by $\mathcal{TAB}_{PH1}^{\mathcal{ALC}+\mathbf{T}}$ and corresponding to a model $\mathcal{M}^{\mathbf{B}}$ of $KB \cup \{\neg F\}$, $\mathcal{TAB}_{PH2}^{\mathcal{ALC}+\mathbf{T}}$ checks whether $\mathcal{M}^{\mathbf{B}}$ is a minimal model of KB by trying to build a model of KB which is preferred to $\mathcal{M}^{\mathbf{B}}$. To this purpose, it keeps track (in K) of the negated box formulas used in $\mathbf{B}$ ($\mathbf{B}^{\square^-}$) in order to check whether it is possible to build a model of KB containing less negated box formulas. The rules of $\mathcal{TAB}_{PH2}^{\mathcal{ALC}+\mathbf{T}}$ are shown in Figure 2. The tableau built by $\mathcal{TAB}_{PH2}^{\mathcal{ALC}+\mathbf{T}}$ closes if it is not possible to build a model smaller than $\mathcal{M}^{\mathbf{B}}$, it remains open otherwise. Since by Definition 3 two models can be compared only if they have the same domain, $\mathcal{TAB}_{PH2}^{\mathcal{ALC}+\mathbf{T}}$ tries to build an open branch containing all the labels appearing in $\mathbf{B}$, i.e. those in $\mathcal{D}(\mathbf{B})$. To this aim, the dynamic rules use labels in $\mathcal{D}(\mathbf{B})$ instead of introducing new ones in their conclusions. The rule ($\sqsubseteq$) is applied to *all the labels of* $\mathcal{D}(\mathbf{B})$ (and not only to those appearing in the branch). The rule ($\square^-$) is applied to a node $\langle S, x : \neg\square\neg C \mid U \mid K, x : \neg\square\neg C\rangle$, that is to say when the negated box formula $x : \neg\square\neg C$ also belongs to the open branch $\mathbf{B}$. Also in this case, the rule introduces a branch on the choice of the individual $v_i \in \mathcal{D}(\mathbf{B})$ to be used in the conclusion. In case a tableau node has the form $\langle S, x : \neg\square\neg C \mid U \mid K\rangle$, and $x : \neg\square\neg C \notin \mathbf{B}^{\square^-}$, then $\mathcal{TAB}_{PH2}^{\mathcal{ALC}+\mathbf{T}}$ detects a clash, called (Clash)$_{\square^-}$: this corresponds to the situation where $x : \neg\square\neg C$ does not belong to $\mathbf{B}$, while the model corresponding to the branch being built contains $x : \neg\square\neg C$, and hence is *not* preferred to the model represented by $\mathbf{B}$. The calculus $\mathcal{TAB}_{PH2}^{\mathcal{ALC}+\mathbf{T}}$ also contains the clash condition (Clash)$_{\emptyset}$. Since each application of ($\square^-$) removes the negated box formulas $x : \neg\square\neg C$ from the set K, when K is empty all the negated boxed formulas occurring in $\mathbf{B}$ also belong to the current branch. In this case, the model built by $\mathcal{TAB}_{PH2}^{\mathcal{ALC}+\mathbf{T}}$ satisfies the same set of $x : \neg\square\neg C$ (for all individuals) as $\mathbf{B}$ and, thus, it is not preferred to the one represented by $\mathbf{B}$.

Let KB be a knowledge base whose corresponding constraint system is $\langle S \mid U\rangle$. Let F be a query and let S' be the set of constraints obtained by adding to S the constraint corresponding to $\neg F$. $\mathcal{TAB}_{PH2}^{\mathcal{ALC}+\mathbf{T}}$ is *sound and complete* in the following sense: an open branch $\mathbf{B}$ built by $\mathcal{TAB}_{PH1}^{\mathcal{ALC}+\mathbf{T}}$ for $\langle S' \mid U\rangle$ is satisfiable in a minimal model of KB iff the tableau in $\mathcal{TAB}_{PH2}^{\mathcal{ALC}+\mathbf{T}}$ for $\langle S \mid U \mid \mathbf{B}^{\square^-}\rangle$ is closed. The termination of $\mathcal{TAB}_{PH2}^{\mathcal{ALC}+\mathbf{T}}$ is ensured by the fact that dynamic rules make use of labels belonging to $\mathcal{D}(\mathbf{B})$, which is finite, rather than introducing "new" labels in the tableau. Also, it is possible to show that the problem of verifying that a branch $\mathbf{B}$ represents a minimal model for KB in $\mathcal{TAB}_{PH2}^{\mathcal{ALC}+\mathbf{T}}$ is in NP in the size of $\mathbf{B}$. The overall procedure $\mathcal{TAB}_{min}^{\mathcal{ALC}+\mathbf{T}}$ is defined as follows:

Definition 5. *Let KB be a knowledge base whose corresponding constraint system is* $\langle S \mid U\rangle$*. Let F be a query and let S' be the set of constraints obtained by adding to S the constraint corresponding to $\neg F$. The calculus $\mathcal{TAB}_{min}^{\mathcal{ALC}+\mathbf{T}}$ checks whether a query F can be minimally entailed from a KB by means of the following procedure:*

- *the calculus $\mathcal{TAB}_{PH1}^{\mathcal{ALC}+\mathbf{T}}$ is applied to $\langle S' \mid U\rangle$;*
- *if, for each branch $\mathbf{B}$ built by $\mathcal{TAB}_{PH1}^{\mathcal{ALC}+\mathbf{T}}$, either: (i) $\mathbf{B}$ is closed or (ii) the tableau built by the calculus $\mathcal{TAB}_{PH2}^{\mathcal{ALC}+\mathbf{T}}$ for $\langle S \mid U \mid \mathbf{B}^{\square^-}\rangle$ is open, then the procedure says YES else the procedure says NO*

In [16] we have shown that $\mathcal{TAB}_{min}^{\mathcal{ALC}+\mathbf{T}}$ is a sound and complete decision procedure for verifying if KB $\models_{\mathcal{ALC}+\mathbf{T}_{min}}^{\mathcal{L}_{\mathbf{T}}} F$, and that the problem is in CO-NEXP^{NP}.

4   Design of DysToPic

In this section we present DysToPic, a multi-engine theorem prover for reasoning in $\mathcal{ALC} + \mathbf{T}_{min}$. DysToPic is a SICStus Prolog implementation of the tableaux calculus $\mathcal{TAB}^{\mathcal{ALC}+\mathbf{T}}_{min}$ introduced in the previous section, wrapped by a Java interface which relies on the Java RMI APIs for the distribution of the computation. The system is designed for scalability and based on a "worker/employer" paradigm: the computational burden for the "employer" can be spread among an arbitrarily high number of "workers" which operate in complete autonomy, so that they can be either deployed on a single machine or on a computer grid.

The basic idea underlying DysToPic is as follows: there is no need for the first phase of the calculus to wait for the result of one elaboration of the second phase on an open branch, before generating another candidate branch. Indeed, in order to prove whether a query F entails from a KB, the first phase can be executed on a machine; every time that a branch remains open after the first phase, the execution of the second phase for this branch can be performed in parallel, on a different machine. Meanwhile, the main machine (worker), instead of waiting for the termination of the second phase on that branch, can carry on with the computation of the first phase (potentially generating other branches). If a branch remains open in the second phase, then F is not minimally entailed from KB (we have found a counterexample), so the computation process can be interrupted early.

4.1   The Whole Architecture

In order to describe the architecture of DysToPic we refer to the *worker-employer* metaphor. The system is characterized by: (i) a single *employer*, which is in charge of verifying the query and yielding the final result. It also implements the first phase of the calculus and uses $\mathcal{TAB}^{\mathcal{ALC}+\mathbf{T}}_{PH1}$ to generate branches: the ones that it cannot close (representing candidate models of KB $\cup \{\neg F\}$), it passes to a *worker*; (ii) an unlimited number of *workers*, which use $\mathcal{TAB}^{\mathcal{ALC}+\mathbf{T}}_{PH2}$ to evaluate the models generated by the *employer*; (iii) a *repository*, which stores all the answers coming from the *workers*. A schema of the architecture of DysToPic is shown in Figure 3.

First, each worker registers to the employer. When checking whether KB $\models_{\mathcal{ALC}+\mathbf{T}_{min}} F$, the employer executes $\mathcal{TAB}^{\mathcal{ALC}+\mathbf{T}}_{PH1}$. If the employer needs to check whether an open branch generated by the first phase represents a minimal model of the KB, then it delegates the execution of the second phase to one of the registered workers, and consequently proceeds with its computation on other branches generated in the first phase. When a worker terminates its execution, it reports its result to the repository.

If every branch has been processed and each worker has answered affirmatively, i.e. each tableaux built in the second phase by $\mathcal{TAB}^{\mathcal{ALC}+\mathbf{T}}_{PH2}$ is open, the employer can conclude that KB $\models_{\mathcal{ALC}+\mathbf{T}_{min}} F$. Otherwise, the employer can

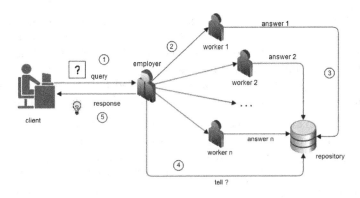

Fig. 3. The architecture of `DysToPic`.

conclude the proof as soon as the first negative answer comes into the repository, since (at least) a worker found a closed tableaux in $\mathcal{TAB}_{PH2}^{\mathcal{ALC}+\mathbf{T}}$ for an open branch (candidate model) generated by the employer, in this case we have that KB $\not\models_{\mathcal{ALC}+\mathbf{T}_{min}} F$. It is worth noticing that the employer has to keep a continuous dialogue with the repository.

The library `se.sics.jasper` is used in order to combine Java and SICStus Prolog to decouple the two phases of the calculus. In detail, the employer handles the query in `Employer.java`, a piece of Java code which presents it to `alct1.pl`, the Prolog core implementing $\mathcal{TAB}_{PH1}^{\mathcal{ALC}+\mathbf{T}}$. Every time that an open branch is generated, `alct1.pl` invokes `Phase1RMIStub.java`, another piece of Java code which will send it to the correct worker. Workers will then have to process the open branches with $\mathcal{TAB}_{PH2}^{\mathcal{ALC}+\mathbf{T}}$, which is implemented in `alct2.pl`.

Concurrency is the main goal of our implementation, since we want the execution of the first phase of the calculus to be independent from the second one. Java natively supports concurrency via multithreading. The employer uses a separate thread (implemented in `Phase1Thread.java`) to perform the current invocation of $\mathcal{TAB}_{PH1}^{\mathcal{ALC}+\mathbf{T}}$ on a query, while its main thread polls the repository waiting for termination (the procedure can be stopped when the first counterexample is found, even if not all of the branches have been explored). During the execution of $\mathcal{TAB}_{PH1}^{\mathcal{ALC}+\mathbf{T}}$, every time that the employer wants to ask a worker to verify a branch, a new thread is spawned. The worker itself makes use of threads: its main thread simply enqueues each request coming from the employer and spawns a new thread which performs $\mathcal{TAB}_{PH2}^{\mathcal{ALC}+\mathbf{T}}$.

4.2   The Implementation of the Tableaux Calculus

Concerning the implementation of the tableaux calculus $\mathcal{TAB}_{min}^{\mathcal{ALC}+\mathbf{T}}$, each machine of the system runs a SICStus Prolog implementation which is strongly related to the implementation of the calculus given by `PreDeLo 1.0`, introduced in [14]. The implementation is inspired by the "lean" methodology of lean $T^A P$, even if it does not follow its style in a rigorous manner. The program comprises a set of clauses, each one implementing a rule or axiom of the tableau calculus. The proof search is provided for free by the mere depth-first search mechanism of Prolog, without any additional mechanism.

DysToPic comprises two main predicates, called `prove` and `prove_phase2`, implementing, respectively, the first and the second phase of the tableau calculus.

Phase 1: The `prove` Predicate. Concerning the first phase of the calculus, executed by the employer, DysToPic represents a tableaux node $\langle S \mid U \rangle$ with two Prolog lists: S and U. Elements of S are either pairs [X, F], representing formulas of the form $x : F$, or triples of the form either [X,R,Y] or [X,<,Y], representing either roles $x \xrightarrow{R} y$ or the preference relation $x < y$, respectively. Elements of U are pairs of the form [[C inc D],L], representing $C \sqsubseteq D^L \in U$ described in Section 3.1.

The calculus $\mathcal{TAB}_{min}^{\mathcal{ALC}+\mathbf{T}}$ are implemented by a top-level predicate

$$\texttt{prove(+ABox,+TBox,[+X,+F],-Tree)}.$$

This predicate succeeds if and only if the query $x : F$ is minimally entailed from the KB represented by TBox and ABox. When the predicate succeeds, then the output term `Tree` matches a Prolog term representing the closed tableaux found by the prover. The top-level predicate `prove/4` invokes a second-level one:

$$\texttt{prove(+S,+U,+Lt,+Labels,+ABOX,-Tree)}$$

having 6 arguments. In detail, S corresponds to ABox enriched by the negation of the query $x : F$, whereas Lt is a list corresponding to the set of concepts $\mathcal{L_T}$. Labels is the set of labels belonging to the current branch, whereas ABOX is used to store the initial ABox (i.e. without the negation of the query) in order to eventually invoke the second phase on it, in order to look for minimal models of the initial KB.

Each clause of the `prove/6` predicate implements an axiom or rule of the calculus $\mathcal{TAB}_{PH1}^{\mathcal{ALC}+\mathbf{T}}$. To search a closed tableaux for $\langle S \mid U \rangle$, DysToPic proceeds as follows. First of all, if $\langle S \mid U \rangle$ is a clash, the goal will succeed immediately by using one of the clauses implementing axioms. As an example, the following clause implements (Clash):

```
prove(S,U,_,_,_,tree(clash)):-
    member([X,C],S),member([X, neg C],S),!.
```

If $\langle S \mid U \rangle$ is not an instance of the axioms, then the first applicable rule will be chosen, e.g. if S contains an intersection [X,C and D], then the clause implementing the $(\sqcap^+)$ rule will be chosen, and DysToPic will be recursively invoked on its unique conclusion. DysToPic proceeds in a similar way for the other rules. The ordering of the clauses is such that the application of the dynamic rules is postponed as much as possible: this implements the strategy ensuring the termination of the calculus described in the previous section. As an example, the clause implementing $(\mathbf{T}^+)$ is as follows:

```
1. prove(S,U,Lt,Labels,ABOX,tree(...,Tree)):-member([X,ti C],S),
2.    (\+(member([X,C],S)); \+(member([X, box neg C],S))),!,
3.    prove([[X,C]|[[X, box neg C]|S]],U,Lt,Labels,ABOX,Tree),!.
```

In line 1, the standard Prolog predicate member is used in order to find a formula of the form $x : \mathbf{T}(C)$ in the list S. In line 2, the side conditions on the applicability of such a rule are checked: the rule can be applied if either $x : C$ or $x : \Box\neg C$ do not belong to S. In line 3 DysToPic is recursively invoked on the unique conclusion of the rule, in which $x : C$ and $x : \Box\neg C$ are added to the list S. The last clause of prove is:

```
prove(...) :- ... , jasper_call(JVM,
        method('employer/Phase1RMIStub', 'solveViaRMI', [static]),...,
        solve_via_rmi(NextWorkerName, 'toplevelphase2(...)' ) ),!.
```

invoked when no other clauses are applicable. In this case, the branch built by the employer represents a model for the initial set of formulas, then toplevelphase2 is invoked on a worker in order to check whether such a model is minimal.

Phase 2: The prove_phase2 Predicate. Given an open branch built by the first phase, the predicate toplevelphase2 is invoked on a worker. It first applies an optimization preventing useless applications of $(\sqsubseteq)$, then it invokes the predicate

prove_phase2(+S,+U,+Lt,+K,+Bb,+Db).

S and U contain the initial KB (without the query), whereas K, Bb and Db are Prolog lists representing K, $\mathbf{B}^{\Box^-}$ and $\mathcal{D}(\mathbf{B})$ as described in Section 3.2. Lt is as for prove/6. Also in this case, each clause of prove_phase2 implements an axiom or rule of the calculus $\mathcal{TAB}_{PH2}^{ALC+\mathbf{T}}$. To search for a closed tableaux, DysToPic first checks whether the current node $\langle S \mid U \mid K \rangle$ is a clash. otherwise the first applicable rule will be chosen, and DysToPic will be recursively invoked on its conclusions. As an example, the clause implementing $(\mathbf{T}^+)$ is as follows:

```
prove_phase2(S,U,Lt,K,Bb,Db) :- select([X,ti C],S,S1),
        prove_phase2([[X,C]|[[X,box neg C]|S1]],U,Lt,K,Bb,Db),!.
```

Notice that, according to the calculus $\mathcal{TAB}_{PH2}^{ALC+\mathbf{T}}$, the principal formula to which the rule is applied is removed from the current node: to this aim, the SICStus Prolog predicate select is used rather than member.

4.3   Performance Testing of DysToPic

We have made an attempt to show how DysToPic performs, especially in comparison with its predecessor PreDeLo 1.0. The performances of DysToPic are promising. We have tested both the provers by running SICStus Prolog 4.1.1 on Ubuntu 14.04.1 64 bit machines. Concerning DysToPic, we have tested it on

4 machines, namely: 1. a desktop PC with an Intel Core i5-3570K CPU (3.4-3.8GHz, 4 cores, 4 threads, 8GB RAM); 2. a desktop PC with an Intel Pentium G2030 CPU (3.0GHz, 2 cores, 2 threads, 4GB RAM); 3. a Lenovo X220 laptop with an Intel Core i7-2640M CPU (2.8-3.5GHz, 2 cores, 4 threads, 8GB RAM); 4. a Lenovo X230 laptop with an Intel Core i7-3520M CPU (2.9-3.6GHz, 2 cores, 4 threads, 8GB RAM).

We have performed two kinds of tests. On the one hand, we have randomly generated KBs with different sizes (from 10 to 100 ABox formulas and TBox inclusions) as well as different numbers of named individuals: in less than 10 seconds, both the provers DysToPic and PreDeLo 1.0 are able to answer in more than the 75% of tests. Notice that, as far as we know, it does not exist a set of acknowledged benchmarks for defeasible DLs. On the other hand, we have tested the two theorem provers on specific examples. As expected, DysToPic is better in than the competitor in answering that a query F is not minimally entailed from a given KB. Surprisingly enough, its performances are better than the ones of PreDeLo 1.0 also in case the provers conclude that F follows from KB, as in the following example:

Example 1. Given TBox $=\{\mathbf{T}(Student)\sqsubseteq\neg IncomeTaxPayer, WorkingStudent\sqsubseteq Student, \mathbf{T}(WorkingStudent) \sqsubseteq IncomeTaxPayer\}$ and ABox$=\{Student(mario), WorkingStudent(mario), Tall(mario), Student(carlo), WorkingStudent(carlo), Tall(carlo), Student(giuseppe), WorkingStudent(giuseppe), Tall(giuseppe)\}$, we have tested both the theorem provers in order to check whether $IncomeTaxPayer$ ($mario$) is minimally entailed from KB=(TBox,ABox). This query generates 1090 open branches in $\mathcal{TAB}_{PH1}^{ALC+\mathbf{T}}$, each requiring the execution of $\mathcal{TAB}_{PH2}^{ALC+\mathbf{T}}$. PreDeLo 1.0 answers in 370 seconds, whereas DysToPic answers in 210 seconds if only two machines are involved (employer + one worker). If 4 workers are involved, DysToPic only needs 112 seconds to conclude its computation.

Example 1 witnesses that the advantages obtained by distributing the computation justify the overhead introduced by the machinery needed for that.

5   Conclusions

We have introduced DysToPic, a multi-engine theorem prover implementing tableaux calculi for reasoning in $\mathcal{ALC}+\mathbf{T}_{min}$. DysToPic implements a distributed version of the calculus $\mathcal{TAB}_{min}^{ALC+\mathbf{T}}$ introduced in [16], exploiting the fact that the two phases characterizing such a calculus can be computed in parallel. We aim at extending DysToPic to the *lightweight* DLs of the DL-Lite and $\mathcal{EL}$ family. Despite their relatively low expressivity, they are relevant for several applications. Extensions of $\mathcal{EL}^{\perp}$ and of DL-Lite$_{core}$ with the typicality operator $\mathbf{T}$ have been proposed in [13], where it has also been shown that minimal entailment is in Π_2^p (for $\mathcal{EL}^{\perp}$, if restricted to a specific fragment). Tableaux calculi performing a two phases computation, similar to $\mathcal{TAB}_{min}^{ALC+\mathbf{T}}$, have been proposed in [11,12].

Acknowledgments. This research is partially supported by INDAM- GNCS Project 2015 "Logiche descrittive e ragionamento non monotono". G.L. Pozzato is supported by the project "ExceptionOWL" (U. di Torino and C. di San Paolo).

References

1. Baader, F., Hollunder, B.: Priorities on defaults with prerequisites, and their application in treating specificity in terminological def. logic. JAR **15**(1), 41–68 (1995)
2. Beckert, B., Posegga, J.: leanTAP: Lean tableau-based deduction. Journal of Automated Reasoning (JAR) **15**(3), 339–358 (1995)
3. Bonatti, P., Lutz, C., Wolter, F.: DLs with circumscription. In: KR, pp. 400–410 (2006)
4. Britz, K., Heidema, J., Meyer, T.: Semantic preferential subsumption. In: KR, pp. 476–484. AAAI Press, Sidney, September 2008
5. Casini, G., Straccia, U.: Rational closure for defeasible description logics. In: Janhunen, T., Niemelä, I. (eds.) JELIA 2010. LNCS, vol. 6341, pp. 77–90. Springer, Heidelberg (2010)
6. Casini, G., Straccia, U.: Defeasible inheritance-based description logics. Journal of Artificial Intelligence Research **48**, 415–473 (2013)
7. Donini, F.M., Nardi, D., Rosati, R.: Description logics of minimal knowledge and negation as failure. ACM ToCL **3**(2), 177–225 (2002)
8. Eiter, T., Lukasiewicz, T., Schindlauer, R., Tompits, H.: Combining answer set programming with DLs for the semantic web. In: KR, pp. 141–151 (2004)
9. Giordano, L., Gliozzi, V., Olivetti, N., Pozzato, G.L.: Preferential description logics. In: Dershowitz, N., Voronkov, A. (eds.) LPAR 2007. LNCS (LNAI), vol. 4790, pp. 257–272. Springer, Heidelberg (2007)
10. Giordano, L., Gliozzi, V., Olivetti, N., Pozzato, G.L.: ALC+T: a preferential extension of description logics. Fundamenta Informaticae **96**, 341–372 (2009)
11. Giordano, L., Gliozzi, V., Olivetti, N., Pozzato, G.L.: A tableau calculus for a nonmonotonic extension of $\mathcal{EL}^\perp$. In: Brünnler, K., Metcalfe, G. (eds.) TABLEAUX 2011. LNCS (LNAI), vol. 6793, pp. 180–195. Springer, Heidelberg (2011)
12. Giordano, L., Gliozzi, V., Olivetti, N., Pozzato, G.L.: A tableau calculus for a nonmonotonic extension of the description logic $DL - Lite_{core}$. In: Pirrone, R., Sorbello, F. (eds.) AI*IA 2011. LNCS (LNAI), vol. 6934, pp. 164–176. Springer, Heidelberg (2011)
13. Giordano, L., Gliozzi, V., Olivetti, N., Pozzato, G.L.: Reasoning abouttypicality in low complexity DLs: the logics $\mathcal{EL}^\perp \mathbf{T}_{min}$ and $DL - Lite_c \mathbf{T}_{min}$. In: IJCAI 2011. pp. 894–899 (2011)
14. Giordano, L., Gliozzi, V., Jalal, A., Olivetti, N., Pozzato, G.L.: Predelo a theorem prover for preferential description logics. In: Baldoni, M., Baroglio, C., Boella, G., Micalizio, R. (eds.) AI*IA 2013. LNCS(LNAI), vol. 8249, pp. 60–72. Springer, Heidelberg (2013)
15. Giordano, L., Gliozzi, V., Olivetti, N., Pozzato, G.L.: Preferential vs rational description logics: which one for reasoning about typicality? In: ECAI 2010, pp. 1069–1070 (2010)
16. Giordano, L., Gliozzi, V., Olivetti, N., Pozzato, G.L.: A NonMonotonic Description Logic for Reasoning About Typicality. Artificial Intelligence **195**, 165–202 (2013)
17. Ke, P., Sattler, U.: Next steps for description logics of minimal knowledge and negation as failure. In: DL 2008. CEUR, vol. 353 (2008)

18. Kraus, S., Lehmann, D., Magidor, M.: Nonmonotonic reasoning, preferential models and cumulative logics. Artificial Intelligence 44(1–2), 167–207 (1990)
19. Krisnadhi, A.A., Sengupta, K., Hitzler, P.: Local closed world semantics: Keep it simple, stupid! In: DL 2011. CEUR, vol. 745 (2011)
20. Lehmann, D., Magidor, M.: What does a conditional knowledge base entail? Artificial Intelligence 55(1), 1–60 (1992)
21. Motik, B., Rosati, R.: Reconciling DLs and rules. J. ACM 57(5) (2010)
22. Straccia, U.: Default inheritance reasoning in hybrid kl-one-style logics. In: IJCAI 1993, pp. 676–681 (1993)

Advances in Multi-engine ASP Solving

Marco Maratea[1], Luca Pulina[2(✉)], and Francesco Ricca[3]

[1] DIBRIS, University degli Studi di Genova, Viale F. Causa 15, 16145 Genova, Italy
marco@dist.unige.it
[2] POLCOMING, University degli Studi di Sassari,
Viale Mancini 5, 07100 Sassari, Italy
lpulina@uniss.it
[3] Dip. di Matematica ed Informatica, University della Calabria, Via P. Bucci,
87030 Rende, Italy
ricca@mat.unical.it

Abstract. Algorithm selection techniques are known to improve the performance of systems for several knowledge representation and reasoning frameworks.This holds also in the case of Answer Set Programming (ASP), which is a rule-based programming paradigm with roots in logic programming and non-monotonic reasoning. Indeed, the multi-engine approach to ASP solving implemented in ME-ASP was particularly effective on the instances of the third ASP competition. In this paper we report about the advances we made on ME-ASP in order to deal with the new standard language ASPCore 2.0, which substantially extends the previous version of the standard language.An experimental analysis conducted on the Fifth ASP Competition benchmarks and solvers confirms the effectiveness of our approach also in comparison to rival systems.

1 Introduction

Algorithm selection [36] techniques are known to improve the performance of solvers for several knowledge representation and reasoning frameworks [24,31, 34,35,37,38,40]. It is well-established in the scientific literature that the usage of these techniques is very useful to deal with empirically hard problems, in which there is rarely an overall best algorithm, while it is often the case that different algorithms perform well on different domains. In order to take advantage of this behavior, these systems are able to select automatically the "best" algorithm/solver on the basis of the characteristics of the instance in input (called *features*). Algorithm selection techniques proved to be particularly effective [5,25,26,31,38] in the case of solvers for Answer Set Programming (ASP) [22,23], which is declarative programming paradigm based on logic programming and non-monotonic reasoning.

The application of algorithm selection techniques to ASP solving was ignited by the release of the portfolio solver CLASPFOLIO ver. 1 [20]. This solver ported to ASP the SATZILLA [40] approach for SAT. Indeed, the main selection technique of CLASPFOLIO was based on *regression*, which tries to estimate solving time to choose the "best" configuration/heuristic of the ASP solver CLASP.

M. Gavanelli et al. (Eds.): AI*IA 2015, LNAI 9336, pp. 179–190, 2015.
DOI: 10.1007/978-3-319-24309-2_14

Later, in [29,30], it was introduced the first multi-engine solver for ASP, called ME-ASP [31]. ME-ASP ports to ASP an approach applied before to QBF [35]. In particular, ME-ASP exploits inductive techniques based on *classification*, i.e., membership to a class, to choose, on a per instance basis, a solver among a selection of black-box heterogeneous ASP solvers that participated to the third ASP Competition [13] and DLV [28], being able to combine the strengths of its component engines. Other proposals [25,38] employ parameters tuning and/or design a solvers schedule. CLASPFOLIO ver. 2 [26] is a framework that includes and can combine several techniques implemented in other ASP solvers based on algorithm selection techniques.

Among all approaches, the one implemented in ME-ASP seems to be very effective in ASP, given it outperforms the other solvers on a broad set of benchmarks encoded in the standard language ASPCore 1.0 [12]. The input language of ME-ASP was, however, limited to ASPCore 1.0, which is the very basic language of the System track of the third ASP Competition [12]. The next editions of this event [2,10] were based on a substantially extended language, called ASPCore 2.0 [9], supporting more expressive language features such as aggregates [14], weak constraints [8] and choice rules [39]. Supporting these additional constructs require a substantial update of the system, including the design of proper (syntactic) features for classification and an update of the engines, and of the consequent inductive model.

In this paper we report about the advances we made on ME-ASP in order to deal with the new standard language ASPCore 2.0. An experimental analysis, conducted on all domains of the fifth ASP Competition and considering the solvers that entered the Single Processor category of the competition, confirms the effectiveness of our approach. In particular, our results show that

- the new features allow to properly classify benchmarks encoded in ASP-Core 2.0;
- ME-ASP performs better than its component engines, and is able to outperform alternative solutions at the state of the art, implemented in CLASPFOLIO ver. 2.2, on the benchmarks of the fifth ASP Competition [10].

The paper is structured as follows. Section 2 introduces needed preliminaries on ASP and classification. Section 3 reviews the key ingredients of a multi-engine approach and explains the choices made in the new version of ME-ASP. Section 4 then presents the results, and the paper ends in Section 5 with some conclusions.

2   Preliminaries

In this section we recall some preliminary notions concerning Answer Set Programming and machine learning techniques for algorithm selection.

2.1   Answer Set Programming

In this section we recall Answer Set Programming syntax and semantics. We refer in particular to the syntax and semantics of the ASPCore 2.0 [9] standard specification that has been employed in ASP competitions [2,10] from 2013. More detailed descriptions and a more formal account of ASP can be found in [8,17,21,23], whereas a nice introduction to ASP can be found in [6]. Hereafter, we assume the reader is familiar with logic programming conventions.

Syntax. The syntax of ASP is similar to the one of Prolog. Variables are strings starting with uppercase letter and constants are non-negative integers or strings starting with lowercase letters. A *term* is either a variable or a constant. A *standard atom* is an expression $p(t_1, \ldots, t_n)$, where p is a *predicate* of arity n and $t_1, \ldots, t_n$ are terms. An atom $p(t_1, \ldots, t_n)$ is ground if $t_1, \ldots, t_n$ are constants. A *ground set* is a set of pairs of the form $\langle consts : conj \rangle$, where $consts$ is a list of constants and $conj$ is a conjunction of ground standard atoms. A *symbolic set* is a set specified syntactically as $\{Terms_1 : Conj_1; \cdots ; Terms_t : Conj_t\}$, where $t > 0$, and for all $i \in [1, t]$, each $Terms_i$ is a list of terms such that $|Terms_i| = k > 0$, and each $Conj_i$ is a conjunction of standard atoms. A *set term* is either a symbolic set or a ground set. Intuitively, a set term $\{X : a(X, c), p(X); Y : b(Y, m)\}$ stands for the union of two sets: The first one contains the X-values making the conjunction $a(X, c), p(X)$ true, and the second one contains the Y-values making the conjunction $b(Y, m)$ true. An *aggregate function* is of the form $f(S)$, where S is a set term, and f is an *aggregate function symbol*. Basically, aggregate functions map multisets of constants to a constant. The most common functions implemented in ASP systems are: #min and #max (undefined for the empty set) computing minimum and maximum, respectively; #count counting the number of terms; and #sum the computes the sum of integers. An *aggregate atom* is of the form $f(S) \prec T$, where $f(S)$ is an aggregate function, $\prec \in \{<, \leq, >, \geq\}$ is a comparison operator, and T is a term called guard. An aggregate atom $f(S) \prec T$ is ground if T is a constant and S is a ground set. A *rule* r has the following form:

$$a_1 \mid \ldots \mid a_n :\!- b_1, \ldots, b_k, \text{not } b_{k+1}, \ldots, \text{not } b_m.$$

where $a_1, \ldots, a_n$ are standard atoms, $b_1, \ldots, b_k$ are atoms, $b_{k+1}, \ldots, b_m$ are standard atoms, and $n, k, m \geq 0$. A literal is either a standard atom a or its negation not a. The disjunction $a_1 \mid \ldots \mid a_n$ is the *head* of r, while the conjunction $b_1, \ldots, b_k, \text{not } b_{k+1}, \ldots, \text{not } b_m$ is its *body*. Rules with empty body are called *facts*. Rules with empty head are called *constraints*. A variable that appears uniquely in set terms of a rule r is said to be *local* in r, otherwise it is a *global* variable of r. An ASP program is a set of *safe* rules. A rule r is *safe* if both the following conditions hold: *(i)* for each global variable X of r there is a positive standard atom ℓ in the body of r such that X appears in ℓ; *(ii)* local variables of r appearing in a symbolic set $\{Terms : Conj\}$ also appear in $Conj$.

The ASPCore 2.0 language used in competitions also includes choice rules, weak constraints and queries. Choice rules [39] are of the form:

$$\{a : l_1, \ldots, l_k\} \geq u :\!- b_1, \ldots, b_k, \text{not } b_{k+1}, \ldots, \text{not } b_m.$$

where a is a an atom and $l_1, \ldots, l_k$ are literals for $k \geq 0$, u is a term, and the body is defined as for standard rules. Intuitively, a choice rule means that, if the body of the rule is *true*, an arbitrary subset of atoms of at least u elements in the head must be chosen as *true*. According to the standard specification [9] we interpret choice rules as a syntactic shortcut for a disjunctive program simulating this behavior.[1]

A *weak constraint* [8] Ω is of the form:

$$:\sim b_1, \ldots, b_k, \textbf{not } b_{k+1}, \ldots, \textbf{not } b_m.[w@l, t_1, \cdots, t_k]$$

where w and l are the weight and level of Ω, and $t_1, \cdots, t_k$ are distinguishing terms. (Intuitively, $[w@l, \bar{t}]$ is read "as weight w at level l for substitution $\bar{t}$", for more details on distinguishing terms see [9]). An ASP program with weak constraints is $\Pi = \langle P, W \rangle$, where P is a program and W is a set of weak constraints. A standard atom, a literal, a rule, a program or a weak constraint is *ground* if no variables appear in it.

A *query* on an ASP program is of the form $q?$, where q is a positive ground atom.

Semantics. Let P be an ASP program. The *Herbrand universe* U_P and the *Herbrand base* B_P of P are defined as usual (see e.g.,[6]). The ground instantiation G_P of P is the set of all the ground instances of rules of P that can be obtained by substituting variables with constants from U_P.

An *interpretation* I for P is a subset I of B_P. A ground literal ℓ (resp., $\textbf{not } \ell$) is true w.r.t. I if $\ell \in I$ (resp., $\ell \notin I$), and false (resp., true) otherwise. An aggregate atom is true w.r.t. I if the evaluation of its aggregate function (i.e., the result of the application of f on the multiset S) with respect to I satisfies the guard; otherwise, it is false.

A ground rule r is *satisfied* by I if at least one atom in the head is true w.r.t. I whenever all conjuncts of the body of r are true w.r.t. I.

A model is an interpretation that satisfies all the rules of a program. Given a ground program G_P and an interpretation I, the *reduct* [16] of G_P w.r.t. I is the subset G_P^I of G_P obtained by deleting from G_P the rules in which a body literal is false w.r.t. I. An interpretation I for P is an *answer set* (or stable model [23]) for P if I is a minimal model (under subset inclusion) of G_P^I (i.e., I is a minimal model for G_P^I) [16].

Given a program with weak constraints $\Pi = \langle P, W \rangle$, the semantics of Π extends from the basic case defined above. Thus, let $G_\Pi = \langle G_P, G_W \rangle$ be the instantiation of Π; a constraint $\Omega \in G_W$ is violated by an interpretation I if all the literals in Ω are true w.r.t. I. An *optimum answer set* O for Π is an answer set of G_P that minimizes the sum of the weights of the violated weak constraints in G_W as a prioritized way.

---

[1] Roughly, choice rules can be seen as a shortcut for $a \mid na \leftarrow b_1, \ldots, b_n, e_1, \ldots, e_m.$, $\leftarrow b_1, \ldots, b_n, \textbf{not } \#count\{a : a, e_1, \ldots, e_m\} \geq k.$ where na is an fresh auxiliary atom that is projected out of the answer.

The semantics of queries is given in terms of cautious reasoning. Given a program P and a query $q?$, the query is true if q is true in all answer sets of P, and is false otherwise.

2.2   Classification for Algorithm Selection

In this work we rely on a per-instance selection algorithm that chooses the best (or a good) algorithm among a pool of available. The selection in our case is of an ASP solver and is made using a set of *features*, i.e., numeric values that represent particular characteristics of a given instance, of a ground ASP program.

In order to make such a selection in an automatic way, we model the problem using *multinomial classification* algorithms, i.e., machine learning techniques that allow automatic classification of a set of instances, given some instance features. In more detail, in multinomial classification we are given a set of patterns, i.e., input vectors $X = \{\underline{x}_1, \ldots \underline{x}_k\}$ with $\underline{x}_i \in \mathbb{R}^n$, and a corresponding set of labels, i.e., output values $Y \in \{1, \ldots, m\}$, where Y is composed of values representing the m classes of the multinomial classification problem. In our modeling, the m classes are m ASP solvers. We think of the labels as generated by some unknown function $f : \mathbb{R}^n \rightarrow \{1, \ldots, m\}$ applied to the patterns, i.e., $f(\underline{x}_i) = y_i$ for $i \in \{1, \ldots, k\}$ and $y_i \in \{1, \ldots, m\}$. Given a set of patterns X and a corresponding set of labels Y, the task of a multinomial classifier c is to extrapolate f given X and Y, i.e., construct c from X and Y so that when we are given some $\underline{x}^\star \in X$ we should ensure that $c(\underline{x}^\star)$ is equal to $f(\underline{x}^\star)$. This task is called *training*, and the pair (X, Y) is called the *training set*.

3   Multi-engine Answer Sets Computation

The key idea at the basis of the application of automated algorithm selection algorithms can be summarized as follows: There is rarely a best solver to solve a given combinatorial problem, while it is often the case that different solvers perform well on different instances. Thus, a method that is able to select a good algorithm among a pool of available ones can perform much better than a static choice. In our framework a number of features of the input are measured, and *multinomial classification* algorithms are used to learn a selection strategy. More in details, the design of a multi-engine ASP solver involves the following steps:

1. Design of cheap-to-compute (syntactic) features that are significant for classifying the instances.
2. Fair design of training and test sets.
3. Selection of solvers that are representative of the state of the art.
4. Induction of a robust selection strategy by applying a classification algorithm.

In this section we report the choices we made in the design of the new version ME-ASP, by instantiating the ingredients we outlined above.

3.1 Features

The design of features is a crucial step of the development: indeed, features must be able to characterize the instances, but also should be cheap to compute, in the sense that they can be extracted very efficiently. Indeed, the overhead introduced by feature computation must be negligible.

The features of ground programs we selected for characterizing our instances are a super-set of those employed in the earlier version of ME-ASP for dealing with ASPCore 1.0. The new set includes features for taking into account the new language constructs of ASPCore 2.0, e.g., number of choice rules, aggregates and weak constraints.

The new features of ME-ASP are divided into four groups (such a categorization is borrowed from [33]):

- **Problem size features**: number of rules r, number of atoms a, ratios r/a, $(r/a)^2$, $(r/a)^3$ and ratios reciprocal a/r, $(a/r)^2$ and $(a/r)^3$;
- **Balance features**: fraction of unary, binary and ternary rules;
- **"Proximity to horn" features**: fraction of horn rules;
- **ASP specific features**: number of true and disjunctive facts, fraction of normal rules and constraints c, number of choice rules, number of aggregates and number of weak constraints.

This final choice of features, together with some of their combinations (e.g., c/r), amounts to a total of 58 features.

3.2 Dataset

In order to train the classifiers, we have to select a pool of instances for training purpose, called the training set. The training set must be broad enough to get a robust model; on the other hand, for reporting a fair analysis, we test the system on instances belonging to benchmarks not "covered" by the training set.

The benchmarks considered for the experiments correspond to the suite of the fifth ASP Competition – see [10] for details about the last event. This is a large and heterogeneous suite of hard benchmarks encoded in ASPCore 2.0, which was already employed for evaluating the performance of state-of-the-art ASP solvers. That suite includes planning domains, temporal and spatial scheduling problems, combinatorial puzzles, graph problems, and a number of application domains, i.e., databases, information extraction and molecular biology field.[2]

The considered pool of benchmarks is composed of 26 domains which are based on both complexity issues and language constructs of ASPCore 2.0. Starting from a total amount of 8572 instances – with *instance* we refer to the complete input program (i.e., encoding+facts) –, we pragmatically randomly split the amount of instances in each domain, using 50% of the total amount for training purpose, and the remaining ones for testing. All the instances were subject to feature selection after grounding them by using GRINGO (ver. 4) [18].

---

[2] An exhaustive description of the benchmark problems can be found in [11].

3.3   Solvers Selection

The selection of solvers has the goal of collecting a pool of engines that are both representative of the state-of-the-art solver (SOTA) and that have "orthogonal" performance (i.e., cover as much as possible of the set of solved instances, with minimal overlap on solved instances).

In order to find the set of training set labels, we have run ASP solvers that entered the Single Processor category of the fifth ASP Competition. In detail, we have run: CLASP [15], several solvers based on translation[3], i.e., LP2SAT3+GLUCOSE, LP2SAT3+LINGELING [27], LP2BV2+BOOLECTOR [32], LP2GRAPH [19], LP2MAXSAT+CLASP and LP2NORMAL2+CLASP [7], and some incarnations of the WASP solver [3,4] (ver. 1, ver. 1.5 and ver. 2, called WASP-1, WASP-1.5 and WASP-2, respectively). In the following, we give more details for each solver. CLASP is a native ASP solver relying on conflict-driven nogood learning, and in this edition includes also the capabilities of CLASPD, an extension of CLASP that is able to deal with disjunctive logic programs. The LP2SAT3* family employs a translation strategy to SAT and resorts to the SAT solvers GLUCOSE and LINGELING for computing the answer sets. The translation strategy mentioned includes the normalization of aggregates as well as the encoding of level mappings for non-tight ground programs: LP2BV2+BOOLECTOR and LP2GRAPH are variants that express the latter in terms of bit-vector logic or acyclicity checking, respectively, supported by a back-end SMT solver. LP2MAXSAT+CLASP competes by translating to a Max-SAT problem and solving with CLASP. LP2NORMAL2+CLASP normalizes aggregates and uses CLASP.

WASP is a native ASP solver based on conflict-driven learning, extended with techniques specifically designed for solving disjunctive logic programs. Unlike WASP-1, which uses a prototype version of DLV [28] for grounding. WASP-2 relies on GRINGO and adds techniques for program simplification and further deterministic inferences. WASP-1.5, instead, combines the two solvers by switching between them depending on whether a logic program is non-HCF or subject to a query.

In order to choose the engines of ME-ASP, we computed the total amount of training instances solved by the state-of-the-art solver (SOTA) i.e., given an instance, the oracle that always fares the best among all the solvers. Looking at the results of the Fifth ASP Competition, we can see that only four solver can deal with the whole set of instances, namely CLASP, LP2NORMAL2+CLASP, WASP1, and WASP1.5. Starting from these results, we look for the minimum number of solvers such that the total amount of instances solved by the pool is the closest to the SOTA solver on the training instances. The result of this procedure allow us to choose as ME-ASP engines three solvers, namely CLASP, LP2NORMAL2+CLASP, and WASP1. Thus, each pattern of the training set is labeled with the solver having the best CPU time on the given instance.

---

[3] We have not considered LP2MIP2 given that we did not receive the license of CPLEX on time.

Table 1. Results of the evaluated solvers. The first column contains the solver names, and it is followed by two columns, reporting the number of solved instances within the time limit (column "Solved"), and the sum of their CPU times in seconds ("Time").

Solver	Solved	Time
ME-ASP	2378	70144.99
CLASP	2253	63385.74
LP2NORMAL2+CLASP	2198	94560.98
CLASPFOLIO	1841	75044.14
WASP1.5	1532	52478.95
WASP2	1407	46939.06
LP2MAXSAT+CLASP	1387	82500.12
LP2GRAPH	1344	72633.53
LP2SAT3+LINGELING	1334	90644.33
WASP1	1313	87193.62
LP2SAT3+GLUCOSE	1305	73893.54
LP2BV2+BOOLECTOR	1011	57498.48

3.4   Classification Algorithm and Training

Concerning the choice of a multinomial classifier, we considered a classifier able to deal with numerical features and multinomial class labels (the solvers). According to the study reported in the original paper on multi-engine ASP solving [29,31], we selected *k-Nearest-neighbor*, NN in the following. NN is a classifier yielding the label of the training instance which is closer to the given test instance, whereby closeness is evaluated using some proximity measure, e.g., Euclidean distance, and training instances are stored in order to have fast look-up, see, e.g., [1]. The NN implementation used in ME-ASP is built on top built of the ANN library (www.cs.umd.edu/ mount/ANN). In order to test the generalization performance, we use a technique known as *stratified 10-times 10-fold cross validation* to estimate the generalization in terms of *accuracy*, i.e., the total amount of correct predictions with respect to the total amount of patterns. Given a training set (X, Y), we partition X in subsets X_i with $i \in \{1, \ldots 10\}$ such that $X = \bigcup_{i=1}^{10} X_i$ and $X_i \cap X_j = \emptyset$ whenever $i \neq j$; we then train $c_{(i)}$ on the patterns $X_{(i)} = X \setminus X_i$ and corresponding labels $Y_{(i)}$. We repeat the process 10 times, to yield 10 different c and we obtain the global accuracy estimate. To tune the parameter k of NN, we repeated the process described above for $k \in [1, 10], k \in \mathbb{N}$. As a result of cross-validation and parameter tuning, we choose $k = 1$, for which we obtained an accuracy greater than 87%.

4   Experiments

We assessed the performance of ME-ASP on the test set, that as described in the previous section contains the half of the instances of the fifth ASP competition that were not comprised in the training set used for generating the inductive model of ME-ASP. All the experiments run on a cluster of Intel Xeon E3-1245 PCs at 3.30 GHz equipped with 64 bit Ubuntu 12.04, granting 600 seconds of CPU time and 2GB of memory to each solver.

In Table 1 we report the performance of ME-ASP compared to the one obtained by the solvers described in Section 3.3. We involved in the analysis also CLASPFOLIO ver. 2.2 [4] for a direct comparison with approaches of algorithm selection.

As a general comment we note that ME-ASP is the solver that solves in absolute terms more instances than any other alternative considered in our analysis, which represents the state of the art in ASP solving.

In detail, comparing ME-ASP with its engines, we can see that it solves 125 instances more than CLASP, 155 instances more than LP2NORMAL2+CLASP, and 1065 instances more than WASP1. This outlines that ME-ASP is consistently better than the component engines, thus confirming that our algorithm selection strategy is effective. This proves empirically that the features of ME-ASP are able to characterize the input programs, and also that the inductive model learned during training is effective in suggesting a good solver for solving a given instance. Indeed, ME-ASP run the best solver 80% of the times, while it makes a suboptimal choice 17% of the times, i.e., it does not predict the best engine, but it runs a solver able to deal the input instance within the time limit. In fact, the SOTA solver composed of the engines of ME-ASP is able to solve 2462 instances, so ME-ASP is able to reach – in terms of solved instances – about 97% of the SOTA solver performance. Regarding the average CPU time per instance, we report that ME-ASP CPU time is about 5% more than the average CPU time per instance of the SOTA solver (29.50 and 28.05, respectively). Finally, we report that feature computation is basically negligible thanks to the selection of cheap-to-compute features, and remains, in average, within the 0.6% of computation time, i.e., 0.19 seconds in average.

Concerning the comparison with CLASPFOLIO, which is the only other system in this comparison that is based on algorithm selection, and represents the state of the art in algorithm selections for ASP, we note that CLASPFOLIO solves 537 instances less than ME-ASP.

Summing up, ME-ASP outperforms any solver that entered the fifth ASP competition, as well as alternative solvers based on algorithm selection, performing very efficiently on this benchmark set.

5 Conclusion

In this paper we presented an extension of the multi-engine ASP solving technique in [31] to deal with the broader set of language features included in the standard language ASPCore 2.0. We implemented an extended version of the

---

[4] CLASPFOLIO has been run with its default setting, and with CLASP ver. 3 as a backend solver. This improved version has been provided by Marius Lindauer, who is thanked.

188 M. Maratea et al.

ME-ASP solver, which is now able to process powerful constructs such as choice rules, aggregates, and weak constraints.

An experimental analysis conducted on the fifth ASP Competition benchmarks and solvers shows that the new version of ME-ASP is very efficient, indeed it outperforms state-of-the-art systems in terms of number of solved instances.

References

1. Aha, D., Kibler, D., Albert, M.: Instance-based learning algorithms. Machine Learning 6(1), 37–66 (1991)
2. Alviano, M., Calimeri, F., Charwat, G., Dao-Tran, M., Dodaro, C., Ianni, G., Krennwallner, T., Kronegger, M., Oetsch, J., Pfandler, A., Pührer, J., Redl, C., Ricca, F., Schneider, P., Schwengerer, M., Spendier, L.K., Wallner, J.P., Xiao, G.: The fourth answer set programming competition: preliminary report. In: Cabalar, P., Son, T.C. (eds.) LPNMR 2013. LNCS, vol. 8148, pp. 42–53. Springer, Heidelberg (2013)
3. Alviano, M., Dodaro, C., Faber, W., Leone, N., Ricca, F.: WASP: a native ASP solver based on constraint learning. In: Cabalar, P., Son, T.C. (eds.) LPNMR 2013. LNCS, vol. 8148, pp. 54–66. Springer, Heidelberg (2013)
4. Alviano, M., Dodaro, C., Ricca, F.: Preliminary report on WASP 2.0. In: Konieczny, S., Tompits, H. (eds.) Proceedings of the 15th International Workshop on Non-Monotonic Reasoning (NMR 2014), pp. 1–5. Vienna, Austria (2014)
5. Balduccini, M.: Learning and using domain-specific heuristics in ASP solvers. AI Communications - The European Journal on Artificial Intelligence 24(2), 147–164 (2011)
6. Baral, C.: Knowledge Representation, Reasoning and Declarative Problem Solving. Cambridge University Press, Tempe, Arizona (2003)
7. Bomanson, J., Janhunen, T.: Normalizing cardinality rules using merging and sorting constructions. In: Cabalar, P., Son, T.C. (eds.) LPNMR 2013. LNCS, vol. 8148, pp. 187–199. Springer, Heidelberg (2013)
8. Buccafurri, F., Leone, N., Rullo, P.: Enhancing Disjunctive Datalog by Constraints. IEEE Transactions on Knowledge and Data Engineering 12(5), 845–860 (2000)
9. Calimeri, F., Faber, W., Gebser, M., Ianni, G., Kaminski, R., Krennwallner, T., Leone, N., Ricca, F., Schaub, T.: ASP-Core-2 input language format (since 2013). https://www.mat.unical.it/aspcomp2013/ASPStandardization
10. Calimeri, F., Gebser, M., Maratea, M., Ricca, F.: The design of the fifth Answer Set Programming competition. ICLP 2014 Technical Communications - CoRR abs/1405.3710 (2014). http://arxiv.org/abs/1405.3710
11. Cabalar, P.: Answer set; Programming? In: Balduccini, M., Son, T.C. (eds.) Logic Programming, Knowledge Representation, and Nonmonotonic Reasoning. LNCS, vol. 6565, pp. 334–343. Springer, Heidelberg (2011)
12. Cabalar, P.: Answer set; Programming? In: Balduccini, M., Son, T.C. (eds.) Logic Programming, Knowledge Representation, and Nonmonotonic Reasoning. LNCS, vol. 6565, pp. 334–343. Springer, Heidelberg (2011)
13. Calimeri, F., Ianni, G., Ricca, F., Alviano, M., Bria, A., Catalano, G., Cozza, S., Faber, W., Febbraro, O., Leone, N., Manna, M., Martello, A., Panetta, C., Perri, S., Reale, K., Santoro, M.C., Sirianni, M., Terracina, G., Veltri, P.: The third answer set programming competition: preliminary report of the system

competition track. In: Delgrande, J.P., Faber, W. (eds.) LPNMR 2011. LNCS, vol. 6645, pp. 388–403. Springer, Heidelberg (2011)

14. Dell'Armi, T., Faber, W., Ielpa, G., Leone, N., Pfeifer, G.: Aggregate functions in disjunctive logic programming: semantics, complexity, and implementation in DLV. In: Proceedings of the 18th International Joint Conference on Artificial Intelligence (IJCAI) 2003, pp. 847–852. Morgan Kaufmann Publishers, Acapulco, August 2003

15. Drescher, C., Gebser, M., Grote, T., Kaufmann, B., König, A., Ostrowski, M., Schaub, T.: Conflict-driven disjunctive answer set solving. In: Brewka, G., Lang, J. (eds.) Proceedings of the Eleventh International Conference on Principles of Knowledge Representation and Reasoning (KR 2008), pp. 422–432. AAAI Press, Sydney (2008)

16. Faber, W., Leone, N., Pfeifer, G.: Recursive aggregates in disjunctive logic programs: semantics and complexity. In: Alferes, J.J., Leite, J. (eds.) JELIA 2004. LNCS (LNAI), vol. 3229, pp. 200–212. Springer, Heidelberg (2004)

17. Faber, W., Leone, N., Pfeifer, G.: Semantics and complexity of recursive aggregates in answer set programming. Artificial Intelligence 175(1), 278–298 (2011). special Issue: John McCarthy's Legacy

18. Gebser, M., Kaminski, R., Kaufmann, B., Schaub, T.: Clingo = ASP + Control: preliminary report. In: Theory and Practice of Logic Programming - Online-Supplement: Proc. of 30th International Conference on Logic Programming (ICLP 2014), pp. 1–9. Cambridge University Press (2014)

19. Gebser, M., Janhunen, T., Rintanen, J.: Answer set programming as sat modulo acyclicity. In: Schaub, T., Friedrich, G., O'Sullivan, B. (eds.) Proceedings of the Twenty-First European Conference on Artificial Intelligence (ECAI 2014). Frontiers in Artificial Intelligence and Applications, vol. 263, pp. 351–356. IOS Press (2014)

20. Gebser, M., Kaminski, R., Kaufmann, B., Schaub, T., Schneider, M.T., Ziller, S.: A portfolio solver for answer set programming: preliminary report. In: Delgrande, J.P., Faber, W. (eds.) LPNMR 2011. LNCS, vol. 6645, pp. 352–357. Springer, Heidelberg (2011)

21. Gelfond, M., Leone, N.: Logic Programming and Knowledge Representation - the A-Prolog perspective. Artificial Intelligence 138(1–2), 3–38 (2002)

22. Gelfond, M., Lifschitz, V.: The stable model semantics for logic programming. In: Logic Programming: Proceedings Fifth Intl. Conference and Symposium, pp. 1070–1080. MIT Press, Cambridge (1988)

23. Gelfond, M., Lifschitz, V.: Classical Negation in Logic Programs and Disjunctive Databases. New Generation Computing 9, 365–385 (1991)

24. Gomes, C.P., Selman, B.: Algorithm portfolios. Artificial Intelligence 126(1–2), 43–62 (2001)

25. Hoos, H., Kaminski, R., Schaub, T., Schneider, M.T.: ASPeed: ASP-based solver scheduling. In: Technical Communications of the 28th International Conference on Logic Programming (ICLP 2012). LIPIcs, vol. 17, pp. 176–187. Schloss Dagstuhl - Leibniz-Zentrum fuer Informatik (2012)

26. Hoos, H., Lindauer, M.T., Schaub, T.: Claspfolio 2: Advances in algorithm selection for answer set programming. TPLP 14(4–5), 569–585 (2014)

27. Janhunen, T.: Some (in)translatability results for normal logic programs and propositional theories. Journal of Applied Non-Classical Logics 16, 35–86 (2006)

28. Leone, N., Pfeifer, G., Faber, W., Eiter, T., Gottlob, G., Perri, S., Scarcello, F.: The DLV System for Knowledge Representation and Reasoning. ACM Transactions on Computational Logic 7(3), 499–562 (2006)

29. Grosan, C., Abraham, A.: Machine learning. In: Grosan, C., Abraham, A. (eds.) Intelligent Systems. ISRL, vol. 17, pp. 261–268. Springer, Heidelberg (2011)
30. Maratea, M., Pulina, L., Ricca, F.: The multi-engine ASP solver ME-ASP. In: del Cerro, L.F., Herzig, A., Mengin, J. (eds.) JELIA 2012. LNCS, vol. 7519, pp. 484–487. Springer, Heidelberg (2012)
31. Maratea, M., Pulina, L., Ricca, F.: A multi-engine approach to answer-set programming. TPLP **14**(6), 841–868 (2014). http://dx.doi.org/10.1017/S1471068413000094
32. Nguyen, M., Janhunen, T., Niemelä, I.: Translating answer-set programs into bit-vector logic. In: Tompits, H., Abreu, S., Oetsch, J., Pührer, J., Seipel, D., Umeda, M., Wolf, A. (eds.) INAP/WLP 2011. LNCS, vol. 7773, pp. 91–109. Springer, Heidelberg (2013)
33. Nudelman, E., Leyton-Brown, K., H. Hoos, H., Devkar, A., Shoham, Y.: Understanding random SAT: beyond the clauses-to-variables ratio. In: Wallace, M. (ed.) CP 2004. LNCS, vol. 3258, pp. 438–452. Springer, Heidelberg (2004)
34. O'Mahony, E., Hebrard, E., Holland, A., Nugent, C., O'Sullivan, B.: Using case-based reasoning in an algorithm portfolio for constraint solving. In: Proc. of the 19th Irish Conference on Artificial Intelligence and Cognitive Science (2008)
35. Pulina, L., Tacchella, A.: A self-adaptive multi-engine solver for quantified boolean formulas. Constraints **14**(1), 80–116 (2009)
36. Rice, J.R.: The algorithm selection problem. Advances in Computers **15**, 65–118 (1976)
37. Samulowitz, H., Memisevic, R.: Learning to solve QBF. In: Proc. of the 22th AAAI Conference on Artificial Intelligence, pp. 255–260. AAAI Press, Vancouver (2007)
38. Silverthorn, B., Lierler, Y., Schneider, M.: Surviving solver sensitivity: an ASP practitioner's guide. In: Technical Communications of the 28th International Conference on Logic Programming (ICLP 2012). LIPIcs, vol. 17, pp. 164–175. Schloss Dagstuhl - Leibniz-Zentrum fuer Informatik (2012)
39. Simons, P.: Extending and Implementing the Stable Model Semantics. Ph.D. thesis, Helsinki University of Technology, Finland (2000)
40. Xu, L., Hutter, F., Hoos, H.H., Leyton-Brown, K.: SATzilla: Portfolio-based algorithm selection for SAT. JAIR **32**, 565–606 (2008)

Defeasible Logic Programming in Satisfiability Modulo CHR

Francesco Santini$^{(\boxtimes)}$

Dipartimento di Matematica e Informatica, Università di Perugia, Perugia, Italy
francesco.santini@dmi.unipg.it

Abstract. We revise some results in Argumentation-based Logic Programming under the umbrella of Satisfiability Modulo CHR (SMCHR), specifically considering Defeasible Logic Programming (DeLP). Strict and defeasible rules in DeLP can be cast to SMCHR rules, which act as conflict "disentanglers" and implement the Theory part. At the same time, we inherit several built-in theory solvers, as SAT, unification, or linear arithmetic ones, which implement the Satisfiability-modulo part. Moreover, we show how to deal with possibilistic extensions of DeLP, i.e., Possibilistic-DeLP, where certainty scores describing the possibility of some events are associated with rules.

1 Introduction

This paper links *Argumentation-based Logic Programming* [9,13,16,20] (*ALP*) to *Satisfiability Modulo Theories* (*SMT*), with the purpose to have declarative and a powerful tool to reason in case of conflict, and reach a justifiable conclusion with a support in Argumentation-based reasoning.

To accomplish such goal, we use *Satisfiability Modulo Constraint Handling Rules* (*SMCHR*) [8], which in turn exploits the declarativeness of a rule-based language as CHR [11,12], and binds it to SMT. Solving CHR constraints in other propositional contexts typically relies on some external machinery. For example, Prolog CHR implementations such as K.U. Leuven CHR system [19] use Prolog's default backtracking search to handle disjunction. The execution algorithm for CHR is based on constraint rewriting and propagation over a global store of constraints. CHR solvers are incremental: when a new constraint c is asserted, we check c and the store against the rules in order to find a match. If there is a match, we fire that rule, possibly generating new constraints in the store. Otherwise c is simply added to the global store.

In the following, we provide a general overview on how to program constraint-propagators on top of solvers (as the *SAT* one), with the purpose to resolve conflicts between arguments, e.g., $arg \wedge \neg arg$. In general, an *unsatisfiable* result (i.e., the asked goal is not satisfiable) from a solver points to an inconsistency in the knowledge base: such conflict can be overcome by writing an ad-hoc propagator to solve it by removing either arg or $\neg arg$ from the constraint store. This decision can be taken by considering qualitative/quantitative preference scores associated

© Springer International Publishing Switzerland 2015
M. Gavanelli et al. (Eds.): AI*IA 2015, LNAI 9336, pp. 191–204, 2015.
DOI: 10.1007/978-3-319-24309-2_15

with arguments, which define a total/partial order among arguments. A conflict resolution-procedure favours the preferred argument between two. In this sense, an example of naturally weighted frameworks is *Possibilistic Defeasible Logic Programming (P-DeLP)* [1], where ALP is mixed with belief scores.

Several different proposals have been crafted to express ALP in ad-hoc logics and settle such conflicts [9,13,16,20]. One of our goals is to offer the features of SMCHR and propagators as a general means to resolve them. We take as an example the *Defeasible Logic Programming* framework *(DeLP)* [13] with the purpose to show how SMCHR can be used to model and solve various reasoning processes in a particular instance of such logics. For example, we are able *i)* to check the "correctness" of an argument structure (following its definition), *ii)* to check if one argument is the counter-argument of another, and *iii)* if it is a proper or a blocking defeater for it [13]. Then we show that similar considerations hold for the possibilistic extension of DeLP, i.e., P-DeLP.

To summarise, the main motivations behind this paper are to:

- have a unifying solving framework in which to solve all the ALP proposals [9, 13,16,20], independently developed;
- link constraint-based representation and solving techniques, as the design of propagators, to help argument-based reasoning in an efficient way. Note in this work we focus on non-Abstract Argumentation Frameworks [9], where, on the contrary, AI-based techniques have been already successfully applied [6], as SAT [5], constraints [2–4], Answer Set Programming (ASP) [10];
- design propagators (which collectively implement a "Theory") on top of different built-in solvers, and then to check their satisfiability (from this, "Satisfiability Modulo Theory"). This unlocks the use of weights, which represent quantitative preferences on arguments, or some additional information to be taken into account during the reasoning. Efficient underlying solvers, as the *bounds* solver or the simplex algorithm (see Sec. 2.2), optimise the search procedure in case of complex constraints over them. This is clearly not possible by using boolean solvers only.

The paper derives from the preliminary work in [18], and is organised as follows: in Sec. 2 we summarise the necessary background-notions to understand SMCHR, its rewriting rules, and underlying solvers. Section 3 shows how SMCHR can represent strict and defeasible rules, and some reasoning processes of DeLP, as checking counter-arguments or defeaters. Such processes can be also reproduced in case of weighted rules, i.e., in P-DeLP. Section 4 reports some of the related works in the literature. Finally, Sec. 5 wraps up the paper and hints directions for future work.

2   CHR and Satisfiability Modulo CHR

2.1   Constraint Handling Rules

Constraint Handling Rules (CHR) [11,12] is essentially a committed-choice language. A program consists of multi-headed guarded rules that rewrite constraints into simpler ones until they are solved. CHR rules define simplification of, and propagation over, multi-sets of relations interpreted as conjunctions of constraint atoms. Simplification rewrites constraints to simpler constraints while preserving logical equivalence (e.g. $X < Y, Y < X <=> false$). Propagation adds new constraints, which are logically redundant but may cause further simplification (e.g. $X \leq Y, Y \leq Z ==> X \leq Z$). Furthermore, simpagation just mixes the features of both simplification and propagation, rewriting and adding new constraints at the same time. Repeatedly applying the rules incrementally solves constraints (e.g. $A \leq B, B \leq C, C \leq A$ leads to $A = B \wedge A = C$). In the following we show the formal syntax of such basic rules: r is the optional unique-name of a rule, each H (and $H_k \backslash H_r$) is the (multi-) head of a rule, and it consists in a conjunction of one or more defined constraints indicated by commas ($H = h_1, \ldots, h_n$), G is the guard being a conjunction of built-in atoms, and B the body being a conjunction of constraints:

- **Simplification:** [r@] H <=> [G|] B.
- **Propagation:** [r@] H ==> [G|] B.
- **Simpagation:** [r@] H$_k$ \ H$_r$ <=> [G|] B.

The @ symbol assigns a name r to a rule. A constraint (also built-in, as =) is a predicate of First-order Logic. Rules are tried and (in case) fired in the order they are written in the program (from top to bottom). For each rule, one of its head constraints is matched against the last constraint added to the store. Matching succeeds if the constraint is an instance of the head. If matching succeeds and the rule has more than one head constraint, the constraint store is searched for partner constraints that match the other head constraints. A guard is a precondition on the applicability of a rule: it is basically a test that either succeeds or fails. If the firing rule is a simplification rule, the matched constraints are removed from the store and the body of the rule is executed by adding the constraints in the body. Similarly for firing a simpagation rule, except that the constraints that match the head-part preceding \ (i.e., H$_k$) are kept in the store. A simpagation rule can be seen as a short hand for H$_k$, H$_r$ ==> B, H$_r$. If the firing rule is a propagation rule, its body is executed without removing any constraint. The rule is remembered with the purpose to not fire it again with the same constraints.

Basically, rules are applied to an initial conjunction of constraints (syntactically, a goal) until exhaustion, i.e., until no more change happens. An initial goal is called *query*. The intermediate goals of a computation are stored in the so-called store. A final goal, to which no more rule is applicable, represents the

```
reflexivity    @ X leq X <=> true.
antisymmetry @ X leq Y, Y leq X <=> X = Y.
transitivity  @ X leq Y, Y leq Z ==> X leq Z.
idempotence  @ X leq Y \ X leq Y <=> true.
```

Fig. 1. Four rules that implement a solver for a less-or-equal constraint.

answer (or result) of a computation. Figure 1 shows four rules to reason on the $\leq$ relation: by posting the goal $A \leq B, B \leq C, C \leq A$ to the store we obtain a final store containing the result $\{A = B, A = C\}$. The reflexivity rule removes not useful information from the store.

2.2   Satisfiability Modulo CHR

SMCHR[1] [7,8] is essentially a Satisability Modulo Theories (SMT) solver where a theory T is implemented in CHR.

SMCHR follows the theoretical operational-semantics of CHR. No assumptions should be made about the ordering of rule applications. The SMCHR system also treats deleted constraints differently from CHR. A deleted constraint stays deleted "forever", i.e., it is not possible to re-generate a copy of the same constraint: for instance, a program $p(x) <=> p(x)$ always terminates.

The following list introduces all the solvers that can be plugged into SMCHR at the time of writing [7,8]:

- *eq*: an equality solver based on union-find. This solver is complete for for systems of equality and dis-equality constraints.
- *linear*: a linear arithmetic solver over the integers based on the simplex algorithm over the rationals. This solver is incomplete if there exists a rational solution for the given goal.
- *bounds*: a simple bounds-propagation solver over the integers. This solver is incomplete.
- *dom*: a simple solver that interprets the constraint int $dom(x, l, u)$ as an integer domain/range constraint $x \in [l..u]$. This, in combination with the *bounds* solver, forms a *Lazy Clause Generation (LCG)* finite domain solver. This solver is incomplete.
- *heaps*: a heap solver for program reasoning based on some of the ideas from *Separation Logic*. This solver is complete.
- *sat*: the boolean satisfiability solver (SAT). This solver is complete and is always enabled by default.

The SMCHR system assumes that loaded solvers are incomplete (the tool can be extended with new solvers): an "UNKNOWN" result indicates that unsatisfiability cannot be proven, meaning that either the goal is satisfiable, or that the solver is incomplete and was unable to prove unsatisfiability. However, if the user knows that a given solver combination is complete, then the answer UNKNOWN can be re-interpreted as a "SAT" response. SMCHR returns the answer "UNSAT" to indicate that the goal is unsatisfiable. Some performance benefits in using the *SAT* solver are that it is possible to inherit all the advantages of no-good clause learning, non-chronological back-jumping, and unit propagation during computation [7].

---

[1] http://www.comp.nus.edu.sg/~gregory/smchr/

3   DeLP Solved in SMCHR

In this section we choose DeLP [13] among all Argument-based Logic Programming frameworks (see Sec. 4) in order to show how SMCHR propagators can effectively model such plethora of systems. DeLP variables are represented with constraint variables. A DeLP program is a set of *i)* facts, *ii)* strict rules, and *iii)* defeasible rules using ground literals L_i.

- *Facts* are ground literals representing atomic information or its strong negation.
- *Strict rules* represent non-defeasible rules, and they are represented as $L_0 \leftarrow L_1, \ldots, L_n$.
- *Defeasible rules* represent tentative information, in the form of rules like $L_0 \leftharpoondown L_1, \ldots, L_n$.

In words, a defeasible rule is used to represent tentative information that may be used if nothing could be posed against it. On the contrary, the information that represents a strict rule, or a fact, is not tentative. A DeLP-program is denoted by a pair (Π, Δ) distinguishing a subset Π of facts and (only two in Fig. 2) strict rules, which represent non-defeasible knowledge, and a subset Δ of defeasible rules (eight in Fig. 2). In Fig. 2 we consider the same example given in [17, Ch. 8]. $\Pi \cup \Delta$ collect information and reason on three rooms a, b, and c, linking their illumination to the day (working or holiday), to the switch position (on/off), to the electricity presence (yes/no), and to the time of the day (day/night): for instance, the first defeasible rule states that "it is reasonable to believe that if the switch of a room is on, then the lights of that room are on". This rule is in Δ because its conclusion can be defeated in case, for instance, there is no electricity.

$$\Pi \begin{cases} night. \\ \sim day \leftarrow night. \\ \sim dark(Y) \leftarrow illuminated(X). \\ sunday. \\ deadline. \end{cases} \quad \begin{array}{l} switch\_on(a). \\ switch\_on(b). \\ switch\_on(c). \\ \sim electricity(b). \\ \sim electricity(c). \\ emergency\_lights(c). \end{array}$$

$$\Delta \begin{cases} light\_on(X) \leftharpoondown switch\_on(X). \\ \sim lights\_on(X) \leftharpoondown \sim electricity(X). \\ lights\_on(X) \leftharpoondown \sim electricity(X), emergency\_lights(X). \\ dark(X) \leftharpoondown \sim day. \\ illuminated(X) \leftharpoondown \sim lights\_on(X), \sim day. \\ working\_at(X) \leftharpoondown illuminated(X). \\ \sim working\_at(X) \leftharpoondown sunday. \\ working\_at(X) \leftharpoondown sunday, deadline. \end{cases}$$

Fig. 2. Π collects DeLP facts and strict rules, while Δ provides defeasible rules.

```
/* Strict rules */
night(x)  ⟹  not day(x);
illuminated(x)  ⟹  not dark(x);

/* Defeasible rules */
switchOn(x)  ⟹  lightsOn(x);
not electricity(x)  ⟹  not lightsOn(x);
not electricity(x) ∧ emergencyLights(x)  ⟹  lightsOn(x);
not day(x)  ⟹  dark(x);
not day(x) ∧ lightsOn(x)  ⟹  illuminated(x);
illuminated(x)  ⟹  workingAt(x);
sunday(x)  ⟹  not workingAt(x);
sunday(x) ∧ deadline(x)  ⟹  workingAt(x);
```

Fig. 3. *argcheck.chr*: a SMCHR propagator coding the DeLP example in Fig. 2.

In Fig. 3 we show a possible encoding to SMCHR of the DeLP program in Fig. 2, where each rule (both in Π and Δ) is encoded in a SMCHR propagation-rule in Fig. 3. Facts, which represent pieces of indefeasible information (i.e., they are in Π as well), are not represented through rules, but with the constraints

$$switchOn(a), switchOn(b), switchOn(c), not\ electricity(b), not\ electricity(c),$$
$$emergencyLights(c), night(a), night(b), night(c), sunday(a), sunday(b),$$
$$sunday(c), deadline(a), deadline(b), deadline(c).$$

Such set F of constraints (or part of it) can be passed to the SMCHR interpreter (in "and") as part of a global goal. If we ask the entire F we obtain an UNSAT result, meaning that not all of them can be warranted, i.e., we have contradictions in our knowledge-base.

A DeLP-query is a ground literal Q that a DeLP program tries to warrant. Our SMCHR can straightforwardly check it. There are several queries that succeed with respect to the program in Fig. 3 because they are warranted, e.g., $illuminated(a) \wedge switchOn(a)$. File `argCheck.chr` stores our "Theory" as a set of propagation rules, and it is shown in Fig. 3. If we call `./smchr --solver argCheck.chr` and we ask the goal $Q = illuminated(a) \wedge switchOn(a)$ then we obtain UNKNOWN, which, being the default *SAT* solver complete, can be reinterpreted as SAT (Q is then warranted). The output of the `smchr` interpreter is shown in Fig. 9, where we can also see the number of generated constraints (three: $lightsOn(a)$, $not\ dark(a)$ and $workingAt(a)$), and other information related to search, as the number of backtracks. Other queries cannot be warranted instead: for instance $Q = switchOn(a) \wedge not\ lightsOn(a)$ returns UNSAT, because $switchOn(a)$ propagates to $lightsOn(a)$, which however conflicts with part of Q. Indeed, also two contradictory constraint in a query, as $day(a) \wedge not\ day(a)$, trivially disagree, and the answer is UNSAT as well.

A derivation using Π only is called a *strict derivation*, that is only the first two rules in Fig. 3. A *defeasible derivation* of a literal Q by using (Π, Δ), denoted by $(\Pi, \Delta) \mathrel{|\!\sim} Q$, is a finite sequence of ground literals in form of $L_1, L_2, \ldots, L_n$ with $L_n = Q$, where *i)*, L_i is a fact in Π, or *ii)* there exists a strict or defeasible

```
LOAD solver "sat"
LOAD solver "argcheck.chr"

> illuminated(a) /\switchOn(a)
UNKNOWN:
illuminated(a) /\
lightsOn(a) /\
not dark(a) /\
switchOn(a) /\
workingAt(a)
```

Fig. 4. Output of Fig. 3 with *illuminated(a)* as input query.

```
/* Strict rules */
night(x) ⟹ not day(x);
illuminated(x) ⟹ not dark(x);

/* Defeasible rules */
not electricity(x) ⟹ not lightsOn(x) ∧ defeasibleNotLightsOn();

/*1*/ defeasibleNotLightsOn(x) ∧ lightsOn(x) ⟹ strictLightsOn(x);
/*2*/ defeasibleNotLightsOn(x) ∧ defeasibleLightsOn(x) ⟹ false;
```

Fig. 5. A SMCHR program to check the validity of an argument structure (see Def. 1) by using (Π, A), where $A = \{\sim lights\_on(X) \leftarrowtail \sim electricity(X)\}$ and Π is supposed as taken from Fig. 2.

rule in (Π, Δ) with head L_i and body $B_1, \ldots, B_k$ and each B is an L_j element of $L_1, L_2, \ldots, L_n$, with $j < i$. A derivation is defeasible if at least one defeasible rule is used. This brings us to define a valid argument structure:

Definition 1 (Argument Structure [13]). *Let H be a ground literal, (Π, Δ) a DeLP-program, and $A \subseteq \Delta$. The pair $\langle A, H \rangle$ is an argument structure if:*[2]

1. *there exists a defeasible derivation for H from (Π, A),*
2. *there is no defeasible derivation from (Π, A) of contradictory literals.*

With respect to Fig. 2, if we consider $A = \{\sim lights\_on(X) \leftarrowtail \sim electricity(X)\}$ and $H = \sim lights\_on(b)$, we can check if $\langle A, H \rangle$ is a valid argument structure by using the SMCHR rules in Fig. 5, which collect all the strict rules in Π plus A, as required by 2) in Def. 1. If we set as goal all the facts F in Π, i.e.,

$$Q = F = night(a) \wedge night(b) \wedge night(c) \wedge switchOn(a) \wedge switchOn(b) \wedge$$
$$switchOn(c) \wedge sunday(a) \wedge sunday(b) \wedge sunday(c) \wedge deadline(a) \wedge deadline(b) \wedge$$
$$deadline(c) \wedge not\ electricity(b) \wedge not\ electricity(c) \wedge emergencyLights(c)$$

---

[2] We do not consider here a third property, i.e., that there is no proper subset A' of A such that A' satisfies 1) and 2)). This leads to have "not useful" information in the support (e.g., r in $(\{r, p, p \to q\}, q)$), but we plan to solve it in future work (see Sec. 5).

then the obtained output is UNKNOWN and the six constraints generated in the store are

$$defeasibleNotLightsOn(b), defeasibleNotLightsOn(c), not\ day(a), not\ day(b),$$
$$not\ day(c), strictLightsOn(b).$$

Having constraint $defeasibleNotLightsOn(b)$ in the store means that we generated $H = \sim lights\_on(b)$ by using at least one defeasible rule. Thus, 1) in Def. 1 is satisfied, as well as 2), since a result of UNKNOWN (i.e., SAT because the solver is complete) means that no contradictory literals have been generated. Note that the same holds for $H = \sim lights\_on(c)$, as it can be appreciated from the same final store of constraints above.

Rule 2 in Fig. 5 is used to have an UNSAT response in case there are two defeasible derivations leading to the contradiction that lights are on and off at the same time. Therefore, we use this rule to check property 2) in Def. 1. Rule 1 in Fig. 5 is used to add a constraint (in this case, $strictLightsOn(b)$) warning that there is also a strict derivation contradicting our defeasible derivation for b. If we wish to remove such a conflict, we only need to add the following (simpagation) rule to Fig. 5, which can remove the constraint $defeasibleNotLightsOn(b)$ from the store:

$$strictLightsOn(x) \setminus defeasibleNotLightsOn(x) <=> strictLightsOn(x);$$

Now we can turn our attention to model and check counter-arguments and defeaters in DeLP:

Definition 2 (Counter-Argument [13]). $\langle B, S \rangle$ *is a counter-argument for* $\langle A, H \rangle$ *at literal* P, *if there exists a sub-argument* $\langle C, P \rangle$ *of* $\langle A, H \rangle$ *such that* P *and* S *disagree, that is, there exist two contradictory literals that have a strict derivation from* $\Pi \cup \{S, P\}$.

Consider two valid argument-structures $\langle \{illuminated(X) \leftarrowtail \sim lights\_on(X), \sim day, light\_on(X) \leftarrowtail switch\_on(X)\}, illuminated(b) \rangle$ and $\langle \{dark(X) \leftarrowtail \sim day\}, dark(b) \rangle$. We can accomplish this check by writing a new SMCHR program with the first two rules of Fig. 3; then we ask a query $Q = F \wedge \{illuminated(b), dark(b)\}$, which contains all the facts F in Π in conjunction with $S \cup P$. As answer we get UNSAT, meaning that one is the counter-argument of the other (obtained by only using strict derivations).

In DeLP the argument comparison criterion between two arguments is modular [13]. For this reason, Def. 3 abstracts away from the comparison criterion, assuming there exists one (denoted by $\succ$):

Definition 3 (Proper/Blocking Defeaters [13]). *Let* $\langle B, S \rangle$ *be a counter-argument for* $\langle A, H \rangle$ *at point* P, *and* $\langle C, P \rangle$ *the disagreement sub-argument. If* $\langle B, S \rangle \succ \langle C, P \rangle$ *(i.e.,* $\langle B, S \rangle$ *is "better" than* $\langle C, P \rangle$*) then* $\langle B, S \rangle$ *is a proper defeater for* $\langle A, H \rangle$. *If* $\langle B, S \rangle$ *is unrelated by the preference relation to* $\langle C, P \rangle$, *(i.e.,* $\langle B, S \rangle \nsucc \langle C, P \rangle$, *and* $\langle C, P \rangle \nsucc \langle B, S \rangle$*) then* $\langle B, S \rangle$ *is a blocking defeater for* $\langle A, H \rangle$.

```
type night(var of atom, num); type illuminated(var of atom, num); type
    dark(var of atom, num);

illuminated(x, y) \ dark(x, z) ⟺ y $> z ∧ r:= (y − z) |   not dark(x,
    r) ∧ pDefeaterIlluminatedDark(x);
illuminated(x, y) \ dark(x, z) ⟺ y $= z ∧ r:= (y − z) |   not dark(x,
    r) ∧ bDefeaterIlluminatedDark(x);
```

Fig. 6. *dark* ∧ *not dark*: resolving conflicts with scores.

In SMCHR, preference can be computed and/or constrained in the guard of a rule, thus allowing us to even represent dynamic preferences, i.e., preferences that are subject to some conditions, as suggested in [16] or in implemented systems as GORGIAS [14]. Therefore, in our approach we opt to associate a score with each argument, and we compute a new numeric result by resolving a conflict, for instance subtracting the strength of a support from another:

$$supp(X) \land \neg supp(Y) <=> X >= Y \land Z := X - Y \mid claim(Z)$$
$$supp(X) \land \neg supp(Y) <=> X < Y \land Z := Y - X \mid \neg claim(Z)$$

In these two rules, contradictory constraints $supp(X)$ and $supp(Y)$ are the same argument but with a different preference score. They are removed from the store through a simplification rule, and the result is the a claim with a different preference score Z, computed in the guard of the rules. If $X \geq Y$ ($\$ >=$ in SMCHR), $supp$ wins over $\neg supp$, otherwise the store contains $\neg claim$. In both of the cases, the final preference Z is the difference between X and Y.

If we import such conflict-resolution method in our running-case (Fig. 2), we can model it via the program in Fig. 6. The type of constraints is defined at the beginning of Fig. 6 in order to let the program correctly manage operations on their arguments. We set the query Q to $dark(b, 2) \land illuminated(b, 3)$. Hence, argument $illuminated(b)$ is associated with a preference value equal to 3, and $dark(b)$ to 2. The program in Fig. 6 states that $illuminated(b)$ propagates to *not dark(b)*, if the first preference score is greater/equal than the preference score of $dark(b)$, and the preference of *not dark(b)* is their difference. The first rule generates a proper defeater in the store when the preference is strictly better. If the two scores are the same, the second rule generates a blocking defeater instead. With our query Q, we fire the first rule and we obtain $illuminated(b, 3) \land$ *not dark(b, 1)* $\land pDefeaterIlluminatedDark(b)$ in the store.

3.1   A Bridge Towards P-DeLP

Finally, we show how SMCHR and constraints have an impact on weighted extensions of ALP (see also Sec. 4). *Possibilistic Defeasible Logic Programming (P-DeLP)* [1] is an extension of DeLP in which defeasible rules are attached with weights, belonging to the real unit interval [0..1], in the following discretised to [0..100] since SMCHR works with integer numbers only. Such score expresses

```
type sw1(num); type sw2(num); type sw3(num); type pumpClog(num);
type pumpFuel(num); type pumpOil(num); type oilOk(num); type fuelOk(num)
  ;
type engineOk(num); type heat(num); type lowSpeed(num);

/* Strict */
pumpClog(x) ==> not fuelOk(x);

/* Defeasible */
sw1(x) ==> x $ <= 60 | pumpFuel(x);
sw1(x) ==> x $ > 60 | pumpFuel(60);
pumpFuel(x) ==> x $ <= 30 | fuelOk(x);
pumpFuel(x) ==> x $ > 30 | fuelOk(30);
sw2(x) ==> x $ <= 80 | pumpOil(x);
sw2(x) ==> x $ > 80 | pumpOil(80);
pumpOil(x) ==> x $ <= 80 | oilOk(x);
pumpOil(x) ==> x $ > 80 | oilOk(80);
oilOk(x) ∧ fuelOk(y) ==> x $ <= y ∧ x $ <= 30 | engineOk(x);
oilOk(x) ∧ fuelOk(y) ==> y $ <= x ∧ y $ <= 30 | engineOk(y);
oilOk(x) ∧ fuelOk(y) ==> 30 $ <= x ∧ 30 $ <= y | engineOk(30);
heat(x) ==> x $ <= 95 | not engineOk(x);
heat(x) ==> x $ > 95 | not engineOk(95);
heat(x) ==> x $ <= 90 | not oilOk(x);
heat(x) ==> x $ > 90 | not oilOk(x);
lowSpeed(x) ∧ pumpFuel(y) ==> x $ <= y ∧ x $ <= 70 | pumpClog(x);
lowSpeed(x) ∧ pumpFuel(y) ==> y $ <= x ∧ y $ <= 70 | pumpClog(y);
lowSpeed(x) ∧ pumpFuel(y) ==> 70 $ <= x ∧ 70 $ <= y | pumpClog(70);
sw2(x) ==> x $ <= 80 | lowSpeed(x);
sw2(x) ==> x $ > 80 | lowSpeed(80);
sw3(x) ∧ sw2(y) ==> x $ <= y ∧ x $ <= 80 | not lowSpeed(x);
sw3(x) ∧ sw2(y) ==> y $ <= x ∧ y $ <= 80 | not lowSpeed(y);
sw3(x) ∧ sw2(y) ==> 80 $ <= x ∧ 80 $ <= y | not lowSpeed(80);
sw3(x) ==> x $ <= 60 | fuelOk(x);
sw3(x) ==> x $ > 60 | fuelOk(80);
```

Fig. 7. SMCHR rules coding the P-DeLP example in [1].

the relative belief or preference strength of arguments. Each fact p_i is associated with a certainty value that expresses how much the relative fuzzy-statement is believed in terms of necessity measures. Weights are aggregated in accordance to $(p_1 \wedge \cdots \wedge p_k \rightarrow q, \alpha)$ iff $(p_1, \beta_1), \ldots, (p_k, \beta_k)$ with $(q, \min(\alpha, \beta_1, \ldots, \beta_k))$. Such computational evaluation can be naturally encoded into SMCHR, as we show in the following example.

The program in Fig. 7 encodes in SMCHR an example provided in [1]. We suppose to have an intelligent agent controlling an engine with three switches *sw1*, *sw2* and *sw3*. These switches regulate different features of the engine, such as the pumping system, speed, etc. Figure 7 shows certain and uncertain knowledge an agent has about how this engine works.

By querying *sw1(100)* we obtain UNKNOWN and *fuelOk(30)*, thus correctly deriving *pumpFuel* with a certainty score equal to 0.3, that is the minimum value among all the constraints in the store: *sw1(100)*, *pumpFuel(60)*, and *fuelOk(30)*.

By switching the first two switches on, i.e., *sw1(100)* ∧ *sw2(100)*, the agent knows that the engine works with a certainty score equal to 0.3. The result is UNKNOWN, and the final constraint store is:

$engineOk(30) \land fuelOk(30) \land lowSpeed(80) \land not\,fuelOk(60) \land oilOk(80) \land$
$pumpClog(60) \land pumpFuel(60) \land pumpOil(80) \land sw1(100) \land sw2(100).$

Moreover, the agent knows that fuel is not properly pumped with a certainty score of 60, and it is pumped properly with a certainty score of 30, at the same time. By defining two rules similar as the ones in Fig. 6, it is possible to leave in the store only that fuel is not properly pumped with a certainty of, e.g., $60 - 30 = 30$.

3.2 Using Solvers Different from SAT

With a small example, in this section we would like to justify the use of solvers different from *SAT*, and consequently justify why it is interesting to keep the Theory and the satisfiability as separated, i.e., why to use SMCHR. In the example in Fig. 8 we show two arguments supporting different claims, that is *claim1* and *claim2*. If they are in conflict, *claim2*, belonging to a different agent, is withdrawn, and *claim3* is added, thus reaching a final (consistent) conclusion. The solution is found by calling ./smchr.macosx --solver ex.chr,bounds: therefore, we add a (incomplete) *bounds* solver, in order to bind the solution variables. The arguments support constraints on the variables, which are object of negotiation: for instance, *int_gt(x,y)* imposes $x > y$, and *int_gt_c(x,c)* imposes that x has to be greater than a constant c. These are two examples of several primitive built-in constraints directly supported by SMCHR. If *claim2* supports that $y > d$ and *claim1* supports $y \not> c$, if $d > c$ a conflict arises. This conflict is resolved by the third rule in Fig. 8: *claim2* is withdrawn from the store and a new argument *claim3* is added, supporting the constraint $y > c - 2$. A possible output is shown in Fig. 2.1, using $Q = claim1(x, y, 5) \land claim2(y, 9)$.

```
type claim1(var of num, var of num, num); type claim2(var of num, num);
    type claim3(var, num);

claim1(x, y, c)  ⟹ not int_gt(x, y) ∧ not int_gt_c(y,c);
claim2(z, d) ⟹ int_gt_c(z,d);
claim1(x, y, c) \ claim2(z, d) ∧ int_gt_c(z,d)  ⟺ d $>= c ∧ k:= c - 2
    | claim3(z,k) ∧ int_gt_c(z,k);
```

Fig. 8. An example of negotiation using the *bounds* propagator.

4   Related Work

In this section we revise some of the most important proposals that combine Argumentation with Logic Programming.

```
> claim1(x,y,5) /\  claim2(y,9)
UNKNOWN:
y > 3 /\
claim1(x,y,5) /\
claim3(y,3) /\
int_lb(y,4) /\
not x > y /\
not y > 5 /\
not int_lb(x,6) /\
not int_lb(y,6)
```

Fig. 9. A possible output for the program in Fig 8, given $Q = claim1(x, y, 2) \wedge claim2(y, 9)$. $int\_lb(y, 6)$ states that 4 is the lower (reachable) bound of y, and not $int\_lb(x, 6)$ and $not\ int\_lb(y, 6)$ are the (unreachable) upper bounds of x and y respectively.

One of the first attempt for integrating Logic Programming and Argumentation is [15], where Donald Nute introduces a formalism called *Logic for Defeasible Reasoning (LDR)*. The proposed language has three different types of rules: *strict*, *defeasible*, and *defeaters*. Even if LDR is not a defeasible formalism, its implementation in d-Prolog is enhanced with comparison criteria between rules.

In his seminal work [9] on Abstract Argumentation, Dung shows how that argumentation can be viewed as a special form of logic programming with negation as failure, e.g. "a logic program can be seen as a schema to generate arguments". Then, he introduces a general logic-programming based method to generate meta-interpreters for argumentation systems.

Two years later, inspired by legal reasoning, Prakken and Sartor [16] present a semantics (given by a fixed point definition) and a proof theory of a system for defeasible reasoning, where arguments are expressed in a logic-programming language with both strong and default negation. Conflicts between arguments are decided with the help of priorities associated with rules; such priorities can be defeasibly derived as conclusions within the system.

In [20] the authors formulate a variety of notions of attack for extended logic programs from combinations of undercuts and rebuts; moreover, they define a general hierarchy of argumentation semantics, which is parametrised

5 Conclusion

We have presented how a constraint propagator as SMCHR can be used to prototype different reasoning problems linked to Argumentation-based Logic Programming. The use of constraints becomes interesting when resolving conflicts depends on relations among arguments and/or their preference value, as in P-DeLP. Such methodology can use different solvers, e.g., *sat* or *bounds*. The ideas in this paper suggest the potentiality of having such a powerful declarative tool, paving the way for Argumentation-based Constraint Logic Programming.

The future goal is to have an automatised SMCHR-based framework where to model also dynamic reasoning over argumentation lines (where each argument structure in a sequence is a defeater of the predecessor), and dialectical trees (where each path from the root to a leaf corresponds to a different acceptable argumentation line). Clearly, arguments can be iteratively added to such tree during a debate. Some reasoning side-procedures, as checking the minimality of an argument support, can be programmed on top of SMCHR by, for instance, embedding SMCHR into an imperative language.

References

1. Alsinet, T., Chesñevar, C.I., Godo, L., Simari, G.R.: A logic programming framework for possibilistic argumentation: Formalization and logical properties. Fuzzy Sets and Systems **159**(10), 1208–1228 (2008)
2. Bistarelli, S., Santini, F.: Conarg: a constraint-based computational framework for argumentation systems. In: Proceedings of the 2011 IEEE 23rd International Conference on Tools with Artificial Intelligence, ICTAI 2011, pp. 605–612. IEEE Computer Society (2011)
3. Bistarelli, S., Santini, F.: Coalitions of arguments: An approach with constraint programming. Fundam. Inform. **124**(4), 383–401 (2013)
4. Bistarelli, S., Rossi, F., Santini, F.: A first comparison of abstract argumentation reasoning-tools. In: ECAI 2014–21st European Conference on Artificial Intelligence. FAIA, vol. 263, pp. 969–970. IOS Press (2014)
5. Cerutti, F., Giacomin, M., Vallati, M.: ArgSemSAT: solving argumentation problems using SAT. In: Computational Models of Argument - Proceedings of COMMA 2014. FAIA, vol. 266, pp. 455–456. IOS Press (2014)
6. Charwat, G., Dvorák, W., Gaggl, S.A., Wallner, J.P., Woltran, S.: Methods for solving reasoning problems in abstract argumentation: A survey. Artificial Intelligence **220**, 28–63 (2015)
7. Duck, G.J.: SMCHR: Satisfiability modulo constraint handling rules. TPLP **12**(4–5), 601–618 (2012)
8. Duck, G.J.: Satisfiability modulo constraint handling rules (extended abstract). In: IJCAI 2013, Proceedings of the 23rd International Joint Conference on Artificial Intelligence (2013)
9. Dung, P.M.: On the acceptability of arguments and its fundamental role in nonmonotonic reasoning, logic programming and n-person games. Artificial Intelligence **77**(2), 321–358 (1995)
10. Egly, U., Gaggl, S.A., Woltran, S.: Answer-set programming encodings for argumentation frameworks. Argument & Computation **1**(2), 147–177 (2010)
11. Frühwirth, T.W.: Theory and practice of constraint handling rules. J. Log. Program. **37**(1–3), 95–138 (1998)
12. Frühwirth, T.W.: Constraint Handling Rules, 1st edn. Cambridge University Press, New York (2009)
13. García, A.J., Simari, G.R.: Defeasible logic programming: An argumentative approach. Theory Pract. Log. Program. **4**(2), 95–138 (2004)
14. Kakas, A., Moraitis, P.: Argumentation based decision making for autonomous agents. In: Proceedings of the Second International Joint Conference on Autonomous Agents and Multiagent Systems, AAMAS 2003, pp. 883–890. ACM (2003)
15. Nute, D.: Defeasible reasoning: a philosophical analysis in Prolog. In: Aspects of Artificial Intelligence, pp. 251–288. Springer (1988)

16. Prakken, H., Sartor, G.: Argument-based extended logic programming with defeasible priorities. Journal of Applied Non-Classical Logics **7**(1–2), 25–75 (1997)
17. Rahwan, I., Simari, G.R.: Argumentation in Artificial Intelligence, 1st edn. Springer Publishing Company, Incorporated (2009)
18. Santini, F.: Argument-based constraint logic-programming in satisfiability modulo CHR. In: 12th International Workshop on Argumentation in Multi-Agent Systems. Informal Proceedings (2015)
19. Schrijvers, T., Demoen, B.: The K.U. Leuven CHR system: Implementation and application. In: First Workshop on Constraint Handling Rules: Selected Contributions, pp. 1–5 (2004)
20. Schweimeier, R., Schroeder, M.: A parameterised hierarchy of argumentation semantics for extended logic programming and its application to the well-founded semantics. Theory Pract. Log. Program. **5**(1–2), 207–242 (2005)

Abstract Solvers for Quantified Boolean Formulas and their Applications

Remi Brochenin$^{(\boxtimes)}$ and Marco Maratea

DIBRIS, University of Genova, Viale F. Causa 15, Genova, Italy
{remi.brochenin,marco.maratea}@unige.it

Abstract. Abstract solvers are a graph-based representation employed in many research areas, such as SAT, SMT and ASP, to model, analyze and compare search algorithms in place of pseudo-code-based representations. Such an uniform, formal way of presenting the solving algorithms proved effective for their understanding, for formalizing related formal properties and also for combining algorithms in order to design new solving procedures.

In this paper we present abstract solvers for Quantified Boolean Formulas (QBFs). They include a direct extension of the abstract solver describing the DPLL algorithm for SAT, and an alternative formulation inspired by the two-layers architecture employed for the analysis of disjunctive ASP solvers. We finally show how these abstract solvers can be directly employed for designing solving procedures for reasoning tasks which can be solved by means of reduction to a QBF.

1 Introduction

Abstract solvers are a relatively new methodology that have been employed in many research areas, such as Propositional Satisfiability (SAT) [1], Satisfiability Modulo Theories (SMT) [1,2], Answer Set Programming (ASP) [3–5], and Constraint ASP [6], to model, analyze and compare solving algorithms in place of pseudo-code-based representations. Abstract solvers are a graph-based representation, where the states of computation are represented as nodes of a graph, the solving techniques as arcs between such nodes, the solving process as a path in the graph and the formal properties of the algorithms are reduced to related graph's properties. Such a uniform, mathematically simple yet formal way of presenting the solving algorithms proved effective for their understanding, for formalizing properties in a clear way and also for combining algorithms for designing new solutions. However, with the notable exception of the recent work on disjunctive ASP [5], up to now this methodology has been employed to solving procedures for reasoning tasks whose complexity is within the first level of the polynomial hierarchy.

In this paper we present, for the first time, abstract solvers for deciding the satisfiability of Quantified Boolean Formulas (QBFs) [7], the prototypical PSPACE-complete problem, thus showing their potential also to analyze "hard" reasoning tasks. The first abstract solution is an extension of the abstract

© Springer International Publishing Switzerland 2015
M. Gavanelli et al. (Eds.): AI*IA 2015, LNAI 9336, pp. 205–217, 2015.
DOI: 10.1007/978-3-319-24309-2_16

solver [1] for describing the DPLL algorithm for SAT [8], enhanced with theory-specific techniques, modeled as additional arcs in the respective graph. The second abstract solver is, instead, based on the two-layers architecture employed for the analysis of disjunctive ASP solvers, which are characterized by a generating layer for finding candidate solutions, and a test layer for checking candidates's minimality: this second solution employs a "multi-layer" architecture whose number of layers depend on the quantifiers alternation in the QBF formula. We also comment on a third viable approach based on compilation into a SAT formula, on which is applied an abstract solver for SAT. For all abstract solutions, correctness results are formalized by means of related graph's properties. We finally show how abstract solvers for QBFs can be directly employed for solving certain reasoning tasks in other areas such as Answer Set Programming [9] and Abstract Argumentation and Dialectical frameworks (see, e.g. [10,11]), whose solutions are obtained by means of reduction to a QBF, thus where an abstract solver for QBF becomes an abstract solver for the (compiled) respective reasoning task.

To sum up, the main contributions of this paper are:

- We present a first abstract solver for QBFs that extends the abstract solver describing the DPLL algorithm for SAT.
- We present a second abstract solver that employs a multi-layer architecture whose idea comes from the two-layers architecture of disjunctive ASP solvers.
- We show a third solution, based on compilation to SAT and an abstract solver for SAT, and correctness results for all mentioned solutions by means of properties of related graphs.
- We show how the afore-presented abstract solvers can be directly employed for solving certain reasoning tasks in other fields.

The paper is structured as follows. Section 2 introduces needed preliminaries. Section 3 then presents the abstract solvers for backtracking-based procedures. Section 4 shows some applications of the introduced abstract solvers. The paper then ends with a discussion of related work and some conclusions in Section 5.

2   Formal Background

Syntax and Semantics. Consider a set P of variables (also called atoms). The symbols $\bot$ and $\top$ denote *false* and *true*, respectively. A literal is a variable a_0 or its negation $\overline{a_0}$. $\overline{\overline{a_0}}$ is the same of a_0. A clause is a finite set of literals. A SAT formula F' in Conjunctive Normal Form (CNF) is a finite set of clauses.

A QBF formula is an expression F of the form:

$$Q_0 X_0 Q_1 X_1 \ldots Q_n X_n F' \tag{1}$$

where:

- every Q_i $(0 \leq i \leq n)$ is a quantifier, either existential $\exists$ or universal $\forall$, and such that for each $0 \leq j \leq n - 1$, $Q_j \neq Q_{j+1}$, i.e. we assume an alternation of quantifiers, and assume that the innermost quantifier is $\exists$;
- $X_0, \ldots, X_n$ are variable groups (equivalently seen as sets) which define a *partition* of P (i.e. the formula is closed) such that (i) each $0 \leq i \leq n$ $X_i \neq \emptyset$, (ii) $\bigcup_{i, 0 \leq i \leq n} X_i = P$, and (iii) for each $0 \leq i, j \leq n$, $i \neq j$, $X_i \cap X_j = \emptyset$.
- F' is a SAT formula over P.

In (1), $Q_0 X_0 Q_1 X_1 \ldots Q_n X_n$ is the *prefix* and F' is the *matrix*. A *level* of a literal l built on a variable $b_i \in X_i$, denoted $level(l)$, is i with $0 \leq i \leq n$. We assume the formula has *max* alternations of quantifiers, so *max* is n. If F is (1) and l is a literal $l = b_i$ or $l = \neg b_i$ for $b_i \in X_i$, then F_l is the QBF:

- whose matrix is obtained from F by substituting (i) b_i with $\top$ and $\neg b_i$ with $\bot$ if $l = b_i$, and (ii) b_i with $\bot$ and $\neg b_i$ with $\top$, otherwise;
- whose prefix is $Q_0 X_0 Q_1 X_1 \ldots Q_i (X_i \setminus b_i) \ldots Q_n X_n$.

The *semantics* of a QBF F can be defined recursively as follows:

1. If the prefix is empty, according to the semantics of propositional logic.
2. If F is $\exists b F'$, $b \in P$, F is satisfiable iff F_b is satisfiable or $F_{\neg b}$ is satisfiable.
3. If F is $\forall b F'$, $b \in P$, F is satisfiable iff both F_b and $F_{\neg b}$ are satisfiable.

Example 1. In the QBF (2) below (from [7]), X_0 is $\{a\}$, X_1 is $\{d\}$, X_2 is $\{b, c\}$, $\exists a \forall d \exists bc$ is the prefix and $\{\{\overline{a}, \overline{d}, b\}, \{\overline{d}, \overline{b}\}, \{b, c\}, \{a, \overline{d}, \overline{c}\}, \{d, b, \overline{c}\}\}$ is the matrix. Note that (2) is unsatisfiable.

$$\exists a \forall d \exists bc \{\{\overline{a}, \overline{d}, b\}, \{\overline{d}, \overline{b}\}, \{b, c\}, \{a, \overline{d}, \overline{c}\}, \{d, b, \overline{c}\}\} \tag{2}$$

Q-DPLL algorithm. An *assignment* to a set X of atoms is a function from X to $\{\bot, \top\}$. A set of literals is called *consistent* if for any literal l it contains it does not contain $\overline{l}$. We identify a consistent set of literals M with an assignment to $At(M)$: $a \in M$ iff a maps to $\top$, and $\neg a \in M$ iff a maps to $\bot$.

Q-DPLL is an extension of DPLL algorithm for SAT for determining the satisfiability of a QBF. As DPLL exhaustively explores the space of assignments, by assigning literals either deterministically or heuristically, to generate classical models of a propositional formula, also Q-DPLL is a classical backtrack-search algorithm that exhaustively explores the space of assignments to test the satisfiability of a QBF. The main difference is that, when a literal whose atom is universally quantified is heuristically chosen, both branches must be explored.

A pseudo-code description of the Q-DPLL algorithm can be found in [7,12]. Solvers implementing this approach are, e.g. EVALUATE [12] and QuBE [13].

Abstract Q-DPLL for SAT. As said before, a QBF without prefix corresponds to a SAT formula, where all atoms are existentially quantified. Some more preliminaries about abstract solvers and related graphs are needed. A *universal literal* is a literal annotated as $l^\forall$. An *existential literal* is a literal annotated as $l^\exists$.

A *decision literal* is a universal literal or an existential literal. A *record* is a string of literals and decision literals. The *terminal states* are *Valid* and *Unsat*. A *state* is a record or a terminal state. The *initial state* is $\emptyset$.

Figure 1 will present the transition rules for the graph describing the Q-DPLL algorithm, but if we restrict to the rules *Unit*, *Decide*, *Backtrack$_\exists$*, *Fail* and *Succeed*, and considering that all atoms are on the same level, we obtain a description of the DPLL algorithm for SAT as presented in, e.g. [5]. Given a SAT formula, *Unit* adds to the current assignment an unassigned literal in a clause where all other literals are contradicted. *Decide* adds to the current assignment an unassigned literal. *Backtrack$_\exists$* restores an inconsistent assignment by going back in the current assignment and flipping the last decision literal. *Fail* determines the formula to be unsatisfiable, i.e. the current assignment is inconsistent but can not be fixed given that it does not contain any decision literal. Finally *Succeed* determines the formula to be satisfiable.

Example 2. Below are two possible paths in the graph $QBF_{\exists a,b,c\{\{a,b\},\{\bar{a},c\}\}}$:

			Initial state :		$\emptyset$
Initial state :	$\emptyset$		*Decide*	$\Longrightarrow$	$a^\exists$
Decide	$\Longrightarrow$	$a^\exists$	*Decide*	$\Longrightarrow$	$a^\exists \bar{c}^\exists$
Unit	$\Longrightarrow$	$a^\exists c$	*Unit*	$\Longrightarrow$	$a^\exists \bar{c}^\exists c$
Decide	$\Longrightarrow$	$a^\exists c\, b^\exists$	*Backtrack$_\exists$*	$\Longrightarrow$	$a^\exists c$
Succeed	$\Longrightarrow$	*Valid*	*Decide*	$\Longrightarrow$	$a^\exists c\, b^\exists$
			Succeed	$\Longrightarrow$	*Valid*

In order to realistically describe the DPLL algorithm, an ordering must be given on the application of the rules, such that a transition rule can not be applied if a rule with higher priority is applicable. In DPLL, the ordering follows how the rules *Unit, Decide, Backtrack$_\exists$, Fail*, and *Succeed* are listed in Figure 1. Thus, the path in Example 2 at the left corresponds to a possible path of the DPLL algorithm, while the path at the right does not, given that *Decide* is applied when *Unit* is applicable.

Finally, we say that a graph G *verifies* a statement S (e.g. G verifies that F is satisfiable) when all the following conditions hold:

1. G is finite and acyclic;
2. Any terminal state in G is either *Failstate* or *Valid*;
3. If a state *Valid* is reachable from the initial state in G then S holds (e.g. F is satisfiable);
4. *Failstate* is reachable from the initial state in G if and only if S does not hold (e.g. F is not satisfiable).

3   Abstract Solvers for QBFs

In this section we introduce the three abstract solvers for deciding the satisfiability of a QBF mentioned in the introduction.

Q-DPLL on a single layer. We show here a description of the Q-DPLL algorithm, within a single layer of computation. As we already wrote, the Q-DPLL algorithm is an extension of the DPLL algorithm for SAT. As a consequence, its transition system updates some of the transition rules for SAT, and introduces further transition rules to take into account the specific problem. First, unit propagate can now be applied subject to further specific conditions. Second, the decision literal must be chosen such that there is no an unassigned literal at a higher level. Third, there are now two types of backtracking, i.e. through an existentially-quantified literal, whose value is switched after a contradiction, or through an universally-quantified literal, whose value is switched after a successful branch. Finally, monotone rules that take into account situations where only a literal, or its negation, is in the formula, are also added. In the following, we will formalize these updates and modular additions.

The graph QBF_F has the states as nodes, as defined in the previous section, and the transitions of Figure 1 as edges.

The rules *Unit*, *Decide* and *Backtrack*$_\exists$ extend the ones for SAT in Section 2. *Unit* adds to the current assignment an unassigned literal in a clause if all other assigned literals are contradicted, *and all other unassigned literals are universally quantified*. *Decide* adds to the current assignment an unassigned literal, *either existentially or universally quantified, such that all atoms at lower levels are assigned*. *Backtrack*$_\exists$ restores an inconsistent assignment by going back in the assignment and flipping the last *existentially-quantified* decision variable.

Specific rules for QBF are *Monotone1*, *Monotone2* and *Backtrack*$_\forall$. The rule *Monotone1* (resp. *Monotone2*) assigns an existential (resp. universal) literal l, and l (resp. $\bar{l}$) appears in some clause while the opposite does not appear in any clause. *Backtrack*$_\forall$ is a counterpart of *Backtrack*$_\exists$: it flips the value of the last universal literal after finding a complete and consistent assignment. For that, it goes back in the current assignment and flips the last decision variable that is universally quantified. Note that this rule is applicable only when all other rules but *Succeed* can not be applied, so that it is triggered only when the current assignment both assigns all the atoms in F since *Decide* does not apply and is consistent since *Fail* does not apply. *Fail* and *Succeed* are the same as for SAT, except that *Succeed* can be triggered only when there is no decision variable that is universally quantified so that *Backtrack*$_\forall$ is not applicable.

Proposition 1. *Given a QBF formula F, the graph QBF_F verifies that F is satisfiable.*

Example 3. Consider the QBF (2) in Example 1, that we call F. A possible path in QBF_F is:

Initial state :	$\emptyset$			
Decide	$\Longrightarrow \overline{a}^\exists$		*Backtrack*$_\exists$	$\Longrightarrow a$
Decide	$\Longrightarrow \overline{a}^\exists \overline{d}^\forall$		*Decide*	$\Longrightarrow a\overline{d}^\forall$
Monotone1	$\Longrightarrow \overline{a}^\exists \overline{d}^\forall b$		*Monotone1*	$\Longrightarrow a\overline{d}^\forall b$
Backtrack$_\forall$	$\Longrightarrow \overline{a}^\exists d$		*Backtrack*$_\forall$	$\Longrightarrow a\,d$
Unit	$\Longrightarrow \overline{a}^\exists d\,\overline{c}$		*Unit*	$\Longrightarrow a\,d\,b$
Unit	$\Longrightarrow \overline{a}^\exists d\,\overline{c}\overline{b}$		*Fail*	$\Longrightarrow Unsat$

Rules

$$Unit \quad L \implies Ll \quad \text{if} \begin{cases} l \text{ does not occur in } L \text{ and} \\ \text{for some clause } C \text{ in the matrix,} \\ l \text{ occurs in } C \text{ and} \\ \text{each other unassigned literal of } C \text{ is universal and} \\ \text{each assigned literal of } C \text{ is contradicted} \end{cases}$$

$$Monotone1 \ L \quad \implies Ll \quad \text{if} \begin{cases} \text{the variable of } l \text{ is existential and} \\ l \text{ occurs in some clause } C \text{ and} \\ \bar{l} \text{ does not occur in any clause } C \end{cases}$$

$$Monotone2 \ L \quad \implies Ll \quad \text{if} \begin{cases} \text{the variable of } l \text{ is universal and} \\ \bar{l} \text{ occurs in some clause } C \text{ and} \\ l \text{ does not occur in any clause } C \end{cases}$$

$$Decide \quad L \quad \implies Ll^Q \quad \text{if} \begin{cases} L \text{ is consistent and} \\ \text{the variable of } l \text{ is unassigned and} \\ \text{the quantifier of the variable of } l \text{ is } Q \text{ and} \\ \text{for all } l' \text{ such that } level(l') < level(l) \\ \text{the variable of } l' \text{ is assigned.} \end{cases}$$

$$Backtrack_\exists \ Ll^\exists L' \implies L\bar{l} \quad \text{if} \begin{cases} Ll^\exists L' \text{ is inconsistent and} \\ l^\exists \text{ is the rightmost existential literal} \end{cases}$$

$$Fail \quad L \quad \implies Unsat \text{ if} \{ L \text{ is inconsistent and existential free}$$

$$Backtrack_\forall \ Ll^\forall L' \implies L\bar{l} \quad \text{if} \begin{cases} \text{no other rule applies except } Succeed \text{ and} \\ l^\forall \text{ is the rightmost universal literal} \end{cases}$$

$$Succeed \quad L \quad \implies Valid \text{ if} \{ \text{ no other rule applies}$$

Fig. 1. The transition rules of the QBF_F graph.

Q-DPLL with multiple layers. In the previous sub-section we presented the abstract solver that describes the Q-DPLL algorithm. Here we present an alternative solution that uses and extends the two-layers architecture employed in abstract solvers for disjunctive ASP [5]. In disjunctive ASP, solvers are mainly organized in the following way:[1] there is a "generate" layer that computes candidate solutions, and a "test" layer that checks whether it is indeed a solution, by checking minimality. In QBF, we extend this architecture by considering that a layer is the solving process "within" the same quantifier level. Within a layer, a SAT problem is solved, and depending from the past search, the current level, and the related quantifier type, the search is directed through levels with some newly added control states. In sum, there is a basic set of transition rules that corresponds to the rules for SAT (plus monotone rules, that can be used also in SAT), which are called Core rules, and another set of rules to direct the search. In the following, we will formalize these concepts.

---

[1] A notable exception is the family of ASP solvers based on translation.

Before introducing the related graph, we need some additional definitions. A stack of records S is a (possibly empty) list of records $S = L_1 :: L_2 :: \ldots :: L_k$ which we can also write as $S = S' :: L_k$, where S' is a stack of records.

An *oracle state* $L_{S,k,Q}$ is made of:

- a record L, which accounts for the current assignment computed;
- a stack of records S, an integer k and a quantifier type Q.

A *control state* $Instr(L_{S,k,Q})$ is made of:

- the action $Instr$ that led to this state, $Instr \in \{Failure, Success, Cont\}$;
- a record L, result of the last computation;
- a stack of records S, results of previous computations;
- an integer k equal to the amount of quantifier alternations that precede the quantifiers currently treated in the last computation;
- a type of quantifier Q in $\{\exists, \forall\}$, corresponding to the type of the quantifiers of the last computation.

Control states guide the search through layers. A state is a control state, an oracle state or a terminal state. Intuitively, the core rules from *Unit* to *Backtrack*, which resemble the ones in the previous sub-section, deal with the computation within a level. Rules *Fail* and *Succeed*, instead, are the rules at the interface between two layers, going from an oracle to a control state. Result processing rules, on the other hand, direct the computation according to the oracle state they are dealing with. In particular, *Fail* (resp. *Succeed*) leads to the actions *Failure* (resp. *Cont*) if the assignment can not (resp. can) be successfully extended at this level, respectively. In consequence of the application of one of these rules, Result processing rules can be triggered. A *Failure* control state on an existential level triggers *Failure*∃ and the result is that the action is unchanged, the level is decremented, the quantifier is changed, and the current record L' becomes the last in the stack. Then, *Failure*∀ is immediately triggered doing similar processing, but leading to an oracle state whose record is inconsistent (and backtrack will be forced). The rational about this behavior is that if we had a failure on an existential level, we must also jump over the previous universal level, because in this branch a solution can not be found. If k becomes 0 and the stack is empty, we can return *Unsat* through *FailureFinal*, meaning the F is unsatisfiable. Instead, a *Cont* control state on an existential level triggers the rule *Continue* that brings (*i*) to an oracle state at the next level whose assignment is added to the queue, the level is incremented, the quantifier is switched and the current assignment is restarted, or (*ii*) to a *Success* state if the maximum level is reached through *FullAssign*. Now, *Success*∃ is triggered and leads to failing level at the upper quantified level, in order to force a backtrack, or *Success*∃' is triggered if the record L at the upper level does contain decision literals: if this happens, the *Success* state is maintained at the upper level, that immediately triggers *Success*∀ that goes one further level up, and the search can proceed as before. If a *Success* is reached at level 0, *Valid* can be returned, meaning that F is satisfiable. The graph $QBF2_F$ has the states as nodes, the transitions of Figure 2 as edges, and $\emptyset_{\emptyset,0,Q}$ as initial state, where Q is the type of the outermost quantifier.

Proposition 2. *Given a QBF formula F, the graph $QBF2_F$ verifies that F is satisfiable.*

Note that this graph seems to be more amenable for being the basis for building new abstract procedures for QBFs, which is one of the main advantage of this methodology. In fact, we can replace the Core rules with any other set of rules that solve the same problem (in this case a SAT problem), and add similar rules as *Fail* and *Succeed* that lead to Result processing rules.

Example 4. Consider the QBF (2) in Example 1, that we call F. A possible path in $QBF2_F$ is:

Initial state :	$\emptyset_{\emptyset,0,Q}$		Backtrack	$\Longrightarrow a_{\emptyset,0,\exists}$
Decide	$\Longrightarrow \overline{a}^{\exists}{}_{\emptyset,0,\exists}$		Succeed	$\Longrightarrow Cont(a_{\emptyset,0,\exists})$
Succeed	$\Longrightarrow Cont(\overline{a}^{\exists}{}_{\emptyset,0,\exists})$		Continue	$\Longrightarrow \emptyset_{a,1,\forall}$
Continue	$\Longrightarrow \emptyset_{\overline{a}^{\exists},1,\forall}$		Decide	$\Longrightarrow \overline{d}^{\exists}{}_{a,1,\forall}$
Decide	$\Longrightarrow \overline{d}^{\exists}{}_{\overline{a}^{\exists},1,\forall}$		Succeed	$\Longrightarrow Cont(\overline{d}^{\exists}{}_{a,1,\forall})$
Succeed	$\Longrightarrow Cont(\overline{d}^{\exists}{}_{\overline{a}^{\exists},1,\forall})$		Continue	$\Longrightarrow \emptyset_{a::\overline{d}^{\exists},2,\exists}$
Continue	$\Longrightarrow \emptyset_{\overline{a}^{\exists}::\overline{d}^{\exists},2,\exists}$		Monotone1	$\Longrightarrow b_{a::\overline{d}^{\exists},2,\exists}$
Monotone1	$\Longrightarrow b_{\overline{a}^{\exists}::\overline{d}^{\exists},2,\exists}$		Succeed	$\Longrightarrow Cont(b_{a::\overline{d}^{\exists},2,\exists})$
Succeed	$\Longrightarrow Cont(b_{\overline{a}^{\exists}::\overline{d}^{\exists},2,\exists})$		FullAssign	$\Longrightarrow Success(b_{a::\overline{d}^{\exists},2,\exists})$
FullAssign	$\Longrightarrow Success(b_{\overline{a}^{\exists}::\overline{d}^{\exists},2,\exists})$		$Success\exists$	$\Longrightarrow \overline{d}^{\exists}\perp_{a,1,\forall}$
$Success\exists$	$\Longrightarrow \overline{d}^{\exists}\perp_{\overline{a}^{\exists},1,\forall}$		Backtrack	$\Longrightarrow d_{a,1,\forall}$
Backtrack	$\Longrightarrow d_{\overline{a}^{\exists},1,\forall}$		Unit	$\Longrightarrow d\, b_{a,1,\forall}$
Unit	$\Longrightarrow d\, \overline{c}_{\overline{a}^{\exists},1,\forall}$		Unit	$\Longrightarrow d\, b\, \overline{b}_{a,1,\forall}$
Unit	$\Longrightarrow d\, \overline{c}\, b_{\overline{a}^{\exists},1,\forall}$		Fail	$\Longrightarrow Failure(d\, b\, \overline{b}_{a,1,\forall})$
Unit	$\Longrightarrow d\, \overline{c}\, \overline{b}\, b_{\overline{a}^{\exists},1,\forall}$		$Failure\forall$	$\Longrightarrow a\perp_{\emptyset,0,\exists}$
Fail	$\Longrightarrow Failure(d\, \overline{c}\, \overline{b}\, b_{\overline{a}^{\exists},1,\forall})$		Fail	$\Longrightarrow Failure(a\perp_{\emptyset,0,\exists})$
$Failure\forall$	$\Longrightarrow \overline{a}^{\exists}\perp_{\emptyset,0,\exists}$		FailureFinal	$\Longrightarrow Unsat$

Solution Based on Variable Elimination. A third solution for solving a QBF is to rely on an approach which directly follows the semantic of a QBF as presented in the previous section, thus considering that, given a QBF F, $\exists aF$ is logically equivalent to $F_a \vee F_{\overline{a}}$, while $\forall aF$ is logically equivalent to $F_a \wedge F_{\overline{a}}$. The expansion of a variable a is obtained by[2]:

1. adding a variable b' for each variable b having $level(b) < level(a)$;
2. quantifying each variable b' in the same way as b and s.t. $level(b') = level(b)$;
3. for each clause C in the scope of a, adding a new clause C' obtained from C by substituting b' to b; and
4. considering the mentioned clause C (resp. C') , those containing $\overline{a}$ (resp. a) are eliminated, while a (resp. $\overline{a}$) is eliminated from the other clauses.

---

[2] Optimizations are possible, e.g. *Unit*, *Monotone*1 and *Monotone*2 are applicable, and a concept of "minimal scope" for a variable can be defined (see, e.g. [7]).

Core rules

$$UnitL_{S,k,Q} \implies Ll_{S,k,Q} \quad \text{if} \begin{cases} l \text{ does not occur in } L \text{ and} \\ \text{for some clause } C \text{ in the matrix,} \\ l \text{ occurs in } C \text{ and} \\ \text{each other unassigned literal of } C \text{ is universal and} \\ \text{each assigned literal of } C \text{ is contradicted} \end{cases}$$

$$Monotone1 \quad L_{S,k,Q} \implies Ll_{S,k,Q} \quad \text{if} \begin{cases} \text{the variable of } l \text{ is existential and} \\ l \text{ occurs in the matrix and} \\ \bar{l} \text{ does not occur in the matrix} \end{cases}$$

$$Monotone2 \quad L_{S,k,Q} \implies Ll_{S,k,Q} \quad \text{if} \begin{cases} \text{the variable of } l \text{ is universal and} \\ \bar{l} \text{ occurs in the matrix and} \\ l \text{ does not occur in the matrix} \end{cases}$$

$$Decide \quad L_{S,k,Q} \implies Ll^{\exists}{}_{S,k,Q} \quad \text{if} \begin{cases} L \text{ is consistent and} \\ \text{neither } l \text{ nor } \bar{l} \text{ occur in } L \text{ and} \\ level(l) = k \end{cases}$$

$$Backtrack \quad Ll^{\exists}L'_{S,k,Q} \implies L\bar{l}_{S,k,Q} \quad \text{if} \begin{cases} Ll^{\exists}L' \text{ is inconsistent and} \\ l^{\exists} \text{ is the rightmost decision literal} \end{cases}$$

$$Fail \quad L_{S,k,Q} \implies Failure(L_{S,k,Q}) \text{ if} \begin{cases} L \text{ is inconsistent and decision free} \end{cases}$$
$$Succeed \quad L_{S,k,Q} \implies Cont(L_{S,k,Q}) \quad \text{if} \begin{cases} \text{no other rule applies} \end{cases}$$

Result processing rules for $k \in \{0..max - 1\}$

$$Continue \quad Cont(L_{S,k,Q}) \implies \emptyset_{S::L,k+1,\overline{Q}}$$
$$FullAssign \quad Cont(L_{S,max,\exists}) \implies Success(L_{S,max,\exists})$$

$$Failure\forall \quad Failure(L_{S::L',k+1,\forall}) \implies L'\perp_{S,k,\exists}$$
$$Failure\exists \quad Failure(L_{S::L',k+1,\exists}) \implies Failure(L'_{S,k,\forall})$$
$$FailureFinal \quad Failure(L_{\emptyset,0,Q}) \implies Unsat$$

$$Success\forall \quad Success(L_{S::L',k+1,\forall}) \implies Success(L'_{S,k,\exists})$$
$$Success\exists \quad Success(L_{S::L',k+1,\exists}) \implies L'\perp_{S,k,\forall}$$
$$\text{if } L' \text{ contains at least a decision literal}$$
$$Success\exists' \quad Success(L_{S::L',k+1,\exists}) \implies Success(L'_{S,k,\forall})$$
$$\text{if } L' \text{ is decision-free}$$
$$SuccessFinal \quad Success(L_{\emptyset,0,Q}) \implies Valid$$

Fig. 2. The transition rules of the $QBF2_F$ graph.

In particular, the process of expanding all universally-quantified variables yields a SAT formula that can be solved with the abstract solver we have seen in Section 2, e.g. by expanding d in (2) we obtain the following SAT formula:

$$\exists abcb'c'\{\{b,c\},\{b,\overline{c}\},\{\overline{a},b'\},\{\overline{b'}\},\{b',c'\},\{a,\overline{c'}\}\}$$

where, e.g. variables b' and c' are added to the prefix (step 1. above) as existentially quantified (step 2.), $\{b',c'\}$ is obtained from $\{b,c\}$ (step 3.) and $\{b,\overline{c}\}$ is $\{d,b,\overline{c}\}$ deprived of d (step 4.). Formal properties for this approach can be stated by relying on the correctness of variable elimination, and formal properties on the abstract graph for SAT.

Of course, a formal result similar to Proposition 1 and 2 could be added. But the correctness of this potential proposition stems directly from the correctness of variable elimination and of abstract solvers for SAT, that are already proved in other papers. Hence, stating such a proposition is not necessary.

4   Applications

Several reasoning tasks in other fields have been solved by a translation to a QBF formula followed by the application of a QBF solver to this formula. Hence, the abstract solvers we have defined can be used to abstract decision procedures for these problems. Indeed, the states and the transition rules from which the transitions are inferred will be identical to those of the graph QBF_F. But this process will be applied to a specific formula F that corresponds to the translation of an instance of the reasoning task to solve. We review here some applications of the abstract solvers we have defined in previous sections.

We will use the graphs of the type QBF_F, but these ideas could equivalently be stated with $QBF2_F$ graphs. Also, in the articles describing each of the considered solvers, where it is possible to find the resulting formulas, the matrix is not in CNF, and the formulas are not necessarily in prenex from. Hence, they are not defined exactly as we defined QBF formulas in this article. As a consequence, each time we define an abstract solver using a formula F we will use $QBF_{NF(F)}$. The function NF converts a non-prenex non-CNF quantified formula F to a QBF formula matching the definition provided in this article, i.e. with prenex form, all the free variables quantified existentially, and matrix in CNF.[3]

Answer Set Programming. Answer set programming is a declarative language representing problems as logic programs, of which solutions are called answer sets. Determining whether a program has any solution is Σ_2^P-complete in the general case of disjunctive answer set programming. In [9] is defined the quantified formula $T_{lp}(P)$ for a program P. The formula $T_{lp}(P)$ is satisfiable if and only if P has at least one answer set. Then, for any program P, the graph $QBF_{NF(T_{lp}(P))}$ verifies that P has at least one answer set.

There is not yet practical ASP solvers built my means of reduction to QBF. The main obstacle is the encoding, and the fact that the defined encoding is in non-prenex non-CNF form, and an efficient transformation in prenex CNF is required. On the other hand, we believe that, once this obstacle is mitigated, our work can help also in the practical interplay between the encoding and the engine solver.

---

[3] Also the solvers that implement approaches based on reduction rely on such conversion, given they employ QBF solvers based on the form used in this article.

Abstract Argumentation Frameworks. Abstract argumentation frameworks are directed graphs which are designed to represent conflicting information. Each vertex of the graph is called an argument. The edges of the graph represent the way arguments can attack each other. The graphs are studied under varied semantics for which a set of arguments is a solution; for instance, in general a set of arguments will have to be conflict-free to be admissible. There are two main types of decision problems generally studied for each of the semantics: knowing whether a given argument belongs to at least a solution (i.e. credulous acceptance), and whether a given argument belongs to all solutions (i.e. skeptical acceptance). The article [14] defines first order formulas that have free variables, designed so that the assignments to the free variables that satisfy the formulas have a strict correspondence to the solutions of an argumentation framework under a given semantics. For instance, the first order formula $\mathcal{PE}(\mathcal{A})$ is defined in this article and satisfying assignments to its free variables correspond to the preferred semantics of the argumentation framework A. Obtaining an abstract procedure from such formulas is possible but would require using the techniques of [11], as explained below.

Abstract Dialectical Frameworks. Abstract dialectical frameworks are a generalization of abstract argumentation frameworks which allow to model more complex interactions between arguments. As a result, decision problems are generally one level higher in the polynomial hierarchy, up to the third level. We are here going to rely on the work of Diller et al. [11] which, for an argument $s \in S$ from an abstract dialectical framework (S, C) and a semantics $\sigma \in \{adm, com, prf, grd, mod, stb\}$, defines a QBF formula which is satisfiable if and only if s is skeptically accepted in (S, C) under σ semantics. A similar formula is defined in the case of credulous acceptance.

For instance, we focus on the case of preferred semantics and skeptical acceptance. The formula is denoted $\forall S_3(\mathcal{E}_{prf}(S, C) \Rightarrow s^{\oplus})$. Note that $\forall S_3$ refers to a quantification over all the elements from a set of variables built from S and used in $\mathcal{E}_{prf}(S, C)$, so that the quantification remains in the first order. Hence, for an abstract dialectical framework (S, C) and $s \in S$, the procedure of [11] for solving skeptical acceptance of s in (S, C) under preferred semantics is abstracted by $QBF_{NF(\forall S_3(\mathcal{E}_{prf}(S,C) \Rightarrow s^{\oplus}))}$. Hence $QBF_{NF(\forall S_3(\mathcal{E}_{prf}(S,C) \Rightarrow s^{\oplus}))}$ verifies that s is skeptically accepted in (S, C) under preferred semantics. Similar graphs can be defined so as to abstract the procedure for other semantics, and for credulous acceptance. For instance, credulous acceptance for preferred semantics would be abstracted by $QBF_{NF(\exists S_3(\mathcal{E}_{prf}(S,C) \wedge s^{\oplus}))}$. In $\forall S_3(\mathcal{E}_{prf}(S, C) \Rightarrow s^{\oplus})$, the formula $\mathcal{E}_{prf}(S, C)$ is very similar to $\mathcal{PE}(\mathcal{A})$ in its function. Using a formula similar to $s^{\oplus}$ which represents the argument s and a set of variables similar to S_3, one could define a formula of the form $\forall S_3(\mathcal{PE}(\mathcal{A}) \Rightarrow s^{\oplus})$ so as to abstract a decision procedure for abstract argumentation frameworks. Assuming we call the obtained formula $pe(A, s)$, the graph $QBF_{pe(A,s)}$ would verify that s is skeptically accepted in A under preferred semantics.

Differently from ASP, efficient approaches based on QBF are implemented for Abstract Dialectical Frameworks [11]. We believe that, in this case, abstract solvers can help to more deeply understand the effectiveness of this approach.

5   Related Work and Conclusions

Abstract solvers have been originally employed in [1] first to describe the DPLL algorithm for SAT, and then by extending this graph to deal with certain SMT logics which can be solved by means of a lazy (i.e. SAT-based) approach to SMT solving. Then, abstract solvers have been applied to Answer Set Programming in several papers. Abstract solvers for backtracking-based ASP solvers for non-disjunctive ASP programs (whose complexity is within the first level of the polynomial hierarchy) have been presented in [15], then extended in (i) [3] to include backjumping and learning techniques, and in (ii) [5] for describing solvers for disjunctive ASP programs (able to express problems up to the second level of the polynomial hierarchy). Another contribution in ASP is presented in [4], where an unifying perspective based on completion (i.e. transforming a logic program into a propositional formula) on some solvers for non-disjunctive ASP programs is given. Finally, abstract solvers for Constraint ASP solvers are presented in [6].

In this paper we have presented, for the first time, abstract procedures for solving reasoning tasks whose complexity is beyond the second level of the polynomial hierarchy, i.e. the satisfiability of QBF, the prototypical PSPACE-complete problem. We have finally shown how these abstract solvers can be used to define abstract procedures for certain reasoning tasks in other fields that can be solved via a translation to a QBF. Other applications are of course possible, e.g. to solving conformant and conditional automated planning problems. However, we do not claim about the efficiency of a new tool built on this basis, given that it usually also requires many iterations of theoretical analysis, practical engineering, and domain-specific optimizations to develop efficient systems. Yet, positive experiences have been already reported for Abstract Dialectial Frameworks [11] and classical planning [16].

Future research includes adding optimization techniques, like backjumping and learning, to our abstract solvers. These are two well-known techniques implemented in several solvers: backjumping is the ability to jump over decision literals that were not directly responsible to the conflict that caused backtracking, while learning adds information (in terms of, e.g. clauses) to the initial formula in order to avoid to follow similar paths in order parts of the search. The presence of two quantifier types enables two different types of transition rules for modeling these techniques, that will be anyway added by means of modular addition of transition rules.

References

1. Leeuwenhoek, R., Oliveras, A., Tinelli, C.: Solving SAT and SAT modulo theories: From an abstract Davis-Putnam-Logemann-Loveland procedure to DPLL(T). Journal of the ACM **53**(6), 937–977 (2006)
2. Barrett, C., Shikanian, I., Tinelli, C.: An abstract decision procedure for a theory of inductive data types. JSAT **3**(1–2), 21–46 (2007)
3. Lierler, Y.: Abstract answer set solvers with backjumping and learning. Theory and Practice of Logic Programming **11**, 135–169 (2011)
4. Lierler, Y., Truszczynski, M.: Transition systems for model generators - a unifying approach. Theory and Practice of Logic Programming **11**(4–5), 629–646 (2011)
5. Brochenin, R., Lierler, Y., Maratea, M.: Abstract disjunctive answer set solvers. In: Schaub, T., Friedrich, G., O'Sullivan, B. (eds.) Proc. of ECAI 2014, vol. 263. Frontiers in Artificial Intelligence and Applications, pp. 165–170. IOS Press (2014)
6. Lierler, Y.: Relating constraint answer set programming languages and algorithms. Artificial Intelligence **207**, 1–22 (2014)
7. Giunchiglia, E., Marin, P., Narizzano, M.: Reasoning with quantified boolean formulas. In: Biere, A., Heule, M., van Maaren, H., Walsh, T. (eds.) Handbook of Satisfiability. Volume 185 of FAIA, pp. 761–780. IOS Press (2009)
8. Davis, M., Logemann, G., Loveland, D.: A machine program for theorem proving. Communications of the ACM **5**(7), 394–397 (1962)
9. Egly, U., Eiter, T., Tompits, H., Woltran, S.: Solving advanced reasoning tasks using quantified boolean formulas. In: Kautz, H.A., Porter, B.W. (eds.) Proc. of AAAI 2000, pp. 417–422, AAAI Press / The MIT Press (2000)
10. Arieli, O., Caminada, M.W.A.: A QBF-based formalization of abstract argumentation semantics. Journal of Applied Logic **11**(2), 229–252 (2013)
11. Diller, M., Wallner, J.P., Woltran, S.: Reasoning in abstract dialectical frameworks using quantified boolean formulas. In: Parsons, S., Oren, N., Reed, C., Cerutti, F. (eds.) Proc. of COMMA 2014. Volume 266 of FAIA, pp. 241–252, IOS Press (2014)
12. Cadoli, M., Giovanardi, A., Schaerf, M.: An algorithm to evaluate quantified boolean formulae. In: Mostow, J., Rich, C. (eds) Proc. of AAAI 1998, pp. 262–267, AAAI Press / The MIT Press (1998)
13. Giunchiglia, E., Narizzano, M., Tacchella, A.: Clause/term resolution and learning in the evaluation of quantified boolean formulas. JAIR **26**, 371–416 (2006)
14. Charwat, G., Dvořák, W., Gaggl, S.A., Wallner, J.P., Woltran, S.: Methods for Solving Reasoning Problems in Abstract Argumentation - A Survey. Artificial Intelligence **220**, 28–63 (2015)
15. Lierler, Y.: Abstract answer set solvers. In: Garcia de la Banda, M., Pontelli, E. (eds.) ICLP 2008. LNCS, vol. 5366, pp. 377–391. Springer, Heidelberg (2008)
16. Cashmore, M., Fox, M., Giunchiglia, E.: Partially grounded planning as quantified boolean formula. In: Borrajo, D., Kambhampati, S., Oddi, A., Fratini, S. (eds.) Proceedings of the Twenty-Third International Conference on Automated Planning and Scheduling, ICAPS 2013, AAAI (2013)

Machine Learning

Learning Accurate Cutset Networks
by Exploiting Decomposability

Nicola Di Mauro$^{(\boxtimes)}$, Antonio Vergari, and Floriana Esposito

University of Bari "Aldo Moro", Bari, Italy
{nicola.dimauro,antonio.vergari,floriana.esposito}@uniba.it

Abstract. The rising interest around tractable Probabilistic Graphical Models is due to the guarantees on inference feasibility they provide. Among them, Cutset Networks (CNets) have recently been introduced as models embedding Pearl's cutset conditioning algorithm in the form of weighted probabilistic model trees with tree-structured models as leaves. Learning the structure of CNets has been tackled as a greedy search leveraging heuristics from decision tree learning. Even if efficient, the learned models are far from being accurate in terms of likelihood. Here, we exploit the decomposable score of CNets to learn their structure and parameters by directly maximizing the likelihood, including the BIC criterion and informative priors on smoothing parameters. In addition, we show how to create mixtures of CNets by adopting a well known bagging method from the discriminative framework as an effective and cheap alternative to the classical EM. We compare our algorithms against the original variants on a set of standard benchmarks for graphical model structure learning, empirically proving our claims.

1 Introduction

Probabilistic Graphical Models (PGMs) [8] provide a powerful formalism to model and reason about rich and structured domains. They capture the conditional independence assumptions among random variables into a graph based representation, sometimes called network (as in Bayesian Networks). Answering *inference* queries in PGMs often results in computing the probability of observing some evidence according the provided graphical structure. However, in general, to compute *exact* inference is a NP-Hard problem, and also some approximate inference routines are intractable in practice [19].

The pursuit for exact and efficient inference procedures led to the recently growing interest in the AI community around *tractable* PGMs. In exchange for inference tractability guarantees, they are less expressive, in the sense that they cannot possibly capture all the conditional probabilistic independences in the data. Tractable PGMs encompass *tree-structured models*, like those learned by the classical Chow-Liu algorithm [3] or by introducing latent variables [2], or even a bound on the treewidth of the model [1]; Bayesian and Markov Networks compiled into Arithmetic Circuits (ACs) [11,12]; and Sum-Product Networks (SPNs) [15] as deep architectures encoding probability distributions by layering

© Springer International Publishing Switzerland 2015
M. Gavanelli et al. (Eds.): AI*IA 2015, LNAI 9336, pp. 221–232, 2015.
DOI: 10.1007/978-3-319-24309-2_17

hidden variables as mixtures of independent components. As the expressiveness of these models increases, the complexity of learning their parameters and structure from data increases as well; as a matter of fact the overall performances degrade as data grows in size.

Cutset Networks (CNets) have been introduced recently in [17] as tractable PGMs with the aim of making learning efficient and scalable. They are weighted probabilistic model trees in the form of OR-trees having tree-structured models as leaves, and non-negative weights on inner edges, resulting into an architecture embedding Pearl's conditioning algorithm [14]. Inner nodes, i.e., conditioning OR nodes, are associated to random variables and outgoing branches represent conditioning on the values for those variables domains.

In [17], well known decision tree learning algorithms are leveraged to build a Cutset Network from data. In a nutshell, iteratively, training instances are split conditioning on the values of the best variable chosen as to maximize the reduction in an approximation of the joint entropy over all variables. While the computation of such a heuristic is cheap, it is not principled in a generative framework where model accuracy is measured as the scored data likelihood. The need to directly estimate the data likelihood is shown when a form of tree post pruning is introduced in [17] as a way to alleviate overfitting. Competitive results against state-of-the-art tractable PGM structure learners [13,18] are achieved when introducing mixtures of Cutset Networks via the Expectation Maximization algorithm (EM).

In this work we introduce a more principled way to learn Cutset Networks. We reformulate the search in the structure space as an optimization task directly maximizing data likelihood in a Bayesian framework. Regularization is achieved through the introduction of the Bayesian Information Criterion (BIC) in the likelihood score and by informative Dirichlet Priors on counting parameters while learning tree-structured models. Therefore, we avoid overfitting without adopting costly techniques like post pruning as in [17]. The direct optimization of the CNet likelihood has been obtained by exploiting its decomposability, leading to a tractable evaluation of the models during learning by limiting computations only on portions of data. We then introduce a very simple yet effective way to learn mixtures of Cutset Networks by exploiting *bagging* [6], opposed to the classical generative use of EM. We empirically verified the gain in terms of likelihood for the learned models with this new proposed approach against the original algorithm variants proposed in [17], with and without pruning, and MT, a solid competitor learning mixtures of trees as proposed in [13], on 18 datasets commonly used as benchmarks for graphical models structure learning.

2   Background

We define $\mathcal{D} = \{\xi_1, \ldots, \xi_M\}$ as a set of M i.i.d. instances over the discrete variables $\mathbf{X} = \{X_1, \ldots, X_n\}$, whose domains are the sets $\mathrm{Val}(X_i) = \{x_i^j\}_{j=1}^{k_i}, i = 1, \ldots, n$. We refer to the value assumed by an instance ξ_m in correspondence of a particular variable X_i as $\xi_m[X_i]$.

2.1   Tree-Structured Models

A *directed tree-structured model* [13] is a Bayesian Network in which each variable has at most one parent. Therefore, the joint probability distribution over $\mathbf{X}$ represented by such models can be written in the form of a factorization as:

$$P(\mathbf{X}) = \prod_{i=1}^{n} P(X_i|\mathrm{Pa}_i) \tag{1}$$

where Pa_i stands for the parent variable of X_i, if present. From Eq. 1 it follows immediately that inference for complete or marginal queries has complexity linear in the number of variables, hence the tractability of tree-structured models.

One classic result in learning tree-structured models is that presented by Chow and Liu in [3], where they prove that maximizing the Mutual Information (MI) among random variables in $\mathbf{X}$ leads to the best tree, in an information-theoretic sense, approximating the underlying probability distribution of $\mathcal{D}$ in terms of the Kullback-Leibler divergence.

Algorithm 1. LearnCLTree($\mathcal{D}$, $\mathbf{X}$, α)

1: **Input:** a set of instances $\mathcal{D}$ over a set of features $\mathbf{X}$; α smoothing parameter
2: **Output:** $\langle \mathcal{T}, \boldsymbol{\theta} \rangle$, a tree $\mathcal{T}$ with parameters $\boldsymbol{\theta}$ encoding a pdf over $\mathbf{X}$
3: $M \leftarrow \mathbf{0}_{|\mathbf{X}| \times |\mathbf{X}|}$
4: **for each** $X_u, X_v \in \mathbf{X}$ **do**
5:     $M_{u,v} \leftarrow$ estimateMutualInformation($X_u, X_v, \mathcal{D}, \alpha$)
6: $\mathcal{T} \leftarrow$ maximumSpanningTree(M)
7: $\mathcal{T} \leftarrow$ traverseTree($\mathcal{T}$)
8: $\boldsymbol{\theta} \leftarrow$ computeFactors($\mathcal{D}, \mathcal{T}$)
9: **return** $\langle \mathcal{T}, \boldsymbol{\theta} \rangle$

Algorithm 1 shows a sketch of the learning process. Firstly, for each pair of variables in $\mathbf{X}$, their MI is estimated from $\mathcal{D}$, optionally by introducing a smoothing factor α (line 5); then a maximum spanning tree is built on the weighted graph induced by the MI as an adjacency matrix (line 6). Rooting the tree in a randomly chosen variable and traversing it leads to the learned Bayesian Network (lines 7-9). In the following we will refer to tree-structured models simply as CLtrees.

CLtrees have been widely employed in AI, both as efficient approximations in density estimation tasks and as the building blocks of more expressive and yet tractable PGMs. A very simple and accurate algorithm to learn mixtures of CLtrees is MT, as presented in [13]. MT learns a distribution of the form: $Q(\mathbf{x}) = \sum_{i=1}^{k} \lambda_i \mathcal{T}_i(\mathbf{x})$, where the tree distributions $\mathcal{T}_i$, are the mixture components and $\lambda_i \geq 0$, such that $\sum_{i=1}^{k} \lambda_i = 1$, are their coefficients. In [13], the best components and weights are the (local) likelihood maxima learned by EM, with k being a parameter fixed in advance.

2.2   Cutset Networks

As introduced in [17], *Cutset Networks* (CNets) are a hybrid of rooted OR trees
and CLtrees, with OR nodes as internal nodes and CLtrees as leaves. Each node
in an OR tree is labeled by a variable X_i, and each edge emanating from it
represents the conditioning of X_i by a value $x_i^j \in Val(X_i)$, weighted by the
probability $w_{i,j}$ of conditioning the variable X_i to the value x_i^j.

Formally, a cutset network is a pair $\langle \mathcal{G}, \gamma \rangle$, where $\mathcal{G} = \mathcal{O} \cup \{\mathcal{T}_1, \dots, \mathcal{T}_L\}$ is
composed of the rooted OR tree, $\mathcal{O}$, plus the leaf CLtrees $\mathcal{T}_l$, and $\gamma = \boldsymbol{w} \cup$
$\{\boldsymbol{\theta}_1, \dots, \boldsymbol{\theta}_L\}$ corresponds to the parameters $\boldsymbol{w}$ of the OR tree and $\boldsymbol{\theta}_l$ of the
CLTrees. The *scope* of a CNet $\mathcal{G}$ (resp. a CLtree $\mathcal{T}_l$), denoted as scope($\mathcal{G}$) (resp.
scope($\mathcal{T}_l$)), is the set of random variables that appear in it. A CNet may be
defined recursively as follows.

Definition 1 (Cutset network). *Given* $\mathbf{X}$ *be a set of discrete variables, a*
Cutset Network *is defined as follows:*

1. *a CLtree, with scope* $\mathbf{X}$*, is a CNet;*
2. *given* $X_i \in \mathbf{X}$ *a variable with* $|Val(X_i)| = k$*, graphically conditioned in an*
   OR node, a weighted disjunction of k *CNets* $\mathcal{G}_i$ *with same scope* $\mathbf{X}_{\setminus i}$ *is a*
   CNet, where all weights $w_{i,j}$, $j = 1, \dots, k$*, sum up to one, and* $\mathbf{X}_{\setminus i}$ *denotes*
   the set $\mathbf{X}$ *minus the variable* X_i*.*

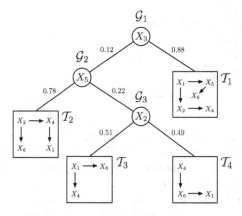

Fig. 1. Example of a binary CNet model. Internal nodes on variables X_i are OR nodes,
while leaf nodes are CLtrees encoding a direct graphical model.

Figure 1 reports a CNet model for binary valued variables, where the internal
nodes denote a conditioning on a variable (i.e., an OR node), while the leaves
correspond to the CLtrees of the model. Note that each node in the model
corresponds to a root of a sub-CNet $\mathcal{G}_i$ or to a CLtree $\mathcal{T}_j$, thus the recursive
definition of a CNnet model.

The log-likelihood function of a CNet may be decomposed as follows.

Proposition 1 (CNet log-likelihood decomposition). *Given a CNet $\langle \mathcal{G}, \boldsymbol{\gamma} \rangle$ over variables $\mathbf{X}$ and a set of instances $\mathcal{D}$, its log-likelihood $\ell_{\mathcal{D}}(\langle \mathcal{G}, \boldsymbol{\gamma} \rangle)$ can be computed as follows:*

$$\ell_{\mathcal{D}}(\langle \mathcal{G}, \boldsymbol{\gamma} \rangle) = \sum_{\xi \in \mathcal{D}} \sum_{i=1,\dots,n} \log P(\xi[X_i]|\xi[\mathrm{Pa}_i]) \tag{2}$$

when $\mathcal{G}$ corresponds to a CLtree. While, in the case of a OR tree rooted on the variable X_i, with $|Val(X_i)| = k$, the log-likelihood is:

$$\ell_{\mathcal{D}}(\langle \mathcal{G}, \boldsymbol{\gamma} \rangle) = \sum_{j=1,\dots,k} M_j \log w_{i,j} + \ell_{\mathcal{D}_j}(\langle \mathcal{G}_j, \boldsymbol{\gamma}_{\mathcal{G}_j} \rangle) \tag{3}$$

where for each $j = 1, \dots, k$, $\mathcal{G}_j$ is the CNet involved in the disjunction with parameters $\boldsymbol{\gamma}_{\mathcal{G}_j}$, and $\mathcal{D}_j$ is a slice of the dataset $\mathcal{D}$ obtained as $\mathcal{D}_j = \{\xi \in \mathcal{D} : \xi[X_i] = x_i^j\}$, $M_j = |\mathcal{D}_j|$ corresponds to the number of instances in $\mathcal{D}_j$, and $\ell_{\mathcal{D}_j}(\langle \mathcal{G}_j, \boldsymbol{\gamma}_{\mathcal{G}_j} \rangle)$ denotes the log-likelihood of the sub-CNet $\mathcal{G}_j$ on the slice $\mathcal{D}_j$.

Proof. For Equation (2):

$$\ell_{\mathcal{D}}(\langle \mathcal{G}, \boldsymbol{\gamma} \rangle) = \sum_{\xi \in \mathcal{D}} \log \prod_{i=1,\dots,n} P(\xi[X_i]|\xi[\mathrm{Pa}_i]) = \sum_{\xi \in \mathcal{D}} \sum_{i=1,\dots,n} \log P(\xi[X_i]|\xi[\mathrm{Pa}_i])$$

While, for Equation (3):

$$\begin{aligned}
\ell_{\mathcal{D}}(\langle \mathcal{G}, \boldsymbol{\gamma} \rangle) &= \sum_{\xi \in \mathcal{D}} \log w_{i,\xi[X_i]} + \log P(\langle \mathcal{G}_{\xi[X_i]}, \boldsymbol{\gamma}_{\mathcal{G}_{\xi[X_i]}} \rangle) \\
&= \sum_{j=1,\dots,k} M_j \log w_{i,j} + \sum_{\xi \in \mathcal{D}_j} \log P(\langle \mathcal{G}_j, \boldsymbol{\gamma}_{\mathcal{G}_j} \rangle) \\
&= \sum_{j=1,\dots,k} M_j \log w_{i,j} + \ell_{\mathcal{D}_j}(\langle \mathcal{G}_j, \boldsymbol{\gamma}_{\mathcal{G}_j} \rangle) \qquad \blacksquare
\end{aligned}$$

Structure Learning. The algorithm to learn CNet structures proposed in [17] performs a greedy top-down search in the OR-trees space. It recursively tries to partition $\mathcal{D}$ into the instance subsets $\{\mathcal{D}_j = \{\xi \in \mathcal{D} : \xi[X_s] = x_s^j\}\}_{j=1}^{|Val(X_s)|}$ over the current scope $\mathbf{X}_{\backslash s}$ by selecting heuristically the best variable X_s maximizing a reformulation of the information gain in a generative context:

$$X_s = \underset{X_i \in \mathbf{X}}{\operatorname{argmax}} \left(\hat{H}_{\mathcal{D}}(\mathbf{X}) - \sum_{x_i^j \in Val(X_i)} \frac{|\mathcal{D}_j|}{|\mathcal{D}|} \hat{H}_{\mathcal{D}_j}(\mathbf{X}_{\backslash i}) \right)$$

where $\hat{H}_{\mathcal{D}_j}(\mathbf{X}) = -\frac{1}{|\mathbf{X}|} \sum_{X_i \in \mathbf{X}} \sum_{x_i^j \in Val(X_i)} P_{\mathcal{D}_j}(x_i^i) \log P_{\mathcal{D}_j}(x_i^j)$ is the average entropy over the single variables in the current scope, limited to the subset $\mathcal{D}_j$, which is introduced as a way to approximate the computation of the joint entropy over the current scope.

Found X_s, the algorithm creates a corresponding inner node C_s whose children will be the nodes $\{C_j\}_{j=1}^{k_s}$ returned by recursive calls on the instance subsets $\{\mathcal{D}_j\}_{j=1}^{k_s}$, with $k_s = |Val(X_s)|$. The weights $\{w_{sj}\}_{j=1}^{k_s}$ are estimated as the proportion of instances falling into each partition. As reported in [17], termination can be achieved when for the current partition $\mathcal{D}$ the number of instances falls under a tunable parameter m, or when the total entropy is less than a threshold λ. In this case a leaf node is added as a CLtree learned on the current instance partition over the current scope according to Algorithm 1. To cope with the risk of overfitting, always in [17], post-pruning based on a validation set is introduced in the form of reduced error pruning [16]. Leveraging this decision tree technique, after a CNet is fully grown, by advancing bottom-up, leaves are pruned and inner nodes without children replaced with a CLtree, if the network validation data likelihood after this operation is higher than that scored by the unpruned network. Following experimental evidence, it appears clear that a search step directly guided by the data likelihood, in this case the pruning stage, is crucial for the accuracy of the learned models; otherwise representing very poor local optima in the terms of likelihood. However, as the same authors report, the additional cost of growing a full network and then traversing it while reevaluating inner nodes, is demanding.

3   dCSN: Decomposability Based CNets Learning

Here, we propose the dCSN algorithm that exploits a different approach from that in [17]: we avoid decision tree heuristics and instead choose the best variable directly maximizing the data log-likelihood. By exploiting the recursive Definition 1, we grow a CNet top-down, allowing further expansion, i.e. the substitution of a CLtree with an OR node, only if it improves the structure log-likelihood, since it is clear to see that maximizing the second term in Equation 3, results in maximizing the global score. In detail, we start with a single CLtree, for variables $\mathbf{X}$, learned from $\mathcal{D}$ and we check whether there is a decomposition, i.e. an OR node applied on as many CLtrees as the values of the best variable X_i, providing a better log-likelihood than that scored by the initial tree. If such a decomposition exists, than the decomposition process is recursively applied to the sub-slices $\mathcal{D}_i$, testing each leaf for a possible substitution. A sketch of the process is shown in Algorithm 2.

In this principled learning framework we do not need to employ post-pruning techniques while we can embed a regularization term in the structure score used in the decomposition process. To penalize complex structures we adopt the Bayesian Information Criterion BIC, we now show how to derive it in our framework and what are its properties. Using another criterion like the BDe [7] could be possible. Following [4], the BIC score of a CNet $\langle \mathcal{G}, \gamma \rangle$ on data $\mathcal{D}$ is defined as: $\mathrm{score}_{\mathrm{BIC}}(\langle \mathcal{G}, \gamma \rangle) = \log P_{\mathcal{D}}(\langle \mathcal{G}, \gamma \rangle) - \frac{\log M}{2}\mathrm{Dim}(\mathcal{G})$, where $\mathrm{Dim}(\mathcal{G})$ is the model dimension, i.e., the number of independent parameters used for the structure representation $\mathcal{G}$, and M is the size of the dataset $\mathcal{D}$. Here, we set $\mathrm{Dim}(\mathcal{G})$ to the number $O_{\mathcal{G}}$ of OR nodes appearing in $\mathcal{G}$.

Given $\mathcal{G}$ and $\mathcal{G}'$ be two CNets, where $\mathcal{G}'$ has been obtained from $\mathcal{G}$ substituting a leaf tree by adding a new sub-CNet rooted in an OR node, then:

$$\text{score}_{\text{BIC}}(\langle\mathcal{G}',\boldsymbol{\gamma}'\rangle) - \text{score}_{\text{BIC}}(\langle\mathcal{G},\boldsymbol{\gamma}\rangle) =$$

$$\ell_{\mathcal{D}}(\langle\mathcal{G}',\boldsymbol{\gamma}'\rangle) - \ell_{\mathcal{D}}(\langle\mathcal{G},\boldsymbol{\gamma}\rangle) - \frac{\log M}{2}(\text{Dim}(\mathcal{G}') - \text{Dim}(\mathcal{G})) =$$

$$\ell_{\mathcal{D}}(\langle\mathcal{G}',\boldsymbol{\gamma}'\rangle) - \ell_{\mathcal{D}}(\langle\mathcal{G},\boldsymbol{\gamma}\rangle) - \frac{\log M}{2}, \qquad (4)$$

since $\text{Dim}(\mathcal{G}') - \text{Dim}(\mathcal{G}) = O_{\mathcal{G}'} - O_{\mathcal{G}} = (O_{\mathcal{G}}+1) - O_{\mathcal{G}} = 1$. Hence, $\mathcal{G}'$ is accepted when $\ell_{\mathcal{D}}(\langle\mathcal{G}',\boldsymbol{\gamma}'\rangle) - \ell_{\mathcal{D}}(\langle\mathcal{G},\boldsymbol{\gamma}\rangle) > \frac{\log M}{2}$. This means that a leaf node may be decomposed (or, a new OR node may be added), when the improvement on the global loglikelihood is greater than $\frac{\log M}{2}$.

The decomposability property of the log-likelihood of a CNet can lead to similar results for the BIC score.

Proposition 2 (CNet BIC score decomposition). *Given a CNet $\langle\mathcal{G},\boldsymbol{\gamma}\rangle$, over variables $\mathbf{X}$ and instances $\mathcal{D}$, made up of $\{T_l\}_{l=1}^{L}$ CLtrees, a decomposition of a tree T_l, having scope $\mathbf{X}_l \subset \mathbf{X}$, with parameters $\boldsymbol{\theta}_l$, with a sub-CNet $\mathcal{G}_i$ rooted in a OR node associated to the variable $X_i \in \mathbf{X}_l$ with parameters $\boldsymbol{\gamma}_i$, leading to a new CNet $\langle\mathcal{G}',\boldsymbol{\gamma}'\rangle$, is accepted iff:*

$$\ell_{\mathcal{D}_l}(\langle\mathcal{G}_i,\boldsymbol{\gamma}_i\rangle) - \ell_{\mathcal{D}_l}(\langle T_l,\boldsymbol{\theta}_l\rangle) > \frac{\log M}{2}. \qquad (5)$$

where $M = |\mathcal{D}|$, and D_l is the subset of D containing only instances associated to the tree T_l.

Proof. Each leaf tree node T_l, $l \in \{1,\ldots,L\}$ is reachable from the root through the path $x_{i_1}^{j_{i_1}}, x_{i_2}^{j_{i_2}}, \ldots, x_{i_P}^{j_{i_P}}$ of length P where $\langle i_1,\ldots,i_P\rangle$ is the sequence of indices for the random variables $\mathbf{X}_{\backslash l} = \mathbf{X} \setminus \mathbf{X}_l$ found in the path. Instances reaching the tree T_l form the set $\mathcal{D}_l = \{\xi \in \mathcal{D} : \xi[X_{i_p}] = x_{i_p}^{j_{i_p}}, \forall p = 1,\ldots,P\}$, that is, for each random variable X_{i_p} in the path, they take the conditioned branch according to their value for X_{i_p}.

If $\ell_{\mathcal{D}_l}(\langle T_l,\boldsymbol{\theta}_l\rangle)$ indicates the *local* log-likelihood of T_l with respect to $\mathcal{D}_i$, then its contribution to the *global* log-likelihood $\ell_{\mathcal{D}}(\langle\mathcal{G},\boldsymbol{\gamma}\rangle)$ corresponds to:

$$\left(M_l \sum_{p=1}^{P} \log w_{i_p,j_{i_p}} \right) + \ell_{\mathcal{D}_l}(\langle T_l,\boldsymbol{\theta}_l\rangle), \qquad (6)$$

where $M_l = |\mathcal{D}_l|$. If we decompose the tree T_l into a sub-CNet $\mathcal{G}_i$, using the CNet log-likelihood decomposition as reported in Equation 3, then the global contribution reported in Equation 6 becomes:

$$\left(M_l \sum_{p=1}^{P} \log w_{i_p,j_{i_p}} \right) + \ell_{\mathcal{D}_l}(\langle\mathcal{G}_i,\boldsymbol{\gamma}_i\rangle), \qquad (7)$$

We have that $\text{score}_{\text{BIC}}(\langle\mathcal{G}',\boldsymbol{\gamma}'\rangle) - \text{score}_{\text{BIC}}(\langle\mathcal{G},\boldsymbol{\gamma}\rangle) = \ell_{\mathcal{D}_l}(\langle\mathcal{G}_i,\boldsymbol{\gamma}_i\rangle) - \ell_{\mathcal{D}_l}(\langle T_l,\boldsymbol{\theta}_l\rangle) - \frac{\log M}{2}$. ∎

Again, instead of recomputing the likelihood on the complete dataset $\mathcal{D}$, due to the decomposability of the likelihood, we can evaluate only the local improvement. Moreover, the decomposition of T_l is independent from all other $T_j, j \neq l$ being their local contributions to the global log-likelihood independent. Hence, it is not significant the order we choose to decompose leaf nodes.

Bayesian Parameter Smoothing. As regards the learning of the CLtrees parameters we adopted a Bayesian approach. To learn the structure of a CLtree T from data $\mathcal{D}$ with parameters $\boldsymbol{\theta}$, the Bayesian approach employs as a scoring function the posterior probability of the graph given the data: $P(\boldsymbol{\theta}|\mathcal{D}) \approx P(\mathcal{D}|\boldsymbol{\theta})P(\boldsymbol{\theta})$. The marginal $P(\mathcal{D}|\boldsymbol{\theta})$ can be expressed in closed form when using the Dirichlet prior over the model parameters $\boldsymbol{\theta}_{X_i|\mathrm{Pa}_i}$, the only distribution that ensures likelihood equivalence, i.e., the hyper-parameters $\alpha_{X_i|\mathrm{Pa}_i}$ of the Dirichlet prior can be expressed as $\alpha_{X_i,\mathrm{Pa}_i} = \alpha q_{X_i,\mathrm{Pa}_i}$, where q is a prior distribution over X, and α, the so called *equivalent sample size* (ESS), is a positive constant independent of i. In this Bayesian approach with the Dirichlet prior, the regularized parameter estimates are:

$$\theta_{x_i|\mathrm{Pa}_i} \approx E_{P(\theta_{x_i|\mathrm{Pa}_i}|\mathcal{D},T)}[\theta_{x_i|\mathrm{Pa}_i}] = \frac{M_{x_i,\mathrm{Pa}_i} + \alpha_{x_i|\mathrm{Pa}_i}}{M_{\mathrm{Pa}_i} + \alpha_{\mathrm{Pa}_i}}, \qquad (8)$$

where $M_{\mathbf{z}}$ is the number of entries in a dataset $\mathcal{D}_{\mathbf{z}}$ having the set of variables $\mathbf{Z}$ instantiated to $\mathbf{z}$. As pointed out in [4], we can use a different Dirichlet prior for each distribution of X_i given a particular value of its parents, leading to choose the regularized parameter estimates as:

$$\hat{\boldsymbol{\theta}}_{X_i|\mathrm{Pa}_i} = \frac{M \cdot P(\mathrm{Pa}_i)P(X_i|\mathrm{Pa}_i)}{M \cdot P(\mathrm{Pa}_i) + \alpha_{X_i|\mathrm{Pa}_i}} + \frac{\alpha_{X_i|\mathrm{Pa}_i}\boldsymbol{\theta}^0(X_i|\mathrm{Pa}_i)}{M \cdot P(\mathrm{Pa}_i) + \alpha_{X_i|\mathrm{Pa}_i}},$$

where $\boldsymbol{\theta}^0(X_i|\mathrm{Pa}_i)$ is the prior estimate of $P(X_i|\mathrm{Pa}_i)$ and $\alpha_{X_i|\mathrm{Pa}_i}$ is the confidence associated with that prior.

In the case of uniform priors, the estimates correspond to the additive of Laplace smoothing. A reasonable choice uses the marginal probability of X_i in the data as the prior probability. This choice is based on the assumption that most conditional probabilities are close to the observed marginal. Thus, we can set $\boldsymbol{\theta}^0(X_i|\mathrm{Pa}_i) = P_D(X_i)$. With fixed $\alpha_{x_i|\mathrm{Pa}_i} = \alpha$, we have: $\hat{\theta}_{X_i|\mathrm{Pa}_i} = \frac{M_{x_i,\mathrm{Pa}_i} + \alpha P_D(X_i)}{M_{\mathrm{Pa}_i} + \alpha}$.

Algorithm 2 reports the pseudocode of dCSN. The dCSN algorithm starts by learning a single CLTree on the whole dataset $\mathcal{D}$ (line 4), and then calls the decomposition procedure on this tree (line 6). The input parameters δ and σ are used for regularization in order to avoid overfitting. σ, resp. δ, is the minimum number of instances, resp. of features, in a slice required to try a decomposition.

Given a CLtree, Algorithm 3 tries to decompose it in a sub-CNet. The aim of dCSN is to attempt to extend the model by replacing one of the CLtree leaf nodes with a new CNet on the same variables. In particular, the **decompose** procedure checks for each variable X_i on the slice $\mathcal{D}$ (line 5), whether the OR decomposition

---

Algorithm 2. $\mathsf{dCSN}(\mathcal{D}, \mathbf{X}, \alpha_f, \delta, \sigma)$

1: **Input:** a set of instances $\mathcal{D}$ over a set of features $\mathbf{X}$; $\alpha_f \in [0,1]$: ESS factor; δ minimum number of instances to decompose, σ minimum number of features to decompose

2: **Output:** a CNet $\langle \mathcal{G}, \boldsymbol{\gamma} \rangle$ encoding a pdf over $\mathbf{X}$ learned from $\mathcal{D}$

3: $\alpha \leftarrow \alpha_f |\mathcal{D}|$

4: $\langle \mathcal{T}, \boldsymbol{\theta} \rangle \leftarrow \mathsf{LearnCLTree}(\mathcal{D}, \mathbf{X}, \alpha)$

5: $\boldsymbol{w} \leftarrow \emptyset$

6: $\langle \mathcal{G}, \boldsymbol{\gamma} \rangle \leftarrow \mathsf{decompose}(\mathcal{D}, \mathbf{X}, \alpha, \mathcal{T}, \boldsymbol{\theta}, \boldsymbol{w}, \delta, \sigma)$

---

associated to that variable (a new CNet) has a log-likelihood better than that of the input CLtree (line 16). If a better decomposition is found, it then recursively (line 21) tries to decompose the sub-CLtrees of the newly introduced CNet. In dCSN α is set to $\alpha_f |\mathcal{D}|$, where $\alpha_f \in [0,1]$ is an input parameter. When we proceed with the decomposition on the slices, α is proportionally reduced, in the procedure decompose, to the number of instances in the slices. In particular, if we initially assume that there are $\alpha = \alpha_f |\mathcal{D}|$ fictitious instances for computing the priors, then we should assume that a proportion $\alpha |\mathcal{D}_i| / |\mathcal{D}|$ falls into the slice $\mathcal{D}_i$, in order to make the priors in $\mathcal{D}_i$ consistent with those in $\mathcal{D}$.

Algorithm 3. $\mathsf{decompose}(\mathcal{D}, \mathbf{X}, \alpha, \mathcal{T}, \boldsymbol{\theta}, \boldsymbol{w}, \delta, \sigma)$

1: **Input:** a set of instances $\mathcal{D}$ over a set of features $\mathbf{X}$; α: ESS; $\mathcal{T}$: the tree structured model to decompose and its parameters $\boldsymbol{\theta}$; δ minimum number of instances to decompose, σ minimum number of features to decompose

2: **Output:** a CNet encoding a pdf over $\mathbf{X}$ learned from $\mathcal{D}$

3: **if** $|\mathcal{D}| > \delta$ and $|\mathbf{X}| > \sigma$ **then**

4:     $\ell_{\mathsf{best}} \leftarrow -\infty$

5:     **for** $X_i \in \mathbf{X}$ **do**

6:         $\mathcal{G}_i \leftarrow \emptyset, \boldsymbol{w}_i \leftarrow \emptyset, \boldsymbol{\theta}_i \leftarrow \emptyset, C_i$ is the OR Node associated to X_i

7:         **for** $x_i^j \in Val(X_i)$ **do**

8:             $\mathcal{D}_j \leftarrow \{\xi \in \mathcal{D} : \xi[X_s] = x_s^j\}$

9:             $w_{ij} \leftarrow |\mathcal{D}_j| / |\mathcal{D}|$

10:            $\langle \mathcal{T}_j, \boldsymbol{\theta}_{ij} \rangle \leftarrow \mathsf{LearnCLTree}(\mathcal{D}_j, \mathbf{X}_{\backslash s}, \alpha w_{ij})$

11:            $\mathcal{G}_i \leftarrow \mathsf{addSubTree}(C_i, \mathcal{T}_j)$

12:         $\boldsymbol{w}_i \leftarrow \boldsymbol{w}_i \cup \{w_{ij}\}, \boldsymbol{\theta}_i \leftarrow \boldsymbol{\theta}_i \cup \{\theta_{ij}\}$

13:         $\ell_i \leftarrow \ell_{\mathcal{D}_i}(\langle \mathcal{G}_i, \boldsymbol{w}_i \cup \boldsymbol{\theta}_i \rangle)$

14:         **if** $\ell_i > \ell_{\mathsf{best}}$ and $\ell_i > \ell_{\mathcal{D}_i}(\langle \mathcal{T}, \boldsymbol{\theta} \rangle)$ **then**

15:            $\ell_{\mathsf{best}} \leftarrow \ell_i, X_{\mathsf{best}} \leftarrow X_i, \mathcal{G}_{best} \leftarrow \mathcal{G}_i, \boldsymbol{\theta}_{best} \leftarrow \boldsymbol{\theta}_i, \boldsymbol{w}_{best} \leftarrow \boldsymbol{w}_i$

16:     **if** $\ell_{\mathsf{best}} - \ell_{\mathcal{D}}(\langle \mathcal{T}, \boldsymbol{\theta} \rangle) > (log|\mathcal{D}|)/2$ **then**

17:         substitute $\mathcal{T}$ with $\mathcal{G}_i$

18:         $\boldsymbol{w} \leftarrow \boldsymbol{w} \cup \boldsymbol{w}_{best}$

19:         **for** $x_b^j \in Val(X_{\mathsf{best}})$ **do**

20:            $\mathcal{D}_j \leftarrow \{\xi \in \mathcal{D} : \xi[X_{\mathsf{best}}] = x_b^j\}$

21:            $\mathsf{decompose}(\mathcal{D}_j, \mathbf{X}_{\backslash \mathsf{best}}, \alpha w_{ij}, \mathcal{T}_j, \boldsymbol{\theta}_j, \boldsymbol{w}, \delta, \sigma)$

Table 1. Datasets used and their number of features and instances.

	$|\mathbf{X}|$	$|T_{train}|$	$|T_{val}|$	$|T_{test}|$		$|\mathbf{X}|$	$|T_{train}|$	$|T_{val}|$	$|T_{test}|$
NLTCS	16	16181	2157	3236	DNA	180	1600	400	1186
MSNBC	17	291326	38843	58265	Kosarek	190	33375	4450	6675
Plants	69	17412	2321	3482	MSWeb	294	29441	3270	5000
Audio	100	15000	2000	3000	Book	500	8700	1159	1739
Jester	100	9000	1000	4116	EachMovie	500	4525	1002	591
Netflix	100	15000	2000	3000	WebKB	839	2803	558	838
Accidents	111	12758	1700	2551	Reuters-52	889	6532	1028	1540
Retail	135	22041	2938	4408	BBC	1058	1670	225	330
Pumsb-star	163	12262	1635	2452	Ad	1556	2461	327	491

Finally, in order to improve the accuracy of the CNet models we adopted a bagging procedure in order to obtain a mixture of CNets. We draw k bootstrapped samples $\mathcal{D}_i$ from the dataset $\mathcal{D}$, sampling $|\mathcal{D}|$ instances with replacements, and on each of those we call dCSN, thus leading to k CNets $\mathcal{G}_i$. The resulting bagged CNet $\mathcal{G}$ corresponds to a weighted sum of all the learned CNets $\mathcal{G}_i$. We set the weights proportional to the likelihood score obtained by each bootstrapped component. In particular, for each instance $\xi \in \mathcal{D}$, the bagged CNet $\mathcal{G}$ would result in the more robust estimation $\hat{P}(\xi : \mathcal{G}) = \sum_{i=1}^{k} \mu_i P(\xi : \mathcal{G}_i)$, where $\mu_i = \ell_{\mathcal{D}}(\langle \mathcal{G}_i, \gamma_i \rangle) / \sum_{j=1}^{k} \ell_{\mathcal{D}}(\langle \mathcal{G}_j, \gamma_j \rangle)$.

4 Experiments

Since the CNet, and the variant CNetP embedding the pruning on validation, as reported in [17] are not publicly available, we implemented them as well as our dCSN and its bagging dCSN-B variant in Python[1]. We were not able to reproduce the results of the mixtures learned with EM as showed in [17], therefore we will just report them (as MCNet). To make the comparison fair in testing the mixture accuracies, we also extended CNet and CNetP by embedding mixtures by bagging, leading to versions CNet-B and CNetP-B respectively. We introduce MT as the last competitor as it is reported to be one of the most competitive tractable PGM [17,18]; for it we used the implementation available in the Libra toolkit [9].

We evaluated the proposed algorithms on an array of 18 datasets that are now standard benchmarks for graphical model structure learners. They have been introduced in [10] and [5] as binarized versions of datasets from different tasks like frequent itemset mining, recommendation and classification. Their names and statistics for their training, validation, test splits are reported in Table 1.

We run both CNet and CNetP with $m = 10$ and $\lambda = 0.01$ fixed to exactly reproduce the original experiments in [17]. We run CNet-B and CNetP-B by learning a number of components k ranging from 5 to 40, with a step of 5.

For dCSN we run a grid search in the space formed by $\alpha_f \in \{.01, .02, .03, .04,$ $.05, .06, .08, .1, .15, .2, .3, .4, .5\}$ and $\delta \in \{200, 300, 400, 500\}$; for dCSN-B we set

---

[1] Source code is available at http://www.di.uniba.it/~ndm/dcsn/.

Table 2. Empirical risk for all algorithms.

	CNet	CNetP	dCSN	CNet-B	CNetP-B	dCSN-B	MT	MCNet
NLTCS	-6.11	-6.06	**-6.04**	-6.09	-6.02	-6.02	**-6.01**	-6.00
MSNBC	-6.06	**-6.05**	-6.05	-6.06	**-6.04**	-6.04	-6.08	-6.04
Plants	**-13.24**	-13.25	-13.35	-12.30	-12.38	**-12.21**	-12.93	-12.78
Audio	-44.58	**-42.05**	-42.06	-42.09	-40.71	**-40.17**	-40.14	-39.73
Jester	-61.71	**-55.56**	**-55.30**	-57.76	-53.17	**-52.99**	-53.06	-52.57
Netflix	-65.61	-58.71	**-58.57**	-63.08	-57.63	**-56.63**	-56.71	-56.32
Accidents	-30.97	-30.69	**-30.17**	-30.25	-30.28	**-28.99**	-29.69	-29.96
Retail	-11.07	**-10.94**	-11.00	-10.99	-10.88	-10.87	**-10.84**	-10.82
Pumsb-star	-24.65	-24.42	**-23.83**	-24.39	-24.19	**-23.32**	-23.70	-24.18
DNA	-90.48	-87.59	**-87.19**	-90.66	-86.85	**-84.93**	**-85.57**	-85.82
Kosarek	-11.19	**-11.04**	-11.14	-10.97	-10.85	-10.85	**-10.62**	-10.58
MSWeb	-10.07	-10.07	**-9.94**	-9.95	-9.91	-9.86	**-9.82**	-9.79
Book	-37.62	-37.35	**-37.22**	-35.88	-35.62	-35.92	**-34.69**	-33.96
EachMovie	**-59.19**	-58.37	-58.47	-54.22	-54.02	**-53.91**	-54.51	-51.39
WebKB	-162.85	-162.17	**-161.16**	-156.79	-156.94	**-155.20**	-157.00	-153.22
Reuters-52	**-88.72**	-88.55	-88.60	-86.22	-86.89	**-85.69**	-86.53	-86.11
BBC	**-262.08**	-263.08	-262.08	-252.01	-257.72	**-251.14**	-259.96	-250.58
Ad	-16.92	-16.92	**-14.81**	-15.94	-16.02	**-13.73**	-16.01	-16.68

instead $\alpha_f \in \{.05, .1\}$ and $\delta \in \{100, 200, 300, 400, 500, 1000\}$, running the algorithm for a number of components k ranging from 5 to 40, with a step of 5. For both dCSN and dCSN-B we fixed $\sigma = 3$. For MT we reproduced the experiment in [18], setting k from 2 to 30 by steps of 2. For all mixture variants, for each mixture configuration, we selected the best one based on the validation likelihood score.

In Table 2 is reported the empirical risk, defined as $1/|\mathcal{D}| \sum_{\xi \in \mathcal{D}} \log P(\xi|\mathcal{G}, \gamma)$ averaged over the set of test instances for all the experiments over the listed datasets. We provide in last column the original scores of MCNet as reported in [17] as a reference. For all the implemented versions we run a pairwise Wilcoxon signed rank test to assess the statistical significance of the scores. In bold are reported the best values, compared to all others, for each dataset. As we can see dCSN is significantly better than CNet and CNetP on 8 datasets, and significantly worse than CNet and CNetP on 1 and 3 datasets, respectively. Considering the bagging version for the mixtures, we see that dCSN-B is significantly better than CNet-B, CNetP-B and MT on 11, 11, and 10 datasets, and significantly worse on 1, 1, and 5 datasets, respectively.

5 Conclusions

Here we proposed a new approach to learn the structure of the recently introduced CNets model. We exploited the decomposable score of CNets to learn their structure and parameters by directly maximizing the likelihood, formulating a score including the BIC criterion and by introducing informative priors on smoothing parameters. Moreover, we presented how to create mixtures of CNets by adopting the bagging method as an alternative to EM. We compared our algorithm against the original variants on a large set of standard benchmarks proving the validity of our claims.

Acknowledgments. This work has been partially founded by the PON02 00563 3489339 project PUGLIA@SERVICE financed by the Italian Ministry of University and Research (MIUR).

References

1. Bach, F.R., Jordan, M.I.: Thin junction trees. In: Advances in Neural Information Processing Systems, vol. 14, pp. 569–576. MIT Press (2001)
2. Choi, M.J., Tan, V.Y.F., Anandkumar, A., Willsky, A.S.: Learning latent tree graphical models. Journal of Machine Learning Research **12**, 1771–1812 (2011)
3. Chow, C., Liu, C.: Approximating discrete probability distributions with dependence trees. IEEE Transactions on Information Theory **14**(3), 462–467 (1968)
4. Friedman, N., Geiger, D., Goldszmidt, M.: Bayesian network classifiers. Machine Learning **29**(2–3), 131–163 (1997)
5. Haaren, J.V., Davis, J.: Markov network structure learning: a randomized feature generation approach. In: Proceedings of the 26th Conference on Artificial Intelligence. AAAI Press (2012)
6. Hastie, T., Tibshirani, R., Friedman, J.: The Elements of Statistical Learning. Springer (2009)
7. Heckerman, D., Geiger, D., Chickering, D.: Learning bayesian networks: the combination of knowledge and statistical data. Machine Learning **20**, 197–243 (1995)
8. Koller, D., Friedman, N.: Probabilistic Graphical Models: Principles and Techniques. MIT Press (2009)
9. Lowd, D., Rooshenas, A.: The Libra Toolkit for Probabilistic Models. CoRR abs/1504.00110 (2015)
10. Lowd, D., Davis, J.: Learning Markov network structure with decision trees. In: Proceedings of the 10th IEEE International Conference on Data Mining, pp. 334–343. IEEE Computer Society Press (2010)
11. Lowd, D., Domingos, P.: Learning arithmetic circuits. CoRR abs/1206.3271 (2012)
12. Lowd, D., Rooshenas, A.: Learning Markov networks with arithmetic circuits. In: JMLR Workshop Proceedings of the 16th International Conference on Artificial Intelligence and Statistics, vol. 31, pp. 406–414 (2013)
13. Meilă, M., Jordan, M.I.: Learning with mixtures of trees. Journal of Machine Learning Research **1**, 1–48 (2000)
14. Pearl, J.: Probabilistic Reasoning in Intelligent Systems: Networks of Plausible Inference. Morgan Kaufmann Publishers Inc., San Francisco (1988)
15. Poon, H., Domingos, P.: Sum-product network: a new deep architecture. In: NIPS 2010 Workshop on Deep Learning and Unsupervised Feature Learning (2011)
16. Quinlan, J.R.: Induction of decision trees. Machine Learning Journal **1**(1), 81–106 (1986)
17. Rahman, T., Kothalkar, P., Gogate, V.: Cutset networks: a simple, tractable, and scalable approach for improving the accuracy of chow-liu trees. In: Calders, T., Esposito, F., Hüllermeier, E., Meo, R. (eds.) ECML PKDD 2014, Part II. LNCS, vol. 8725, pp. 630–645. Springer, Heidelberg (2014)
18. Rooshenas, A., Lowd, D.: Learning sum-product networks with direct and indirect variable interactions. In: JMLR Workshop and Conference Proceedings of the 31st International Conference on Machine Learning, pp. 710–718 (2014)
19. Roth, D.: On the hardness of approximate reasoning. Artificial Intelligence **82**(1–2), 273–302 (1996)

Common-Sense Knowledge for Natural Language Understanding: Experiments in Unsupervised and Supervised Settings

Luigi Di Caro[✉], Alice Ruggeri, Loredana Cupi, and Guido Boella

Department of Computer Science, University of Turin,
Corso Svizzera 185, Torino, Italy
{dicaro,ruggeri,cupi,boella}@di.unito.it

Abstract. Research in Computational Linguistics (CL) has been grow-ing rapidly in recent years in terms of novel scientific challenges and commercial application opportunities. This is due to the fact that a very large part of the Web content is textual and written in many languages. A part from linguistic resources (e.g., WordNet), the research trend is mov-ing towards the automatic extraction of semantic information from large corpora to support on-line understanding of textual data. An example of direct outcome is represented by common-sense semantic resources. The main example is ConceptNet, the final result of the Open Mind Common Sense project developed by MIT, which collected unstructured common-sense knowledge by asking people to contribute over the Web. In spite of being promising for its size and broad semantic coverage, few applications appeared in the literature so far, due to a number of issues such as inconsistency and sparseness. In this paper, we present the results of the application of this type of knowledge in two different (supervised and unsupervised) scenarios: the computation of semantic similarity (the keystone of most Computational Linguistics tasks), and the automatic identification of word meanings (Word Sense Induction) in simple syntactic structures.

1 Introduction

Recent Computational Linguistics advances are fully oriented towards the auto-matic extraction of semantic information through big and multilingual data anal-yses, since semantics help tasks such as disambiguation, summarization, entail-ment, question answering, and so forth. This explains the fortunate and growing area of semantic resources, often constructed with automatic approaches, when manual building of ontologies is not feasible on large scale. Semantic informa-tion extraction is currently approached by distributional analysis of linguistic items over specific contexts [1] or by starting from seeds and patterns to build ontologies from scratch [2]. In some cases, linguistic items are substituted by super-senses (i.e., top-level hypernyms) [3].

In recent years, the need and the opportunity of automatically extracting semantic information has been answered by Big Data. This led to the construc-tion of very large semantic resources, such as ConceptNet [4], i.e., a semantic

© Springer International Publishing Switzerland 2015
M. Gavanelli et al. (Eds.): AI*IA 2015, LNAI 9336, pp. 233–245, 2015.
DOI: 10.1007/978-3-319-24309-2_18

graph that has been directly created from collection of unstructured common-sense knowledge asked to people contributing over the Web. In contrast with linguistic resources such as WordNet [5], ConceptNet contains semantic information that are more related to common-sense facts. For this reason, it has a wider spectrum of semantic relationships though a much more sparse coverage. For instance, among the more unusual types of relationships (24 in total), it contains semantic relations like "*ObstructedBy*" (i.e., referring to what would prevent it from happening), " and *CausesDesire*" (i.e., what does it make you want to do). In addition, it also has classic relationships such as "*is_a*" and "*part_of*" as in other linguistic resources.

ConceptNet is a resource based on common-sense rather than linguistic knowledge and it contains much more function-based information (e.g., all the actions a concept can be associated with) contained in even complex syntactic structures. While it has been recognized as a very promising type of knowledge for many computational tasks, it is not significantly used yet due to its complexity. More in detail, ConceptNet has the following main problems:

1. *Specificity.* It contains very specific semantic information (e.g., $< knowledge - CapableOf - openhumanmind >$) that are difficult to integrate in automated tasks;
2. *Completeness.* It is not complete (due to the methodology used to build it), since semantic features are arbitrarly associated to only few of all the possible relevant concepts (e.g., ConceptNet contains $< jazz-IsA-styleofmusic >$) but not $< rock - IsA - styleofmusic >$);
3. *Correctness.* It contains pragmatics statements which are not semantically correct, e.g., $< cat - Antonym - dog >$;
4. *Relativity.* It has semantic features such as $< dog - HasProperty - small >$, which is not always true.

This paper presents an application of ConceptNet (as common-sense knowledge) in two different scenarios: the computation of similarity scores at word-level (one of the key task in Computational Linguistics) and the identification of the different meanings that can be associated with words in sentences. Although the tasks are of different type (supervised and unsupervised respectively), they are based on the same idea of replacing words with ConceptNet semantic information, to measure whether this knowledge has the potential to serve computational tasks even without any particular approach for data alignment, noise removal, semantic information propagation, etc.

2   Background and Related Work

Word Sense Disambiguation (WSD) [6] is maybe the most crucial Natural Language Processing task, since its aim is to capture the correct meaning to be associated to a word in a specific context. This allows to interpret the correct sense of a word, in order to understand how similar is with respect to other words. This permits a set of comparisions between texts, which is useful for many computational tasks such as Information Retrieval, Named Entity Recognition, Question Answering, and so forth. Generally speaking, systems are usually asked to compute similarity scores between pieces of texts at different granularity (word, sentence, discourse) [7].

In order to evaluate the similarity between texts, semantic resources are often used to consider a larger semantic basis to make more accurate comparisons. While linguistic resources such as WordNet and VerbNet constitute a highly-precise source of information, they often cover very few semantic relations usually focusing on taxonomic relations. Common-sense knowledge, on the other side, represents a much larger set of semantic features, which, however, is affected by noise and lack of completness.

If we consider the objects, agents and actions as *terms* in text sentences, we can try to extract their meaning and semantic constraints by using the idea of *affordances* [8]. The affordances of an object can be seen as the set of functionalities that it naturally communicates to the agents through its shape, size, and other phisical characteristics.

For instance, let us think to the sentence *"The squirrel climbs the tree"*. In this case, we need to know what kind of subject *"squirrel"* is to figure out (and visually imagine) how the action will be performed. Let us now consider the sentence *"The elephant climbs the tree"*. Even if there is no change in the grammatical structure of the sentence, the agent of the action creates some semantic problem. In fact, in order to climb a tree, the subject needs to fit to our mental model of *"something that can climb a tree"*. In addition, this also depends on the mental model of *"tree"*. Moreover, different agents can be both correct subjects of an action whilst they may produce different meanings in terms of how the action can be performed. A study of these language dynamics can be of help for many NLP tasks, e.g., Part-Of-Speech tagging as well as more complex operations such as dependency parsing and semantic relation extraction. Some of these concepts are latently studied in different disciplines related to statistics. Distributional Semantics (DS) [9] represents a class of statistical and linguistic analysis of text corpora that tries to estimate the validity of connections between subjects, verbs, and objects by means of statistical sources of significance.

3   A Large Common-sense Knowledge: ConceptNet

The Open Mind Common Sense[1] project developed by MIT collected unstructured common-sense knowledge by asking people to contribute over the Web. In this paper, we started focusing on ConceptNet [4], that is a semantic graph that has been directly created from it. In contrast with linguistic resources like WordNet [5], ConceptNet contains semantic information that are more related to common-sense facts.

---

[1] http://commons.media.mit.edu/

Table 1. Some of the existing relations in ConceptNet, with example sentences in English.

Relation	Example sentence
IsA	NP is a kind of NP.
LocatedNear	You are likely to find NP near NP.
UsedFor	NP is used for VP.
DefinedAs	NP is defined as NP.
HasA	NP has NP.
HasProperty	NP is AP.
CapableOf	NP can VP.
ReceivesAction	NP can be VP.
HasPrerequisite	NP—VP requires NP—VP.
MotivatedByGoal	You would VP because you want VP.
MadeOf	NP is made of NP.
...	...

For this reason, it has a wider spectrum of semantic relationships though a much more sparse coverage. For instance, among the more unusual types of relationships (24 in total), it contains information like "*ObstructedBy*" (i.e., referring to what would prevent it from happening), "and *CausesDesire*" (i.e., what does it make you want to do). In addition, it also has classic relationships like "*is_a*" and "*part_of*" as in most linguistic resources (see Table 1). ConceptNet is a resource based on common-sense rather than linguistic knowledge since it contains much more function-based information (e.g., all the actions a concept can be associated with) contained in even complex syntactic structures.

4   Supervised Experiment: Semantic Similarity

From a computational perspective, being words ambiguous, the disambiguation process (i.e., Word Sense Disambiguation) is one of the most studied tasks in Computational Linguistics. To make an example, the term *count* can mean many things like *nobleman* or *sum*. Using contextual information, it is often possible to make a choice. Again, this choice is done by means of comparisons among contexts, that are still made of words. In other terms, we may state that the *computational* part of almost all computational linguistics research is about the calculus of matching scores between linguistic items, i.e., *words similarity*.

4.1   Description of the Experiment

The experiment starts from the transformation of a word-word-score similarity dataset into a context-based dataset in which the words are replaced by sets of semantic information taken from ConceptNet. The aim is to understand if common-sense knowledge may represent a useful basis for capturing the similarity between words, and if it may outperform systems based on linguistic resources such as WordNet.

4.2   Data

We used the dataset SimLex-999 [10] that contains one thousand word pairs that were manually annotated with similarity scores. The inter-annotation agreement is 0.67 (Spearman correlation). We leveraged ConceptNet to retrieve the semantic information associated to the words of each pair, then keeping the simple intersection[2]. Then, we applied the same approach with the semantic information in WordNet. Note that we did not make any disambiguation of the words, so we used all the semantic data of all the possible senses of the words contained in the similarity dataset.

4.3   Running Example

Let us consider the pair *rice-bean*. ConceptNet returns the following set of semantic information for the term *rice*:

[hasproperty-edible, isa-starch, memberof-oryza, atlocation-refrigerator, usedfor-survival, atlocation-atgrocerystore, isa-food, isa-domesticate-plant, relatedto-grain, madeof-sake, isa-grain, isa-traditionally, receivesaction-eatfromdish, isa-often, receivesaction-cook, relatedto-kimchi, atlocation-pantry, atlocation-ricecrisp, relatedto-sidedish, atlocation-supermarket, receivesaction-stir, isa-staplefoodinorient, hasproperty-cookbeforeitbeeat, madeof-ricegrain, partof-cultivaterice, receivesaction-eat, derivedfrom-rice, isa-cereal, relatedto-white, hasproperty-white, hascontext-cook, relatedto-whitegrain, relatedto-food]

Then, the semantic information for the word *bean* are:

[usedfor-fillbeanbagchair, atlocation-infield, atlocation-can, usedfor-nutrition, usedfor-cook, atlocation-atgrocerystore, memberof-leguminosae, usedfor-makefurniture, usedfor-grow, atlocation-foodstore, isa-legume, usedfor-count, isa-domesticateplant, hasproperty-easytogrow, partof-bean, atlocation-cookpot, isa-vegetableorperson-brain, atlocation-beansoup, atlocation-soup, atlocation-pantry, usedfor-plant, isa-vegetable, atlocation-container, usedfor-supplyprotein, atlocation-jar, usedfor-useasmarker, atlocation-field, derivedfrom-beanball, usedfor-shootfrompeashooter, atlocation-coffee, usedfor-fillbag, receivesaction-grindinthis, usedfor-beanandgarlicsauce, atlocation-beancan, usedfor-makebeanbag, usedfor-eat]

Finally, the intersection produces the following set (*semantic intersection*):

[atlocation-atgrocerystore, isa-domesticateplant, atlocation-pantry]

Then, for each non-empty intersection, we created one instance of the type:

---

[2] In future works we will study more advanced types of matching.

<*semantic intersection*>, <*similarity score*>

and compute a standard *term-document* matrix, where the *term* is a semantic term within the set of semantic intersections and the *document* dimension represents the instances (i.e., the word pairs) of the original dataset. After this preprocessing phase, the *score* attribute is discretized into two bins:

- *non-similar* class - range in the dataset [0, 5]
- *similar* class - range in the dataset [5.1, 10]

4.4 Results

The splitting of the data into two clusters allowed us to experiment a classic supervised classification system, where a Machine Learning tool (a Support Vector Machine, in our case) has been used to learn a binary model for automatically classifying *similar* and *non-similar* word pairs. The result of the experiment is shown in Table 2. Noticeably, the classifier based on ConceptNet data has been able to reach a quite good accuracy (65.38% of correctly classified word pairs), considering that the inter-annotation agreement of the original data is only 0.67 (Spearman correlation). The use of WordNet produced a total F-measure of 0.582 (0.723 for the *non-similar* class and only 0.143 for the *similar* class). This demonstrates the potentiality of the semantic information contained in ConceptNet even if not structured and disambiguated as in WordNet.

Table 2. Classification results in terms of Precision, Recall, and F-measure with ConceptNet (CN) and WordNet (WN). With the use of WordNet, the system achieved 0.582 of total F-measure (0.723 for the non-similar class and only 0.143 for the similar class), while ConceptNet carried to higher accuracy levels.

Class	Prec. (CN)	Recall (CN)	F (CN)	Prec. (WN)	Recall (WN)	F (WN)
non-similar	0,697	0,475	0,565	0,574	0,978	0,723
similar	0,633	0,815	0,713	0,745	0,079	0,143
weighted total	**0,664**	**0,654**	**0,643**	0,649	0,582	0,467

Note that with ConceptNet, *similar* word pairs are generally easier to identify with respect to *non-similar* ones. On the other side, WordNet resulted to be not sufficient to generalize over *similar* word pairs.

5   Unsupervised Experiment: Word Sense Induction

In this section, we present an approach for automatic inducing senses (or meanings) related to the use of short linguistic constructions of the type *subject-verb-object*. As in the supervised experiment, we used the same approach, that is to replace words with the semantic information obtained from a semantic resource such as ConceptNet. The previous experiment demonstrated that the use of common-sense knowledge in this approach may outperform standard resources such as WordNet.

Instead of focusing on single words (as in the supevised setting), we experimented on a simple sentence-level task. The reason is twofold: 1) subject-verb-object are easily minable with very simple NLP parsers, and 2) they represent complete meanings (entire scenes, or actions). A *sense* is intended as in the well-known tasks named Word Sense Disambiguation (WSD) [6] and Word Sense Induction (WSI) [11], i.e., the meaning that a word assumes in a specific context. While WSD focuses on the classification of the meaning of a word among a given set of possible senses, WSI automatically finds senses as *clusters* of different contexts in which a word appears. Obviously, WSI systems are more complex to evaluate (as all clustering methods in general), eventhough [12] proposed a *pseudo-word* evaluation mechanism which is able to simulate the classic WSD scenario.

5.1   Description of the Experiment

The experiment is composed by three different phases: (1) the data pre-processing step with the generation of two transactional databases (transactions of items, as in the fields of Frequent Itemset Mining and Association Rules [13]) that we also call *semantic itemsets*; (2) the extraction of frequent, closed, and *diverse* itemsets (we will briefly introduce the meaning of all these names in the next paragraphs); and finally (3) the creation of *semantic verb models*, that generalize and automatically induce *senses* from entire linguistic constructions at sentence-level.

For a specific input verb v, we parse all the subject-verb-object (SVO) triples in the dataset[3] that have a higher frequency than a set threshold[4], taking into consideration morphological variations of v. Then, for each SVO triple, we replaced the subject-term and the object-term with the relative semantic features contained in ConceptNet. Table 3 shows an example of the information collected in this phase. Then, we associate each semantic information to a unique *id* and construct two transactional databases: one for the semantic information of the subjects, and one for the objects.

Once the transactional databases are built for a specific verb "v", we use techniques belonging to the field of Frequent Itemset Mining to extract *frequent patterns*, i.e, semantic features that frequently co-occur in our transactional databases[5].

This is done for both the transactional databases (subject and object databases associated to the verb 'v'). Since our aim is to capture all the linguistic *senses*, i.e., the different meanings connectable to the use of a specific

---

[3] More details about the used dataset are illustrated in a next section.

[4] In our experiments we considered SVO triples that occur at least 50 times in the whole ClueWeb09 corpus, in order to remove noisy data.

[5] In our experimentation, we used the library called SPMF for finding closed frequent itemsets[6], applying the CHARM algorithm [14].

Table 3. An example of subject- and object-terms semantic transformation for one triple of the verb *"to learn"* (*student-learns-math*). This represents one row of the two transactional databases.

Subject-term	Subject semantic features	Object-term	Object semantic features
student	*CapableOf*-study, *AtLocation*-at_school, *IsA*-person, *Desires*-learn, *PartOf*-class, *CapableOf*-read_book)	**math**	*IsA*-subject, *HasProperty*-useful_in_business, *UsedFor*-model_physical_world, ...

verb, we also need to obtain itemsets that cover all the items that are found in frequent itemsets. In other words, we want to extract *diverse itemsets*, i.e., a minimal set of frequent and closed itemsets that cover all the frequent items.

In the final phase, once obtained the frequent and diverse itemsets for both the two transactional databases, we connect all the subject-itemsets with all the object-itemsets, weighting the connection according to the their co-occurrences in the same triples of the original dataset.

The *semantic verb model* constructed for a specific verb *"v"* is thus a set of weighted connections between *frequent and diverse* semantic features belonging to the subjects of *"v"* and *frequent and diverse* semantic features of the objects of *"v"*. On the one hand, this is a way to summarize the semantics suggested by the verb occurrences. On the other hand, it is also a result that can be used to generate new data by querying existing semantic resources with such semantic subject- and object-itemsets. Still, this can be done without looking for words similar to frequent subjects and objects, but by finding new subjects and objects that, even if not similar in general, have certain semantic information that fill the specific context.

The resulting models are automatically calculated, and they are very concise, since in all the large and sparse semantic space only few features are relevant to certain meanings (headed by the verb). This is also in line with what stated in [15] where the authors claimed that semantics is actually structured by low-dimensionality spaces that are covered up in high-dimensional standard vector spaces.

5.2 Data

We used a dataset of *subject-verb-object* (SVO) triples generated as part of the NELL project[7]. This dataset contains a set of 604 million triples extracted from

---

[7] http://rtw.ml.cmu.edu/resources/svo/

the entire dependency-parsed corpus *ClueWeb09* (about 230 billion tokens)[8]. The dataset also provides the frequency of each triple in the parsed corpus. The aim of the evaluation is two-folds: to demonstrate that the proposed approach is able to induce senses with a better efficacy than a classic word-based strategy, and to show that the resulting semantic model also drammatically reduces the dimensionality of the feature space, yet without mixing features.

5.3   Running Example

In order to better explain the whole data flow, we list some steps on an example verb (*to sing*). Within the dataset, we extracted 82 different triples (after morphologic and NER normalizations). The cardinality of subjects and objects is 95. The total number of semantic relations retrieved from ConceptNet is 8786. Then, with a minimum support of 0.05 (i.e., 5%), the output model of our approach is constituted by 4 diverse itemsets for the objects and 24 for the subjects, with an average itemset cardinality of 18.5 and 12.6 respectively, covering more than 50% of the semantic features of all the input triples. Thus, the size of the feature space for modeling the senses of the verb *to sing* goes from 95 words to 28 (4 + 24) semantic features (clusters of ConceptNet semantic information). In the next section we show the aggregate results on an extensive set of experiments using the *pseudo-word* evaluation proposed by [12].

The original dataset has several issues we needed to solve in order to guarantee the significance of the evaluation. In detail, we removed very low-frequency triples containing mispelled words as well as meaningless sequences of characters that were not filtered out during the generation of the data. We manually tested different cutoffs to see the quality of the triples of the top-10 verbs in the corpus, selecting $m = 50$ as minimum frequency of the *subject-verb-object* triples in the *ClueWeb09* corpus.

Then, since the subject and object terms of the triples were not normalized, we merged all linguistic variations according to their morphological roots. Still, we integrated a Named Entity Recognition (NER) module to transform proper names into simple generic semantic classes, like people and organizations[9].

5.4   Results

Word Sense Induction systems are usually evaluated by using the pseudo-word strategy presented in [12]. The process proceeds as follows: first, two random words are selected, merged, and replaced by a unique token (for instance, the words *dog* and *apple* are replaced by the single pseudo-word *dogapple* in the corpus). Then, the WSI system is evaluated in terms of how well it separates the *dogapple*-instances into two correct clusters based on their context (as in a standard Word Sense Disambiguation scenario). In our case, since we use a large corpus of *subject-verb-object* instances, we evaluate our approach on verbs,

---

[8] http://lemurproject.org/clueweb09/
[9] We used the Stanford NLP toolkit available at http://www-nlp.stanford.edu/

Table 4. Average feature space cardinality for the three approaches (word-based ww, semantics-featured sf, semantic-model sm) on 100 random verb pairs within the top-100 frequent verbs.

F. Sp. Size (ww)	Min supp.	Sem. Loop	F. Sp. Size (sf)	F. Sp. Size (sm)
145,70	10%	1	4701,65	13,65
147,30	10%	2	2346,25	52,20
147,25	10%	3	1856,85	101,00

F. Sp. Size(ww)	Min supp.	Sem. Loop	F. Sp. Size (sf)	F. Sp. Size (sm)
146,90	20%	1	5159,70	4,55
145,10	20%	2	2232,40	14,10
143,50	20%	3	1863,80	41,75

F. Sp. Size (ww)	Min supp.	Sem. Loop	F. Sp. Size (sf)	F. Sp. Size (sm)
147,30	30%	1	4604,00	-
149,20	30%	2	2313,85	6,85
149,10	30%	3	1805,75	25,35

defining the context as their *subject-object* pairs. In particular, we extracted the top-100 frequent verbs in the dataset. From them, we randomly selected 100 verb pairs and replaced them with their pseudo-word. Some examples of pairs were *take-do*, *find-give*, *look-try*. On these evaluation data, we compared three approaches:

ww The ww-approach (*word-based*) constructs a dataset using subject and object terms as features.

sf The sf-approach (*semantic-featured*) constructs a dataset using the ConceptNet semantic information associated to subjects and objects as features, without mining frequent and diverse patterns from them.

sm The sm-approach (*semantic verb model*) constructs a dataset using the generated diverse itemsets as feature space.

For the ww-approach, for each verb pair under evaluation, we build a verb matrix $M_{I,J}^{v}$ where the rows represent the I-triples *subject-verb-object* containing the verb v while the columns $J = J_s \bigcup J_o$ are the union of subject and object terms (J_s and J_o respectively). For example, the triple *Robert-eats-vegetables* becomes an M^{eat}-row $<s\text{-}person{=}1, o\text{-}vegetable{=}1>$ with only two non-zero features (notice that *Robert* has been substituted with the feature representing its associated Named Entity *person*, whereas *vegetables* has been linguistically-normalized to *vegetable*). The creation of the sf and the sm models is done in the same way. For the sf-approach we replaced subject and object terms with their ConceptNet semantic information, while for the sm-approach we replaced the features of the sf model with the generated diverse itemsets.

In this section we show and discuss the comparison of the three different models (ww, sf and sm) in terms of dimensionality reduction, classification accuracy and clustering adherence. We also experimented the usefulness of the

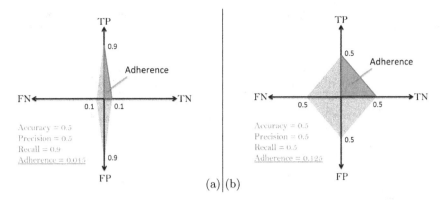

Fig. 1. An example sketch of the clustering evaluation results for the *ww*-approach (left figure (a)) and the *sf*- and *sm*-approach (right figure (b)). Accuracy and precision values are identical. Recall is superior for *ww*, in spite of the inability of the clustering process to separate the instances (almost all instances are in one single cluster). Clustering adherence permits to give light to the better intersection between clusters and classes obtained with our proposed approaches *sf* and *sm*.

recursive usage of ConceptNet. In detail, instead of simply replacing a word w with its semantic terms $< s_1, s_2, ..., s_n >$, we continue substituting each s_i with $< s_{i1}, s_{i2}, ..., s_{in} >$. We call this reiteration *semantic loop*.

Dimensionality Reduction. As shown in Table 4, the average size of the feature space is around 146 for the *ww*-approach, that is the average number of terms that fill the subject and object slots for the top-100 verbs in the collection (after normalization). When such terms are replaced by their ConceptNet semantic information (*sf*), the feature space becomes much larger. Actually, we only kept the top-1000 most common terms plus the terms that are as common as the least common term (i.e. ties are not broken). Note that while the semantic loop increases, the number of features for *sf* decreases. This is due to the fact that also ties on the 1000th term decrease with higher semantic loops, so the cutoff is able to operate earlier. Finally, the semantic model *sm* produces a very reduced feature space, compared to the *ww* and *sf* models.

Clustering. we used K-Means (with K=2) on the three models, then matching the resulting two clusters with the two actual *verb1* and *verb2* groupings. For each pair of instances i and j, it can happen one of the following 4 cases:

- i and j are clustered in the same cluster and they belong to the same class (true positive TP)
- i and j are clustered in the same cluster but they actually belong to different classes (false positive FP)
- i and j are clustered in different clusters but they actually belong to the same class (false negative FN)

Table 5. Average clustering adherence for the three approaches (word-based *ww*, semantics-featured *sf*, semantic-model *sm*) on 100 random verb pairs within the top-100 frequent verbs.

Clust. Adh. (ww)	Min supp.	Sem. Loop	Clust. Adh. (sf)	Clust. Adh. (sm)
7,90%	10%	1	13,57%	21,15%
7,46%	10%	2	15,69%	22,84%
5,71%	10%	3	15,95%	25,65%
Clust. Adh. (ww)	Min supp.	Sem. Loop	Clust. Adh. (sf)	Clust. Adh. (sm)
6,78%	20%	1	8,22%	22,38%
7,76%	20%	2	15,92%	24,48%
6,61%	20%	3	20,02%	26,10%
Clust. Adh. (ww)	Min supp.	Sem. Loop	Clust. Adh. (sf)	Clust. Adh. (sm)
8,46%	30%	1	11,40%	-
7,28%	30%	2	15,52%	22,46%
7,84%	30%	3	19,88%	25,35%

- i and j are clustered in different clusters and they belong to different classes (true negative TN)

The result of the clustering on the *ww*-approach is quite worthless in almost all the cases. In particular, the first cluster uses to contain one or two instances while the rest of the data remains in the second cluster. Thus, the computation of standard accuracy values carries to the same problem of the classification case (standard accuracy values are not meaningful). To overcome this inconvinient, we propose a novel measure, i.e., the *clustering adherence*. In words, its aim is to see how the feature space lets identify some coherent overlapping between the returned clusters and the actual classes in terms of TF and TN only (only in Information Retrieval systems FP and FN are actually important as TP and TN). Figure 1 shows the geometric representation of the measure. TP and TN form an *area of correctness*, that can be compared to its maximum value (when FP and FN are zero). Even if the three approaches have similar accuracy values, they present different clustering adherence. In particular, the clustering on the *ww*-model produces very low TP-TN areas with respect to *sf* and *sm*. Table 5 shows the complete results.

6   Conclusions and Future Works

Common-sense knowledge, differently from top-down semantic resources, has the potential to impact computational tasks thanks to the high number of semantic relations that is usually unfeasible to create manually. Nevertheless, the complexity in terms of specificity, completeness, correctness, and relativity makes it difficult to be used in numerous tasks. In this paper, we experimented two uses of common-sense knowledge in both an unsupervised and a supervised setting, by replacing words with the information associated to them in ConceptNet, a large common-sense semantic resource. The results highlights the power of this approach even before adequate management of the above-mentioned problematic common-sense type of semantic data. In future work, we will leverage techniques for data filtering and compression to improve quality in ConceptNet, evaluating the impact in main computational tasks.

References

1. Padó, S., Lapata, M.: Dependency-based construction of semantic space models. Computational Linguistics **33**(2), 161–199 (2007)
2. Navigli, R., Velardi, P., Faralli, S.: A graph-based algorithm for inducing lexical taxonomies from scratch. In: Proceedings of the Twenty-Second International Joint Conference on Artificial Intelligence-Volume Volume Three, pp. 1872–1877. AAAI Press (2011)
3. Lenci, A.: Carving verb classes from corpora. In: Simone, R., Masini, F. (eds.) Word Classes, p. 7. John Benjamins, Amsterdam, Philadelphia (2010)
4. Speer, R., Havasi, C.: Representing general relational knowledge in ConceptNet 5. In: LREC, pp. 3679–3686 (2012)
5. Miller, G.A.: Wordnet: a lexical database for english. Communications of the ACM **38**(11), 39–41 (1995)
6. Stevenson, M., Wilks, Y.: Word-sense disambiguation. The Oxford Handbook of Comp. Linguistics, 249–265 (2003)
7. Manning, C.D., Schütze, H.: Foundations of statistical natural language processing. MIT press (1999)
8. Gibson, J.J.: The Theory of Affordances. Lawrence Erlbaum (1977)
9. Baroni, M., Lenci, A.: Distributional memory: A general framework for corpus-based semantics. Computational Linguistics **36**(4), 673–721 (2010)
10. Hill, F., Reichart, R., Korhonen, A.: Simlex-999: Evaluating semantic models with (genuine) similarity estimation (2014). arXiv preprint arXiv:1408.3456
11. Denkowski, M.: A survey of techniques for unsupervised word sense induction. Language & Statistics II Literature Review (2009)
12. Schutze, H.: Dimensions of meaning. In: Supercomputing 1992., Proceedings, pp. 787–796. IEEE (1992)
13. Borgelt, C.: Frequent item set mining. Wiley Interdisciplinary Reviews: Data Mining and Knowledge Discovery **2**(6), 437–456 (2012)
14. Zaki, M.J., Hsiao, C.J.: Charm: an efficient algorithm for closed itemset mining. In: SDM, Vol. 2, pp. 457–473. SIAM (2002)
15. Karlgren, J., Holst, A., Sahlgren, M.: Filaments of meaning in word space. In: Macdonald, C., Ounis, I., Plachouras, V., Ruthven, I., White, R.W. (eds.) ECIR 2008. LNCS, vol. 4956, pp. 531–538. Springer, Heidelberg (2008)

An AI Application to Integrated Tourism Planning

Francesca Alessandra Lisi$^{(\boxtimes)}$ and Floriana Esposito

Dipartimento di Informatica, Università degli Studi di Bari "Aldo Moro", Bari, Italy
{francesca.lisi,floriana.esposito}@uniba.it

Abstract. Integrated Tourism can be defined as the kind of tourism which is explicitly linked to the localities in which it takes place and, in practical terms, has clear connections with local resources, activities, products, production and service industries, and a participatory local community. In this paper we report our experience in applying Artificial Intelligence techniques to Integrated Tourism planning in urban areas. In particular, we have modeled a domain ontology for Integrated Tourism and developed an Information Extraction tool for populating the ontology with data automatically retrieved from the Web. Also, we have defined several Semantic Web Services on top of the ontology and applied a Machine Learning tool to better adapt the automated composition of these services to user demands. Use cases of the resulting service infrastructure are illustrated for the Apulia Region, Italy.

1 Introduction

Integrated Tourism can be defined as the kind of tourism which is explicitly linked to the localities in which it takes place and, in practical terms, has clear connections with local resources, activities, products, production and service industries, and a participatory local community. The goal of Integrated Tourism is twofold. For the various interests, requirements and needs the aim is to be fused together into a composite, integrated strategic tourism plan. For tourism the aim is to be planned with the intention of being fused into the social and economic life of a region and its communities. Although there is evidence that some tourism destinations have developed without conscious strategic and integrated planning, many of them have experienced unforeseen consequences (either physical, or human, or marketing or organizational impacts) which have led to their deterioration. Integrated Tourism has turned out to be crucial in the sustainable development of rural areas (so-called Integrated Rural Tourism) [19]. However, the integrated approach can be beneficial also to urban areas as testified by recent progress in *Urban Tourism* research [1]. Urban tourism is the practice of taking a vacation and visiting an inner-city area, as opposed to visiting such places as historical sites and natural features. Such cities market tourist attractions located within the city itself. This can include points of interest such as churches, a particular city block or feature, and amazing architectural creations. Urban Tourism is complex, difficult to pin down and define, and depends on

M. Gavanelli et al. (Eds.): AI*IA 2015, LNAI 9336, pp. 246–259, 2015.
DOI: 10.1007/978-3-319-24309-2_19

many factors such as the size of the town, its history and heritage, its morphology and its environment, its location, its image, etc. In the case of Integrated Urban Tourism, tourism is being seen as a cornerstone of a policy of urban development that combines a competitive supply able to meet visitors' expectations with a positive contribution to the development of towns and cities and the well-being of their residents.

In this paper, we report our experience in applying Artificial Intelligence (AI) techniques to Integrated Urban Tourism within the *Puglia@Service*[1] project. In particular, the techniques employed are borrowed from the AI areas of Knowledge Engineering (KE), Information Extraction (IE) and Machine Learning (ML). The resulting e-tourism application is intended to be compliant with the principles and the standards of the so-called *Semantic Web* [2]. The Semantic Web provides a common framework that allows data to be shared and reused across application, enterprise, and community boundaries, mainly by means of shared vocabularies called *ontologies* [23]. In *Puglia@Service*, ontologies come into play in the definition of *Semantic Web Services* [9]. More precisely, Web Services are enriched with semantic annotations involving concepts taken from undelying domain ontologies. This will enable users and software agents to automatically discover, invoke, compose, and monitor Web resources offering services, under specified constraints, for Integrated Tourism in the Apulia Region, Italy.[2] Moreover, a distinguishing feature of the application described in this paper is that the automated composition of Semantic Web Services can be improved by learning from users' feedback. Preliminary and partial reports of the work done within *Puglia@Service* can be found in [12,14].

The paper is structured as follows. Section 2 provides some background information. In particular, Section 2.1 summarizes the goals of *Puglia@Service* whereas Section 2.2 reviews relevant literature on e-tourism. Section 3 shortly describes a domain ontology, named *OnTourism*, which models the relevant features of Integrated Tourism. Section 4 briefly presents a IE tool, named WIE-ONTOUR, which has been developed for populating *OnTourism* with data automatically retrieved from the Web. Section 5 illustrates some of the Semantic Web Services which have been defined on top of *OnTourism*. Section 6 introduces a ML tool, named FOIL-$\mathcal{DL}$, which has been run on data provided by users in order to automatically generate new axioms for *OnTourism*. Section 7 concludes the paper with final remarks and directions of future work.

2 Background

2.1 Integrated Tourism Services in Apulia

Puglia@Service is an Italian PON Research & Competitivity project aimed at creating an innovative service infrastructure for the Apulia Region, Italy. The research conducted in *Puglia@Service* falls within the area of *Internet-based*

---

[1] http://www.ponrec.it/open-data/progetti/scheda-progetto?ProgettoID=5807
[2] http://en.wikipedia.org/wiki/Apulia

Service Engineering, *i.e.* it investigates methodologies for the design, development and deployment of innovative services. Concerning this area, the project will have an impact on the Apulia regional system at a strategic, organizational and technological level, with actions oriented to service innovation for the "sustainable knowledge society". The reference market of *Puglia@Service* is represented by the so-called *Knowledge Intensive Services* (KIS), an emerging category of the advanced tertiary sector, and transversal to the other economic sectors, that is supposed to play a prominent role within the restructuring process which will follow the world economic crisis.

Objective of the project is to promote a new service culture over the Apulia Region, marking a discontinuity point in the local development model, and guiding the transition of the region towards the "smart territory" paradigm where the territory is intended to be a multiplayer system able to improve, by means of an adequate technological and digital infrastructure, its attitude to innovation as well as its skills in managing the knowledge assets of the regional stakeholders. The arrangement of the new service model into the regional context will regard new generation services for Public Administration and for Integrated Tourism.

As for the application to Integrated Tourism (*Puglia@Service.Tourism*), which is of interest to this paper, the project addresses some of the issues analyzed in a report entitled "Sustainable Tourism and Local Development in Apulia Region" (2010)[3] and prepared by the Local Economic and Employment Development (LEED) Programme and the Tourism Committee of the Organisation for Economic Co-operation and Development (OECD) in collaboration with Apulia Region. The region as a touristic destination needs a better management in spite of the recent growth of visitors and the high potential. In particular, the report emphasizes the lack of an adequate ICT infrastructure and little use of new technologies.

Puglia@Service.Tourism encompasses an intervention on the Apulia tourism system, based on the definition of an Internet-based service model which increases the capability of KIS to create value for the region and for the tourist. Here, the tourist is not only "service user" but also "information supplier". In particular, the application will require the development of methods and technologies enabling an interaction model between the tourist and the territory with the ultimate goal of local development along three directions: Culture, Environment and Economy. For the purposes of this paper, we shall focus only on the Environment dimension along which *Puglia@Service.Tourism* aims at promoting forms of tourism with a low environmental impact centered around the notion of eco-compatible mobility. This will contribute to the achievement of a twofold goal. On one side, the tourist will benefit from decision support facilities during his/her tours, *e.g.* he/she will receive suggestions about sites of interest and public transportation means suitable to reach a certain destination. On the other side, the territory will benefit from the environmental sustainability of local tourism. The reduced environmental impact of eco-mobility together with the need for a more efficent transportation system in touristic places can be

---

[3] http://www.oecd.org/cfe/leed/46160531.pdf

obtained by combining sensoring tools and applications with rewarding mechanisms that encourage tourists and citizens to make eco-compatible choices. A possible scenario is described in the following. Once arrived in a touristic destination, the tourist could use his/her smartphone/PDA in order to obtain a suggestion about specific itineraries compliant with his/her profile and the information about the context. The tourist will be informed about the availabiliy of alternative transportation means and will be offered some credits for the green options (byking, trekking, car pooling, car sharing, etc.). In order to support this scenario, the *Puglia@Service.Tourism* infrastructure should deal with multi-dimensional information useful to suggest a touristic strategy which should meet users' expectations and preferences (in culture, enogastronomy, shopping, relax, etc.); environmental conditions, both meteorological and natural; multi-modal transportation means; availability of car pooling and car sharing services; transfer time between sites of interest. The "fingerprint" of tourists visiting an area in a given time span can be anonymized and employed to improve continuously the user profiling with the choices made by tourists with the same profile. To this aim it is necessary to track the trajectories of citizens and tourists by means of localization and wireless communication technologies (traces from mobile phones, PDA, vehicles with GPS, etc.).

It is straightforward to notice that Internet-based Service Engineering for KIS in Integrated Tourism should strongly rely on some Web technology enabling an automated service composition, just like Semantic Web Services.

2.2   State of the Art of e-Tourism

Most research on the application of Information and Communication Technology (ICT) to the tourism industry (so-called e-tourism) has been conducted by specializing technologies originally conceived for e-commerce (see [5] for a comprehensive yet not very recent review). However, e-tourism has been considered particularly challenging since the very beginning due to the technical issues raised by *interoperability*. Werthner and Klein [24] defined interoperability as the provision of a well-defined and end-to-end service which is in a consistent and predictable way. This generally covers not merely technical features but also in the case of electronic market environments, contractual features and a set of institutional rules. Interoperability enables partners to interact electronically with each other by the most convenient method and to deliver the right information at the right time to the right user at the right cost.

Ontologies have played a crucial role in solving the interoperability problem. Using a domain ontology a mediator software system (such as HARMONISE [7,18]) effectively "'translates"' partners' data and allows them to communicate electronically. Maedche and Staab [16,17] showed that Semantic Web technologies can be used for tourism applications to provide useful information on text and graphics, as well as generating a semantic description that is interpretable by machines. Dogac *et al.* [8] describe how to deploy semantically enriched travel Web services and how to exploit semantics through Web service registries.

They also address the need to use the semantics in discovering both Web services and Web service registries through peer-to-peer technology. Hepp *et al.* [10] investigate the use of ontological annotations in tourism applications. They show, based on a quantitative analysis of Web content about Austrian accommodations, that even a perfect annotation of existing Web content would not allow the vision of the Semantic Web to become a short-term reality for tourism-related e-commerce. Also, they discuss the implications of these findings for various types of e-commerce applications that rely on the extraction of information from existing Web resource, and stress the importance of Semantic Web Services technology for the Semantic Web. Siorpaes and Bachlechner [22] develop a system based on a fast and flexible Semantic Web backbone with a focus on e-tourism. The major benefits of this approach are its simplicity, modularity, and extensibility. In [11], Jakkilinki *et al.* describe the underlying structure and operation of a Semantic Web based intelligent tour planning tool. The proposed tour planner has inbuilt intelligence which allows it to generate travel plans by matching the traveller requirements and vendor offerings stored in conjunction with the travel ontology. Bousset *et al.* [3] present a decision support system which combines tools to assist in the analysis of the views, concerns and planned strategies of a wide range of tourism stakeholders in the face of given trends in tourists' expectations. Ricca *et al.* [21] present a successful application of logic programming for e-tourism: the iTravel system. The system exploits two technologies: (i) a system for ontology representation and reasoning, called OntoDLV; and, (ii) a semantic IE tool. The core of iTravel is an ontology which models the domain of tourism offers. The ontology is automatically populated by extracting the information contained in the tourism leaflets produced by tour operators. A set of specifically devised logic programs is used to reason on the information contained in the ontology for selecting the holiday packages that best fit the customer needs. An intuitive web-based user interface eases the task of interacting with the system for both the customers and the operators of a travel agency. Brilhante *et al.* [4] propose TripBuilder, a user-friendly and interactive system for planning a time-budgeted sightseeing tour of a city on the basis of the points of interest and the patterns of movements of tourists mined from user-contributed data. The knowledge needed to build the recommendation model is entirely extracted in an unsupervised way from two popular collaborative platforms: Wikipedia and Flickr. TripBuilder interacts with the user by means of a friendly Web interface that allows her to easily specify personal interests and time budget. The sightseeing tour proposed can be then explored and modified.

3 Modeling Domain Knowledge in *OnTourism*

Modeling the knowledge concerning the domain of Integrated Tourism is the first of the KE activities we have performed within *Puglia@Service.Tourism*. Domain ontologies for tourism are already available, *e.g.* the *travel*[4] ontology is centered around the concept of *Destination*. However, it is not fully satisfactory from the

---

[4] http://www.protege.cim3.net/file/pub/ontologies/travel/travel.owl

viewpoint of Integrated Tourism. For instance, it lacks concepts modeling the reachability of places. Therefore, we have decided to build a domain ontology, named *OnTourism*,[5] more suitable for the project objectives and compliant with the OWL2 standard.[6] It consists of 379 axioms, 205 logical axioms, 117 classes, 9 object properties, and 14 data properties, and has the DL expressivity of $\mathcal{ALCOF}(\mathbf{D})$ according to Protégé [7] ontology metrics. [8]

The main concepts forming the terminology of *OnTourism* model the sites (class *Site*), the places (class *Place*), and the distances between sites (class *Distance*). Sites of interest include accommodations (class *Accommodation* with subclasses *Bed_and_Breakfast* and *Hotel* among the others), attractions (class *Attraction* with several subclasses, *e.g.*, *Church*), stations (class *Station*), and civic facilities (class *Civic*). The terminology encompasses the amenities (class *Amenity* with subclasses like *Wheelchair_Access*) and the services (class *Service* further specialized in, *e.g.*, *Car_Rental* and *Bike_Rental*) offered by hotels. Also, it models the official 5-star classification system for hotel ranking (class *Rank* with instances *1_star*, *2_stars*, and so on) as well as a user classification system for accommodation rating (class *Rate* with instances *Excellent*, *Very_Good*, *Average*, *Poor*, *Terrible*). Finally, the terminology includes landscape varieties (class *Landscape* with instances *City*, *Country*, *Lake*, *Mountain*, *River*, and *Sea*) and transportation means (class *Transportation_Mean* with instances *Bike*, *Car*, *Foot*, and *Public_transit*). Distances are further classified into *Distance_by_car* and *Distance_on_foot* according to the transportation means used.

The object properties in *OnTourism* model the relationship between a site and a distance (*hasDistance*), the relationship between a distance and the two sites (*isDistanceFor*), and the relationship between a site and the place where the site is located at (*isLocatedAt*) (see Figure 1). Also, for each accommodation, it is possible to specify the amenities available (*hasAmenity*) and the services provided (*provides*). The user rating allows to classify accommodations into five categories (from *Excellent_Accommodation* to *Terrible_Accommodation*). In the case of hotels, the ranking (*hasRank*) is the starting point for the definition of five categories (from *Hotel_1_Star* to *Hotel_5_Stars*).

The data properties in *OnTourism* allow to refer to sites by *name* and to places by *address*, *zipcode*, *city*, and *country*. Details about accommodations are the number of rooms (*numberOfRooms*) and the average price of a room (*hasPrice*). Distances between sites have a numerical value in either length or time units (*hasLengthValue*/*hasTimeValue*). Note that each of these numerical values would be better modeled as attribute of a ternary relation. However, only binary relations can be represented in OWL. The concept *Distance* and the properties *hasDistance*, *isDistanceFor* and *hasLengthValue*/*hasTimeValue* are necessary to simulate a ternary relation by means of binary relations.

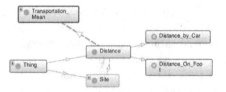

Fig. 1. Portion of the *OnTourism* ontology modeling the distances between sites.

4   Populating *OnTourism* with WIE-OnTour

At the ontology population step, the ontology is augmented with instances of concepts and properties. This KE activity is often perfomed by relying on some IE tool. IE aims at the automated extraction of structured information from unstructured and/or semi-structured machine-readable documents. In most of the cases this activity concerns processing texts by means of natural language processing. The proliferation of the Web has intensified the need for developing IE systems that help people to cope with the enormous amount of data that is available online, thus giving raise to Web Information Extraction (WIE) [6]. WIE tools typically exploit the HTML/XML tags and layout format that are available in online text. As a result, less linguistically intensive approaches have been developed for IE on the Web using wrappers, which are sets of highly accurate rules that extract a particular page's content. Wrappers typically handle highly structured collections of web pages, such as product catalogues and telephone directories. They fail, however, when the text type is less structured, which is also common on the Web.

WIE-ONTOUR is a wrapper-based WIE tool implemented in Java and conceived for the population of *OnTourism* with data concerning accommodations (in particular, those in the categories "Hotel" and "B&B") available in the web site of TripAdvisor[9]. Note that the tool does not generate instances of the classes *Amenity* and *Service* because it is not necessary for our purposes. However, amenities and services can be implicitly declared. For instance, the fact that a hotel, say *hotel_10*, offers a bike rental service is modeled by means of an axiom stating that *hotel_10* is instance of the class ∃*hasAmenity.Bike_Rental*. The tool is also able to compute distances of the extracted accommodations from sites of interest (*e.g.*, touristic attractions) by means of Google Maps[10] API. Finally, the tool supports the user in the specification of sites of interest. A snapshot of WIE-ONTOUR is shown in Figure 2.

Instantiations of *OnTourism* for the main destinations of urban tourism in Apulia have been obtained with WIE-ONTOUR. Here, we consider an instantiation for the city of Bari (the capital town of Apulia). It contains 34 hotels, 70 B&Bs, 106 places, 208 distances for a total of 440 individuals. Information about

---

[9] http://www.tripadvisor.com/
[10] http://maps.google.com/

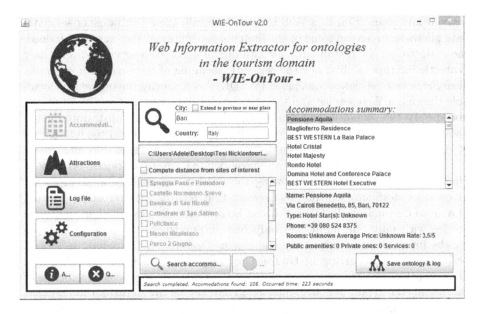

Fig. 2. Web Information Extraction for the city of Bari, Italy, with WIE-ONTOUR.

the rank, the amenities, the services, and the average room price has been added in the ontology for each extracted accommodation. The distances are provided in time and length on foot and have been computed with respect to *Basilica di San Nicola*[11] and *Cattedrale di San Sabino*[12] (both instances of *Church*). The restriction to foot distances is due to the aforementioned preference of Integrated Tourism for eco-mobility.

5   Building Semantic Web Services Upon *OnTourism*

Semantic Web Services are the server end of a client-server system for machine-to-machine interaction via the Web [9]. As a component of the Semantic Web, they are defined with mark-up languages which make data machine-readable in a detailed and sophisticated way. In particular, *OWL-S*[13] is an ontology which provides a standard vocabulary that can be used together with the other aspects of OWL to create service descriptions. The use of OWL-S makes it easy for programmers to combine data from different sources and services without losing meaning. Web services can be activated "behind the scenes" when a Web browser makes a request to a Web server, which then uses various Web services to construct a more sophisticated reply than it would have been able to do on its own. Semantic Web Services can also be used by automatic programs that run

---

[11] Basilica of St. Nicholas: http://en.wikipedia.org/wiki/Basilica_di_San_Nicola
[12] Cathedral of St. Sabinus: http://en.wikipedia.org/wiki/Bari_Cathedral
[13] http://www.w3.org/Submission/OWL-S/

without any connection to a Web browser. Overall, the interchange of semantic data allows to overcome some of the limits of conventional Web services. Indeed, the mainstream XML standards for interoperation of Web services specify only syntactic interoperability, not the semantic meaning of messages. For example, *Web Services Description Language* (WSDL)[14] can specify the operations available through a Web service and the structure of data sent and received but can not specify the semantic meaning of the data or semantic constraints on the data. This requires programmers to reach specific agreements on the interaction of Web services and makes automatic Web service composition difficult.

In *Puglia@Service.Tourism*, we have defined several services on top of *OnTourism* (and *travel*) by following the OWL-S approach. For example, let us consider *destination_attractions_service* that returns the attractions located in a given destination. In OWL-S it can be semantically described as an atomic service with only one input and only one output where the parameter types for the input and the output are the classes *Destination* (belonging to *travel*) and *Attraction* (occurring in *OnTourism*) respectively. Several specializations of *destination_attractions_service* have been considered, one for each subclass of the parameter types. For example, *city_churches_service* returns the churches (output parameter of type *Church*) located in a given city (input parameter of type *City*). When executed for the city of, *e.g.*, Bari, the service will query the underlying domain ontologies (more precisely, their instance level) to retrieve each *Church* that *isLocatedAt* some *Place* in Bari, *e.g. Basilica di San Nicola* and *Cattedrale di San Sabino*. Note that these instances will be returned also by *destination_attractions_service* because they are inferred to be instances of *Attraction*. As a further case, *near_attraction_accomodations_service* returns all the accommodations (output parameter of type *Accommodation*) near a given attraction (input parameter of type *Attraction*). Note that closeness can be defined on the basis of the distance between sites (class *Distance*) either in a crisp way (*i.e.*, when the distance value is under a fixed threshold) or in a fuzzy way (*i.e.*, through grades of closeness). In both ways, however, the computation should consider the transportation means used (*Distance_by_car* vs. *Distance_on_foot*) as well as the measure units adopted (*hasLengthValue* vs. *hasTimeValue*).

In *Puglia@Service.Tourism*, we have chosen to define only OWL-S atomic services in order to exploit the notorious advantages of the WSDL grounding. Composite services can be automatically obtained by applying service composition methods such as the one described in [20]. The simplest form of composite service is based on the control construct of *Sequence*. For example, the services *city_churches_service* and *near_attraction_accomodations_service* can be executed in sequence by having the output of the former as input to the latter. Note that the type mismatch is only apparent since *Church* is a subclass of *Attraction*. One such service composition satisfies, *e.g.*, the user request of knowing the accommodations around *Basilica di San Nicola* and *Cattedrale di San Sabino* in Bari. Considering that Bari is a major destination of religious tourism in Apulia, this composite service effectively supports the demand from pilgrims who prefer to

---

[14] http://www.w3.org/TR/2007/REC-wsdl20-primer-20070626/

find an accommodation in the neighborhood of places of worship so that they can practise their own religions at any hour of the day. Also, if the suggested accommodations are easy to reach (*i.e.*, at foot distance) from the site of interest, the service will bring benefit also to the city, by reducing the car traffic. In a more complex scenario, disabled pilgrims might need a wheelchair-accessible accommodation. The service composition mechanism should then append a further specialized service, say *wheelchairaccess_hotels_service*, which returns the hotels (output parameter of type *Hotel*) with disabled facilities (input parameter of type *Wheelchair_Access*). Indeed, the resulting composite service can be considered more compatible with the special needs of this user profile.

6 Improving Services by Learning from Users' Feedback

In *Puglia@Service.Tourism*, the automated composition of Semantic Web Services will be enhanced by exploiting users' feedback. The idea is to apply ML tools in order to induce ontology axioms which can be used for discarding those compositions that do not reflect the preferences/expectations/needs of a certain user profile. Here, we illustrate this idea with an application scenario which builds upon the accommodation rating provided by TripAdvisor's users. More precisely, we consider the task of accommodation finding which consists of distinguishing good accommodations from bad ones according to the amenities available, the services offered, the location, and the distance from sites of interest. In order to address this classification problem, we need ML tools able to deal with the inherent incompleteness of Web data and the inherent vagueness of concepts such as closeness. One such tool is FOIL-$\mathcal{DL}$.

FOIL-$\mathcal{DL}$ implements a method for learning a set of *fuzzy* General Concept Inclusion (GCI) axioms from positive and negative examples for a target class A_t in any *crisp* OWL 2 ontology [15]. In particular, the GCIs inducible with FOIL-$\mathcal{DL}$ are $\mathcal{EL}(\mathbf{D})$ axioms of the form

$$C \sqsubseteq A_t \, , \tag{1}$$

where the left-hand side is defined according to the following syntax

$$C \longrightarrow \top \mid A \mid \exists R.C \mid \exists T.\mathbf{d} \mid C_1 \sqcap C_2 \, . \tag{2}$$

Here, A is a primitive concept, R is an object property, T is a data property, C, C_1, C_2 are complex concepts, and $\mathbf{d}$ can be any of the following membership functions widely used in fuzzy set theory

$$\mathbf{d} := ls(a, b) \mid rs(a, b) \mid tri(a, b, c) \, , \tag{3}$$

where *e.g.* $ls(a, b)$ is the left-shoulder membership function.

For instance, the 5 fuzzy sets that can be automatically generated by FOIL-$\mathcal{DL}$ for the data property *hasLengthValue* of *OnTourism* are graphically shown in Figure 3. Here, we have used the membership functions $ls(a, b)$ for the first

Fig. 3. Fuzzy sets (*VeryLow* (VL), *Low* (L), *Fair* (F), *High* (H), and *VeryHigh* (VH)) derived from the concrete domain used as range of the data property *hasLengthValue* of *OnTourism* in the case of (a) *Distance_by_car* and (b) *Distance_on_foot*.

fuzzy set, $tri(a, b, c)$ for the subsequent three ones, and $rs(a, b)$ for the fifth one. Also, we have distinguised between the two cases of *Distance_by_car* and *Distance_on_foot*. Notably, in the latter case, a very low distance does not exceed 900 meters, an average distance is about 1500 meters, and so on. New concepts such as *hasLengthValue_VeryLow* are obtained from these fuzzy sets and can appear in the GCI axioms induced by FOIL-$\mathcal{DL}$. For example, the axiom

$$Hotel\_3\_Stars \sqcap \exists hasDistance.(\exists isDistanceFor.(Train\_Station) \sqcap$$
$$hasLengthValue\_VeryLow) \sqcap hasPrice\_Fair \sqsubseteq Good\_Hotel$$

translates into the following sentence in natural language: 'A 3-star hotel with a very low distance from a train station and a fair room price is a good hotel'.

More formally, given a crisp $\mathcal{DL}$[15] ontology $\mathcal{K}$ (*background theory*), an atomic concept A_t (*target concept*), a set $\mathcal{E} = \mathcal{E}^+ \cup \mathcal{E}^-$ of *crisp* concept assertions labelled as either positive or negative examples for A_t (*training set*), and a set $\mathcal{L}_\mathcal{H}$ of *fuzzy* $\mathcal{EL}(\mathbf{D})$ GCI axioms (the *language of hypotheses*) of the form (1), the goal of FOIL-$\mathcal{DL}$ is to find a set $\mathcal{H} \subset \mathcal{L}_\mathcal{H}$ (a *hypothesis*) of axioms satisfying the following properties: (i) $\forall e \in \mathcal{E}^+, \mathcal{K} \cup \mathcal{H} \models e$ (*completeness*), and (ii) $\forall e \in \mathcal{E}^-, \mathcal{K} \cup \mathcal{H} \not\models e$ (*consistency*). Further details about FOIL-$\mathcal{DL}$ are reported in [15].

As an illustration of the potential usefulness of FOIL-$\mathcal{DL}$ in Integrated Tourism planning, we discuss here a couple of experiments concerning the filtering of results returned by the services reported in Section 5. We set up a learning problem with the class *Bad_Accommodation* (abbr. *BA*) as target concept. Ratings from TripAdvisor users have been exploited for distinguishing good accommodations from bad ones by using WIE-ONTOUR (see Section 4). Out of the 104 accommodations, 57 with a higher percentage (say, over 70%) of positive users' feedback are asserted to be instances of *Good_Accommodation*, whereas 15 with a lower percentage (say, under 50%) are asserted to be instances of *Bad_Accommodation*. The latter, of course, play the role of positive examples in our learning problem. Syntactic restrictions are imposed on the form of the learnable GCI axioms. More precisely, conjunctions can have at most 3 conjuncts and at most 2 levels of nesting are allowed in existential role restrictions. The two trials differ as for the alphabet underlying the language of hypotheses.

---

[15] $\mathcal{DL}$ stands for any DL.

In the first experiment, we have not considered the distances of the accommodations from the sites of interest (*i.e.*, we have forbidden the use of *hasDistance* in $\mathcal{L}_{\mathcal{H}}$). With this configuration, FOIL-$\mathcal{DL}$ returns just the following GCI with confidence 0.5:

$$B\&B \sqcap \exists hasAmenity.(Pets\_Allowed) \sqcap \exists hasAmenity.(Wheelchair\_Access) \sqsubseteq BA$$

The GCI suggests that B&Bs are not recommended even though they provide disabled facilities. It can be used to filter out from the result set of *wheelchairaccess_accommodations_service* those accommodations which are classified as bad.

In the second experiment, conversely, we have considered the distances of the accommodations from the sites of interest (*i.e.*, we have enabled the use of *hasDistance* in $\mathcal{L}_{\mathcal{H}}$). With this configuration, FOIL-$\mathcal{DL}$ returns the following GCI with confidence 1.0:

$$\exists hasAmenity.(Bar) \sqcap \exists hasAmenity.(Wheelchair\_Access) \sqcap$$
$$\exists hasDistance.(\exists isDistanceFor.(B\&B) \sqcap \exists isDistanceFor.(Church)) \sqsubseteq BA$$

The GCI strenghtens the opinion that B&Bs are not recommendable accommodations for disabled people whatever their distance from the churches is.

As a further experiment, we have restricted our analysis of accommodations in Bari to only B&Bs. Starting from 12 positive examples and 39 negative examples for *Bad_Accommodation*, FOIL-$\mathcal{DL}$ returns the following two GCIs with confidence 0.154 and 0.067 respectively:

$$\exists hasAmenity.(Pets\_Allowed) \sqcap \exists hasAmenity.(Wheelchair\_Access) \sqsubseteq BA$$
$$\exists hasAmenity.(Bar) \sqcap \exists hasAmenity.(Wheelchair\_Access) \sqsubseteq BA$$

which confirm that B&Bs should not be recommended to disabled tourists.

7   Conclusions and Future Work

In this paper we have reported our ongoing work on the use of AI techniques for supporting Integrated Tourism services in the urban areas of the Apulia region within the *Puglia@Service* project. More precisely, we have shortly described *OnTourism*, a domain ontology for Integrated Tourism. Also, we have briefly presented WIE-ONTOUR, a WIE tool which has been developed for populating *OnTourism* with data automatically retrieved from the Web sites of TripAdvisor and Google Maps. Moreover, we have illustrated the semantic descriptions in OWL-S of some Integrated Tourism services built on top of *OnTourism*. Finally, we have outlined an application scenario for FOIL-$\mathcal{DL}$, a ML tool able to deal with incomplete data and vague concepts, whose findings can be used to enhance the automated composition of OWL-S services.

Though developed for the purposes of the project, the technical solutions here described are nevertheless general enough to be reusable for similar applications in other geographical contexts. Overall, the application shows the added value of having ontologies and ontology reasoning behind an Interned-based service infrastructure. Note that by ontology reasoning we intend also non-standard forms such as the inductive inference supported by FOIL-$\mathcal{DL}$. The use of ML algorithms in our application makes a great difference from the related works mentioned in Section 2.2, including iTravel [21] which is the closest in design and implementation. However, at the current stage, the application suffers from some limits. First, WIE-ONTOUR is hard-coded. More precisely, the tool needs to be continuously maintained to keep up with updates of the structure of Web sites which are inspected for extracting the data. Second, FOIL-$\mathcal{DL}$ can not deal with big data. Tests over larger cities show that the system does not scale well.

For the future we intend to carry on the work on the application of FOIL-$\mathcal{DL}$ to the automated service composition. Notably, we shall consider the problem of learning from the feedback provided by specific user profiles. Also, in order to overcome the limits of FOIL-$\mathcal{DL}$, we are investigating the possibility of reducing the size of data by means of fuzzy granulation [13]. This should have a positive impact especially on the number of instances for the distance relation, and consequently on the computational cost of the learning process.

Acknowledgments. This work was partially funded by the Italian PON R&C 2007-2013 project PON02_00563_3489339 *"Puglia@Service*: Internet-based Service Engineering enabling Smart Territory structural development".

References

1. Ashworth, G., Page, S.J.: Urban tourism research: Recent progress and current paradoxes. Tourism Management **32**, 1–15 (2011)
2. Berners-Lee, T., Hendler, J., Lassila, O.: The Semantic Web. Scientific American, May 2001
3. Bousset, J.P., Skuras, D., Tešitel, J., Marsat, J.B., Petrou, A., Fiallo-Pantziou, E., Kušová, D., Bartoš, M.: A decision support system for integrated tourism development: Rethinking tourism policies and management strategies. Tourism Geographies **9**(4), 387–404 (2007)
4. Brilhante, I., Macedo, J.A., Nardini, F.M., Perego, R., Renso, C.: Tripbuilder: a tool for recommending sightseeing tours. In: de Rijke, M., Kenter, T., de Vries, A.P., Zhai, C.X., de Jong, F., Radinsky, K., Hofmann, K. (eds.) ECIR 2014. LNCS, vol. 8416, pp. 771–774. Springer, Heidelberg (2014)
5. Buhalis, D., Law, R.: Progress in information technology and tourism management: 20 years on and 10 years after the internet - the state of eTourism research. Tourism Management **29**(4), 609–623 (2008)
6. Chang, C.H., Kayed, M., Girgis, M.R., Shaalan, K.F.: A survey of web information extraction systems. IEEE Transactions on Knowledge and Data Engineering **18**(10), 1411–1428 (2006)
7. Dell'Erba, M., Fodor, O., Höpken, W., Werthner, H.: Exploiting semantic web technologies for harmonizing e-markets. J. of IT & Tourism **7**(3–4), 201–219 (2005)
8. Dogac, A., Kabak, Y., Laleci, G., Sinir, S.S., Yildiz, A., Kirbas, S., Gurcan, Y.: Semantically enriched web services for the travel industry. SIGMOD Record **33**(3), 21–27 (2004)
9. Fensel, F., Facca, F.M., Simperl, E., Toma, L.: Semantic Web Service. Springer, Heidelberg (2011)

10. Hepp, M., Siorpaes, K., Bachlechner, D.: Towards the semantic web in e-tourism: can annotation do the trick? In: Ljungberg, J., Andersson, M. (eds.) Proceedings of the Fourteenth European Conference on Information Systems, ECIS 2006, Göteborg, Sweden, pp. 2362–2373 (2006)
11. Jakkilinki, R., Georgievski, M., Sharda, N.: Connecting destinations with an ontology-based e-tourism planner. In: Sigala, M., Mich, L., Murphy, J. (eds.) Information and Communication Technologies in Tourism, ENTER 2007, Proceedings of the International Conference in Ljubljana, Slovenia, pp. 21–32. Springer (2007)
12. Lisi, F.A., Esposito, F.: Semantic web services for integrated tourism in the apulia region. In: Giordano, L., Gliozzi, V., Pozzato, G.L. (eds.) Proceedings of the 29th Italian Conference on Computational Logic, Torino, Italy, June 16–18, 2014. CEUR Workshop Proceedings, vol. 1195, pp. 178–193. CEUR-WS.org (2014)
13. Lisi, F.A., Mencar, C.: Towards fuzzy granulation in OWL ontologies. In: Ancona, D., Maratea, M., Mascardi, V. (eds.) Proceedings of the 30th Italian Conference on Computational Logic, Genova, Italy, June 30 – July 2, 2015. CEUR Workshop Proceedings. CEUR-WS.org (2015)
14. Lisi, F.A., Esposito, F.: Supporting integrated tourism services with semantic technologies and machine learning. In: Horridge, M., Rospocher, M., van Ossenbruggen, J. (eds.) Proceedings of the ISWC 2014 Posters & Demonstrations Track a Track within the 13th International Semantic Web Conference, ISWC 2014, Riva del Garda, Italy, October 21, 2014. CEUR Workshop Proceedings, vol. 1272, pp. 361–364. CEUR-WS.org (2014)
15. Lisi, F.A., Straccia, U.: A FOIL-like method for learning under incompleteness and vagueness. In: Zaverucha, G., Santos Costa, V., Paes, A. (eds.) ILP 2013. LNCS, vol. 8812, pp. 123–139. Springer, Heidelberg (2014)
16. Mädche, A., Staab, S.: Applying semantic web technologies for tourism information systems. In: Wöber, K., Frew, A., Hitz, M. (eds.) Information and Communication Technologies in Tourism 2002, pp. 311–319. Springer, Vienna (2002)
17. Mädche, A., Staab, S.: Services on the move - towards P2P-enabled semantic web services. In: Frew, A., Hitz, M., O'Connor, P. (eds.) Information and Communication Technologies in Tourism 2003. Springer (2003)
18. Missikoff, M., Werthner, H., Hoepken, W., Dell'Erba, M., Fodor, O., Formica, A., Taglino, F.: Harmonise - towards interoperability in the tourism domain. In: Frew, A., Hitz, M., O'Connor, P. (eds.) Information and Communication Technologies in Tourism 2003, pp. 58–66. Springer, Vienna (2003)
19. Oliver, T., Jenkins, T.: Sustaining rural landscapes: The role of integrated tourism. Landscape Research 28(3), 293–307 (2003)
20. Redavid, D., Iannone, L., Payne, T.R., Semeraro, G.: OWL-S atomic services composition with SWRL rules. In: An, A., Matwin, S., Raś, Z.W., Ślezak, D. (eds.) Foundations of Intelligent Systems. LNCS (LNAI), vol. 4994, pp. 605–611. Springer, Heidelberg (2008)
21. Ricca, F., Dimasi, A., Grasso, G., Ielpa, S.M., Iiritano, S., Manna, M., Leone, N.: A logic-based system for e-tourism. Fundamenta Informaticae 105(1–2), 35–55 (2010)
22. Siorpaes, K., Bachlechner, D.: Ontour: tourism information retrieval based on YARS. In: Demos and Posters of the 3rd European Semantic Web Conference (ESWC 2006), Budva, Montenegro, 11th 14th June, 2006 (2006)
23. Staab, S., Studer, R. (eds.): Handbook on Ontologies, 2nd edn., International Handbooks on Information Systems. Springer (2006)
24. Werthner, H., Klein, S.: Information Technology and Tourism - A Challenging Relationship. Springer Verlag, Wien (1999)

Testing a Learn-Verify-Repair Approach
for Safe Human-Robot Interaction

Shashank Pathak[1], Luca Pulina[2][(⊠)], and Armando Tacchella[3]

[1] iCub Facility, Istituto Italiano di Tecnologia (IIT), Via Morego, 30, Genova, Italy
Shashank.Pathak@iit.it
[2] POLCOMING, Università degli Studi di Sassari, Viale Mancini 5, Sassari, Italy
lpulina@uniss.it
[3] DIBRIS, Università degli Studi di Genova, Via Opera Pia 13, Genova, Italy
Armando.Tacchella@unige.it

Abstract. Ensuring safe behaviors, i.e., minimizing the probability that a control strategy yields undesirable effects, becomes crucial when robots interact with humans in semi-structured environments through adaptive control strategies. In previous papers, we contributed to propose an approach that (*i*) computes control policies through reinforcement learning, (*ii*) verifies them against safety requirements with probabilistic model checking, and (*iii*) repairs them with greedy local methods until requirements are met. Such learn-verify-repair work-flow was shown effective in some — relatively simple and confined — test cases. In this paper, we frame human-robot interaction in light of such previous contributions, and we test the effectiveness of the learn-verify-repair approach in a more realistic factory-to-home deployment scenario. The purpose of our test is to assess whether we can verify that interaction patterns are carried out with negligible human-to-robot collision probability and whether, in the presence of user tuning, strategies which determine offending behaviors can be effectively repaired.

1 Introduction

Human-Robot Interaction (HRI) — see, e.g., [8] — is a broad topic in modern robotics, covering tasks which require a robot and a human to share a common workspace and engage in physical activities, as well as other forms of interaction. Safety in the context of HRI is recognized as a standing challenge in robotics [9]. One of the most important reasons is the fact that, outside controlled factory or laboratory settings, the environment is usually unstructured and potentially hostile. In turn, this increases the chance of inadequate perception and thus the risk of undesirable behaviors. Since robots are capable of inflicting severe physical damage, when interaction with humans happens outside "safety bubbles" the problem of preserving the environment against potential odds arises even if the behavior of the robot is fully scripted. The problem of ensuring safety is even more challenging when the robot is also expected to learn a task [5]. While there are attempts to solve this problem, e.g., using learning by demonstration, or

© Springer International Publishing Switzerland 2015
M. Gavanelli et al. (Eds.): AI*IA 2015, LNAI 9336, pp. 260–273, 2015.
DOI: 10.1007/978-3-319-24309-2_20

ensuring safety via a separate procedure that accounts for collision probabilities with the human, they often require restrictions on the environment, or the robot, or the kind of interactions the human is expected to perform.

In previous papers [14,15], we contributed to define a methodology based on a learn-verify-repair work-flow that seeks to approach the problem of learning safe control policies in a principled, yet widely applicable way. In particular, in [15] it is shown how such policies can be (*i*) computed using reinforcement learning starting from minimal assumptions about the environment, (*ii*) verified using probabilistic model checking, and (*iii*) repaired using — largely heuristic — verification-based local methods. For a simple reach-avoid task, it is shown that restrictions introduced in previous literature such as reactive-only safety and learning-by-demonstration of safe tasks can be lifted using the learn-verify-repair work-flow. In [14], the approach is formalized in the context of probabilistic model repair, and a greedy local repair algorithm is shown to be sound and complete.

The main contribution of this work is to consider HRI as an essential ingredient of the intended application scenario, and put the learn-verify-repair approach to the test using a simulated, yet challenging and realistic, case study. In particular, we consider the case in which a robot endowed with manipulation abilities is taught a cooperative task at factory. The task is assumed to be complex enough to require learning, e.g., because some parameters of the system are not known exactly [13], and critical enough so that safety of the learning results must be verified. At the factory stage, only learning and verification are required, because it is assumed that technicians are expert enough to tune learning until safety requirements are fulfilled. Once the robot is shipped to the end-user, she has the possibility to customize the factory-learned policy in order to make the robot fit her needs. At this stage, we would like the learn-verify-repair approach to be able to detect inappropriate, i.e., unsafe, customizations and repair customized policies so that safety is preserved. Since the repair procedure is automated, the end-user need not to be an expert in learning. Still, she might be forced to accept some compromise made by the repair procedure which is oblivious of her preferences. In the end, we need to check whether the right balance between user satisfaction and uncompromising safety can be struck.

The rest of the paper is structured as follows. In Section 2 the models, the algorithms and the techniques to be used throughout the paper are introduced. In Section 3, the mathematical modeling of safe HRI is described. The main contributions are detailed in Section 4, describing the scenario considered, and Section 5, describing the experimental results. Section 6 concludes the paper with some final remarks.

2   Preliminaries

2.1   Reinforcement Learning

The formal framework underlying Reinforcement Learning (RL) is that of Markov decision processes — see, e.g. [17]. In particular, a *(stationary, discrete time) Markov decision process* is a collection of objects $\mathcal{D} = \{T, S, A_s, p(\cdot|s, a), r(s, a)\}$ where T is a set of points in time called *decision epochs* such that $T \subseteq \mathbb{N}^+$; S is a finite, time-independent set of *states* which the system occupies at each decision epoch; A_s is a finite time-independent set of *actions* allowable in some state $s \in S$ where $A = \bigcup_{s \in S} A_s$ collects all actions; $p(j|s, a)$ is a (non-negative) *transition probability function* such that $\sum_{j \in S} p(j|s, a) = 1$, denoting the probability that the system is in state $j \in S$ at time $t + 1$ when action $a \in A_s$ is accomplished in state s at time t; $r(s, a)$ is a real-valued bounded function, i.e., $|r(s, a)| \leq M$ for some finite $M \in \mathbb{R}$; r is defined for all $s \in S$ and $a \in A_s$, and it denotes the value of the *reward* received when executing action a in state s. It is assumed that the process is *infinite horizon*, i.e., the set of decision epochs is $T \equiv \{1, 2, \ldots\}$. A *(stationary) policy* is a procedure for action selection in each state. Policies can be *deterministic*, i.e., return a specific action for each state, or *probabilistic*, i.e., return the probability of executing an action in a state. The *expected total discounted reward* (also the *value*) of policy π can be defined for every state $s \in S$ as

$$V_\gamma^\pi(s) = \lim_{k \to \infty} E_s^\pi \{ \sum_{t=1}^{k} \gamma^{t-1} r(X_t, Y_t) \} \qquad (1)$$

where X_t and Y_t are random variables to take values in S and A, respectively; $E_s^\pi \{\ldots\}$ denotes the expectation with respect to the policy π conditioned to $X_1 = s$; the parameter $\gamma \in [0, 1)$ is a real constant called *discount factor* that determines the relative value of delayed versus immediate rewards, the former being weighted (exponentially) less than the latter. Given the assumption that $|r(s, a)| \leq M$ for some finite $M \in \mathbb{R}$, it is always guaranteed that the limit in (1) exists. In the following, γ is omitted and $V^\pi(s)$ denotes the value of policy π in state s. Given any two policies π and π', it is the case that π *dominates* π', denoted as $\pi \geq \pi'$, exactly when $V^\pi(s) \geq V^{\pi'}(s)$ for all $s \in S$. An *optimal policy* π^* is a policy such that $\pi^* \geq \pi$ for every policy π. Given a Markov decision process $\mathcal{D}$ and some optimality criterion on policies as the one in (1), the *Markov decision problem* (MDP) is the problem of finding an optimal policy in $\mathcal{D}$.

Reinforcement learning is a family of methods to find approximate solutions to MDPs — see [3] and also [20] for a more recent account wherefore this section borrows notation and concepts. The kind of RL algorithms discussed hereafter are part of the *model-free* family, wherein the transition probability function of the underlying Markov decision process is not assumed to be known. To enable computation/approximation of optimal policies in a model-free fashion, the concept of *quality* (also *Q-value*) $Q(s, a) \in \mathbb{R}$ of each action $a \in A$ in each state $s \in S$ is defined as

$$Q_\gamma^\pi(s, a) = \lim_{k \to \infty} E_{s,a}^\pi \{ \sum_{t=1}^{k} \gamma^{t-1} r(X_t, Y_t) \} \qquad (2)$$

where $E_{s,a}^\pi \{\ldots\}$ denotes the expectation with respect to the policy π conditioned to $X_1 = s$ and $Y_1 = a$. While several model-free RL algorithms exists, two well known and well understood ones are *Q-learning* [19], and SARSA [18]. Both algorithms learn on-line while interacting with the environment (or a simulator) and iteratively improve on estimates of Q-values based on previous ones and the immediate rewards received by the environment. More specifically, both algorithms use the following *update equation*

$$Q(s_t, a_t) \leftarrow Q(s_t, a_t) + \alpha \delta_\gamma(s_t, a_t, r_{t+1}, s_{t+1}, a_{t+1}) \qquad (3)$$

where $\alpha \in (0,1)$ is a *step size parameter* and $\delta_\gamma(\cdot)$ is the *temporal difference equation*. In Q-learning, $\delta_\gamma(\cdot)$ is defined as

$$\delta_\gamma(s_t, a_t, r_{t+1}, s_{t+1}, a_{t+1}) = \left[r_{t+1} + \gamma \max_a Q(s_{t+1}, a) - Q(s_t, a_t) \right]. \qquad (4)$$

while in SARSA (State-Action-Reward-State-Action) it is defined as

$$\delta_\gamma(s_t, a_t, r_{t+1}, s_{t+1}, a_{t+1}) = \left[r_{t+1} + \gamma Q(s_{t+1}, a_{t+1}) - Q(s_t, a_t) \right]. \qquad (5)$$

In both (4) and (5) the parameter γ is the discount factor as defined for (2). The main difference between Q-learning and SARSA is that the former is an *off-policy* algorithm: it will converge to an optimal policy[1] regardless of the *exploration policy*, i.e., the policy followed while learning. On the other hand, SARSA is *on-policy*, i.e., an exploration policy must be available to choose an action at each step to compute $\delta_\gamma(\cdot)$.

The exploration policy for SARSA — as well as a final policy for both algorithms — can be computed from Q-values. In particular, a probabilistic policy can be obtained with a *softmax criterion* [3]. The probability π_Q of executing action a in state s is given by

$$\pi_Q(s, a) = (e^{\frac{Q(s,a)}{\beta}}) / (\sum_{\alpha \in A} e^{\frac{Q(s,\alpha)}{\beta}}) \qquad (6)$$

The index Q refers to the fact that π is computed with current Q-values, and $\beta \in (0, +\infty)$ is a *temperature parameter*: for $\beta \to \infty$ (high temperature), all actions are equally probable, whereas for $\beta \to 0$ (low temperature), the probability of the action with the highest value tends to 1, i.e, the policy becomes greedy (deterministic). An alternative way to interpolate between probabilistic and deterministic policies is using ϵ-*greedy* policies with $\epsilon \in [0,1]$. A policy $\pi_Q(s, a)$ is ϵ-greedy if, given some state s, it chooses the action a such $a = \text{argmax}_a \{Q(s,a)\}$ with probability $1 - \epsilon$, and it chooses an action uniformly at random with probability ϵ. In the following, when no ambiguity arises, we will drop the subscript Q and leave the dependency of π on Q implicit.

2.2 Probabilistic Model Checking

Following [11], given a set of atomic propositions AP, a *Deterministic Time Markov Chain (DTMC)* is defined as a tuple $(W, \overline{w}, \mathbf{P}, L)$ where W is a finite

---

[1] Under the assumptions that each state-action pair is visited an infinite number of times, and the learning parameter α is decreased appropriately [19].

set of *states*, and $\overline{w} \in W$ is the *initial state*; $\mathbf{P} : W \times W \rightarrow [0,1]$ is the *transition probability matrix*; and $L : W \rightarrow 2^{AP}$ is the *labeling function*. An element $\mathbf{P}(w, w')$ gives the probability of making a transition from state w to state w' with the requirement that $\sum_{w' \in W} \mathbf{P}(w, w') = 1$ for all states $w \in W$. A terminating (absorbing) state w is modeled by letting $\mathbf{P}(w, w) = 1$. The labeling function maps every state to a set of propositions from AP which are true in that state.

Temporal properties about DTMCs are expressed with Probabilistic Computational Tree Logic (PCTL) formulas, whose syntax is defined considering the set Σ of *state formulas*, and the set Π of *path formulas*. Σ is defined inductively as: (i) if $\varphi \in AP$ then $\varphi \in \Sigma$; (ii) $\top \in \Sigma$; if $\alpha, \beta \in \Sigma$ then also $\alpha \wedge \beta \in \Sigma$ and $\neg \alpha \in \Sigma$; and (iii) $\mathcal{P}_{\bowtie p}[\psi] \in \Sigma$ where $\bowtie \in \{\leq, <, \geq, >\}$, $p \in [0,1]$ and $\psi \in \Pi$, where $\mathcal{P}_{\bowtie p}[\psi]$ is the *probabilistic path operator*.

The set Π contains exactly the expressions of type $X\alpha$ (*next*), $\alpha \mathcal{U}^{\leq k} \beta$ (*bounded until*) and $\alpha \mathcal{U} \beta$ (*until*) where $\alpha, \beta \in \Sigma$ and $k \in \mathbb{N}$. We also abbreviate $\top \mathcal{U} \beta$ as $\mathcal{F} \beta$ (*eventually*) and $\top \mathcal{U}^{\leq k} \beta$ as $\mathcal{F}^{\leq k} \beta$ (*bounded eventually*). A path τ is defined as a non-empty sequence of states $w_0, w_1, \ldots$ where $w_i \in W$ and $\mathbf{P}(w_i, w_{i+1}) > 0$ for all $i \geq 0$.

Given a DTMC $\mathcal{M} = \{W, \overline{w}, \mathbf{P}, L\}$ and a state $w \in W$, let $\mathcal{M}, w \models \varphi$ denote that $\varphi \in \Sigma$ is satisfied in w. Similarly, we write $\mathcal{M}, \tau \models \psi$ for a path τ of $\mathcal{M}$ satisfying a formula $\psi \in \Pi$. The semantics of path formulas is

- $\tau \models X\alpha$ exactly when $\tau(1) \models \alpha$;
- $\tau \models \alpha \mathcal{U}^{\leq k} \beta$ exactly when there is some $i \leq k$ such that $\tau(i) \models \beta$ and $\tau(j) \models \alpha$ for all $j < i$;
- $\tau \models \alpha \mathcal{U} \beta$ exactly when there is some $k \geq 0$ such that $\tau \models \alpha \mathcal{U}^{\leq k} \beta$.

Accordingly, the semantics of state formulas is

- $w \models a$ with $a \in AP$ exactly when $a \in L(w)$;
- $w \models \top$ for all $w \in W$; $w \models \alpha \wedge \beta$ exactly when $w \models \alpha$ and $w \models \beta$; $w \models \neg \alpha$ exactly when $w \models \alpha$ does not hold;
- $w \models \mathcal{P}_{\bowtie p}[\psi]$ exactly when $Prob_w(\{\tau \in Path_w \mid \tau \models \psi\}) \bowtie p$.

Intuitively, $Prob_w(\ldots)$ is the probability that the paths in $Path_w$ that also satisfy ψ are taken. $Prob_w(\ldots)$ is provably measurable for all possible PCTL path formulas, and it can be always be computed from $\mathbf{P}$ — see [11] for details.

Given a DTMC $\mathcal{M} = \{W, \overline{w}, \mathbf{P}, L\}$ and a PCTL formula φ, we define *probabilistic model checking* (PMC) as the problem of deciding whether $\mathcal{M}, \overline{w} \models \varphi$. As shown in [7], PMC is decidable in time linear in the size of $\mathcal{M}$. There is a growing number of tools that solve PMC – see http://www.prismmodelchecker.org/other-tools.php – among which we selected MRMC [10].

3   Safe HRI Through Learn-Verify-Repair Work-Flow

Before delving into the details of the test proposed, it is useful to discuss some basic assumptions regarding the learn-verify-approach work-flow in the context of HRI. As mentioned before, the scenarios of interest are those in which the human and the robot interact in a limited and shared physical space to accomplish some cooperative task. In these scenarios, the problem of safety becomes relevant and it must be placed in context with HRI. According to Dautenhahn, human-robot interaction could be classified as human-centered, robot-centered or robot-cognition centered [8]. Considering the latter perspective would bring to the table questions that do not admit viable engineering solutions yet, such as what exactly constitutes safety and how the robot can perceive and recognize (un)safe situations. For all the purposes of this paper, it is sufficient to consider safety either as a human-centered or robot-centered perspective. In both cases, unsafe events can be identified a priori by means of standard engineering approaches, e.g., risk analysis. Furthermore, the choice of expressing safety in formal logic is inherently close to the question of ethics in HRI, especially when a deontic ethical code should be given to the system — see, e.g., [4]. In the following, the assumptions that (i) safety can be identified a priori and (ii) it can be described by a (probabilistic, temporal) logic formula, are always taken for granted.

Under the assumption that the environment — including the human interacting with the robot — can be modeled as a Markov decision process, and a suitable policy to interact successfully with such environment can be computed by RL, the problem of modeling HRI boils down to the problem of describing all possible interactions between the robot and the environment. From a mathematical point of view, when a policy π is combined with some underlying Markov decision process $\mathcal{D}$, the resulting model $\mathcal{M}_{\mathcal{D},\pi}$ is a DTMC. Intuitively, since the robot can assign probability to each action according to the policy π, the transition probability $\mathbf{P}(s,s')$ from any given state s to some other s' can be computed by marginalizing on actions. If safety is defined according to some PCTL formula φ, then checking whether $\mathcal{M}_{\mathcal{D},\pi}, s_0 \models \varphi$ holds — given some known initial state s_0 — will solve the problem of assessing whether HRI is safe. In theory, the approach is efficient since model checking a PCTL property over a DTMC is polynomial in the size of the DTMC [2]. In practice, however, one should always consider the fact that the state space will be large, unless very simple domains are considered.

As we mentioned before, in [14,15] the problem of verifying PCTL properties is considered together with the problem of "fixing" issues before deploying the system being analyzed. In [15] a classical reach-avoid task is considered, involving safety as obstacle avoidance. In [14] a greedy method for repairing a DTMC $\mathcal{M}$ in such a way that some property ϕ holds within a desired level probability σ is considered, i.e., if $\mathcal{M} \models \mathcal{P}_{<\sigma}[\phi]$ is not satisfied, find M' such that $\mathcal{M}' \models \mathcal{P}_{<\sigma}[\phi]$ holds. Here, a merger that improves on both previous contributions is proposed. Greedy model repair methods are considered in a context where the robot is interacting with a human, not just objects as in [15], and a realistic interaction pattern is considered, rather than simple examples as in [14]. The exercise which is accomplished here can be broadly categorized as repairing unsafe HRI, in the sense that (i) unsafety of HRI was established using PCTL model checking to

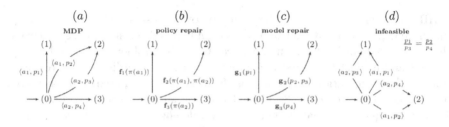

Fig. 1. Model repair and policy repair. (a): interaction of agent and environment modeled as a Markov decision process with actions $a_i \in \mathcal{A}$ and transition probabilities $p_i \in \mathbb{R}_{[0,1]}$. (b): repairing the policy, edge probabilities are linear functions $\mathbf{f}$ of variables $\pi_{a_i} \in \mathbb{R}_{[0,1]}$. (c): repairing the model, edge probabilities are linear functions $\mathbf{g} \mapsto \mathbb{R}$ of variables $p_i \in \mathbb{R}_{[0,1]}$. (d): a pathological case where model repair is infeasible.

show that a property like those mentioned in the previous subsection is violated, and (ii) modifying the robot policy is performed with the aim of making the property satisfied. The probabilistic model $\mathcal{M}_{\mathcal{D},\pi}$ which is to be repaired with respect to some PCTL property φ emerges from the combination of stochasticity of the environment — a Markov decision process $\mathcal{D}$ — and the behavior of the agent — the policy π. Since for any DTMC $\mathcal{M}$ which violates a property φ it is possible to define a repaired DTMC $\hat{\mathcal{M}}$ [14] repair itself could be seen as repair of the Markov decision process to yield another process $\hat{\mathcal{D}}$ or of the policy followed by the agent to yield another policy $\hat{\pi}$. We call the former as *model repair* while latter as *policy repair*. The difference between the two procedures is exemplified in Figure 1. It should be noted that the notion of model repair is weaker than the one used by others authors. For instance, Chen et al. define model repair over the complete set of feasible policies [6]. Irrespective of the kind of repair, the constituent steps involve appropriate assignment to the variables which can indeed be modeled as multi-variable optimization problem. An important issue to note here is the feasibility of such a repair. Given the constraint of preserving the topology of the graph, a model repair would usually be feasible. However, a pathological case could be as shown in Fig. 1, under condition $\frac{p_1}{p_2} = \frac{p_3}{p_4}$. This would mean that no matter how one changes policy (and hence a_1 and a_2), the edges of resulting DTMC would have same probability values.

Another related issue shown in Fig. 2 regards the proper change in the *on-policy* learning of the agent, after an automated repair is performed. This is interesting because it would enable the agent to interleave learning with repair. As mentioned before we looked into two most popular policies of agents in such setups: π_ϵ and π_β, defined over *Q-value* of state s. Since $Q(s,a) \xrightarrow{\pi_\epsilon} \mathcal{T}(a|s)$ is not an invertible mapping, a naïve approach is to impose an additional constraint of minimal change in Q-values. Similarly, it can be seen that $Q(s,a) \xrightarrow{\pi_\beta} \mathcal{T}(a|s)$ is not invertible too; nevertheless, simple constraints such as maintaining the same highest Q-value would fetch a unique solution.

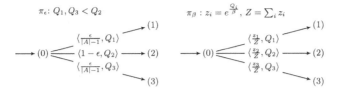

Fig. 2. Relation between policy, Q-value and *repair* in case of policy repair of DTMC $\mathcal{M}$. The repaired node (0), its outgoing edges as well as children nodes, (1), (2) and (3) are shown here.

4   Challenging the Learn-Verify-Repair Work-Flow

Potentially, HRI may involve a wide variety of scenarios each characterized by different safety constraints and learning objectives. While, at least in principle, the approach summarized in the previous Section can deal with many such scenarios, in order to obtain a quantitative assessment of the learn-verify-repair work-flow it is useful to focus on a specific scenario which contains all the basic ingredients found in more complex ones. In particular, the case of a single robot interacting with a single human across a common workspace is considered. It is assumed that the robot observes the human while she is accomplishing a given task, which at some point, requires the robot to chip in and, e.g., finalize the task alone or help the human to do so. For the case in point, the task to be accomplished by the robot is to touch a ball that the user is moving according to a fixed pattern on a table. The task must be learned by the robot, but RL is run offline in a simulator to avoid risk of injuries to the human during the trial-and-error process which characterizes RL.

As shown in Figure 3 two different flows of activities are considered. The first one — Figure 3 (left) — is thought to happen at the end of the production stage (factory), where the robot is configured, trained and checked by experts to accomplish a given task. The second one — Figure 3 (right) — is thought to happen during the deployment stage (e.g., household), where the user is allowed to (*i*) *calibrate* the robot, i.e., adapt its behavior to the contingencies of the environment to be found at the user's place, and to (*ii*) *modify* the robot's behavior, i.e., customize the robot according to specific preferences. For the case considered, an example of calibration could refer to the pattern being executed in a slightly different way by the user, whereas an example of behavior modification might be trying to make the robot move faster than the factory setting, at the expense of some accuracy. In all the phases, safety of HRI must be ensured.

Considering the activity flow performed at the factory — Figure 3 (left) — it can be observed that the trainer of the robot is probably some expert technician with programmer-level access to the robot internals. Furthermore, it is assumed that she has complete and fine grained control over the various phases of the process. Because of this, she may (*i*) configure, (*ii*) train and (*iii*) run simulations to compute policies until a policy complying with the safety (and effectiveness)

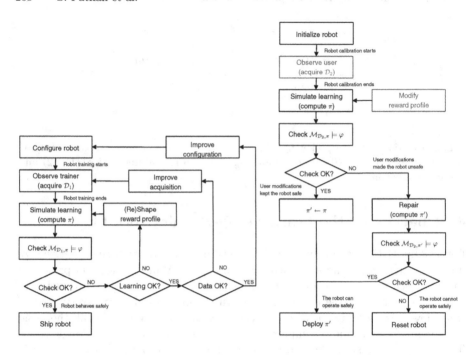

Fig. 3. Experimental setup: learning, verification and repair activities at the factory (left) and at the user place (right). The boxes in red denote activities in which the end user can calibrate the robot or modify its behavior.

requirements is achieved. At this stage, verification is required to ensure that the robot indeed meets safety requirements, but it is the technician's responsibility to "tune" phases $(i) - (iii)$ in order to achieve the stated objectives. It is worth observing that (iii) requires a great deal of expertise in *reward shaping*, i.e., the process of assigning rewards at each step of the RL process in order to make the algorithm converge to a close-to optimal policy. Failure to shape rewards in a sensible way may result in sub-optimal or, worse, unsafe policies, while the precise relationship between rewards and the result of learning remains largely elusive — see, e.g., [1].

At the user's site — Figure 3 (right) — the person interacting with the robot is now a potentially inexperienced end-user. While she is not granted programmer-level access to the robot internals, she might still enjoy a certain level of customizability of the robot. In particular, calibration and modification of robot's behavior might be purposefully granted to the end-user in order to improve interaction. As it can be seen from the figure, calibration amounts to observe the end-user and adapt the behavior of the robot accordingly, i.e., learn a new policy π. Even if learning occurs using the same reward profile that was set at the factory, the learned policy π might not meet the safety requirements, i.e., it could be the case that $\mathcal{M}_{\mathcal{D}_2,\pi} \not\models \varphi$, where φ is the PCTL formula representing

Table 1. The table shows statistics about the probability of reaching the target state p_T after learning in the factory settings. Mean and standard deviation for this probability are computed across different sequences of episodes.

Learning Policy	$\hat{p}_T$				$\tilde{p}_T$			
	N_{20}	N_{25}	N_{35}	N_{50}	N_{20}	N_{25}	N_{35}	N_{50}
SARSA π_β	0.128	0.876	1	1	0.034	0.014	0	0
Q-learning π_β	0.879	0.921	0.952	0.957	0.036	0.023	0.018	0.018
SARSA π_ϵ	0.033	0.040	0.060	0.107	0.020	0.020	0.023	0.033
Q-learning π_ϵ	0.492	0.579	0.555	0.613	0.033	0.039	0.043	0.034

safety requirements. In this case, since the end-user is not allowed programmer-level access, the only way to make the policy viable is to repair it automatically. If it is assumed that the end-user is able also to modify the reward profile — e.g., using apprenticeship learning as in [1] — then it is also the case that an acquisition of the user behavior very close to the one set at the factory level, might end up in a different policy, potentially unsafe. Also in this case, repair might fix the resulting policy to guarantee safety. If, in spite of repaired policy π', the safety requirements are not met, then the robot must be reset to factory settings.

5   Experimental Results

Our experimental setup involves a simulator of the humanoid iCub [12] wherein, for simplicity, the human-robot interaction is restricted to grab a ball moved in a planar pre-determined trajectory — an ellipse in our case. Even if the results we present are obtained with a simulator, there are some elements which we acquired from the real setup, namely (i) the sequence of ball movements and (ii) the sequence of the human elbow movements. In this setting, the state space S is nine dimensional and it is composed of two coordinates for the ball on the plane, three for the elbow of the human and four for the arm of the robot. The action space A is also four dimensional and it corresponds to a planned movement of the robot's hand plus its orientation. For simulation, this is sufficient to obtain reasonable results. In the real setup, a low-level controller [16] takes care of converting a planned destination in the Cartesian space to a smooth movement pattern. Therefore, the results that we obtain in simulation should be reproducible in the real setup without a lot of additional effort. In our experiments, safety always involves avoiding collision with the human forearm. While the forearm position is not directly observed, an estimation is possible knowing the ball position and the elbow position. The movement of the robot is deemed unsafe when its hand enters a cylinder of suitable radius whose vertical axis passes through the ball position and the elbow position.

Initially, the "factory" setting is considered, learning is conducted for k-thousands of episodes (denoted by N_k), and safety is assessed as reachabil-

Table 2. Times (in seconds) for verifiably safe HRI for different learnings. Here, learning algorithm is one of {Q-learn, SARSA}, exploration strategy is either of {π_β, π_ϵ}. Subscript ro denotes case where rewarding is oblivious of safety condition, else no subscript is used.

Learning Policy	Verify				Repair			
	N_{20}	N_{25}	N_{35}	N_{50}	N_{20}	N_{25}	N_{35}	N_{50}
$SARSA_{ro}\ \pi_\epsilon$	786.368	794.780	831.873	878.995	1.491	1.527	1.512	1.568
$SARSA_{ro}\ \pi_\beta$	73.498	65.778	68.816	70.247	1.401	1.542	1.585	1.569
Q-learn $\quad \pi_\epsilon$	22.479	95.632	538.107	209.435	1.081	1.424	1.816	2.022
Q-learn $\quad \pi_\beta$	82.120	197.988	179.962	287.755	1.692	1.763	2.440	2.006
SARSA $\quad \pi_\epsilon$	2.151	7.222	5.824	4.357	0.666	1.228	1.141	1.100
SARSA $\quad \pi_\beta$	116.691	127.224	64.851	63.908	1.281	1.404	1.310	1.300

ity to unsafe states, i.e., collision with the human.[2] A number of runs is made for each sequence of episodes, to account for randomization occurring in the learning algorithms. The main parameter of interest is the reachability to the target state p_T, i.e., the probability of the robot grasping the ball without hitting the human. In this simple setup, this is always complementary to hitting the human, so the probability of unsafe states is always $1 - p_T$. Since many runs are considered, we report mean $\hat{p}_T$ and standard deviation $\tilde{p}_T$ of p_T over such runs. Table 1 shows the progress of learning under a choice of learning algorithms — Q-learning or SARSA — and exploration strategies — ϵ-greedy or softmax-β. The progress of learning is monitored by computing p_T via model checking. With the progress of learning, p_T increases and thus safety improves. However, different learning approaches have different convergence rates: learning via SARSA and exploring the actions through softmax policy performs best in this regard, reaching optimal results over 35000 episodes. Other methods do not converge to the optimal policy even after 50000 episodes. This state of affairs corresponds with well-known results in the literature [3]. While the best case scenarios do not require further investigation with formal methods — the probability of colliding with the human is made arbitrarily small by the learning process — it is worth asking the question of how repair could be useful in cases where an optimal policy cannot be achieved.

In Table 2, a set of results related to repairing of policies is considered. In the table, it is possible to observe that even sub-optimal policies can be repaired. Here, the main bottleneck is the time for verification, whereas time to repair remains fairly constant across different settings. Since verification entails a global analysis, its runtime depends substantially on the topology of the model, whereas repair is a local method, fairly insensitive to the size and structure of the overall model [14]. In practice, repair is up to two orders of magnitude faster than verification. As it would be expected, both verification and repair

---

[2] All the experiments are performed on a Intel i5 PC with 8GB of RAM running Linux Ubuntu 14.04.

are independent of the approach chosen to implement learning. The results of Table 2 tell us that verification must be performed as an off-line process because it is fairly time consuming even is such a simple setup. On the other hand, repair has the potential of being performed as an on-line activity. The question which now arises is whether repair can be useful at the user's site to correct customizations that turn out to be not safe enough.

Assessing whether repair can be useful to correct haphazard customizations is the purpose of the experiment reported in Table 3. Here we consider again the learn-verify-repair work-flow, but this time with slightly different settings of rewards as they may result after modification at the "end-user" site. Here, the variation of the rewarding schema is modeled as noise, which has one of the following forms:

$$\mathbf{R} = \begin{cases} \mathbf{R_0} + \mathcal{U}[\mu - \frac{\delta}{2}, \mu + \frac{\delta}{2}] & \text{additive-uniform} \\ \mathbf{R_0} * \mathcal{U}[\mu - \frac{\delta}{2}, \mu + \frac{\delta}{2}] & \text{multiplicative-uniform} \\ \mathbf{R_0} + \mathcal{N}[\mu, \sigma] & \text{additive-normal} \\ \mathbf{R_0} * \mathcal{N}[\mu, \sigma] & \text{multiplicative-normal} \end{cases} \quad (7)$$

When observed against the reachability to the target state, effect of noise shows no discernible pattern, i.e., the magnitude of rewards does not have a direct bearing over the convergence of learning. As expected, reachability to the target state increases with the increase in episodes, and if learning is carried out for many episodes, there is no point in applying repair at all. Actually, this is true if we assume, as we did, that learning is simulated in the robot's internal computing resources and not executed on-line with a human subject. In this case, it would not be possible to reach the number of episodes required for convergence of RL, but the learning policy could be too suboptimal for repair to produce any viable result. In this direction, further experiments with the real robot are required to establish whether repair can improve on the results of learning.

6   Conclusions

In this paper, the learn-verify-repair approach proposed in [15] and later refined in [14] is put to the test considering a realistic usage scenario, wherein the robot is trained at factory by experts, and customized at the user's site. The goal of the approach tested is to decouple safety assessment from the robotic platform, low level control laws and learning algorithms. In practice, framing safety as formal verification problem and harnessing probabilistic model checking is a potential enabler for both behavioral safety assessment and policy repair. The experiments that we have presented in this paper show that the former aim can be fulfilled, at least on problems of relatively small scale, whereas the second it is feasible and scalable, but still not entirely useful in cases where learning alone can reach high levels of safety. These are the cases in which safety and effectiveness "agree with each other", i.e., the more effective is the robot to grasp the ball, the safer it will be. Therefore, more investigation on the scalability of verification and the

Table 3. Variation of reachabilities to the target state obtained by the policy learned through N_i episodes. Rewards is randomized by applying an additive noise which is either uniformly random across width δ of the current reward vectors $\mathbf{R}_0$ or is normal with std. deviation δ. Noise levels low, medium and high stand for δ being 5%, 50% and 100% of $\mathbf{R}_0$. Five random runs are made to obtain the learned policy, and subscript μ and σ signify mean and standard deviation of reachabilities across such runs.

type	level	$\hat{p}_T$				$\tilde{p}_T$			
		N_{20}	N_{25}	N_{35}	N_{50}	N_{20}	N_{25}	N_{35}	N_{50}
uniform	low	0.126	0.724	0.991	0.998	0.034	0.014	0.003	0.003
uniform	med	0.487	1	1	1	0.021	0	0	0
uniform	hig	0.309	1	1	1	0.028	0	0	0
normal	low	0.678	1	1	1	0.019	0	0	0
normal	med	0.155	1	1	1	0.031	0	0	0
normal	hig	0.501	0.857	1	1	0.024	0.007	0	0

usefulness of repair might be required in out-of-factory contexts. However the results herewith presented suggest that the learn-verify-repair methodology can be useful in practice for debugging reward profiles and repairing close-to-optimal policies.

References

1. Abbeel, P., Ng, A.Y.: Apprenticeship learning via inverse reinforcement learning. In: Proceedings of the twenty-first international conference on Machine learning, p. 1. ACM (2004)
2. Baier, C., Katoen, J.P., et al.: Principles of model checking, vol. 26202649. MIT press, Cambridge (2008)
3. Barto, A.G.: Reinforcement learning: An introduction. MIT press (1998)
4. Bringsjord, S., Arkoudas, K., Bello, P.: Toward a general logicist methodology for engineering ethically correct robots. IEEE Intelligent Systems **21**(4), 38–44 (2006)
5. Calinon, S., Sardellitti, I., Caldwell, D.G.: Learning-based control strategy for safe human-robot interaction exploiting task and robot redundancies. In: 2010 IEEE/RSJ International Conference on Intelligent Robots and Systems (IROS), pp. 249–254. IEEE (2010)
6. Chen, T., Hahn, E.M., Han, T., Kwiatkowska, M., Qu, H., Zhang, L.: Model repair for Markov decision processes. In: 2013 International Symposium on Theoretical Aspects of Software Engineering (TASE), pp. 85–92. IEEE (2013)
7. Courcoubetis, C., Yannakakis, M.: The complexity of probabilistic verification. Journal of the ACM (JACM) **42**(4), 857–907 (1995)
8. Dautenhahn, K.: Socially intelligent robots: dimensions of human-robot interaction. Philosophical Transactions of the Royal Society B: Biological Sciences **362**(1480), 679–704 (2007)
9. Haddadin, S., Albu-Schäffer, A., Hirzinger, G.: Safe physical human-robot interaction: measurements, analysis & new insights. In: International Symposium on Robotics Research (ISRR2007), Hiroshima, Japan, pp. 439–450 (2007)

10. Katoen, J., Zapreev, I., Hahn, E., Hermanns, H., Jansen, D.: The ins and outs of the probabilistic model checker mrmc. Performance Evaluation **68**(2), 90–104 (2011)
11. Kwiatkowska, M., Norman, G., Parker, D.: Stochastic model checking. In: Bernardo, M., Hillston, J. (eds.) SFM 2007. LNCS, vol. 4486, pp. 220–270. Springer, Heidelberg (2007)
12. Metta, G., Natale, L., Nori, F., Sandini, G., Vernon, D., Fadiga, L., von Hofsten, C., Rosander, K., Lopes, M., Santos-Victor, J., et al.: The iCub Humanoid Robot: An Open-Systems Platform for Research in Cognitive Development. Neural networks: the official journal of the International Neural Network Society (2010)
13. Ng, A., Coates, A., Diel, M., Ganapathi, V., Schulte, J., Tse, B., Berger, E., Liang, E.: Autonomous inverted helicopter flight via reinforcement learning. In: Experimental Robotics IX, pp. 363–372 (2006)
14. Pathak, S., Ábrahám, E., Jansen, N., Tacchella, A., Katoen, J.-P.: A greedy approach for the efficient repair of stochastic models. In: Havelund, K., Holzmann, G., Joshi, R. (eds.) NFM 2015. LNCS, vol. 9058, pp. 295–309. Springer, Heidelberg (2015)
15. Pathak, S., Pulina, L., Metta, G., Tacchella, A.: Ensuring safety of policies learned by reinforcement: reaching objects in the presence of obstacles with the icub. In: 2013 IEEE/RSJ International Conference on Intelligent Robots and Systems, Tokyo, Japan, November 3–7, 2013, pp. 170–175. IEEE (2013)
16. Pattacini, U., Nori, F., Natale, L., Metta, G., Sandini, G.: An experimental evaluation of a novel minimum-jerk cartesian controller for humanoid robots. In: 2010 IEEE/RSJ International Conference on Intelligent Robots and Systems, October 18–22, 2010, Taipei, Taiwan, pp. 1668–1674 (2010)
17. Puterman, M.L.: Markov decision processes: discrete stochastic dynamic programming, vol. 414. John Wiley & Sons (2009)
18. Rummery, G., Niranjan, M.: On-line Q-learning using connectionist systems. University of Cambridge, Department of Engineering (1994)
19. Watkins, C.J., Dayan, P.: Q-learning. Machine Learning **8**(3–4), 279–292 (1992)
20. Wiering, M., Van Otterlo, M.: Reinforcement learning. In: Adaptation, Learning, and Optimization, vol. 12. Springer (2012)

An Approach to Predicate Invention
Based on Statistical Relational Model

Stefano Ferilli[1,2]([✉]) and Giuseppe Fatiguso[1]

[1] Dipartimento di Informatica, Università di Bari, Bari, Italy
stefano.ferilli@uniba.it
[2] Centro Interdipartimentale per la Logica e sue Applicazioni,
Università di Bari, Bari, Italy

Abstract. Predicate Invention is the branch of symbolic Machine Learning aimed at discovering new emerging concepts in the available knowledge. The outcome of this task may have important consequences on the efficiency and effectiveness of many kinds of exploitation of the available knowledge. Two fundamental problems in Predicate Invention are how to handle the combinatorial explosion of candidate concepts to be invented, and how to select only those that are really relevant. Due to the huge number of candidates, there is a need for automatic techniques to assign a degree of relevance to the various candidates and select the best ones. Purely logical approaches may be too rigid for this purpose, while statistical solutions may provide the required flexibility.

This paper proposes a new Statistical Relational Learning approach to Predicate Invention. The candidate predicates are identified in a logic theory, rather than in the background knowledge, and are used to restructure the theory itself. Specifically, the proposed approach exploits the Markov Logic Networks framework to assess the relevance of candidate predicate definitions. It was implemented and tested on a traditional problem in Inductive Logic Programming, yielding interesting results.

1 Introduction

In the context of symbolic Machine Learning, Predicate Invention (PI) deals with the problem of discovering new emerging concepts in the available knowledge. Invention of suitable predicates is so relevant not just because it allows to identify hidden concepts in the available knowledge, but also because it may have important consequences on the efficiency and effectiveness of many kinds of tasks involved in the exploitation of such a knowledge. E.g., it allows theory restructuring in order to have more compact and comprehensible theories; it may help in using and linking multi-domain knowledge; it may suggest how to shift the language towards higher-level representations; it may enrich the available knowledge and improve the comprehension of a domain by human experts; etc. Associated to inductive learning, in particular, PI would allow to obtain a compression of the learned theory or to catch exceptions in the learned clauses.

© Springer International Publishing Switzerland 2015
M. Gavanelli et al. (Eds.): AI*IA 2015, LNAI 9336, pp. 274–287, 2015.
DOI: 10.1007/978-3-319-24309-2_21

Despite the problem has been investigated for many years, limited results have been obtained due to its intrinsic complexity.

Two fundamental problems in PI are how to handle the combinatorial explosion of candidate concepts to be invented, and how to select only those that are really relevant. Indeed, a huge number of possible concepts might be defined with the given knowledge, most of which are just casual or otherwise irrelevant aggregations of features. So, proper filtering must be carried out in order to keep the most significant and relevant concepts only. Due to the large number of candidates, manual selection by domain experts is infeasible, and automatic techniques are needed to assign a degree of relevance to the various candidates and select the best ones.

PI can be carried out on two different kinds of sources: on a background knowledge made up of facts, or on a general theory made up of rules. In a Machine Learning setting, both are available: the theory is the model obtained as a result of running some learning system on the facts that describe the observations and the examples. In some sense, such a model has already performed a kind of selection of relevant information in the background knowledge (in this case, information that may characterize a set of classes and/or discriminate them from other classes). So, working on the theory should make the predicate invention problem somehow easier while still accounting also for the background knowledge.

Inductive Logic Programming (ILP) is the branch of Machine Learning interested in dealing with logic-based (especially first-order) representation formalisms. However, using a purely logical approach, ILP techniques are sometimes too rigid to deal with noisy data, where they may be affected by overfitting. Statistical approaches may provide the flexibility that is required to overcome these problems. Statistical Relational Learning (SRL) is the branch of ILP aimed at extending its classical, purely logical approach to handle probabilities. Using SRL to approach the PI problem resulted in the Statistical Predicate Invention (SPI) area of research, interested in the discovery of new concepts from relational data by means of statistical approaches.

This paper proposes *Weighted Predicate Invention* (WPI), a SPI approach that focuses on the discovery of tacit relations inside the theories learned by a traditional ILP system.

Albeit working in the SPI setting, it aims at preserving the ILP perspective on the PI problem, but combining it with statistical learning as a guidance to avoid the problem of noisy concepts. Thus, the proposed approach should fit both the specific perspectives of ILP and SRL. As regards the former, it invents new concepts based on the analysis of the structure of a first-order theory in order to find commonalities between the formulas in the theory and then inventing predicates by inter-construction. As to the latter, the role SRL plays in WPI is crucial but quite different from previous SPI proposals. Markov Logic Networks (MLNs) are used to estimate whether a candidate concept to be invented is actually useful with respect to the logic theory in which it was found, by comparing the weight of first-order formulas in MLNs with and without the subsumption of an invented predicate we can estimate if an that predicate should be introduced in the initial theory or not. After a predicate is invented it can be incorporated into the theory, and the process can be carried out again, until no new predicate can be invented. This results in a restructuring of the initial first-order theory which now contains new possible concepts and has a more compact form.

After discussing related works in the next section, we recall some preliminaries in Section 3. Then, we describe our approach to the problem of Statistical Predicate Invention in Section 4, and provide experimental results on a standard ILP dataset in Section 5. Finally, Section 6 concludes the paper and proposes future work.

2   Related Works

Historically, the problem of PI originated in the ILP field, where typical approaches have been based on the analysis of the structure of clauses. In the classical vision of PI in ILP, predicates are invented by using different techniques based on the analysis of the structure of clauses. The goal is to obtain a compression of a first-order theory, or to improve the performance of inductive learning tasks, or to catch exceptions in learned clauses [9]. The problem has been explored analyzing first-order formulas, trying to apply restructuring operators such as inter-construction, looking for commonalities among clauses, as in [9,22], or intra-construction, looking for differences among clauses, as in [9,13]. Many of these works introduced PI in the context of learning systems such as Golem and Cigol [9]. Some results in this field also came from the work of Pazzani [19], in which the search for a predicate to invent is on a second-order instantiation of the logic theory. Other approaches to PI rely on the idea of finding tacit concepts that are useful to catch irregularities in the patterns obtained for a given induction problem [12,20].

Another stream of research that can be considered as related to PI was carried out in the pure SRL setting. Specifically, the proposal in [4] works in statistical learning. It consists of the search for hidden variables in a Bayesian network, by looking for structural patterns, and if a subsequent development using a clustering method to group observed variables and find hidden ones for each analyzed group. A task in SRL that is very similar to the task of PI in ILP is known as Hidden Variables Discovery. The approaches to PI proposed in this stream focus on the generation of concepts starting from facts. This makes them more relevant to the field of Hidden Variable Discovery, which is pure SRL, than they are to the original ILP concept of PI. So, the research ended up with proposing solutions that are more related to standard Parameter Learning, and specifically Structure Learning [8], than they are to PI. This caused the current cost/benefit balance of these works not clearly positive, although they are SPI are promising.

Recently some works better merged together the ideas of PI with those coming from Hidden Variable Discovery. In [8] attention is moved to the problem of SPI in a context of MLNs using a multiple clustering approach to group both relations and constants in order to improve Structure Learning techniques for Markov Logic. Another approach [2] invents predicates in a statistical context and then exploits the invented predicate in the FOIL learning system. In [16] a particular version k-means is used to cluster separately constants and relations. Another proposal [15] involves the application of a number of techniques to cluster multi-relational data.

Only in the last few years there have been some developments in the field of SPI. Some of the proposed approaches use some version of relational clustering to aggregate concepts and relations into new ones [7]. E.g., [6] aims at avoiding the limitations of the purely ILP approach (that is prone to overfitting on noisy data, which causes the generation of useless predicates) using a statistical bottom-up only approach to invent predicates only in a statistical-relational domain.

3   Preliminaries

Our approach to SPI works in the first-order logic setting. Specifically, it uses Datalog (a function-free fragment of Prolog) as a representation language. The basic elements in this setting are atoms, i.e. claims to which a truth value can be assigned. An *atom* takes the form of a predicate applied to its arguments. The number of arguments required by a predicate is called its *arity*. The arguments of a predicate must be *terms*, that in Datalog can be only variables or constants. A *literal* is either an atom (positive literal) or a negation of an atom (negative literal). A *clause* is a disjunction of literals; if such a disjunction involves at most one positive literal then it is called a Horn clause, its meaning being that, if all atoms corresponding to negative literals (called the *body* of the clause) are true, then the positive literal (called the *head* of the clause) must be true. A clause may be represented as a set of literals. An atom, a literal or a clause is called *ground* if its terms are all constants. More information on this setting can be found in [1, 10]

A clause made up of just the head is a *fact*, a clause made up of just the body is a *goal*, while a clause made up of a head and a non-empty body is a *rule*. A literal in a rule is linked if at least one of its terms is present also in other literals; a rule is linked if all of its literals are linked. We adopt the Object Identity (OI) assumption, by which terms (even variables) that are denoted by different symbols in a clause must refer to distinct objects[1]. So, a variable in a clause cannot be associated to another variable or constant in the same clause, nor can two variables in the same clause be associated to the same term.

So, for instance, $p(X, a, b, Y)$ is an atom (and also a positive literal), and $\neg p(X, a, b, Y)$ is a (negative) literal, where p is the predicate, the arity of p is n, and X, a, b, Y are terms (specifically, X, Y are variables and a, b are constants). $\{p(t_1)\}$ and $\{p(t_1), \neg q(t_1, t_2), \neg r(t_2)\}$ are (Horn) clauses. The former is a fact,

---

[1] While not causing a loss in expressive power [18], OI allows to define a search space for first-order logic machine learning systems that fulfills desirable properties ensuring efficiency and effectiveness. The ILP system we will use in the experiments works under OI.

and the latter is a (linked) rule, with head $p(t_1)$ and body $\{q(t_1, t_2), r(t_2)\}$. Under OI, in a clause $\{p(X), \neg q(X, Y)\}$ it is implicitly assumed that $X \neq Y$.

The vocabulary on which we build our representations is a triple $\mathcal{L} = \langle P, C, V \rangle$ where P is a set of predicate symbols, C is a set of constant symbols and V is a set of variable symbols, all possibly infinite. So, $C \cup V$ is the set of terms in our vocabulary. In the following, predicates and constants will be denoted by lowercase letters, and variables by uppercase letters.

A Markov Logic Network [17] consists of a set of first-order formulas associated to weights that represent their relative strength. Given a finite[2] set of constants C, the set of weighted first-order formulas can be used as a template for constructing a Markov Random Field (MRF) by grounding the formulas by the constants in all possible ways. The result is a graph in which nodes represent ground atoms, and maximal cliques are groundings of first-order formulas, also called 'features' in the MRF. The joint probability for atoms is given by:

$$P(X = x) = \frac{1}{Z} exp \left(\sum_{i=1}^{|F|} w_i n_i(x) \right) \tag{1}$$

where Z is a normalization constant, $|F|$ is the number of first order formulas in the set, $n_i(x)$ is the number of true groundings of formula F_i given the set of constants C, and w_i is the weight associated to formula F_i. In the Markov Logic framework two types of parameters learning can be distinguished: Structure Learning and Weight Learning. The former consists in inducing first-order formulas given a set of ground atoms as evidence and a query predicate, by maximizing the joint probability on the MRF based on the evidence. The latter consists in estimating the weight of each first-order formula given a first-order theory and the facts in the evidence, by maximizing the likelihood of a relational database using formula (1). The weight of a formula captures the importance of the formula itself, i.e. its inclination to be true in all possible worlds defined by the set of constants observed in the evidence. A complete description of inference and parameters learning techniques are discussed in [3].

4   Weighted Predicate Invention

As already pointed out, PI aims at generating of new symbols that define latent concepts in relational data or in a logic theory, defined in terms of the previously available symbols. Approaches to PI can be bottom-up (starting from ground atoms) or top-down (starting from a first-order theory). We propose a top-down approach to SPI based on using Markov Logic Networks, called *Weighted Predicate Invention (WPI)*. WPI is motivated by the observation that concepts that are latent in a logic theory may be useful in accomplishing tasks related to the

---

[2] This is a requirement set by MLNs. In our setting, the step that uses MLNs works on a given theory in which the set of constants used is finite, even if they were drawn from a possibly infinite set.

given problem, but not all such concepts are really significant. While this intuition is not novel [9], it is one of the major problems that PI has encountered so far. Our approach distinguishes significant concepts to be invented by their weight, obtained by considering all possible worlds according to evidence data.

In this context we chose the MLN framework, which is a well-known approach to combining FOL and probability. In MLNs, each rule is assigned a weight, that can be learned exploiting a convex optimization problem by the use of gradient descent techniques as in [11]. Discriminative Weight Learning aims at finding the weights that maximize the product of the rules' prior probabilities and data likelihood; so, weights determine the best characterization of a MLN in the worlds defined by the training set. Since the structure of first order rules plays a central role in the computation of weights, weights are particularly suited as a metric to validate the effectiveness of invented concepts.

The basic idea underlying WPI is to analyze an inductively learned theory consisting of a set of first-order rules under OI. It finds common patterns between the rules' bodies working in different steps. First a bipartite graph is created, whose nodes are distinguished between predicates and rules appearing in the theory. Then, a pattern is searched in the graph and checked for matching with subsets of literals in the body of rules. Specifically, th pattern consists of predicate nodes that may be involved in the definition of the predicate to be invented. This problem is complex due to the indeterminacy that characterizes the first-order logic setting. Indeed, it may not have a solution at all. If a matching is found, a candidate rule to be invented is built, and checked for validation before definitely including it in the existing first-order theory. The validation process is based on the use of Weight Learning in the MLN framework to assign a weight to rules in first-order theory. We produce two MLN, weights in the latter must be non-decreasing with respect the former to consider the new rule validated. In this case we can add the invented rule into the first-order theory.

WPI can be applied iteratively, inventing several predicates that allow to obtain more compact versions of the theory. Every invented predicate subsumes the body of rules involved in its invention.

4.1   Searching for a Pattern

Call R the set of first-order rules in the theory. Given a rule $r_i \in R$, let us denote by b_i the set of literals in the body of r_i, and by c_i a subset of b_i. Also, call P the set of all predicates in bodies of rules in the theory, where a literal l is represented as a pair $l = (p, V)$ with p its predicate and V the list of its arguments. We define a bipartite graph $G = \langle R \cup P, E \rangle$ where $E \subseteq R \times P$ is the set of edges such that $\{r, p\} \in E$ iff $\exists l = (p, V) \in R$. We call *upper nodes* the elements of R and *lower nodes* the elements of P. So, every rule is connected to all the predicate symbols appearing in its body.

To find commonalities in the bodies of clauses we consider all possible pairs $I = (\pi, \rho)$, where $\pi \subseteq P$ is a set of lower nodes (made up of at least two elements) that are connected to the same upper-node and $\rho \subseteq R$ is the set of rules in the theory that include π. Among all possible such pairs, we look for one that maximizes (wrt set inclusion) π. Predicates appearing in such I's will be used to form a candidate pattern to define a predicate to be invented.

Consider a theory made up of three rules, $R = \{r_1, r_2, r_3\}$, where:

$$r_1 : q(X) : - \qquad a(X), b(Y), b(W), c(X,Y), d(Y,W).$$
$$r_2 : q(X) : - a(X), b(W), c(X,Y), c(Y,W), g(X), h(Z,Y).$$
$$r_3 : q(X) : - \qquad a(X), f(Z,Y), h(X,Y).$$

The following predicates are available in each rule:

$$r_1 \rightarrow \quad \{a/1, b/1, c/2, d/2\}$$
$$r_2 \rightarrow \{a/1, b/1, c/2, g/1, h/2\}$$
$$r_3 \rightarrow \quad \{a/1, f/2, h/2\}$$

After the building the bipartite graph, the maximal intersection of lower-nodes is $I = (\pi, \rho)$ where $\pi = \{a/1, b/1, c/2\}$ and $\rho = \{r_1, r_2\}$.

4.2 Candidate Selection

Given a pattern, we aim at finding a subset of literals in the theory rules that matches it. Clearly, only the rules in ρ are involved in this operation. For each rule $r_i \in \rho$, consider the subset of its literals that are relevant to the pattern, $c_i = \{l \in r_i | l = (p, V), p \in \pi\}$. Note that c_i may not be linked. For each predicate p_j in the pattern, call n_j the minimum number of occurrences across the c_i's. We try to invent predicates defined by n_j occurrences of predicate p_j, for $j = 1, \ldots, |\pi|$. The underlying rationale is that this should ensure higher chance to find that set of literals in the rules of the theory. In the previous example the minimum number of literals for all predicates in the pattern $\{a/1, b/1, c/2\}$ is one, thus the best subset of literals to match is $\{a(\cdot), b(\cdot), c(\cdot, \cdot)\}$.

Now, for each c_i, consider the set Γ_i of all of its subsets that include exactly n_j occurrences of each predicate p_j in π. We call each element of a Γ_i a *configuration*, and look for a configuration that is present in all Γ_i's (modulo variable renaming). More formally, we look for a $\overline{\gamma}$ s.t. $\forall i = 1, \ldots, |\rho| : \exists j_i \in \{1, \ldots, |\Gamma_i|\}$ s.t. $\overline{\gamma} \equiv \gamma_{ij_i} \in \Gamma_i$. In the example we have $\Gamma_1 = \{\gamma_{11}, \gamma_{12}\}$ and $\Gamma_2 = \{\gamma_{21}, \gamma_{22}\}$ where:

$$\gamma_{11} = \{a(X), b(Y), c(X,Y)\}, \gamma_{12} = \{a(X), b(W), c(X,Y)\}$$
$$\gamma_{21} = \{a(X), b(W), c(X,Y)\}, \gamma_{22} = \{a(X), b(W), c(Y,Z)\}$$

The existence of a solution is not guaranteed. In case it does not exist, we proceed by removing one occurrence of a predicate and trying again, until subsets two literals are tried (inventing a predicate defined by a single literal would be nonsense, since it would just be a synonym). For the proposed example the best configuration is $\gamma_{12} \equiv \gamma_{21}$.

The selected configuration $\overline{\gamma}$ becomes the body of the rule $\overline{r}$ that defines the predicate to be invented. If i is the name of the invented predicate, then the head

Algorithm 1. Finding matching clause

```
 1: function CANDIDATE_SELECTION(π, ρ)
 2:     bestConfig ← emptyList();
 3:     maxIntersection ← 0;
 4:     maxLiteralsInConfig ← 0;
 5:     for all pred ∈ π do
 6:         maxLiteralsInConfig ← maxLiteralsInConfig + minUsed(pred);
 7:     Γ ← ⟨⟩;
 8:     for all rule ∈ ρ do                              ▷ A list of configurations
 9:         Γ ← ⟨getAllConfigs(rule, π))|Γ⟩;
                              ▷ candidateConfig is a list of configurations from every rule.
10:     for all candidateConfig in cartesianProduct(Γ) do
11:         intersection ← setIntersection(candidateConfig);
12:         if |intersection| > maxIntersection then
13:             maxIntersection ← |intersection|;
14:             bestConfig ← candidateConfig;
15:         if maxIntersection = maxLiteralsInConfig then
16:             break;
17:     return bestConfig;
```

of such a rule is obtained by applying i to a list of arguments corresponding to the different variables in $\overline{\gamma}$. In the example the rule would be

$$i(X, Y, W) \ :- \ a(X), b(W), c(X, Y).$$

After adding this rule to the theory, for each rule $r_i \in \rho$ the configuration $\overline{\gamma_i} \in \Gamma_i$ equivalent to $\overline{\gamma}$ can in principle be removed and replaced by the corresponding instance of predicate i. The whole procedure is sketched in Algorithm 1.

4.3 Candidate Validation

Let us call the new rule r_0, that defines the invented predicate, the *invented rule*. To prevent the invention of useless predicates, a validation step must be run that determines whether the invented predicate is actually relevant. Such a validation should be based on an estimator of the relevance of a rule in the context defined by the given theory and the set of evidence facts in the background knowledge. Given such an estimator, the idea is that the introduction of the invented rule in the original theory must not decrease the relevance of the existing rules.

We propose to use the weights learned by the weight learning functionality of the MLN framework as an estimator, and proceed by building two MLNs. For each MLN we apply Discriminative Weight Learning to maximize the likelihood of the training dataset using a learning process based on Diagonal Newton as in [11]. The former simply adds the invented rule to the initial theory. The latter also applies the invented rule to the existing rules, replacing the subset of literals in each rule, that match the invented rule body, with the head of the invented rule properly instantiated. In the previous example, one would get:

$$r_0 : i(X, Y, W) : - \qquad a(X), b(W), c(X, Y).$$
$$r_1 : q(X) : - \qquad b(Y), d(Y, W), i(X, Y, W).$$
$$r_2 : q(X) : - \; c(Y, W), g(X), h(Z, Y), i(X, Y, W).$$
$$r_3 : q(X) : - \qquad a(X), f(Z, Y), h(X, Y).$$

In the former, the invented rule is disjoint from the rest of the graph, because the invented predicate is not present in the other rules. This causes the weights of the other rules not to change. In the latter, the body of some rules in the original theory has changed so that the invented rule is no more disjoint in the graph. So, we expect a variation of the rule weights in the two cases. Comparing the two weights, we consider the invented predicate as relevant if the weight in the latter template is greater than the weight in the former.

We run Discriminative Weight Learning on both templates, obtaining two sets of weighted first-order rules with respect to the evidence facts and the query predicate for the problem defined by theory. Call $w'_0, w'_1, \ldots, w'_k$ the weights of rules in the former MLN, and $w''_0, w''_1, \ldots, w''_k$ the weights of rules in the latter MLN. Then, we pairwise compare the weights of rules, and specifically the weight of the invented rule and the weights of the rules in ρ involved in the predicate invention process. The invented rule is considered as validated if no weight after the application of the invented predicate is less than it was before:

$$\forall i = 0, \ldots, k : w'_i \le w''_i$$

Otherwise, if the introduction of the invented rule in the initial theory causes a decrease in the relative importance of any rule, then the invented rule is not added to the theory. Given the new theory, WPI can be run again on it in order to invent, if possible, further predicates that define implicit concepts. Iterating this procedure yields a wider theory restructuring.

4.4   Discussion

A typical problem of PI approaches is the risk of combinatorial explosion for the search space of the groups of literals that define the invented predicate. Instead of analyzing this problem from a theoretical or structural viewpoint, in this paper we propose an operational model that avoids the invention of trivial or useless concepts. Nevertheless, the computational complexity forces us to consider the advantages taken by the use of WPI in a context of *learning for restructuring* for possible applications that are discussed in the final section.

The main cause of the possible combinatorial explosion is in the variable number of literals per predicate for each rule in ρ. The higher the number of literals per predicate, the higher the number of possible configurations γ_i. Another limitation is in the cost of evaluating Discriminative Weight Learning twice for every predicate we can invent, in fact this kind of parameters learning consists in numerical optimization problem on the MRF instantiated by the correspondent MLN [3].

5   Experiments

The WPI approach was implemented to test its effectiveness, both as regards predicate invention and as regards theory restructuring. To manage MLNs and Discriminative Weight Learning, WPI relies on Tuffy [14], an implementation of the MLN framework based on a Relational DBMS to scale up learning. Compared to Alchemy [21], Tuffy exploits a simplified syntax for MLNs and provides a Discriminant Weight Learning technique to learn weights starting from a template, the data and a query predicate. Then, an experiment was devised. To simulate a totally automatic setting, the theories for the experiment were learned using InTheLEx [5], a fully incremental, non-monotonic, multi-strategy learner of First-Order Logic (Datalog) theories from positive and negative examples, based on the OI assumption.

A toy dataset was exploited in the experiment, for a twofold reason. First, because of the computational complexity of the SPI approach. Second, because it allows an easier insight into the results. Specifically, we considered the classical *Train Problem*, well-known in ILP, in its extended version given by Muggleton. It includes 20 examples of Eastbound or Westbound trains, with the goal to predict Eastbound ones. Due to the small size of this dataset we applied a

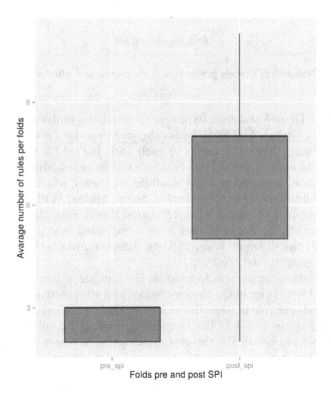

Fig. 1. Number of rules per fold in the theory before and after using WPI

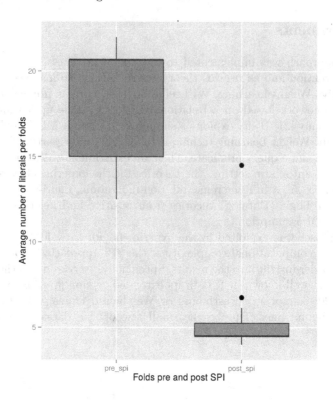

Fig. 2. Number of literals per rule in folds before and after using WPI

Leave-One-Out Cross-Validation technique to avoid the problem of overfitting. This resulted in a total of 20 folds, each using one example for testing purposes and the remaining 19 for training. For each fold, InTheLEx was run on the training set to inductively learn a theory, which was to be tested on the remaining example. Of course, each fold resulted in a different theory, which in turn allowed different possibilities to predicate invention. So, we applied WPI to each learned theory and compared the resulting restructured theory with the one originally provided by InTheLEx. Experimental results confirmed that different theories were learned in the different folds, yielding different invented predicates and, thus, different restructured theories.

The quantitative comparison focused on the number of new (invented) concepts, the number of rules in the theories before and after restructuring, and the number of literals per rule in these theories. A statistical analysis of the outcomes is reported in Figures 1 and 2. The boxplot reported in Figure 1 clearly shows that the number of rules in the theories significantly increases on average after invention and restructuring. However, the number of invented predicates/rules shows some variability in the different folds, including folds in which no predicate could be invented at all. Applying the invented predicates to the initial theory should result in a compression of the theory itself, because the definition of the invented predicate is removed from the other rules and replaced by a single literal that is an instance of the invented predicate. The significant compression obtained in the experiment can be appreciated from the boxplot in Figure 2, concerning the number of literals per rule. So, overall, as expected, the number of rules increases on average, but their size decreases on average. Specifically, 4.25 new concepts were invented on average in each of the 20 folds, which more than doubled the size of the theories on average. However, the average number of literals per rule dropped from 18.41 to 5.30 on average, which (also considering the increase in number of rules) yields an average compression factor of 28.79%.

From a qualitative point of view, we looked for possible interesting new concepts emerging from the inductively learned theories. An interesting result is, for example, the invention in many folds of the concept that any railway car in the train is somehow connected to the locomotive: $\{car(Car), has\_car(Train, Car)\}$. This concept catches a general intuition about trains, and factorizing it makes the theory more understandable.

6 Conclusions and Future Works

Predicate Invention is the branch of symbolic Machine Learning aimed at discovering new emerging concepts in the available knowledge. The outcome of this task may have important consequences on the efficiency and effectiveness of many kinds of exploitation of the available knowledge. Two fundamental problems in Predicate Invention are how to handle the combinatorial explosion of candidate concepts to be invented, and how to select only those that are really relevant. Due to the huge number of candidates, there is a need for automatic techniques to assign a degree of relevance to the various candidates and select the best ones. Purely logical approaches may be too rigid for this purpose, while statistical solutions may provide the required flexibility. This paper proposed a new Statistical Relational Learning approach to Predicate Invention. The candidate predicates are identified in a logic theory, rather than in the background knowledge, and are used to restructure the theory itself. Specifically, the proposed approach exploits the Markov Logic Networks framework to assess the relevance of candidate predicate definitions. It was implemented and tested on a traditional problem in Inductive Logic Programming, yielding interesting results in terms of invented predicates and consequent theory restructuring.

Future work will include: optimizing the implementation; devising more informative heuristics for selecting and choosing candidate definitions for the predicate to be invented; running more extensive experiments, also on real-world datasets. Specifically, we want to investigate the implications of using WPI on the benchmark Mutagenesis dataset, to check whether interesting implicit concepts can be found in that domain.

Acknowledgments. This work was partially funded by the Italian PON 2007-2013 project PON02_00563_3489339 'Puglia@Service'.

References

1. Ceri, S., Gottlob, G., Tanca, L.: Logic Programming and Databases. Springer-Verlag (1990)
2. Craven, M., Slattery, S.: Relational learning with statistical predicate invention: Better models for hypertext. Machine Learning **43**(1/2), 97–119 (2001)
3. Domingos, P., Kok, S., Lowd, D., Poon, H., Richardson, M., Singla, P.: Markov logic (2008)
4. Elidan, G., Friedman, N.: Learning hidden variable networks: The information bottleneck approach. Journal of Machine Learning Research **6**, 81–127 (2005)
5. Esposito, F., Semeraro, G., Fanizzi, N., Ferilli, S.: Multistrategy Theory Revision: Induction and Abduction in INTHELEX. Machine Learning **38**(1/2), 133–156 (2000)
6. Kemp, C., Tenenbaum, J.B., Griffiths, T.L., Yamada, T., Ueda, N.: Learning systems of concepts with an infinite relational model. In: AAAI, pp. 381–388. AAAI Press (2006)
7. Kok, S., Domingos, P.: Toward statistical predicate invention. In: Open Problems in Statistical Relational Learning, SRL 2006 (2006)
8. Kok, S., Domingos, P.: Statistical predicate invention. In: Ghahramani, Z. (eds.) ICML. ACM International Conference Proceeding Series, vol. 227, pp. 433–440. ACM (2007)
9. Kramer, S.: Predicate invention: A comprehensive view. Technical Report TR-95-32, Oesterreichisches Forschungsinstitut fuer Artificial Intelligence, Wien, Austria (1995)
10. Lloyd, J.W. : Foundations of Logic Programming, 2nd edn. Springer-Verlag (1987)
11. Lowd, D., Domingos, P.: Efficient weight learning for Markov logic networks. In: Kok, J.N., Koronacki, J., Lopez de Mantaras, R., Matwin, S., Mladenič, D., Skowron, A. (eds.) PKDD 2007. LNCS (LNAI), vol. 4702, pp. 200–211. Springer, Heidelberg (2007)
12. Muggleton, S.: Predicate invention and utilization. J. Exp. Theor. Artif. Intell **6**(1), 121–130 (1994)
13. Muggleton, S., Buntine, W.: Machine invention of first-order predicates by inverting resolution. In: MLC 1988, pp. 339–352 (1988)
14. Niu, F., Ré, C., Doan, A., Shavlik, J.W.: Tuffy: Scaling up statistical inference in markov logic networks using an RDBMS (2011). CoRR, abs/1104.3216
15. Perlich, C., Provost, F.: Aggregation-based feature invention and relational concept classes. In: KDD 2003, pp. 167–176. ACM Press (2003)
16. Popescul, A., Ungar, L.H.: Cluster-based concept invention for statistical relational learning. In: KDD, pp. 665–670. ACM (2004)
17. Richardson, M., Domingos, P.: Markov logic networks. Machine Learning **62**(1–2), 107–136 (2006)
18. Semeraro, G., Esposito, F., Malerba, D., Fanizzi, N., Ferilli, S.: A logic framework for the incremental inductive synthesis of datalog theories. In: Fuchs, N.E. (ed.) LOPSTR 1997. LNCS, vol. 1463, pp. 300–321. Springer, Heidelberg (1998)

19. Silverstein, G., Pazzani, M.: Relational cliches: constraining constructive induction during relational learning. In: Proceedings of the Sixth International Workshop on Machine Learning. Kaufmann, Los Altos (1989)
20. Srinivasan, A., Muggleton, S., Bain, M.: Distinguishing exceptions from noise in non monotonic learning. In: Rouveirol, C. (eds.) ECAI 1992 Workshop on Logical Approaches to Machine Learning (1992)
21. Sumner, M., Domingos, P.: The alchemy tutorial, July 26, 2013
22. Wogulis, J., Langley, P.: Improving efficiency by learning intermediate concepts. In: IJCAI 1989, pp. 657–662. Morgan Kaufmann (1989)

Empowered Negative Specialization in Inductive Logic Programming

Stefano Ferilli[1,2], Andrea Pazienza[1](✉), and Floriana Esposito[1,2]

[1] Dipartimento di Informatica, Università di Bari, Bari, Italy
{stefano.ferilli,andrea.pazienza,floriana.esposito}@uniba.it
[2] Centro Interdipartimentale per la Logica e sue Applicazioni,
Università di Bari, Bari, Italy

Abstract. In symbolic Machine Learning, the incremental setting allows to refine/revise the available model when new evidence proves it is inadequate, instead of learning a new model from scratch. In particular, *specialization* operators allow to revise the model when it covers a negative example. While specialization can be obtained by introducing negated preconditions in concept definitions, the state-of-the-art in Inductive Logic Programming provides only for specialization operators that can negate single literals. This simplification makes the operator unable to find a solution in some interesting real-world cases.

This paper proposes an empowered specialization operator for Datalog Horn clauses. It allows to negate conjunctions of pre-conditions using a representational trick based on predicate invention. The proposed implementation of the operator is used to study its behavior on toy problems purposely developed to stress it. Experimental results obtained embedding this operator in an existing learning system prove that the proposed approach is correct and viable even under quite complex conditions.

1 Introduction

Supervised Machine Learning approaches based on First-Order Logic (FOL) representations are particularly indicated in real-world tasks in which the relationships among objects play a relevant role in the definition of the concepts of interest. However, the learned theory is valid only until there is evidence to the contrary (i.e., new observations that are wrongly classified by the theory). In such a case, either a new theory is to be learned from scratch using the new batch made up of both the old and the new examples, or the existing theory must be incrementally revised to account for the new evidence as well. If positive and negative examples are provided in a mixed and unpredictable order to the learning system, two different refinement operators are needed: a *generalization* operator to refine a hypothesis that does not account for a positive example, and a *specialization* operator to refine a hypothesis that erroneously accounts for a negative example.

The focus of this paper is on supervised incremental inductive learning of logic theories from examples, and specifically on the specialization operator.

© Springer International Publishing Switzerland 2015
M. Gavanelli et al. (Eds.): AI*IA 2015, LNAI 9336, pp. 288–300, 2015.
DOI: 10.1007/978-3-319-24309-2_22

In these operators the addition of negative information may allow to learn a broader range of concepts. Following the theoretical study in [5], this paper defines an operational search space for a specialization operator that may involve multiple literals in a negation, and provides an algorithm to compute it. An implementation of the operator was embedded in the incremental learning system InTheLEx [4] and tested on purposely developed datasets.

Section 2 introduces our logic framework and the state-of-the-art specialization operator for it. Section 3 describes our new proposal and Section 4 evaluates its efficiency and effectiveness. Finally, Section 5 concludes the paper.

2 Preliminaries

Inductive Logic Programming (ILP) aims at learning logic programs from examples. In our setting, examples are represented as clauses, whose body describes an observation, and whose head specifies a relationship to be learned, referred to terms in the body. Negative examples for a relationship have a negated head. A learned program is called a *theory*, and is made up of *hypotheses*, i.e. sets of program clauses all defining the same predicate. A hypothesis *covers* an example if the body of at least one of its clauses is satisfied by the body of the example. The *search space* is the set of all clauses that can be learned, ordered by a generalization relationship.

In ILP, a standard practice to restrict the search space is to impose biases on it [10]. We are concerned with *hierarchical* (i.e., non-recursive) theories made up of linked Datalog⁻ clauses. *Datalog* [1,6] is a sublanguage of Prolog in which a term can only be a variable or a constant. The missing expressiveness of function symbols can be recovered by *flattening* [12]. Datalog⁻ extends pure Datalog by allowing the use of negation in the body of clauses. In the following, we will denote by $body(C)$ and $head(C)$ the set of literals in the body and the atom in the head of a Horn clause C, respectively. We adopt the Object Identity (OI) assumption: *within a clause, terms denoted by different symbols must be distinct*. This notion can be viewed as an extension of both Reiter's *unique-names* assumption [11] and Clark's Equality Theory [9] to the variables of the language. $Datalog^{OI}$ is the resulting language. It has the same expressive power as Datalog [14], but causes the classical ordering relations among clauses to be modified, thus yielding a new structure of the corresponding search spaces for the refinement operators.

The ordering relation defined by the notion of θ-subsumption under OI upon Datalog clauses [3,13] is θ_{OI}-subsumption, denoted by $\leq_{OI}$. Requiring that terms are distinct, θ_{OI}-subsumption maps each literal of the subsuming clause onto a single, different literal in the subsumed one. So, equivalent clauses under $\leq_{OI}$ must have the same number of literals, and the only way to have equivalence is through variable renaming. Indeed, under OI, substitutions[1] are required to be injective.

---

[1] Substitutions are mappings from variables to terms [16]. Given a substitution σ by which a clause $C \leq_{OI} E$, σ^{-1} denotes the corresponding antisubstitution, i.e. the inverse function of σ, mapping some terms in E to variables in C.

The canonical inductive paradigm requires the learned theory to be complete and consistent. For hierarchical theories, the following definitions are given (where E^- and E^+ are the sets of all the negative and positive examples, resp.):

Definition 1 (Inconsistency)

- A *clause* C *is* inconsistent *wrt* $N \in E^-$ *iff* $\exists \sigma$ *injective substitution s.t.* $body(C).\sigma \subseteq body(N) \wedge \neg head(C).\sigma = head(N)$
- A *hypothesis* H *is* inconsistent *wrt* N *iff* $\exists C \in H$: C *is inconsistent wrt* N.
- A *theory* T *is* inconsistent *iff* $\exists H \subseteq T$, $\exists N \in E^-$: H *is inconsistent wrt* N.

Definition 2 (Incompleteness)

- A hypothesis H *is* incomplete *wrt* $P \in E^+$ *iff* $\forall C \in H$: $not(P \leq_{OI} C)$.
- A theory T *is* incomplete *iff* $\exists H \subseteq T$, $\exists P \in E^+$: H *is incomplete wrt* P.

When the theory is to be learned incrementally, it becomes relevant to define operators that allow a refinement of *too weak* or *too strong* programs [7]. *Refinement operators* are the means by which wrong hypotheses in a logic theory are changed in order to account for new examples with which they are incomplete or inconsistent. A refinement operator, applied to a clause, returns another clause that subsumes (upward refinement) or is subsumed by (downward refinement) the given clause.

Refinement operators have several applications in the automatic synthesis of logic theories. They were introduced by Shapiro [15], who used them for refining discarded hypotheses. The mathematics of refinement operators in themselves were studied by Laird [8], who described both "downward" refinement (of which Shapiro's operators were examples) and "upward" refinement. Tinkham [17] applies these ideas to allow patterns in previously-seen theories to guide the synthesis of new theories: a generalization (upward refinement) operator and a specialization (downward refinement) operator are used to form generalizations of known theories; these generalizations, together with the specialization operator, are then used to synthesize new theories.

In the following, we will assume that logic theories are made up of clauses that have only variables as terms, built starting from observations described as conjunctions of ground facts (i.e., variable-free atoms). This restriction simplifies the refinement operators for a space ordered by θ_{OI}-subsumption defined in [3, 13], and the associated definitions and properties.

Definition 3 (Refinement operators under OI) *Let C be a Datalog clause.*

- $D \in \rho_{OI}(C)$ *(downward refinement operator) when* $body(D) = body(C) \cup \{l\}$, *where l is an atom s.t. $l \notin body(C)$.*
- $D \in \delta_{OI}(C)$ *(upward refinement operator) when* $body(D) = body(C) \setminus \{l\}$, *where l is an atom s.t. $l \in body(C)$.*

The research on incremental approaches is not very wide, due to the intrinsic complexity of learning in environments where the available information about the concepts to be learned is not completely known in advance, especially in a FOL setting. Thus, [3,13] still represent the state-of-the-art for this research.

When a negative example is covered, a specialization of the theory must be performed. Starting from the current theory, the misclassified example and the previous positive examples, the specialization algorithm outputs a revised theory. In our framework, specializing means adding proper literals to a clause that is inconsistent with respect to a negative example, in order to avoid its covering that example. The possible options for choosing such a literal might be so large that an exhaustive search is infeasible.

According to Definition 3, only positive literals can be added. These literals should characterize all the past positive examples and discriminate them from the current negative one. Thus, the operator tries to specialize the clause by adding to it a literal that is present in all positive examples but not in the negative one. To satisfy the property of maximal generality, the operator must add as few atoms as possible. If no (set of) positive literal(s) is able to characterize the past positive examples and discriminate the negative example that causes inconsistency, the addition of a negative literal to the clause body might restore consistency. Such a literal should be able to discriminate the negative example from all the past positive ones. These literals cause the need to extend the representation language of clauses from Datalog to Datalog¬.

Consider a clause C that is inconsistent with respect to a negative example N, and the positive examples $P_1, \ldots, P_n$ that are θ_{OI}-subsumed by C. The specialization operator should restore consistency, returning a refinement C' of C which still θ_{OI}-subsumes $P_1, \ldots, P_n$, but not N.

Let us first define the *residual* of an example with respect to a clause, consisting of all the literals in the example that are not involved in the θ_{OI}-subsumption test, after having properly turned their constants into variables:

Definition 4 (Residual) *Let C be a clause, E an example, and σ_j a substitution s.t. $body(C).\sigma_j \subseteq body(E)$ under OI.*
A residual of E wrt C under the mapping σ_j, denoted by $\Delta_j(E, C)$, is:

$$\Delta_j(E, C) = body(E).\underline{\sigma}_j^{-1} - body(C)$$

where $\underline{\sigma}_j^{-1}$ is the *extended antisubstitution* of σ_j, defined on the whole set $consts(E)$, that associates *new* (fresh) variables to the constants in E that do not have a corresponding variable according to σ^{-1}. Each substitution by which a clause subsumes an example yields a distinct residual.

Let us define the target space for the negative literals to be added by the operator. Extending the specialization operator proposed in [3], [5] defined a new operator that selects a literal that is present in all residuals of the negative example and that is not present in any residual of any positive example:

$$l \in \mathbf{S}'_c = \neg(\overline{\mathbf{S}} - \overline{\mathbf{P}})$$

where:

$$\overline{\mathbf{S}} = \bigcap_{j=1,\ldots,m} \Delta_j(N, C)$$

$$\overline{\mathbf{P}} = \bigcup_{\substack{i=1,\ldots,n \\ j_i=1,\ldots,n_i}} \Delta_{j_i}(P_i, C)$$

and, given a set of literals $\varphi = \{l_1, \ldots, l_n\}$, $n \geq 1$: $\neg\varphi = \{\neg l_1, \ldots, \neg l_n\}$.

However, this *space of consistent negative downward refinements* does not ensure completeness with respect to the previous positive examples, as shown in the following example[2].

Example 1. ex:wrongspsoperator Consider the real world task of classifying **edible mushrooms**. A mushroom m is described by the following features: a stem s, a cap c, spores p, gills g, dots d. Two positive examples: $P_1 = m :- s, c, p, g.$ and $P_2 = m :- s, c, d.$ produce as a least general generalization the clause $C_1 = m :- s, c$. Then, the negative example $N_1 = m :- s, c, p, g, d.$ arrives. The residuals are $\Delta(P_1, C_1) = \{p, g\}$, $\Delta(P_2, C_1) = \{d\}$, and $\Delta(N_1, C_1) = \{p, g, d\}$. No specialization by means of positive literals can be obtained. Switching to the space of negative literals, no single literal from the negative residual, if negated, generates a clause that is complete with all previous positive examples:

$$C_2' = m :- s, c, \neg p. \qquad C_2'' = m :- s, c, \neg g. \qquad C_2''' = m :- s, c, \neg d.$$

where $P_1 \not\leq_{OI} C_2'$, $P_1 \not\leq_{OI} C_2''$ and $P_2 \not\leq_{OI} C_2'''$. Indeed, $\mathbf{S}_c' = (\{p, g, d\}) - (\{p, g\} \cup \{d\}) = \{p, g, d\} - \{p, g, d\} = \emptyset$. So, in this case no single (positive or negative) literal can restore completeness and consistency of the theory.

3 Extended Negative Downward Refinement

When the negation of a single atom is insufficient to restore consistency while preserving completeness, it might be the case that negating a conjunction of atoms resolves the problem. The atoms in the conjunction must be all present in the negative example, but at least one of them must not be present in any positive example. Unfortunately, these solutions are not permitted in the representation language, since only literals may appear in the body of clauses. A possible solution is to invent a new predicate defined by the conjunction, and to negate a single atom that is a suitable instance of this predicate. By a resolution step, the meaning of the resulting theory would be preserved.

Example 2. In the previous example, either $C_2' = m :- s, c, \neg(p, d).$ or $C_2'' = m :- s, c, \neg(g, d).$ would be correct refinements of C_1 wrt $\{P_1, P_2, N_1\}$. So, we might invent a new predicate n, defined as $n :- p, d.$ or $n :- g, d.$, and specialize C_1 in $C_1' = m :- s, c, \neg n.$. I.e., an edible mushroom must not have both spores and dots.

So, the extension comes into play when no single literal is sufficient to restore correctness of the theory. In particular, we would like to find a minimal (in terms of number of elements) such set. A formal definition of this operator is given

---

[2] For the sake of readability, in the following we will often switch to a propositional representation. This means that the residual is unique for each example, so the subscript in $\Delta_i(\cdot, \cdot)$ is no more necessary.

in [5]. Here, we show a possible way to define and restrict the search space, and implement a preliminary version of the operator for analysis purposes. As regards the implementation, the constraint is to determine a minimal subset of the residual, useful to specialize the clause for creating the invented predicate definition. This becomes much more relevant, but much more challenging as well, when the number of literals in the residual is very large, with very few thereof being sufficient to reach the objective. Obviously, generating and testing all possible subsets of the residual by increasingly larger cardinality, until the discriminant one is found, would ensure finding out the minimal solution. However, in the worst case this approach would require exponential time in the cardinality of the residual. We want our knowledge about the positive examples to guide our search. Specifically, we observe that no subset of literals that is present in the residual of a positive example can be used for specialization, unless other literals are included as well, otherwise that positive example would be uncovered. In the simplest case, no pair of literals that appear in the residual of a positive example can be used as a solution (i.e., as the conjunction of literals to be negated) unless another literal, not present in the residual, is added as well. So, we start considering the pairs of literals in the negative residual that are not present in any positive residual. If none of them is a solution, we must consider triplets of literals, each of which may contain at most two literals that are in the same positive residual. If again none of them is a solution, we must consider 4-tuples of literals, each of which may contain at most three literals that are in the same positive residual. And so on.

The proposed heuristic is based on the same principle but starts from the opposite perspective, proceeding top-down from the whole negative residual and progressively deriving smaller subsets thereof. Two subsets are derived from each given set of literals, based on a pair of literals in it: each subset includes one of such literals but not the other. So, the question is how to select the pairs of literals that guide the splits. Based on the previous observations, using a pair of literals belonging to different positive residuals would increase the chances of uncovering positive examples, because the remaining literals in each resulting subset tend to be concentrated in less positive residuals. Conversely, using a pair of literals that are in the same positive residual leaves in the resulting subsets more literals that belong to different positive residuals, because each of them guarantees that those two positive examples are not uncovered. So, we use the list of pairs of literals found in positive residuals as a guide for building the space of possible definitions of the invented predicate.

Algorithm 1 generates the search space for candidate definitions of an invented predicate based on this heuristic. The search space is represented as a binary tree T, made up of vertexes V and edges E, where each vertex is a candidate definition. The length of the definition (number of literals) decreases as the depth of the vertex increases, up to definitions made up of two literals (due to the strict inclusion test, the IF statement in the inner loop does not generate offspring from nodes that include just two literals). The tree is built by scanning the set R of pairs of literals that appear in at least one positive residual, and

---

Algorithm 1. Search space generation

Require: Δ: negative residual; $\{R_i\}$ = residuals of positive examples

   $R \leftarrow \cup_i \{\{l', l''\} \mid l', l'' \in R_i, l' \neq l''\}$

   $T = (V, E) \leftarrow (\{\Delta\}, \emptyset)$

   $Q \leftarrow \{\Delta\}$

   while $R \neq \emptyset$ **do**

      $Q' \leftarrow \emptyset$

      extract $\{l', l''\}$ from R

      for all $L \in Q$ **do**

         if $\{l', l''\} \subset L$ /* if L is a pair it is not split */ **then**

            $L_l \leftarrow L \setminus \{l'\}, L_r \leftarrow L \setminus \{l''\}$

            $Q' \leftarrow Q' \cup \{L_l, L_r\}$

            $V \leftarrow V \cup \{L_l, L_r\}; E \leftarrow E \cup \{(L, L_l), (L, L_r)\}$

         else

            $Q' \leftarrow Q' \cup \{L\}$

         end if

      end for

      $Q \leftarrow Q'$

   end while

Ensure: T

---

appending two children to each vertex in the current frontier Q (i.e., the current leaves) that includes the considered pair of literals. Each of the children removes from its parent a literal of the current pair, based on the proposed heuristic. At the beginning only the root, associated to the whole negative residual Δ, is present. Then, each loop uses an element of R to grow the offspring of the current frontier, obtaining a new frontier Q' that replaces Q in the next round.

To ensure that a minimal solution is found, one must first check if any of the vertexes made up of pairs of literals is able to restore consistency while still ensuring completeness. If no such vertex exists, the level immediately above, whose vertexes are made up of triplets of literals, is scanned, and so on until a suitable set of literals is found (in the worst case, the whole negative residual, associated to the root of the tree, is attempted), or the specialization fails. The proposed search space can be explored in depth or in breadth. In the former case, one must traverse the tree until the leaves are reached, considering only leaves associated to pairs of literals. If the whole tree has been scanned without finding a solution, these leaves are removed and the search is started again, this time focusing on the leaves associated to triplets only. And so on. In the case of a breadth search, the tree levels are explored until the 2-literal level is reached. Then, such a level is scanned looking for a complete and consistent solution. If no such solution exists, the level immediately above is tried, and so on.

In our implementation we adopted the breadth solution, but to save memory space we deleted the previous level as soon as its offspring is generated. In terms of Algorithm 1, the statement that updates V and E is dropped, so that only the vertexes in Q are available for processing. The consequence of this choice is that, if no solution can be found at the level of interest in a given round of the loop, the tree must be re-generated from the root up to the previous level in the next round of the loop. This has clearly a significant impact on runtime, but we wanted to try this solution to have an idea of the algorithm's behavior when the search space becomes too large for being stored in memory.

However, we also implemented a quick optimization to speed up the algorithm in some easy cases. Specifically, a preliminary step is introduced to check if the specialization predicate can be created by just taking a pair of literals that is present in the negative residual but not in any positive residual. Any element in this difference is a pair of literals that, being present in the negative example, if negated prevents it from being covered, and, never appearing in any positive example, does not cause any positive example to be uncovered. So, it can be used to define the invented predicate to be used for specialization without generating the whole search space up to the two-literal leaves. If the difference is empty, then no pair of literals can uncover the negative example while still covering the positive examples, so the algorithm proceeds with the normal computation for groups made up of at least three literals.

The proposed solution is as follows. After determining the set R of all unordered pairs of literals that are present in any positive residual, it is scanned and, for each pair $\{l', l''\}$ in it, the negative residual generates two shorter subsets, one containing only l' and the other containing only l''. This operation results in a binary tree, in which each branch selects a different literal for each possible solution. For each subsequent pair of literals that is considered, all leaves that contain both those literals are in turn split into two subsets. The splits go on until the tree reaches depth $|\Delta| - N$, involving solutions made up of N literals. At the beginning of the algorithm execution $N = 3$, because the solutions made up of just two literals have already been considered (and discarded) in the pre-processing step. Then, N is progressively increased by 1.

Example 3. Consider a hypothesis: $C = h :\text{-} a, b.$, three positive examples: $P_1 = h :\text{-} a, b, c, d, e.$ $P_2 = h :\text{-} a, b, e, f, g.$ $P_3 = h :\text{-} a, b, c, e, f.$, and a negative one: $N = h :\text{-} a, b, c, d, e, f, g.$ The unordered pairs of literals in the residual of the negative example are computed: $\Delta(N, C) = \{c, d, e, f, g\} \rightarrow$
$S = \{\{c, d\}, \{c, e\}, \{c, f\}, \{c, g\}, \{d, e\}, \{d, f\}, \{d, g\}, \{e, f\}, \{e, g\}, \{f, g\}\}$.
For each positive residual, all unordered pairs of literals are determined:
$\Delta(P_1, C) = \{c, d, e\} \rightarrow \{\{c, d\}, \{c, e\}, \{d, e\}\}$
$\Delta(P_2, C) = \{e, f, g\} \rightarrow \{\{e, f\}, \{e, g\}, \{f, g\}\}$
$\Delta(P_3, C) = \{c, e, f\} \rightarrow \{\{c, e\}, \{c, f\}, \{e, f\}\}$
Their union is $R = \{\{c, d\}, \{c, e\}, \{d, e\}, \{e, f\}, \{e, g\}, \{c, f\}, \{f, g\}\}$, and $S \setminus R = \{\{c, g\}, \{d, f\}, \{d, g\}\}$. Any element in this difference can be used to define the invented predicate to be used for specialization.

Let us add another positive example so that no two-literal solutions exist: $P_4 = h :\text{-} a, b, c, d, f, g.$ Now the set of pairs of literals from the positive residuals is:
$R = \{\{c, d\}, \{c, e\}, \{d, e\}, \{e, f\}, \{e, g\}, \{c, f\}, \{f, g\}, \{c, g\}, \{d, f\}, \{d, g\}\}$
and the difference between the two combinations is empty:
$S \setminus R = \{\{c, d\}, \{c, e\}, \{c, f\}, \{c, g\}, \{d, e\}, \{d, f\}, \{d, g\}, \{e, f\}, \{e, g\}, \{f, g\}\} \setminus$

$\{\{c, d\}, \{c, e\}, \{d, e\}, \{e, f\}, \{e, g\}, \{c, f\}, \{f, g\}, \{c, g\}, \{d, f\}, \{d, g\}\} = \emptyset$.
We must use the general procedure to look for solutions that include at least 3 literals. Considering the first pair $\{c, d\}$, the two subsets are $L_l = \{c, e, f, g\}$ and $L_r = \{d, e, f, g\}$. This ensures that literals c and d are not together in any candidate final solution. The second pair is $\{c, e\}$, that is a subset of L_l (but not of L_r). So, L_l is split into $L_{ll} = \{c, f, g\}$ and $L_{lr} = \{e, f, g\}$. Since at this round we are interested in solutions involving three literals, these leaves are tried. Here the solution is indeed given by $L_{1r} = \{e, f, g\}$.

4   Evaluation

The performance of the operational procedure proposed in the previous section was evaluated while changing different parameters. First we studied the increase in runtime for finding a solution for an increasing number of literals in the residuals. Then, we studied the increase in runtime for finding a solution for both an increasing number of literals in the residuals and an increasing number of positive examples. Finally, we estimated the average runtime per literal, and then the results of the previous two phases will be compared and appropriate conclusions will be drawn. All experiments were run on a laptop endowed with a 2.5 GHz Intel i5 processor and 2GB RAM, running YAP Prolog 6.2.2 [2] on Ubuntu Linux 12.04.

The host learning system for our tests is InTheLEx [4]. Due to its behavior and to the peculiarities of its operators, InTheLEx places much burden on the generalization operator, which provides very refined clauses and clauses that are close to the theoretical least general generalization. While this makes it able to handle cases, e.g., of learning from positive examples only, this also causes the specialization operator to be fired very seldom, and often when a solution actually does not exist. Among the cases in which it is fired, in real-world domains it is often sufficient in its basic setting to find a solution. This results in a very limited chance that the proposed specialization operator based on predicate invention is used. This is however not bad. First, because specialization by negative literals is actually a last resort in learning a theory. Second, because predicate invention is a delicate activity that requires some kind of selection and cannot be performed extensively.

On 16 real-world datasets concerning document processing, each including 353 layout descriptions of first pages of scientific papers, the empowered negative specialization was necessary in so few cases that no statistical evaluation of its performance could be run. So, real-world datasets seem unsuitable to thoroughly test our methodology. For this reason, we purposely created two series of datasets that ensure that the specialization through predicate invention is fired, and that allow to suitably tune both the number of literals in the residuals and the number of positive examples as needed to test the approach in increasingly stressing conditions for the operator.

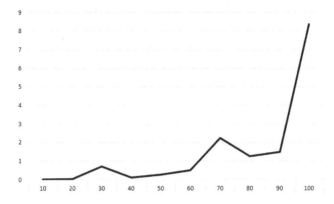

Fig. 1. Runtime (in minutes, y-axis) by number of literals in the residuals (x-axis).

4.1 Efficiency vs Residual Length

The first test involved 50 datasets, 5 for each of 10 different settings. Each dataset included 7 positive and 1 negative examples. The i-th setting ($i = 1, \ldots, 10$) produces datasets that involve $10i$ literals in the residual of the negative example: i.e., 10 literals for the first setting, 20 literals for the second setting, and so on. This is obtained as follows: a clause is created, and its body instantiated in all (positive and negative) examples. Then, $10i$ additional literals are appended to the negative example (to form its residual), and each positive residual is obtained by randomly selecting 80% of the literals from the negative residual, ensuring that each positive residual is different from the others. This should result in a larger search space for the predicate invention-based specialization algorithm. Each of the 5 datasets for each setting involves different combinations of literals in the positive examples, to provide more varied testbeds.

InTheLEx was run on each dataset, fed with all the positive examples and the clause generalizing them. The averages for the various settings are summarized in Figure 1 which shows the trend for larger and larger negative residuals that is clearly increasing, albeit not monotonically. Another interesting observation concerns the change in slope of the curve that appears around the 90-literals case. Up to 60 literals, a solution is found in less than one minute on average. Runtime is still acceptable (never above 3 minutes) from 70 to 90 literals. Instead, a sudden increase happens for 100 literals (even if the average is still below 9 minutes). A possible explanation is that a larger number of literals may cause a significant extension of the search space, resulting in much higher runtime if a solution made up of a few literals is not found. Indeed, an observation of the learned theories revealed that the number of literals that make up the invented predicate is less than half the number of literals in the residual on average.

Table 1. Runtimes for increasingly larger negative residuals

Setting	1	2	3	4	5	6	7	8	9	10
Positive Examples	7	9	11	13	15	17	19	21	23	25

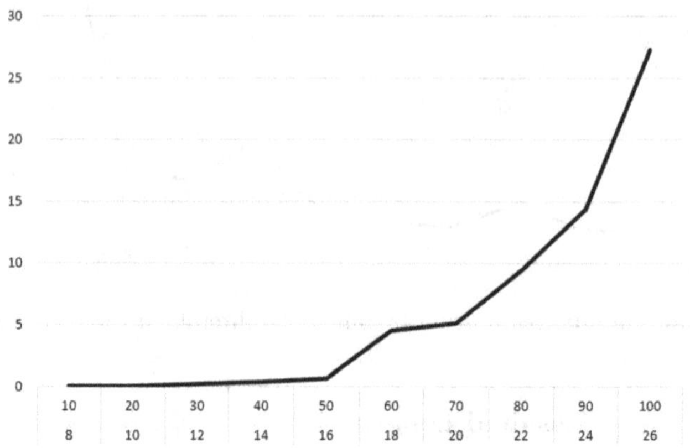

Fig. 2. Runtime (in minutes, y-axis) by number of literals in the residuals and number of examples for each setting (x-axis).

4.2   Efficiency vs Residual Length and Number of Positive Examples

Another 50 artificial datasets were created for the second kind of evaluation, using the same strategy as before except that also the number of positive examples was linearly increased in each setting, as reported in Table 1. Again, each positive example is obtained as an instance of the initial clause by adding a residual made up of 80% randomly selected literals from the negative residual, ensuring that no two combinations are the same.

The corresponding results of runtime for InTheLEx are graphically summarized in Figure 2. Up to 50 literals (and 15 positive examples) runtime is still below one minute. This is pretty good, because the number of positive examples at 50 literals in the residual is more than doubled compared to the previous experiment. For the 60 and 70 settings runtime increases around 5 minutes, and stays below 15 minutes up to the 90 literals setting, where the number of positive examples is three times as much as the previous experiment. For the 100 literals & 25 positive examples setting, runtime is around 27 minutes.

Comparing the two graphics in Figures 1 and 2, we note that in both cases runtime is below one minute up to the 50 literals setting. This means that neither the increase in the number of examples, nor that in the number of literals, significantly affects performance. The two curves part away starting from the 60 literals setting. Both go up, but the latter much more than the former. Summing up, both parameters affect runtime only after a given complexity is reached, and the number of examples has a heavier impact than the number of literals. One possible explanation is that, after that complexity level is reached, the search

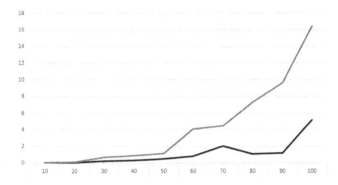

Fig. 3. Runtime (y-axis) per literal (x-axis) trend comparison.

space becomes inherently more complex, not allowing easy/small solution by its very structure. Or, at least, this is what seems to happen in the randomly generated toy problems. It will be interesting to check whether in real-world domains the non-randomness in descriptions avoids such a behavior.

The average runtime per literal for the various settings in the two experiments is reported in Figure 3. As expected, due to more literals causing more combinations to be potentially tested, the runtime is not constant. However, again, the number of examples seems to be the cause of the largest increase in time. While the number of possible literal combinations is determined by the negative residual, the branching factor in the tree is determined by the number of (positive) examples. Thus, once again it seems that the hard part is somehow related to the dataset structure: indeed, this behavior can be explained by the fact that the more positive examples, the more branches must be generated before finding the solution. Manual checking of sample cases in each setting has confirmed that the solution found by the proposed procedure is always minimal.

5   Conclusions and Future Work

Incremental supervised Machine Learning approaches using First-Order Logic representations are mandatory when tackling complex real-world tasks, in which relationships among objects play a fundamental role. A noteworthy framework for these approaches is based on the space of Datalog Horn clauses under the Object Identity assumption, which ensures the existence of refinement operators fulfilling desirable requirements, unless they have some limitations that this paper aims at overcoming. After recalling the most important elements of the framework and of the current downward operator, this paper proposed and implemented an algorithm for multi-literal negation-based specialization. The efficiency of the operator, integrated in the InTheLEx learning system, was tested using purposely devised experiments.

Future work includes a study of the possible connections of the extended operator with related fields of the logic-based learning, such as deduction, abstraction and predicate invention. Experiments aimed at assessing the efficiency and effectiveness of the operator in real-world domains are also planned.

Acknowledgments. This work was partially funded by the Italian PON 2007-2013 project PON02_00563_3489339 'Puglia@Service'.

References

1. Ceri, S., Gottlöb, G., Tanca, L.: Logic Programming and Databases. Springer-Verlag, Heidelberg (1990)
2. Costa, V.S., Rocha, R., Damas, L.: The YAP Prolog system. Theory and Practice of Logic Programming **12**(1–2), 5–34 (2012)
3. Esposito, F., Laterza, A., Malerba, D., Semeraro, G.: Locally finite, proper and complete operators for refining datalog programs. In: Michalewicz, M., Raś, Z.W. (eds.) ISMIS 1996. LNCS, vol. 1079, pp. 468–478. Springer, Heidelberg (1996)
4. Esposito, F., Semeraro, G., Fanizzi, N., Ferilli, S.: Multistrategy Theory Revision: Induction and Abduction in INTHELEX. Machine Learning Journal **38**(1/2), 133–156 (2000)
5. Ferilli, S.: Toward an improved downward refinement operator for inductive logic programming. In: Atti del 11th Italian Convention on Computational Logic (CILC-2014), vol. 1195, pp. 99–113. Central Europe (CEUR) Workshop Proceedings (2014)
6. Kanellakis, P.C.: Elements of relational database theory. In: van Leeuwen, J. (ed.) Handbook of Theoretical Computer Science, Formal Models and Semantics, vol. B, pp. 1073–1156. Elsevier Science Publishers (1990)
7. Komorowski, J., Trcek, S.: Towards refinement of definite logic programs. In: Raś, Z.W., Zemankova, M. (eds.) ISMIS 1994. LNCS, vol. 869, pp. 315–325. Springer, Heidelberg (1994)
8. Laird, P.D.: Inductive inference by refinement. In: Proc. of AAAI-1986, Philadelphia, PA, pp. 472–476 (1986)
9. Lloyd, J.W.: Foundations of Logic Programming, 2nd edn. Springer-Verlag, Berlin (1987)
10. Nédellec, C., Rouveirol, C., Adé, H., Bergadano, F., Tausend, B.: Declarative bias in ILP. In: de Raedt, L. (ed.) Advances in Inductive Logic Programming, pp. 82–103. IOS Press, Amsterdam, NL (1996)
11. Reiter, R.: Equality and domain closure in first order databases. Journal of the ACM **27**, 235–249 (1980)
12. Rouveirol, C.: Extensions of inversion of resolution applied to theory completion. In: Inductive Logic Programming, pp. 64–90. Academic Press (1992)
13. Semeraro, G., Esposito, F., Malerba, D.: Ideal refinement of datalog programs. In: Proietti, M. (ed.) LOPSTR 1995. LNCS, vol. 1048, pp. 120–136. Springer, Heidelberg (1996)
14. Semeraro, G., Esposito, F., Malerba, D., Fanizzi, N., Ferilli, S.: A logic framework for the incremental inductive synthesis of datalog theories. In: Fuchs, N.E. (ed.) LOPSTR 1997. LNCS, vol. 1463, pp. 300–321. Springer, Heidelberg (1998)
15. Shapiro, E.Y.: Inductive inference of theories from facts. Technical Report Research Report 192, Yale University (1981)
16. Siekmann, J.H.: An introduction to unification theory. In: Banerji, R.B. (ed.) Formal Techniques in Artificial Intelligence - A Sourcebook, pp. 460–464. Elsevier Science Publisher (1990)
17. Tinkham, N.L.: Schema induction for logic program synthesis. Artif. Intell. **98**(1–2), 1–47 (1998)

Semantic Web

GENOMA: GENeric Ontology Matching Architecture

Roberto Enea, Maria Teresa Pazienza, and Andrea Turbati[(✉)]

DII, ART Group, University of Rome, Tor Vergata, Rome, Italy
roberto.enea@gmail.com, {pazienza,turbati}@info.uniroma2.it

Abstract. Even though a few architectures exist to support the difficult ontology matching task, it happens often they are not reconfigurable (or just a little) related to both ontology features and applications needs.

We introduce GENOMA, an architecture supporting development of Ontology Matching (OM) tools with the aims to reuse, possibly, existing modules each of them dealing with a specific task/subtasks of the OM process. In GENOMA flexibility and extendibility are considered mandatory features along with the ability to parallelize and distribute the processing load on different systems. Thanks to a dedicated graphical user interface, GENOMA can be used by expert users, as well as novice, that can validate the resulting architecture.

We highlight as main features of developed architecture:

- to select, combine and set different parameters
- to evaluate the matching tool applied to big size ontologies
- efficiency of the OM tool
- automatic balancing of the processing load on different systems

Keywords: Ontology matching · Evaluation · Architecture

1 Introduction

As a matter of fact ontology development became a very frequent task for either expert or novice users. Ontologies can be written using different standards: among others, the most frequently used is RDF[1] (Resource Description Framework) together with its two extensions: RDFS[2] (RDF Scheme), and OWL[3](Web Ontology Language). In creating a new ontology generally two possible approaches are adopted:

1. starting from a shared vocabulary (as the FOAF[4] ontology when describing people-related terms) and then adding the new domain specific data
2. defining everything from scratch.

Consequently, using more than one ontology in a new task can be extremely difficult, since different resources[5] (identified by different lexicalizations) of distinct ontologies

---

[1] http://www.w3.org/TR/2004/REC-rdf-primer-20040210/
[2] http://www.w3.org/TR/2014/REC-rdf-schema-20140225/
[3] http://www.w3.org/TR/2004/REC-owl-ref-20040210/
[4] http://www.foaf-project.org/

© Springer International Publishing Switzerland 2015
M. Gavanelli et al. (Eds.): AI*IA 2015, LNAI 9336, pp. 303–315, 2015.
DOI: 10.1007/978-3-319-24309-2_23

can be used to identify the same concept. While humans naturally deal with such an ambiguity and try to understand the similarity between these resources, systems cannot easily do the same. This means that even when using a formal representation for an ontology (with a given serialization), the richness of the natural language is still an issue!

In this paper, we present GENOMA, an architecture that helps in the development of Ontology Matching (O.M.) systems, which are tools that infer different types of similarities among two or more ontologies. By first, we introduce the most relevant tools for O.M., we compare them thus providing a motivation for our architecture (section 2). In the following section 3, and in its subsections, we introduce our new architecture, describing its main features. Finally, with section 4, we provide a small summary of what we achieved with our research activity. Description of Ontology Matching instances is not the objective of this paper.

2   Ontology Matching Tools: An Overview

Ontology Matching is the branch of the Ontology Engineering that aims in finding similarities between resources of two or more ontologies. It tries not just to find similarities between pair of resources, but also to align ontologies and then to merge them in a new ontology, which is composed of all resources, found in the input ones. While to recognize similarities between ontologies several techniques and algorithms have been defined, the basic process [1] is always the same: the input to the framework consists in two or more ontologies from which, using a matcher (that can be composed of a single module or be built on top of several minor modules), it is produced an alignment matrix. The matcher(s) can be configured using specific parameters as well as some external linguistic resources (such as WordNet[6], for example). The resulting alignment matrix is NxM (where N is the number of resources of the first ontology, M is the number of the resources of the second ontology) and contains the similarity values between these resources. A_{ij} is the similarity value obtained from a particular algorithm between the i-resource of the first ontology and the j-resource of the second ontology.

Main differences among existing O.M. tools are:

- the size of the ontologies they are able to manage;
- the formal language in which the ontologies should be written (mostly RDF and one of its serialization, such as RDF/XML or N-Triples);
- which resources they are able to compare (classes, instances, properties, ...);
- the natural languages in which the two ontologies should be written (or if they are able to compare ontologies written in different languages);
- The cardinality of the output alignment for each resource it returns (1:1 or n:m);

---

[5] In this paper, we use the term resources to refer to any element inside an ontology (classes, instances or properties) unless otherwise specified

[6] http://wordnet.princeton.edu/

- open source or proprietary (a common problem when dealing with any software tools);
- if they use external data or not (this can affect their license as well);
- adoption of just syntactic matching or also some sort of semantic matching approach;

To have more details on the possible features of the matching process, please refer to [2], in our paper we are interested in discussing just how to combine them in a unified architecture and in evaluating the similarity metrics used in the matching tool. The ideas behind the similarity value are that:

- each resource has the maximum similarity value with itself;
- each similarity value is greater than or equal to zero (since normally they are normalized values, they are between 0 and 1);
- considering the same two resources, the similarity vales remains the same irrespective of which ontology is considered as the first one in the comparison.

Once these three properties are upheld, the similarity value can be computed using any kind of techniques: string-based, language-based, structure-based, scheme-based, extensional-based, relational-based, probabilistic-based and semantic-based.

Let us introduce what are considered the state of the art Ontology Matching (OM) tools in order to better understand some of the existing solution for this task.

2.1   Cupid

Cupid [3] is scheme-based OM tool. It was developed in the context of a collaboration of the University of Washington, the Microsoft Research and the University of Leipzig. It implements an algorithm, which uses both a linguistic matcher (using an external thesauri) and a structure-based matcher. Input ontologies are considered as graphs. The implemented algorithm is divided in three steps. By first, it uses the linguistic matcher to computer the similarities values considering the external thesauri. Then it adopts a structure-based matcher to obtain the similarities values. Finally, it combines the values obtained in previous phases using a weighted sum and then it produces the alignment matrix.

2.2   COMA and COMA++

COMA [4] (Combination of Matching Algorithm) adopt a parallel approach in using matchers. It was developed at the University of Leipzig and it is composed of an extendible library of matchers, a framework which can be used to combine the matchers in a single algorithm and a platform for the evaluation of the results. COMA is provided with six elementary matchers, five hybrid matchers and one matcher that uses the other matchers. Most of these matchers are string-based and linguistic, supporting external resources. COMA, offers a more flexible architecture than Cupid and considers the user feedback.

COMA++ [5] is the newer version of COMA that introduces an optimization in reusing the alignment, a better implementation of the base matchers and a user interface.

2.3 iMAP

iMap [6] can be considered as an atypical Ontology Matching tools, since it works on scheme matching and not on what is normally defined as an ontology. This means that iMap tries to find complex matches in scheme, such as "our-price = price * (1 + tax-rate)"[7] or it is able to discover that the field *address* should be matched with the field *street* and the field *number*. The tool iMAP deals with the problem of matching as a search problem in a, possibly, enormous space. Its modules are called *searchers*, and they work in parallel, each one searching for a specific pattern. For example, a searcher analyzes only string fields, while another focuses on fields containing integers. The operations each searcher performs, depend on the data type they are looking for.

The algorithm implemented in iMAP is divided in three distinct phases:

3. All the searchers, which are working in parallel, return their results, which are the candidate for the next step. Since the number of returned results could be extremely large, a *bean search* [7] approach is adopted.
4. For each attribute of the target scheme, all the candidates of the first scheme are considered and only the ones with specific characteristics are retained.
5. Results of the previous steps are put inside a similarity matrix using the Similarity Estimator Module. This matrix is composed of the couples [target attribute, candidate attributes]. Then the requested alignment is extracted from a filtered version of this matrix.

2.4 GLUE

GLUE [8], developed by the Washington University uses machine learning techniques to discover similarities one-to-one between two distinct taxonomies. The machine learning algorithm used, can be divided into three steps:

1. It learns the join distribution function of the classes of the two taxonomies. It uses two modules implementing the following base techniques: *content learning* (based on Bayes Learning) and *name learning* (similar to the first one). A meta-matcher performs a linear combination of the results of the other two modules. The weights for the linear combination are decided by the user.
2. The system computes the similarity between pair of classes adopting a joint distributed function provided by the user. From this, a similarity matrix is produced.
3. The alignment between the two taxonomies is obtained by applying several filtering operations.

GLUE is the evolution of a previous tool, LSD [9].

---

[7] http://pages.cs.wisc.edu/~anhai/projects/schema-matching.html

2.5   Falcon-AO

Falcon-AO [10] was developed by the China Southeast University. It adopt a divide-et-impera approach to perform the matching between two ontologies written in RDFS or OWL. It was designed to manage large ontologies (thousands of resources) using partitioning techniques. The algorithm identifies three phases:

1. It divides the ontologies into smaller structures to take advantages of the hierarchy, through the use of the property *rdf:subClassOf*, and the use of the agglomerative clustering algorithm ROCK [11].
2. It compares the clusters, obtained in the previous step, focusing on the anchors, which are linked resources, obtained using string-based comparison. More anchors there are between two clusters, more similar those clusters are assumed to be. Only clusters with a number of anchors over a specific threshold are considered possible candidate.
3. Two matchers, one linguistic and the other structural, are adopted to generate the similarities matrices; then these matrices are aggregated into the resulting alignment matrix. If the linguistic matcher is able to obtain good results alone, the structural one, which is more complex, is not used.

2.6   ASMOV

ASMOV [12], Automated Semantic Mapping of Ontologies with Verification, was developed during a collaboration between Infotech Software and the University of Miami. It receives two ontologies and an optional alignment and returns a many-to-many alignment among the resources of the two ontologies, mainly classes and properties. It uses an iterative process. In its main phase, after a preprocessing step, it uses lexical, structural and extensional matchers to calculate, iteratively, the similarity between pairs of resources that are then aggregated in a weighted mean. It uses also external resources, such as WordNet and UMLS[8]. Then, it verifies the consistency of the proposed alignments, using five pattern typologies, which can generate inconsistencies. Its main application is the integration of bioinformatics data.

2.7   O.M. Tools Comparison and Motivation for an O.M. Architecture

In the previous sections, we have roughly sketched a few existing O.M. tools and architectures. Table 1 summarizes these features (according to the ones specified in the beginning of Section 2).

Apart from these features, Falcon offers the possibility to set the values for some parameters, and only COMA++ offers the possibility to decide which matchers to use, without deciding how they can communicate each other and no one even suggests the possibility to deploy his tool on a distributed environment.

---

[8] http://www.nlm.nih.gov/research/umls/

Table 1. O.M. tools characteristics

	Falcon	ASMOV	Cupid	iMap	GLUE	COMA++
Specific for big ontologies	Yes	No	No	No	No	No
Input	RDF/RDFS/ OWL	OWL	XML	XML, DB Schemas	XML, DB Schemas	RDF/RDFS/ OWL
Resources matched	Classes + Properties + Instances	Classes + Properties	XML Elements	XML, DB schema Elements	XML, DB schema Elements	Classes + Properties
Specific Natural languages	No	English	English	No	No	English
Output alignment	1:1	m:n	1:1	1:1	1:1	1:1
Open source	Yes	No	No	No	No	Yes
Linguistic resources	No	WordNet	WordNet	No	No	WordNet
Type of matching	Linguistic Structural	Linguistic Structural Extensional	Linguistic Structural	Parallel searcher + similarity estimator	Statistical Approach, Machine learning	Linguistic

After analyzing previous O.M. tools, it emerges that there is no tool that is always better than the others, it depends on several factors. Then there is no architecture that can be classified as the best one or a matching techniques which is always correct. Each time is left to the application developers the decision on what matcher/architecture to choose. And it is not an easy task! That's why we decided to develop a new generic architecture, helping experts or even beginners, to try different combinations and decide which one is the best given a specific application context. Another important feature of our novel architecture is the possibility either to reuse existing matchers, or to develop new ones, always in the context of our architecture. This feature can appear normal nowadays in a context in which extendibility is a core peculiarity of almost any tool or system. Generally, in the scientific environment, the developed tools are close systems (sometimes they are even proprietary one or open one, but their code is too complex and not sufficiently explained to be easily understood and extended).

A strong requirement for our architecture is the development of an easy user interface to help users to assemble the matcher in any way they want to. Finally, since ontologies are becoming bigger and bigger, our architecture should be able to scale out by being deployed on several machine to avoid performance issues.

3   Gen.O.M.A.

Gen.O.M.A (Generic Ontology Matching Architecture) or GENOMA is a sort of meta architecture which helps in the development of new specific architectures for the ontology matching task. In this section by first we introduce the architecture (section 3.1), then the provided user interface, which guide the user in the creation of each new architecture and its validation (section 3.2). Then we describe some of the existing matchers modules (section 3.3) and we conclude this section by providing examples of tested and validated architectures (section 3.4), by showing their peculiarities and differences.

3.1   Architecture

In every matching system, the main problem to overcome is improving system's computational efficiency. This is particularly important when evaluating large ontologies, which are the ones used in real case matching scenarios. GENOMA[9] offers the possibility to be executed in a distributed environment; to achieve this feature, it uses Java RMI[10] (Remote Method Invocation) which enables the execution of each matchers on a given server.

In Fig. 1 we show the architecture adopted by GENOMA. To handle ontologies, the OWL ART API[11] are adopted. The main component of the ontology matching process is the *Matching Composer* (sometimes called just *Composer*). This module coordinates the whole process. It parses the configuration file, handled by the module *Configuration*, of the selected engine[12] (more information about this configuration are given in section 3.2), then launches the desired matchers, using the relative parameters (the only mandatory parameters are the 2 ontologies addresses; the parameters that are specific to each matcher are optional, since they all have default values). Each matcher returns to the Composer its own processed similarity matrix, which is then passed to the next matcher, or a notification if an error occurred during ontology processing. Once all the desired matchers have completed their task by returning the similarity matrix, the Composer produces a detail log and the alignment file, that is an xml file, containing the final similarity matrix, filtered using the threshold selected by the user before starting the matching process (the *Filtered Similarity matrix* in Fig. 1). A matcher can uses external linguistic resources, for example by adopting Ontoling[13] to access DICT[14] and WordNet.

---

[9] It can be download from: https://bitbucket.org/aturbati/ontology-matching-architecture
[10] https://docs.oracle.com/javase/tutorial/rmi/
[11] http://art.uniroma2.it/owlart/
[12] An engine is the implementation of a matching architecture produced by the user interface of GENOMA.
[13] http://art.uniroma2.it/software/OntoLing/
[14] www.dict.org

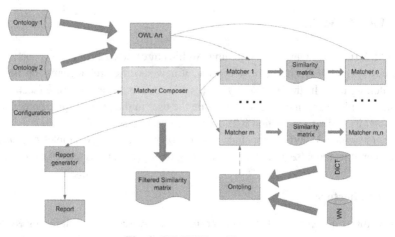

Fig. 1. GENOMA architecture

3.2   User Interface

To help the user in creating, visualizing and validating an architecture for an O.M. tool, GENOMA provides an easy to use and complete User Interface[15] (see Fig. 2 and hereafter for details). Since the beginning, the user can decide whether to use an already existing configuration or to create a new architecture, using the provided matcher (plus the ones possibly developed by himself).

When selecting an existing configuration, the user specifies the two ontologies to be compared, an optional alignment file (useful when evaluating an O.M. architecture), and the Engine (a file specifying all the information describing a given architecture). Once these information have been provided, the matching process starts. The user can monitor the entire process (its status, any error, the log file and the alignment file produced by the selected architecture): in fact all these details are stored in a MySQL[16] DB, to help retrieve them and guarantee their persistency.

The user interface is even more useful when deciding to assemble a new architecture. The GUI (shown in Fig. 2) provides the user what he needs to create each new specific architecture. The user can either save or load a previous created architecture, as well as can validate the current architecture (see Section 3.3 to understand the validation provided by GENOMA). One important feature is the possibility to save the current work as an image, to help showing the created architecture or sharing it with colleagues to discuss it.

The creation of the new architecture is completely interactive and mainly mouse oriented, to help novices in this delicate task. By selecting the desired matcher, the user just need to specify the parameters values (or he can accept the default ones) and

---

[15]The User Interface is deployed as a web application in Tomcat.
[16] www.mysql.com/

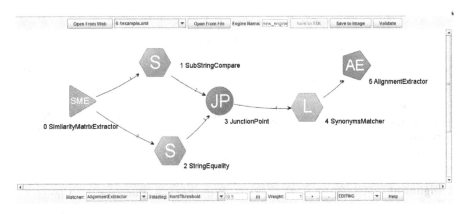

Fig. 2. User Interface when creating an architecture

then link the matchers in several possible deployment configurations (see Section 3.4). All this is achieved easily by using a Java applet and just a couple clicks of the mouse[17]!

3.3 Provided Base Matchers

All the elements that can be considered when assembling an architecture in the user interface, at the moment, are divided into, three main classes:

- Structural elements (known as system elements);
- String-based matchers;
- Language-based matchers.

The elements of these classes all are deployed in the same manner. The system elements are: *SimilarityMatrixExtractor* (SME), *JuctionPoint* and *AlignmentExtractor*.

The Similarity Matrix Extractor is the first element in every architecture created by using GENOMA. It can extract the similarity matrix, obtained from another architecture, so it can start the current architecture by using results obtained in the context of a previous execution.

The Junction Point is used to combine two similarity matrices, obtained from the parallel processing (see Section 3.4). Its configuration parameter consists in which aggregation function to use. The default value is the triangular norm, while other value are possible, by expanding its implementation.

The Alignment Extractor executes the inverse operation of the SME: given a similarity matrix, it generates the alignment. It also apply a filter operation before generating the alignment in case the user has provided it.

The other classes, String-based and Language-based matchers, provide the matching part in the architecture. The former considers only the words associated to a resource (for example thought the property *rdfs:label* or using the name of the resource itself)

---

[17] Just to see how easy to use it is, interested persons could access it at
http://artemide.art.uniroma2.it:8080/GenomaWeb-1.0/

Table 2. Matchers list

Name	Symbol	Type	Matcher Functionality	Number of edges
Similarity Matrix Extractor	SME	System	Extraction the similarity matrix	0 input N output
Junction Point	JP	System	Aggregation similarity matrices	2 input N output
Alignment Extractor	AE	System	Extraction the alignment from the similarity matrix	1 input 0 output
String Equality Matcher	S	String-based	Comparing Strings	1 input N output
SubString Compare Matcher	S	String-based	Comparing Substrings	1 input N output
N-grams Compare Matcher	S	String-based	N-grams comparison	1 input N output
String Hamming Distance Matcher	S	String-based	Hamming distance in the comparison	1 input N output
Synonyms Matcher	L	Language-based	Comparison using synonyms	1 input N output
Cosynonyms Matcher	L	Language-based	Comparison using synonyms and then counting the number of synonyms they have in common	1 input N output

threating it as sequence of characters, while the latter performs an analysis based on the language used, for example considering synonyms obtained from a linguistic resource, such as WordNet. The provided language based matchers, in the current release, work only on the English language; the use of the framework OntoloLing and Linguistic Watermark [13] with the associated resource, DICT, could be used to analyze other languages as well.

The system is able to automatically validate each architecture developed by the user, by checking that all the constrains are respected (for example that each architecture starts with a SME, ends with a Alignment Extractors).

In Table 2, the existing matchers are presented and commented.

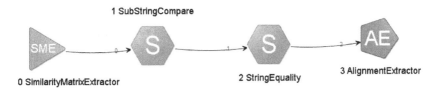

1 SubStringCompare

0 SimilarityMatrixExtractor

2 StringEquality

3 AlignmentExtractor

Fig. 3. Matchers in series

3.4   Examples of Architectures

Matchers can be deployed in series or in parallel. In this section, we present two base configurations to better understand what is possible to achieve by using GENOMA. When the user is defining an architecture, the first decision he has to make is whether to adopt a parallel or a series given two matchers.

In Fig. 3, we have an example of two matchers, one SubStringCompare and one StringEquality, in series, while in Fig. 4 the same two matchers are deployed in parallel.

When two matchers are in series, the resulting similarity matrix of the first matcher is passed as input matrix to the second one. The resulting similarity values, after the second matcher, are the normalized weighted sum of the two separated values (each matcher has a weight in the interval [0, 1]). This means that the result is the weighted sum of all the local results of the single matchers in the series.

When the matchers are in parallel, a JunctionPoint is needed immediately after them. This component performs the aggregation of each result by using the algorithm selected during the configuration phase.

The easy to learn and complete user interface enables the construction of more complex architectures using the same approach: just by combining several matchers in series and in parallel. For example in Fig. 5 we have an architecture composed of a parallel execution of two matchers (a StringHanningDistance and a NgramCompare), which are then deployed in series with another matcher (a SynonymsMatcher). Even more complex architectures, can be deployed very easily just by selecting matchers of interest.

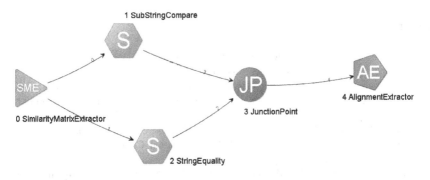

1 SubStringCompare

0 SimilarityMatrixExtractor

2 StringEquality

3 JunctionPoint

4 AlignmentExtractor

Fig. 4. Matchers in parallel

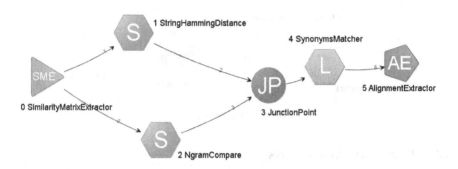

Fig. 5. Parallel and series example

4   Conclusion

In this article, we have described GENOMA, an architecture that helps developing, deploying and validating complex and totally customizable O.M. architecture and tools. These architectures can be changed at any given time, to find the better one given a specific domain or application needs, leaving the user the complete freedom of experimenting while defining its architecture.

Several implemented and tested matchers are provided with GENOMA, to give the possibility to immediately test it. Since it is open source and well documented, implementing new matchers, to extend its capability, is possible.

Regarding the deployment of the produced O.M. architecture, it can be on the same machine or distributed among several servers to obtain better performance by sharing the complexity required by this task.

Thanks to all these features, using GENOMA enables every user (either expert or novice) to rapidly define an O.M. architecture, to test it and possibly extends the modules functionalities by developing new matchers.

Acknowledgements. This research has been partially founded by the european project Sema-Grow.

References

1. Euzenat, J., Shvaiko, P.: Ontology Matching. Springer (2007)
2. Euzenat, J., Shvaiko, P.: Ontology Matching. Springer (2013)
3. Madhavan, J., Bernstein, P.A., Rahm, E.: Generic schema matching with Cupid. In: Proceedings of the 27th International Conference on Very Large Data Bases, San Francisco, CA, USA, pp. 48–58 (2001)
4. Do, H.-H., Rahm, E.: COMA, a system for flexible combination of schema matching approaches. In: Proceedings of the 28th International Conference on Very Large Data Bases, Hong Kong, China, pp. 610–621 (2002)
5. Aumueller, D., Do, H.H., Massmann, S., Rahm, E.: Schema and ontology matching with COMA ++. In: Proceedings of the 2005 ACM SIGMOD International Conference on Management of Data, Baltimore, Maryland, pp. 906–908 (2005)

6. Dhamankar, R., Lee, Y., Doan, A., Halevy, A., Domingos, P.: iMAP: discovering complex semantic matches between database schemas. In: 23rd International Conference on Management of Data (SIGMOD), pp. 383–394 (2004)
7. Russell, S.J., Norvig, P.: Artificial Intelligence: a Modern Approach. Pearson Education, Englewood Cliffs (2003)
8. Doan, A., Madhavan, J., Domingos, P., Halevy, A.: Learning to map between ontologies on the semantic web. In: Proceedings of the 11th International Conference on World Wide Web, Honolulu, Hawaii, USA, pp. 662–673 (2002)
9. Doan, A., Madhavan, J., Domingos, P., Halevy, A.: Ontology matching: a machine learning approach. In: Handbook on Ontologies, pp. 173–235. Springer, Berlin (2004)
10. Hu, W., Qu, Y.: Falcon-AO: A practical ontology matching system. Journal of Web Semantics, 237–239, September 2008
11. Guha, S., Rastogi, R., Shim, K.: ROCK: a robust clustering algorithm for categorical attributes. In: Proceedings of the 15th International Conference on Data Engineering, Washington, DC, USA, pp. 512–521 (1999)
12. Jean-Mary, Y.R., Shironoshita, E.P., Kabuka, M.R.: Ontology matching with semantic verification. Journal of Web Semantics, 235–251 (2009)
13. Pazienza, M.T., Stellato, A., Turbati, A.: Linguistic watermark 3.0: an RDF framework and a software library for bridging language and ontologies in the semantic web. In: 5th Italian Semantic Web Workshop on Semantic Web Applications and Perspectives, (SWAP 2008), FAO-UN, Rome, Italy, December 15–17, 2008

Open Data Integration Using SPARQL and SPIN: A Case Study for the Tourism Domain

Antonino Lo Bue[(⊠)] and Alberto Machì

ICAR-CNR Sezione di Palermo, Via Ugo La Malfa 153, 90146 Palermo, Italy
{lobue,machi}@pa.icar.cnr.it

Abstract. Open Data initiatives from governments and public agencies in Europe have made large amounts of data available on the web. Linked Data principles can help to improve Open Data integration, disclosing the connections between datasets, leveraging a powerful usage of data and enabling innovative ways to improve citizens' life. In this work we present a novel approach for Open Data integration, based on SPARQL federated queries formalized as SPIN rules. This methodology allows the interlinking and enrichment of heterogeneous Open Data, using the distributed knowledge of the Linked Open Data cloud. We present a case study on integrating and publishing, via enrichment and interlinking techniques, tourism domain datasets as Linked Open Data.

Keywords: Semantic web · Linked open data · SPARQL · Interlinking

1 Motivation

Recent developments in e-government and open government initiatives, aiming for openness, transparency and efficiency has fostered the release of governmental open data across Europe. Unfortunately, due to a lack of homogenous guidelines about data delivery and information management, agencies deliver open data using heterogeneous processes and data structures, failing to elicit interconnections among data.

Semantic web technologies, in particular Linked (Open) Data (LOD) [1,2], are opening new ways for data integration and reuse, making data interoperable at a semantic and technological level. A notable example of a successful story of Linked Open Data usage for eGovernment is the UK government portal (data.gov.uk)[3].

The EU delivered in 2013 the Digital Agenda[1], asking each local government to align with the *European Interoperability Framework* (*EIF*) - which describes the technical, semantic and legal interoperability guidelines and processes. The same concepts were also part of the *ISA* and *ISA2*[2] programs, which aim to establish a program on interoperability solutions for European public administrations, businesses and citizens. As a consequence of EU Digital Agenda and *ISA* programmes, first steps on Open Government in Italy were done with the *Guidelines for the Development of*

---

[1] http://ec.europa.eu/digital-agenda/
[2] http://ec.europa.eu/isa/

© Springer International Publishing Switzerland 2015
M. Gavanelli et al. (Eds.): AI*IA 2015, LNAI 9336, pp. 316–326, 2015.
DOI: 10.1007/978-3-319-24309-2_24

the *Heritage Public Information*[3] published in 2014 and the *Guidelines for Semantic Interoperability*[4] (*SPC guidelines*). The *SPC guidelines* suggest, in particular, a workflow for Open Data selection, creation and publishing, identifying 7 steps for publishing Linked Open Data:

(1) Dataset selection,
(2) Cleaning,
(3) Analysis and model creation,
(4) Enrichment,
(5) Interlinking,
(6) Validation,
(7) Publication.

Similar to SPC workflow, the W3C has identified a set of *Best Practices for Publishing Linked Data on the Web*. A group note of 9 January 2014[5] outlines the process of development and delivery of open government data as Linked Open Data in 10 steps:

(1) Prepare Stakeholders,
(2) Select a Dataset,
(3) Model the Data,
(4) Specify Appropriate License
(5) The Role of "Good URIs" for Linked Data,
(6) Standard Vocabularies
(7) Convert Data to Linked Data,
(8) Provide Machine Access to Data,
(9) Announce to the Public,
(10) Social Contract of a Linked Data Publisher

Point 3 of *SPC guidelines* and point 4 of W3C *Best Practices for Publishing Linked Data on the Web* describe the model creation phase. Modeling process, in the context of RDF, refers to the act of capturing the context of data, defining the relationships among them and describing the vocabulary used through ontologies.

In order to produce 5 stars Linked Open Data, source data need to be linked to other datasets. This phase is explained on point 7 of W3C Best Practices and point 5 of *SPC guidelines*. For example being able to connect data from a government authority with *DBpedia*, is a way to show the value of adding content to the Linked Data Cloud.

In this paper we present a case study on integrating, via mapping and interlinking techniques, heterogeneous Open Data and publishing Linked Open Data for the tourism domain.

---

[3] http://www.agid.gov.it/sites/default/files/linee_guida/
    patrimoniopubblicolg2014_v0.7finale.pdf
[4] http://www.agid.gov.it/sites/default/files/documentazione_trasparenza/cdc-spc-gdl6-
    interoperabilitasemopendata_v2.0_0.pdf
[5] http://www.w3.org/TR/ld-bp/

2   Related Work

The term Linked Data refers to data published on the web following 4 principles:

- Use URIs as names for things;
- Use HTTP URIs so that people can look up those names;
- When someone looks up a URI, provide useful information, using the standards;
- Include links to other URIs so that people can discover more things.

These principles promote linking between entities and allow these entities to be connected, enriched and machine-readable.

RDF[6] standard is the basic language that enables the formal representation of information as triples (*subject, predicate, object*). OWL[7] is a semantic language, built on top of RDF, which allow the formal representation of ontologies. SPARQL[8] Query Language is used to query and retrieve information stored in the RDF format thus allowing and facilitating access to the so-called Web of Data.

The SPARQL inferencing notation (SPIN)[9] is a standard language that enables the representation of SPARQL rules and constraints in ontologies [4,5]. SPIN rules can be serialized into RDF, and this allows encapsulating interlinking rules inside the domain ontology model itself and using the model as the unique reference for the integration process.

Data linking is usually performed by comparing the properties of a given dataset with homologous properties of the target dataset. A complete survey of data linking methods by Ferrara et als. is available at [6]. A vast literature is exploring the use of geographic primitives in Linked Data, applying geometric distance measures [7,8,9,10] as well as string matching algorithms. Other methodologies for Linked Data interlinking are based on ontology alignment [11] or machine learning approaches [12].

A number of projects have been developed, providing different approaches to compare entities or semi-automatically define links between different datasets [13]. One of the most popular declarative systems is Silk [15], a logical rule system with built-in functions for establishing links between data items within different linked data sources. The Datalift project [14] provides services for converting data formats to raw RDF data and maintaining vocabularies of keywords suggested to the user in mapping raw RDF resources to domain ontology resources via SPARQL Construct queries. Its interlinking module relies on Silk. A limit of Silk is that it doesn't support rule chaining and is based on a non-RDF mapping language.

Examples of research initiatives producing Linked Data for the tourism domain are the TourMISLOD project [16], which provides a core source of European tourism statistics, such as Linked Data, and OpeNER Linked dataset [17], which provides accommodation data, and other information, such as a short description and location

---

[6]   http://www.w3.org/RDF/

[7]   http://www.w3.org/2001/sw/wiki/OWL

[8]   http://www.w3.org/TR/rdf-sparql-query/

[9]   http://www.w3.org/Submission/2011/SUBM-spin-overview-20110222/

information in Tuscany. Both approaches focus mainly on accommodation modeling and instance building, missing to describe Point of Interests (Cultural, Environmental), Places and Events for the tourism domain.

3   Mapping and Interlinking Open Data

The goal of our research is two-fold: (1) demonstrate that Open Data can be integrated as Linked Open Data by inferring interlinking patterns from the distributed knowledge of the Linked Open Data cloud and identifying equivalent resources to integrate; (2) formalize a methodology for including in the mapping of a domain ontology, interlinking information to be used as a unique reference for the integration/publishing process.

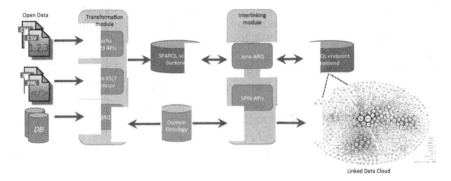

Fig. 1. System overview

Our system prototype, developed as a framework of Java servlets, consists of two main modules. The *transformation module* takes as input Open Data, serialized in Comma Separated Value (CSV) format, in XML or described in a relational database. It generates RDF triples according to a domain ontology and stores the results in a backend store. The *interlinking module* reads the data from the backend store and takes as configuration input a domain ontology containing interlinking rules (defined in SPIN) to be executed. Rules discover links between the dataset instances and the Linked Open Data cloud, using the cloud knowledge itself to infer the linking patterns.

For each Open Data source, all the integration information is stored as RDF triples. Thus, the selected domain ontology:

— defines the semantic model as a vocabulary of OWL classes, properties and axioms,
— formalizes the interlinking methods of each class of the model using SPIN rules.

Defining rules at class levels allows triggering of the SPIN rules dynamically, as soon as a new source instance is added to the backend. This is particularly useful when dealing with frequently updated Open Data as, for example, in the transport domain, or when dealing with data describing events in the tourism domain.

A description of the implementation and a demo of the system are available at `http://slab.icar.cnr.it/opendata/index.html`

3.1   The Transformation Module

The first step of Open Data integration as semantic data, requires transforming source datasets in RDF triples. In order to handle different source formats, we have developed the transformation module as a wrapper that encapsulates existing technologies and services that transform tabular data, XML documents, and relational database tables into RDF statements.

Tabular data expressed in CSV format are managed by a 2-step process: firstly data are translated into generic RDF triples using Apache Any2310 APIs, and then obtained RDF raw triples are re-mapped to the domain ontology vocabulary via a SPIN Construct rule. Complex conversions are handled by user-defined functions called inside the SPIN rule. The final results are stored in the backend server via the SPARQL Update protocol[11].

XML data are transformed by a SAXON[12] service in a single step process, using domain specific XSLT transformation style sheets, which map the XML source data into RDF triples according to the domain ontology vocabulary.

A detailed example of the XML to RDF conversion is available at `http://slab. icar.cnr.it:8080/xslt/transformation_module.jsp`.

Relational database source data translation into RDF is performed by a single step process which exploits the D2RQ APIs[13], using the domain ontology vocabulary in a customized mapping script expressed in the D2RQ mapping language.

3.2   Interlinking Rules in SPIN

To express interlinking, we exploit SPARQL federated queries to merge facts about the same entity extracted from different sources on the LOD cloud, and SPIN functions in order to implement *backward chaining* reasoning.

In SPARQL federated queries, *sub queries* can be directed to different SPARQL endpoints via the 'SERVICE' directive and the query processor merges results from sub queries.

Each *sub query* result (p,q,t) can be seen as part of a Horn clause $(p \wedge q \wedge ... \wedge t) \rightarrow (u,v,z)$ that asserts one or more triples about the data, like, for example, an `owl:sameAs` relation between the local resource and a LOD dataset resource.

Each data source can contribute information, providing facts to the inference rule that asserts the interlinking predicate.

---

[10]   https://any23.apache.org/

[11]   http://www.w3.org/TR/sparql11-update/

[12]   http://sourceforge.net/projects/saxon/

[13]   http://d2rq.org/

SPIN allows defining personalized functions using JavaScript[14] code or SPARQL code. Calling a SPARQL function inside the SPARQL interlinking rule implements *backward chaining*.

Furthermore, multi-step SPARQL processes can be implemented exploiting the execution order features of SPIN. An ordered set of rules can be then applied to a graph of instances.

The following SPIN rule gives an example of an interlinking rule where the instance of a hotel is interlinked with the *Geonames* description of the municipality it belongs to, discovered via reverse geocoding of its coordinates on the *GADM*[15] *Linked Data service* and inference of the *Geonames* equivalent municipality on the *DBpedia* dataset. The SPARQL variable ?this is a special SPIN construct that refers to any instance of the class in which the rule or constraint is defined.

```
1    INSERT {
2            ?municipality a etLite:Municipality .
3            ?municipality owl:sameAs ?geonames .
4            ?this etLite:inMunicipality ?municipality .
6    }
7    WHERE {
8            ?this geo:lat ?lat .
9            ?this geo:long ?long .
10           BIND (IRI(CONCAT("http://gadm.geovocab.org/services/withinRegion?",
11                  "lat=", ?lat, "&long=", ?long, "#point")) AS ?GADMservice) .
12           SERVICE <http://slab.icar.cnr.it:8891/sparql> {
13                   OPTIONAL {
14                           ?GADMservice <http://gadm.geovocab.org/spatial#PP> ?GADMplace.
15                           BIND (IRI(REPLACE(str(?GADMplace), "_", "/")) AS ?GADMplace_loc).
16                   } .
17                   GRAPH <http://slab.icar.cnr.it/graph/GADM> {
18                           ?GADMplace_loc rdfs:label ?GADMlabel .
19                           ?GADMplace_loc a <http://gadm.geovocab.org/ontology#Level3> .
20                           ?GADMplace_loc <http://gadm.geovocab.org/ontology#in_country> ?country .
21                   }
22           } .
23           SERVICE <http://dbpedia.org/sparql> {
24                   ?dbpedia owl:sameAs ?GADMplace .
25                   ?dbpedia owl:sameAs ?geonames .
26                   FILTER STRSTARTS(str(?geonames), "http://sws.geonames.org") .
27           } .
28           BIND (IRI(CONCAT("http://slab.icar.cnr.it/testKB/",REPLACE(?GADMlabel, " ", "_")))
29                   AS ?municipality) .
30   }
```

This SPARQL federated query takes as input latitude and longitude values from the backend data and asserts an owl:sameAs predicate linking the input resource to the Geonames LOD dataset. The query requests a federated access to 2 services, the GADM Linked Data web service (actually wrapped into the local backend endpoint) and the DBpedia endpoint.

Latitude and longitude of the source instance are used to dereference from the *GADM* service the resource of administrative level 3 (city) whose boundary the coordinates are located in. The obtained *sub query* result is used to run a query against the *DBpedia* endpoint in order to get the *Geonames* reference for that resource by evaluating the DBpedia equivalent resource owl:sameAs predicate.

The outcome of this interlinking rule is that an instance of a hotel, mapped from the open dataset, containing as property the string label of the hotel city - e.g. "Rome" and a pair of coordinates is enriched with a semantic geo-reference with the *Geo-*

---

[14] http://spinrdf.org/spinx.html
[15] http://gadm.geovocab.org/

names resource describing the city of Rome
(http://sws.geonames.org/3169070).

The following SPIN rule illustrates another example of a federated query that infers linking information ,namely equivalence of a museum instance with a resource on the *Linked Open Data Museum*s dataset and also infers its cultural topics.

```
1    INSERT {
2        ?this owl:sameAs ?lodm .
3        ?this skos:subject ?lodm_sub .
4        ?this skos:subject ?dbpedia_broader
5    }
6    WHERE {
7        ?this geo:lat ?lat1 .
8        ?this geo:long ?long1 .
9        ?this etCore:localityName ?mibac_loc_name .
10
11       SERVICE <http://slab.icar.cnr.it:8891/sparql> {
12           GRAPH <http://slab.icar.cnr.it/graph/linkedopendata-musei> {
13               ?lodm rdfs:label ?lodm__loc_label .
14               ?lodm <http://www.w3.org/2006/vcard/ns#latitude> ?lat2 .
15               ?lodm <http://www.w3.org/2006/vcard/ns#longitude> ?long2 .
16               ?lodm skos:subject ?lodm_sub
17           }.
18       FILTER (<bif:haversine_deg_km>(xsd:float(?lat1), xsd:float(?long1), xsd:float(?lat2),
19               xsd:float(?long2)) < 0.1) .
20       FILTER ilo:levenshtein(str(?mibac_loc_name), str(?lodm_loc_label), 0.5).
21           }.
22       SERVICE <http://dbpedia.org/sparql> {
23           ?lodm_sub skos:broader ?dbpedia_broader
24           }.
26   }
```

This interlinking rule matches a source instance using latitude, longitude and label data in 2 sub queries. The first sub query is executed against the *Linked Open Data Museums* dataset, asserting the owl:sameAs equivalence of nearby instances and inferring the skos:subject information about the instance. The second sub query uses the result of the first sub query to search on DBpedia for any other broader skos:subject related to the skos:subject found in the first sub query. The rule makes use of a custom implementation of the SPIN *Levenshtein* function, for string matching on labels, and also exploits Virtuoso[16] geospatial *Haversine* distance function in order to find geospatial equivalence between the geolocation attributes of museums in both datasets.

3.3   Mapping and Interlinking Patterns

In order to produce a correct mapping between the source datasets and the domain ontology semantics used in the target RDF graphs, mapping schemes are conceived by a domain expert with the help of an ontology expert and then translated into proper mapping rules (using SPARQL or XSLT or D2RQ mapping language).

As well as for mapping definition, interlinking/validation rule definition is a complex task that requires cooperation between a domain expert that defines and validates the semantics of each interlinking pattern, and a SPARQL/SPIN expert in order to translate the interlinking patterns into SPIN rules and/or functions.

At the beginning, defining interlinking patters for a domain and validating these is a time consuming task, despite that the usage of a well know standard as SPARQL

---

[16] http://virtuoso.openlinksw.com/dataspace/doc/dav/wiki/Main/

can facilitate the task. Re-use of existing functions or simple interlinking/validation rules is possible. General-purpose interlinking/validation patterns can be defined in an interlinking ontology (which can be seen as an upper ontology) and then reused by simply importing that ontology when defining a new domain ontology; in this case, the overall effort can be reduced because only domain-specific patterns must be defined and implemented.

4   The Tourism Case Study

The integration services developed are currently being tested within a tourism domain case study focused on the italian region Calabria in the framework of the Italian Project PON *SmartCities Dicet-InMoto-Orchestra*, stream Information Mobility and Tourism[17].

We are currently testing the *transformation module* using an ontology for the tourism and cultural domain named *eTourism*[18] that describes services (e.g. Hotels, B&B), points of interest (e.g. Museums, Archeological parks, Libraries) and events. Mapping and Interlinking services were tested against different source data types and different source formats ranging from tabular CSV, to XML and relational databases. In particular we have collected and mapped different datasets:

- Open Data from the Italian Minister of Cultural Heritage and Tourism (MiBACT)[19] describing cultural places (e.g. museums) and cultural events;
- Open Data from regional administration of Calabria and Chamber of commerce about cultural and environmental resources
- Scraped web data about accomodation services extracted from Tripadvisor, Booking.com, Venere.com and other websites;
- Open Data from UNESCO[20] about World heritage sites and European Union Interest Sites;

In order to test the *interlinking module* and publish such data as Linked Open Data, we have developed specific mapping schemes to translate the heterogeneous data sources into RDF triples conforming with the *eTourism* ontology, while for each ontology class we have defined a set of interlinking rules using SPIN.

The following table is a summary of all types of SPIN rules created so far and exploited for our case study:

---

[17] http://www.progettoinmoto.it
[18] http://slab.icar.cnr.it/eTourismLite/
[19] http://dbunico20.beniculturali.it/DBUnicoManagerWeb/
[20] http://www.unesco.beniculturali.it/index.php?it/121/open-data

Table 1. Interlinking patterns implemented via SPIN rules for the case study. The table shows requested source properties, LOD datasets exploited, types of linking rule and inferred triples.

Source information	Exploited LOD Dataset	Built-in Function	Inferred triples	Type
[opendata:place] geo:lat, geo:long	GADM, DBpedia, Geonames		[opendata:place] owl:sameAs [geonames:municipality] (ADM3) | [geonames:region] (ADM2) | [geonames:country (ADM1)	Semantic reverse geocoding, schema mappings
[opendata:museum] rdfs:label | foaf:name	DBpedia	levenshtein string distance	[opendata:museum] owl:sameAs [dbpedia:museum]	String distance
[opendata:museum] rdfs:label | foaf:name & geo:lat, geo:long	Linked Open Data Musei, DBpedia	haversine distance wrapper, levenshtein string distance	[opendata:museum] owl:sameAs [lodm:museum], [opendata:museum] skos:subject [dbpedia:subject]	Geographical distance, string distance, schema mappings
[opendata:hotel] | [opendata:museum] geo:lat, geo:long	Linked Geo Data	haversine distance wrapper	[opendata:museum] | [opendata:hotel] owl:sameAs [lgd:hotel]	Geographical distance
[opendata:place] opendata:plainTextAddress	GADM, DBpedia, Geonames	google geocoding wrapper	[opendata:place] geo:lat, geo:long, owl:sameAs [geonames:municipality] (ADM3)	Geocoding, Semantic reverse geocoding, schema mappings

LOD datasets used for the case study interlinking rules are:

- *GADM* (Global Administrative Areas) linked data service to evaluate the administrative area that contains a specific pair of latitude and longitude coordinates (e.g. `lat=41.9, long=12.4` coordinates are contained inside the "city of Rome" administrative area)
- *Geonames* and *DBpedia* resources linked, via `owl:sameAs` relation, to infer semantic references to cities, regions and states and enrich the original dataset with inferred data (e.g. DBpedia's `dbpedia-owl:abstract` descriptions), as well as to include multilingual labels (e.g. Geonames' `gn:alternateName` labels);
- *CulturaItalia Linked Data* to enhance data about museums, public archives and libraries open data assets;
- *Linked Geo Data* resource linked via `owl:sameAs` relation, to hotels and museum open data assets

We have also implemented a set of ad hoc SPIN functions for matching techniques and constraints validation:

- Textual similarity functions as Levensthein and Jaro-Winkler algorithms (*re-using existing Javascript implementations under SPINx Functions*);
- Data validation functions as RDF syntax violation, missing values check and coordinate separator fix (*developed using SPARQL predicates as SPIN Functions*). For example, below two SPIN validation constraint show how to validate latitude values:

```
# Latitude is not bound
1    ASK WHERE {
2        NOT EXISTS {
3            ?this geo:lat ?lat }
4    }
```

```
# Latitude is empty
1    ASK WHERE {
2        ?this geo:lat ?lat .
3        FILTER (str(?lat) == "") .
4    }
```

- Virtuoso RDF geometry[21] extensions wrapper to compare and match geospatial coordinates (e.g. the Haversine distance between 2 geospatial points) (*reused from available Virtuoso functions*);
- Google geocoding wrapper that takes RDF triples describing an address and import geocoded latitude and longitude data from Google as RDF predicates using the W3C Basic Geo vocabulary[22] (*rewritten as SPARQL predicates using Google API 3.0 methods*).

5   Conclusion and Future Work

This paper presented an approach for mapping, enriching and interlinking open data as linked open data, using federated queries over the LOD cloud expressed as SPIN rules. A major advantage of the proposed methodology is the encapsulation of the interlinking rules inside the domain ontology model. Two examples of interlinking SPIN rules in the tourism domain were given.

Matter of current research is the evaluation of the methodology and the formal expression of a modular interlinking ontology, (using SPIN templates and VoID descriptions[23]) in order to suggest dataset to link to and interlinking patterns.

Acknowledgements. This research is funded by Italian project PON *SmartCities Dicet-InMoto-Orchestra*, stream Information Mobility and Tourism.

References

1. Berners-Lee, T.: Design issues: Linked data (2006)
2. Bizer, C., Heath, T., Berners-Lee, T.: Linked data-the story so far. International Journal on Semantic Web and Information Systems **5**(3), 1–22 (2009)
3. Shadbolt, N., O'Hara, K.: Linked data in government. IEEE Internet Comput. **17**(4), 72–77 (2013)
4. Callahan, A., Dumontier, M.: Evaluating scientific hypotheses using the SPARQL inferencing notation. In: Simperl, E., Cimiano, P., Polleres, A., Corcho, O., Presutti, V. (eds.) ESWC 2012. LNCS, vol. 7295, pp. 647–658. Springer, Heidelberg (2012)
5. Fürber, C., Hepp, M.: Using SPARQL and SPIN for data quality management on the semantic web. In: Abramowicz, W., Tolksdorf, R. (eds.) BIS 2010. LNBIP, vol. 47, pp. 35–46. Springer, Heidelberg (2010)
6. Ferrara, A., Nikolov, A., Scharffe, F.: Data linking for the semantic web. Semantic Web: Ontology and Knowledge Base Enabled Tools, Services, and Applications **169** (2013)

---

[21] http://docs.openlinksw.com/virtuoso/rdfsparqlgeospat.html#rdfsparqlgeospat
[22] http://www.w3.org/2003/01/geo/
[23] http://www.w3.org/TR/void/

7. Szekely, P., et al.: Exploiting semantics of web services for geospatial data fusion. In: Proceedings of the 1st ACM SIGSPATIAL International Workshop on Spatial Semantics and Ontologies. ACM (2011)
8. Zhang, M., Yuan, J., Gong, J., Yue, P.: An Interlinking Approach for Linked Geospatial Data. ISPRS-International Archives of the Photogrammetry, Remote Sensing and Spatial Information Sciences 1(2), 283–287 (2013)
9. Zhang, Y., Chiang, Y.Y., Szekely, P., Knoblock, C.A.: A semantic approach to retrieving, linking, and integrating heterogeneous geospatial data. In: Joint Proceedings of the Workshop on AI Problems and Approaches for Intelligent Environments and Workshop on Semantic Cities, pp. 31–37. ACM, August 2013
10. Harth, A., Gil, Y.: Geospatial data integration with linked data and provenance tracking. In: W3C/OGC Linking Geospatial Data Workshop (2014)
11. Scharffe, F., Euzenat, J.: Linked data meets ontology matching: enhancing data linking through ontology alignments. In: Proc. 3rd International Conference on Knowledge Engineering and Ontology Development (KEOD), pp. 279–284 (2011)
12. Isele, R., Bizer, C.: Learning linkage rules using genetic programming. In: Proceedings of the Sixth International Workshop on Ontology Matching, pp. 13–24 (2011)
13. Bizer, C., Schultz, A.: The R2R Framework: Publishing and Discovering Mappings on the Web. COLD **665** (2010)
14. Scharffe, F., Atemezing, G., Troncy, R., Gandon, F., Villata, S., Bucher B., Vatant, B.: Enabling linked data publication with the Datalift platform. In: Proc. AAAI Workshop on Semantic Cities (2012)
15. Volz, J., Bizer, C., Gaedke, M., Kobilarov, G.: Silk-A Link Discovery Framework for the Web of Data. LDOW **538** (2009)
16. Sabou, M., Arsal, I., Braşoveanu, A.M.: TourMISLOD: A tourism linked data set. Semantic Web **4**(3), 271–276 (2013)
17. Bacciu, C., et al.: Accommodations in tuscany as linked data. In: LREC 2014, pp. 3542–3545 (2014)

Natural Language

Bootstrapping Large Scale Polarity Lexicons through Advanced Distributional Methods

Giuseppe Castellucci[1]$^{(\boxtimes)}$, Danilo Croce[2], and Roberto Basili[2]

[1] Department of Electronic Engineering, University of Roma Tor Vergata,
Via Del Politecnico 1, 00133 Roma, Italy
castellucci@ing.uniroma2.it
[2] Department of Enterprise Engineering, University of Roma Tor Vergata,
Via Del Politecnico 1, 00133 Roma, Italy
{croce,basili}@info.uniroma2.it

Abstract. Recent interests in Sentiment Analysis brought the attention on effective methods to detect opinions and sentiments in texts. Many approaches in literature are based on hand-coded resources that model the *prior* polarity of words or multi-word expressions. The development of such resources is expensive and language dependent so that they cannot fully cover linguistic sentiment phenomena. This paper presents an automatic method for deriving large-scale polarity lexicons based on Distributional Models of Lexical Semantics. Given a set of heuristically annotated sentences from Twitter, we transfer the sentiment information from sentences to words. The approach is mostly unsupervised, and experiments on different Sentiment Analysis tasks in English and Italian show the benefits of the generated resources.

Keywords: Polarity lexicon generation · Distributional semantics

1 Introduction

Opinion Mining [19] aims at tracking the opinions expressed in texts with respect to specific topics, e.g. products or people. In particular, Sentiment Analysis (SA) deals with the problem of deciding whether an excerpt of text, e.g. a sentence or a phrase, is expressing a trend towards specific feelings. In SA, polarity lexicons of positive and negative words have been defined to support the development of automatic systems that match phrases or sentences with the entries of the lexicon, see for example [27,30]. In these resources, entries are associated to their prior polarity, i.e. whenever they tend to evoke a positive or negative sentiment. For example, *"good"* can be associated to a prior positive sentiment in contrast to *"sad"*, considered negative in every domain. These lexicons are often hand-compiled, as [25] or [12]. However, from a linguistic point of view, a priori membership of words to polarity classes can be considered too restrictive, as sentiment expressions are often topic dependent, e.g. the occurrences of the word *mouse* are mostly neutral in the consumer electronics domain, while it can

M. Gavanelli et al. (Eds.): AI*IA 2015, LNAI 9336, pp. 329–342, 2015.
DOI: 10.1007/978-3-319-24309-2_25

be negatively biased in a restaurant domain. Accounting for topic specific phenomena would require manual revisions. Moreover, if these resources exist for English they scarce for others, and compiling a new one can be expensive.

In this paper, we propose an efficient and unsupervised methodology to derive large-scale polarity lexicons, by exploiting the extra-linguistic information within Social Media data, e.g. the presence of emoticons in messages. The approach is based on Distributional Models of Lexical Semantics, by exploiting the equivalence in sentences and words representations available in some distributional models (e.g. the dual LSA space for words and texts introduced in [14]). As sentences can be clearly related to polarity, a classifier can always be trained in such spaces and used to transfer sentiment information from sentences to words. Specifically, we train polarity classifiers by observing sentences and we classify words to populate a polarity lexicon. Annotated messages are derived from Twitter[1] and their polarity is determined by simple heuristics. It means that words in specific domains can be related to sentiment classes by looking at their semantic closeness to emotionally biased sentences. The resulting approach is highly applicable, as the distributional model can be acquired without any supervision, and the provided heuristics do not have any bias with respect to languages or domains.

This paper is an extension of our previous work presented in [5]. Here, we demonstrate the effectiveness of the proposed approach, by evaluating its portability to multiple and different languages. In particular, we generated polarity lexicons for English and Italian that will be released to the research community[2]. Their contribution is measured against different SA tasks in the two languages. In particular, our evaluation is based on Twitter Sentiment Analysis, as recently it has been the focus of highly participated challenges, as the recent SemEval [18,21] and Evalita[3] [2] tasks demonstrate. Moreover, we prove that the acquired lexicon is also beneficial in context-aware sentiment recognition tasks recently introduced by [28], where tweets are seen not in isolation, but are immersed in a stream. We show that our large-scale lexicon can be easily adopted to extend a contextual model for Sentiment Analysis in Twitter, where tweets preceding a given message are also used to characterize the sentiment expressed.

In the remaining, related works are discussed in Section 2. Section 3 presents the proposed methodology, while Section 4 describes the experimental evaluations. Conclusions are derived in Section 5.

2   Related Work

Polarity lexicon generation has been tackled in many researches and three main areas can be pointed out.

---

[1] http://www.twitter.com
[2] http://sag.art.uniroma2.it/demo-software/distributional-polarity-lexicon/
[3] http://www.evalita.it

Manually Annotated Lexicons. Earlier works are based on manual annotations of terms with respect to emotional categories. For example, in [25] sentiment labels are manually associated to 3600 English terms. In [12] a list of positive and negative words are manually extracted from customer reviews. The MPQA Subjectivity Lexicon [30] contains words, each with its prior polarity (positive or negative) and discrete strength (strong or weak). The NRC Emotion Lexicon [17] is composed by frequent English nouns, verbs, adjectives, and adverbs annotated through Amazon Mechanical Turk with respect to eight emotions (e.g. joy, sadness, trust) and sentiment. However, the manual development and maintenance of lexicons may be expensive, and coverage issues can arise.

Lexicons Acquired Over Graphs. Graph based approaches exploit an underlying semantic structure that can be built upon words. In [8] the WordNet [16] synset glosses are exploited to derive three scores describing the positivity, negativity and neutrality of the synsets. The work in [20] generates a lexicon as graph label propagation problem. Each node in the graph represents a word. Each weighted edge encodes a relation between words derived from WordNet [16]. The graph is constructed starting from a set of manually defined seeds. The polarity for the other words is determined by exploiting graph-based methods.

Corpus-Based Lexicons. Statistics based approaches are more general as they mainly exploit corpus processing techniques. For example, [27] proposed a minimally supervised approach to associate a polarity tendency to a word by determining if it co-occurs more with positive words than negative ones. More recently, [31] proposed a semi-supervised framework for generating a domain-specific sentiment lexicon. Their system is initialized with a small set of labeled reviews, from which segments whose polarity is known are extracted. It exploits the relationships between consecutive segments to automatically generate a domain-specific sentiment lexicon. In [13] a minimally-supervised approach based on Social Media data is proposed by exploiting hashtags or emoticons related to positivity and negativity, e.g., #happy, #sad, :) or :(. They compute a score, reflecting the polarity of each word, through a Point wise Mutual Information based measure between a word and an emotion. In [23] word contexts are adopted to generate sentiment orientation for words. In particular, the sentiment of context words, available in an already built lexicon, is shown to contribute in deriving the sentiment orientation of a target word. As a result, the so-called *SentiCircle* is derived for each target word by considering the contexts in which they appear. The approach here presented can be seen as more general, as it does not rely on any existing lexicon, but it could be used to build a *SentiCircle*.

3   Polarity Lexicon Generation Through Distributional Approaches

In order to rely on comparable representations for words and sentences, Distributional Models (DM) of Lexical Semantics are exploited. DMs are intended to acquire semantic relationships between words, mainly by looking at the words usage. The foundation for these models is the *Distributional Hypothesis* [11], i.e. words that are used and occur in the same contexts tend to purport similar meanings. Although DMs are similar in nature, as they all derive vector representations for words from more or less complex corpus processing stages, quite different methods have been proposed to derive them.

Main approaches estimate semantic relationships in terms of vector similarity. Different relationships can be modeled, e.g. *topical* similarities if vectors are built considering the occurrence of a word in documents or *paradigmatic* similarities if vectors are built considering the occurrence of a word in the context of another word [22]. In such models, words like *run* and *walk* are close in the space, while *run* and *read* are projected in different subspaces. These representations can be derived mainly in two ways: *counting* the co-occurrences between words, e.g. [14], or *predicting* word representations in a supervised setting. Despite the specific algorithms used for the space acquisition, all these approaches allow to derive a projection function $\Phi(\cdot)$ of words into a geometrical space, i.e. the d-dimensional vector representation for a word $w_k \in \mathbb{W}$ is obtained through $\boldsymbol{w_k} = \Phi(w_k)$. Geometrical regularities will be exploited to determine the prior sentiment for words, i.e. our assumption is that polarized words lie in specific subspaces. However, in DMs opposite polarity words are often similar, as they share the same contexts. In the following, we discuss how we can transfer known sentence polarity to single words by exploiting those subspaces. similarity on shared contexts between words. Polarized words often share the same contexts, thus a DM will have similar representations for opposite polarity words, e.g. *awesome* and *ugly*. Our aim is to leverage on DMs, because of their ability to represent the semantics relationships between words, but we also want to consider the sentiment of those words to obtain a new sentiment related representation.

3.1 Lexicon Generation Through Classification

The semantic similarity (closeness) established by traditional DMs does not correspond well with emotional similarity. Sentiment or emotional differences between words must be captured into representations that are able to coherently express the underlying sentiment. In this perspective, a discriminant function can be derived through machine learning over these representations. Let us consider a space $\mathbb{R}^d$ where some geometrical representation of a set of annotated examples can be derived. In general, a linear classifier can be seen as a separating hyper plane $\theta \in \mathbb{R}^d$ that is used to classify a new example represented in the same space. Each θ_i corresponds to a specific dimension, or feature i that has been extracted from the annotated examples. After a learning stage, the magnitude of each θ_i reflects the importance of the feature i with respect to a target phenomenon. In this sense, when applied on distributional vectors of word semantics, linear classifiers are expected to learn the regions useful to discriminate examples with respect to the target classes. If these classes reflect the sentiment expressed by words, a classifier should find those subspaces better correlating examples with the sentiment. In this way, any set of words $w_i \in \mathbb{W}$ associated with their prior polarity could be used to train a sentiment classifier.

In fact, given a set of seed words whose prior polarity is known, their projection in the Word Space model $w_k^{seed} = \Phi(w_k^{seed})$ is sufficient to train the linear classifier. This would find what dimensions in $\mathbb{R}^d$ are related to the different polarities. Classification thus corresponds to transferring the knowledge about sentiment implicit in the seed words to the other remaining words.

A number of limitations affect this view. First, the definition and annotation of seed words could be expensive and certainly not portable across natural languages. Second, lexical items do change emotional flavor across domains and the knowledge embodied by the seed lexicons may not generalize when different domains are faced. We suggest to avoid the selection of lexical seeds and emphasize the role of distributional models: the representations of both sentences and words are here capitalized to automatize the development of portable sentiment lexicons. We propose to make use of sentences as seeds of the classifier training, as these embody sentiment in a more explicit (and unambiguous) manner: for example, sentences including strong sentiment markers can be cheaply gathered, thus providing a large scale seed resource. As these sentences and words (i.e. candidate entries for the polarity lexicon) lie into the same space (i.e. sentences and semantically related words belong to the same subspaces), we aim to train a classifier over sentences and apply it to a very large lexicon. Subspaces strongly related to a sentiment class can be here used to project it over the lexicon.

In details, we have words $w_k \in \mathbb{W}$ and their vector representation $w_k \in \mathbb{R}^d$ obtained by projecting them in a Word Space, i.e. $w_k = \Phi(w_k)$. We also have a training set $\mathbb{T}$, including sentences associated to a polarity class. In order to project an entire sentence in the same space, we apply a simple but effective linear combination operator. For each sentence $t \in \mathbb{T}$, we derive the vector representation $t \in \mathbb{R}^d$ by summing all the word vectors composing the sentence, i.e. $t = \sum_{w_i \in t} \Phi(w_i)$. It is one of the simpler, but still expressive, method that is used to derive a representation that accounts for the underlying meaning of a sentence, as discussed in [14]. Having projected an entire sentence in the space, we can find all the dimensions of the space that are related to a sentiment class. Sentence representations are fed to a linear learning algorithm that induces a discriminant function f expected to capture the sentiment related subspaces by properly weighting each dimension i of the original space. The lexicon is generated by applying f to the entire $\mathbb{W}$. As we deal with multiple sentiment classes, f can be seen as m distinct binary functions $(f_1, \dots, f_m)$, one for each sentiment class. Each word $w_k \in \mathbb{W}$ is classified with all the f_i, thus receiving m distinct scores s_i^k, each one reflecting the classifier confidence in the membership of w_k to class i. Each s_i^k is normalized through a softmax function[4], obtaining the final polarity score o_i^k: each w_k can be represented both with its distributional representation, i.e. $w_k = \Phi(w_k)$, and its sentiment representation, i.e. o^k.

Generating a Dataset Through Emoticons. An annotated dataset of sentences $\mathbb{T}$ is needed to acquire a linear classifier that emphasizes specific subspaces. Although different dataset of such kind exists, our aim is to use a general method-

---

[4] $o_i^k = e^{s_i^k} / \sum_{j=1}^m e^{s_j^k}$

ology that can enable the use of this technique in different domains or languages. We are going to use heuristic rules to select sentences by exploring Twitter messages and the emoticons that can be found in them. The method is based on a Distant Supervision approach [10]. In order to derive messages belonging to the positive or negative classes, we select Twitter messages whose last token is a smile either positive, e.g. :) or :D or negative, e.g. :(or :-(. Neutral messages are filtered by looking at those messages that end with a url, as in many cases these are written by newspaper accounts and they use mainly non-polar words to announce an article. In order to have a more accurate dataset, we further filter out those messages that contain elements of other classes, i.e. if a message ends with a positive smile and it contains either a negative smile or a url it will be discarded. It is worth nothing that if a more fine-grained emoticon classification is available (see [26]), it will be possible to derive a dataset made by even more heterogeneous data, i.e. observe finer grain phenomena.

4   Evaluating Polarity Lexicons

In this Section, details about the acquisition of polarity lexicons are provided, and different Sentiment Analysis tasks are evaluated with these resources.

Word Vectors Generation. As discussed in Section 3.1 distributional representation for words are needed. We generated word vectors according to a Skip-gram model [15] through the word2vec[5] tool. In particular, we derived 250 dimensional word vectors[6], by using a corpus of more than 20 million tweets downloaded during the last months of the 2014 year, for the English language. For Italian, we adopted a corpus of more than 2 million of tweets downloaded during the 2013 summer. We processed each tweet with a custom version of the Chaos parser [4]: lemmatization and part-of-speech (pos) tagging are applied to derive *lemma::pos* input for the word vector generation. We obtained $188,635$ and $16,579$ words, for English and Italian respectively, that have been classified to generate the polarity lexicon.

Dataset Generation. We applied the heuristics described in Section 3.1 to the same datasets used for word vector generation. Then, for both languages, we filtered these data by randomly selecting $7,000$ tweets.

Linear Classifier. Support Vector Machines (SVM) [29] are among the most effective classifiers applied in many different fields. In NLP, they have been used for their capability to learn both linear and non-linear function (by exploiting the notion of kernel function [24]). In this paper, we learn a linear function that can separate data between three sentiment classes of interest. We adopted the LibLinear [9] formulation of SVM to acquire the classifiers.

Distributional Polarity Lexicon Generation. We represented each sentence in the training set $\mathbb{T}$ by linearly combining word vectors[7] considering verbs,

---

[5] https://code.google.com/p/word2vec/
[6] word2vec settings are: *min-count=50, window=5, iter=10* and *negative=10*.
[7] In order not to be biased by the query terms, the last token is never considered.

nouns, adjectives and adverbs. As we are dealing with three sentiment classes, i.e. positive, negative and neutral, a One-Vs-All (OVA) strategy is adopted to derive the optimal classifiers. Classifier hyper-parameters are tuned by splitting the training set $\mathbb{T}$ with a 80/20 split: for each parameter configuration, a learning phase is performed on the 80% of the data and the accuracy, i.e. the percentage of correctly classified examples, is computed over the remaining 20%. The best configuration is then used to train a classifier over the whole dataset $\mathbb{T}$. The acquired classifier is used to derive the lexicon by classifying the words of the distributional model, thus deriving the polarity scores as described in Section 3.1. In Tables 1 and 2 an excerpt of the English and Italian lexicons can be found, where *pos*, *neg* and *neu* refer respectively to *positivity*, *negativity* and *neutrality*. In both cases, the approach seems able to transfer the polarity to words, given the sentence-based classifiers. Qualitatively, it seems that polar words tend to lie in specific subspaces, which is captured by the linear classification strategy.

Table 1. Example of polarity lexicon terms and relative sentiment scores (English language).

term	pos	neg	neu
good::j	0.74	0.11	0.15
:)	0.86	0.04	0.10
bad::j	0.12	0.80	0.08
pained::v	0.13	0.74	0.13
#apple::h	0.14	0.16	0.70
article::n	0.16	0.09	0.75

Table 2. Example of polarity lexicon terms and relative sentiment scores (Italian language).

term	pos	neg	neu
ottimo::j	0.77	0.08	0.15
:)	0.73	0.08	0.19
sofferenza::n	0.16	0.58	0.26
soffrire::v	0.08	0.65	0.27
#apple::h	0.17	0.12	0.71
articolo::n	0.19	0.05	0.76

Tasks Settings. Experiments reported in the rest of the Section are all performed by exploiting the kernelized formulation of the Support Vector Machine (SVM) algorithm [29]. SVM and kernel functions [24] have been shown to achieve state-of-the-art results in many NLP tasks (see for example [7]). Kernels allow representing data at an abstract level, while their computation still refer to core informative properties. Moreover, kernel functions can be combined, e.g. the contribution of kernels can be summed, in order to account at the same time for different linguistic properties. In the targeted tasks, multiple kernels are combined to verify the contribution of each representation: in particular, one kernel will be made dependent on the automatically generated polarity lexicon.

As presented in Section 3, a $m = 3$-dimensional vector o^k is available for each word w_k in the vocabulary, each expressing a positivity, negativity and neutrality score for w_k. In order to represent an entire sentence t for SVM, we propose to adopt a very simple feature representation by summing up all the polarity lexicon vectors o^k corresponding to the words w_k in t[8], i.e. $t = \sum_{w_k \in t} o^k$. This should be able to capture when many words agree with respect to the polarity; the dimension associated to a particular sentiment should have a higher score.

---

[8] We apply a normalization on the resulting vector t.

Obviously, this basic representation has some limitations, e.g. it doesn't account for the scope of negation.

4.1 Sentiment Analysis in Twitter

In recent years, the interest in mining the sentiment expressed in the Web is growing, and different Twitter based challenges have been proposed in the Computational Linguistics area. We want to test the automatically generated lexicon in the 2013 and 2014 SemEval tasks [18,21] as well as in the 2014 Evalita Italian challenge on Twitter [2]. In particular, we concentrate on the task of assigning a sentiment to a target tweet. For example, the tweet *"Porto amazing as the sun sets... http://bit.ly/c28w"* should be recognized as *positive*, while *"@knickfan82 Nooo ;(they delayed the knicks game until Monday!"* as *negative*.

All the evaluations are addressed with two basic feature representations: a Bag-Of-Word (BOW) and a Word Space (WS). The former captures directly the lexical information, whereas each binary dimension expresses the presence (or absence) of a particular word in a sentence. The latter relies on a word space representation and represents a smoothing of the lexical overlap measure between messages: it models paradigmatic relations among words. The WS representation of a sentence is obtained by summing the vectors of all its verbs, nouns, adjectives and adverbs. We can further augment the representation with the polarity scores related to the Distributional Polarity Lexicon (DPL). Again, only verbs, nouns, adjectives and adverbs are considered.

The SVM learning algorithm is adopted along with kernel functions, which are applied on different representations. Every information (e.g. the polarity lexicon) is represented independently through a kernel function: the overall normalized sum of the different kernels is then adopted as the overall kernel function.

Table 3. Twitter Sentiment Analysis 2013 results. *Best-System* refers to the top scoring system in SemEval 2013.

Kernel	F1-pn	F1-pnn
BOW	59.72	63.53
BOW+SBJ	61.46	64.95
BOW+DPL	60.78	64.09
BOW+WS	66.12	68.56
BOW+WS+SBJ	65.20	67.93
BOW+WS+DPL	66.40	68.68
Best-System	69.02	-

Table 4. Twitter Sentiment Analysis 2014 results. *Best-System* refers to the top scoring system in SemEval 2014.

Kernel	F1-pn	F1-pnn
BOW	58.74	61.38
BOW+SBJ	60.82	62.85
BOW+DPL	62.49	64.01
BOW+WS	65.20	66.35
BOW+WS+SBJ	64.29	66.13
BOW+WS+DPL	66.11	67.07
Best-System	70.96	-

In Tables 3 and 4 the experimental outcomes for the 2013 and 2014 SemEval datasets are reported, as well as the Best-System in the two challenges. Performance measures are the *F1-pn* and the *F1-pnn*. The former is the arithmetic mean between the F1 measures of the positive and negative classes, i.e. the official score adopted by the SemEval challenges. The latter is the arithmetic mean between the F1 measures of the positive, negative and neutral classes. The WS representation is based on the same word space used to generate the polarity lexicon. Here, we compare the contribution of DPL with a well-known lexicon, i.e. the Subjectivity Lexicon by [30]. It is composed by words manually annotated with subjective polarity information (positive, negative, neutral) and a strength (weak or strong) value. For each tweet we generate a new feature representation (SBJ) where each dimension refers to a polarity value with its relative strength, as found in the message. For example, the SBJ representation of *"Getting better!"* is a feature vector whose the only non-zero element is strong_pos. In this scenario, only linear kernel functions are considered. In Table 3 results are shown for the 2013 test dataset, which is composed by $3,814$ examples. First, the baseline performance achievable with a linear kernel applied to the simple BOW representation is shown. Then, the combination of the other representations is experimented. When applying the word space, an improvement can be noticed. It means that distributional representations are useful to capture the semantic phenomena behind sentiment-related expressions, even in short texts. When combining also DPL further improvements are obtained for both performance measures. It seems that DPL effectively acts as a smoothing of the contribution of the pure lexical semantics information (WS). It is noticeable that the BOW+WS+DPL system would have ranked in 2nd position in the 2013 ranking. Same trends are observable for the 2014 test set, as shown in Table 4. In this case, we were not able to rely on a complete test set, as, at the time of this experimentation, some of the messages were no longer available for download. Our evaluation is carried out on $1,562$ test examples, while the full test set was composed by $1,853$. It makes a direct comparison with the in challenge systems impossible, but it still can give an idea of the achievable performances. Again, performances are measured with the BOW and WS representation combined with SBJ and DPL. As it can be noticed, the use of distributed word representations is also beneficial in this scenario, as demonstrated by the BOW+WS row of Table 4. Again, when using the acquired polarity lexicon improvements in both the performance scores is noticeable.

The impact of the Italian lexicon has been measured against the data provided by the Evalita 2014 Sentipolc [2] challenge. Here, Twitter messages are annotated with respect subjectivity, polarity and irony. We selected only those messages annotated with polarity and that were not expressing any ironic content. Thus, our evaluations are pursued on $2,566$ and $1,175$ messages, used respectively for training and testing. In this scenario, linear kernel, 2-degree polynomial kernel (poly) and 1-gamma Gaussian kernel (rbf) are adopted. In Table 5 and 6 performance measures for this setting are reported, respectively with linear kernel and with non-linear kernels. Again, the F1 mean between the positive and negative classes ($F1\text{-}pn$), as well as the mean between all the involved classes are reported ($F1\text{-}pnn$). We compare DPL with another Italian polarity lexicon, called SENTIX in [3]. It consists of words automatically annotated with

4 sentiment scores, i.e. *positive, negative, polarity* and *intensity*. In our evaluation, features correspond to the sum of the four scores across words appearing in
a message (STX kernel). In the linear kernel case, the benefits of using a polarity lexicon for augmenting the BOW representation is more evident, and the
improvements in using the two resources is very similar. When adopting the WS
representation, performances increase, and when using also the DPL lexicon, it
seems that the interaction with the WS features is beneficial in deciding whether
a tweet is *positive* or *negative*, as demonstrated by the 66.04 *F1-Pn* measures.
A different outcome can be noticed when adopting a polynomial kernel over the
BOW representation and a Gaussian kernel over the WS representation. In this
case, the combination of a linear kernel over the polarity lexicon seems not to be
beneficial. Instead, applying a Gaussian kernel also over the DPL lexicon allows
to further push the performances about of 1 point in *F1-Pn*. The same benefit
is not measured when adopting this strategy with the STX representation.

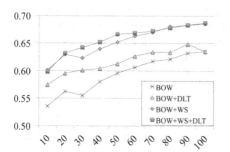 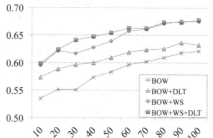

Fig. 1. SemEval 2013 learning curve.   **Fig. 2.** SemEval 2014 learning curve.

In order to verify the generalization capability of the acquired lexicon, the
above evaluations have been carried out in a scenario where a reduced number
of training example is available. To this end, Figure 1 and 2 show the learning
curves, i.e. the *F1-pnn* evaluated at different training set sizes, respectively for
the SemEval 2013 and SemEval 2014 tasks. As it is expected, the pure lexical
information used in the BOW model achieves a lower F1 score and it is noticeable when few labeled examples are used. When DPL is used to augment the
pure lexical information (BOW) it is able to reach higher performances with only
10% of the training data, resulting in about 4 absolute point gain in the *F1-pnn*
measure. A more informative generalization is achieved when the Wordspace is
used (WS). When DPL is used in combination with the BOW+WS representation this phenomenon is alleviated, but starting from 20 − 30% of the training
data, performances are always higher.

The same trends can be noticed for the experiments in Italian. Figure 3
shows the learning curves obtained by the SVM when linear kernels are used,
while in Figure 4 non-linear kernels are shown. The *F1-pnn* is plotted at different
percentages of the training set. In the linear setting the contribution of the lexicon over the pure Bag-of-Words representation is immediately evident even with

Table 5. Twitter Polarity Classification in Italian (linear kernel case).

Kernel	F1-pn	F1-pnn
BOW	61.58	57.97
BOW+STX	62.45	58.04
BOW+DPL	62.35	58.30
BOW+WS	65.48	61.13
BOW+WS+STX	64.82	60.64
BOW+WS+DPL	66.04	60.99

Table 6. Twitter Polarity Classification in Italian (non-linear kernel case).

Kernel	F1-pn	F1-pnn
pol_{BOW}	65.52	60.29
$pol_{BOW}+rbf_{WS}$	68.52	63.24
$pol_{BOW}+rbf_{WS}+STX$	66.05	61.18
$pol_{BOW}+rbf_{WS}+DPL$	68.45	63.14
$pol_{BOW}+rbf_{WS}+rbf_{STX}$	66.72	61.53
$pol_{BOW}+rbf_{WS}+rbf_{DPL}$	**69.17**	**63.40**

only 10% of the training data: this means that the DPL representation is beneficial to generalize pure lexical information. When augmenting the BOW+WS representation the lexicon contribution is more evident starting at $40-50\%$ of the training data. This is due to the fact that the WS representation is able to generalize data in poor training conditions, but it is not completely sufficient to disambiguate different sentiment phenomena when more data are available.

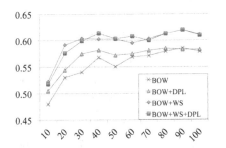

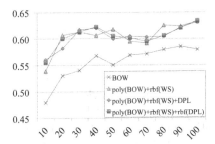

Fig. 3. Twitter Polarity Classification in Italian (linear kernel case).

Fig. 4. Twitter Polarity Classification in Italian (non-linear kernel case).

4.2 Sentiment Analysis in Twitter with Contextual Information

Another setting for Twitter Sentiment Analysis in English considers the *Contextual Information* of a message, as defined in [28]. A context is a temporally ordered sequence of messages where the last element is the target tweet, which can be classified by considering also the additional information of its preceding messages. In [28], a sequence labeling algorithm is adopted to relate sentiment information of sequences with the sentiment of a target message. In particular, the classification is carried out through the SVM-HMM [1] algorithm. It learns a model isomorphic to a k-order Hidden Markov Model (HMM), and a sequence is classified by applying the Viterbi algorithm to find the sequence of HMM states explaining the given (contextual) observations. These models have been demonstrated to be very effective, resulting in improvements with respect to alternatives where tweets are classified in isolation. In the following, the same experimental settings of [28] are adopted. Two different contexts are considered: *Conversation*, where a target tweet is immersed in a stream defined by the temporally ordered sequences of reply messages. The *Hashtag* context, instead, includes all temporally preceding messages sharing at least one hashtag with the target tweet. In both cases, contexts of different size have been considered. The sentiment class for messages related to the context (i.e. different from the target) is assigned by a multi-classifier trained on a Bag-Of-Word (BOW) and Word Space (WS) representations.

Table 7. Twitter Conversation Context.

S	System		+SBJ	+DPL
	BOW	65.7	65.6	67.0
3	BOW+WS	66.5	67.6	68.0
	BOW+WS+USP	68.9	68.3	68.5
	BOW	65.2	64.1	65.5
6	BOW+WS	67.1	69.1	67.1
	BOW+WS+USP	65.4	67.5	67.1
	BOW	62.3	65.1	65.7
∞	BOW+WS	67.0	67.9	68.2
	BOW+WS+USP	67.9	69.0	68.6

Table 8. Twitter Hashtag Context.

S	System		+SBJ	+DPL
	BOW	64.1	64.7	65.6
3	BOW+WS	67.8	67.6	68.3
	BOW+WS+USP	69.3	66.6	**70.3**
	BOW	63.7	63.7	64.9
6	BOW+WS	68.9	65.6	67.9
	BOW+WS+USP	69.1	67.9	67.9
	BOW	65.1	65.3	64.3
31	BOW+WS	67.6	67.7	67.9
	BOW+WS+USP	67.1	67.8	67.7

In Tables 7 and 8 performances for the context-based settings are reported at different context sizes (S). BOW, WS and User Sentiment Profile (USP) representations are used as basic features. WS is based on the Word Space used to generate the lexicon. The USP models the sentiment attitude of the user, acquired within the previous messages in its timeline, as defined in [28]. In addition, the Distributional Polarity Lexicon (DPL) and the Subjectivity Lexicon (SBJ) representations are adopted as for the previous tasks. We report the *F1-pnn* of the classification of the target tweet, as in [28]. In Table 7, it can be noticed that the adoption of polarity lexicons is beneficial for the classification of tweets also in conversation streams. In particular, the adoption of DPL is more evident when augmenting the BOW representation with small context size, i.e. 3. When using also WS or USP, improvements are less prominent, and the difference between the SBJ and DPL is negligible for larger context sizes. It means that the distributional polarity lexicon is able to overcome data sparsity issues when less information is available, while its contribution is less important within richer contexts. When considering the hashtag context, performances trends are the same. Even in this setting, larger contexts seem to provide useful information for the sequence classification. Lexicons contribution (SBJ and DPL) is more evident when less data are available. In fact, at context size 3, augmenting the system BOW+WS+USP with DPL allows to obtain the state-of-the-art on this dataset, as compared to the best configuration (*69.32*) as measured by [28].

5   Conclusions

In this paper, a novel unsupervised methodology to generate large-scale polarity lexicons is presented. Emotion related aspects are observed over annotated sentences and used to transfer the sentiment information to lexical items. This transfer is made possible as both sentences and words lie in the same space, characterized by the underlying Distributional Model, where training a linear classifier is straightforward. The method is quite general, as it does not rely on any hand-coded resource, but mainly uses simple cues, e.g. emoticons, for generating a large corpus of labeled sentences. Moreover, it is largely applicable to resource poor languages. The generated lexicon has been shown beneficial on diverse Sentiment Analysis tasks for two languages. Future investigations will address both other languages and the integration of more complex grammatical features. In fact, the experimented classification algorithms were not sensitive to negation or other grammatical markers nor to ironic phenomena in the texts. The proposed approach should be carefully adopted to deal with ironic phenomena. Distributional polarity vectors capture the main usage of words, but not ironic or metaphoric senses of them. It should be interesting to verify if an approach similar to the one suggested in [6] could be beneficial when adopting distributional polarity vectors as features. In fact, in [6] a specific feature set based on Distributional Models has been adopted to capture when a word is used *"out-of-context"*, suggesting an ironic use of it. A similar strategy could be adopted or extended by considering also distributional polarity vectors.

References

1. Altun, Y., Tsochantaridis, I., Hofmann, T.: Hidden Markov support vector machines. In: Proc. of ICML (2003)
2. Basile, V., Bolioli, A., Nissim, M., Patti, V., Rosso, P.: Overview of the evalita 2014 sentiment polarity classification task. In: Proc. of the 4th EVALITA (2014)
3. Basile, V., Nissim, M.: Sentiment analysis on Italian tweets. In: Proc. of the 4th WS: Computational Approaches to Subjectivity, Sentiment and Social Media Analysis (2013)
4. Basili, R., Pazienza, M.T., Zanzotto, F.M.: Efficient parsing for information extraction. In: ECAI, pp. 135–139 (1998)
5. Castellucci, G., Croce, D., Basili, R.: Acquiring a large scale polarity lexicon through unsupervised distributional methods. In: Biemann, C., Handschuh, S., Freitas, A., Meziane, F., Métais, E. (eds.) NLDB 2015. LNCS, vol. 9103, pp. 73–86. Springer, Heidelberg (2015)
6. Castellucci, G., Croce, D., De Cao, D., Basili, R.: A multiple kernel approach for twitter sentiment analysis in Italian. In: 4th EVALITA 2014 (2014)
7. Croce, D., Moschitti, A., Basili, R.: Structured lexical similarity via convolution kernels on dependency trees. In: Proc. of EMNLP (2011)
8. Esuli, A., Sebastiani, F.: Sentiwordnet: a publicly available lexical resource for opinion mining. In: Proc. of 5th LREC, pp. 417–422 (2006)
9. Fan, R.E., Chang, K.W., Hsieh, C.J., Wang, X.R., Lin, C.J.: Liblinear: A library for large linear classification. JMLR **9**, 1871–1874 (2008)
10. Go, A., Bhayani, R., Huang, L.: Twitter sentiment classification using distant supervision. Processing, 1–6 (2009)
11. Harris, Z.: Distributional structure. In: Katz, J.J., Fodor, J.A. (eds.) The Philosophy of Linguistics. Oxford University Press (1964)

12. Hu, M., Liu, B.: Mining and summarizing customer reviews. In: Proc. of 10th Int. Conf. on Knowledge Discovery and Data Mining, pp. 168–177. ACM (2004)
13. Kiritchenko, S., Zhu, X., Mohammad, S.M.: Sentiment analysis of short informal texts. JAIR **50**, 723–762 (2014)
14. Landauer, T., Dumais, S.: A solution to plato's problem: The latent semantic analysis theory of acquisition, induction and representation of knowledge. Psychological Review **104** (1997)
15. Mikolov, T., Chen, K., Corrado, G., Dean, J.: Efficient estimation of word representations in vector space. CoRR abs/1301.3781 (2013). http://arxiv.org/abs/1301.3781
16. Miller, G.A.: Wordnet: A lexical database for english. Commun. ACM **38**(11), 39–41 (1995)
17. Mohammad, S.M., Turney, P.D.: Emotions evoked by common words and phrases: using mechanical turk to create an emotion lexicon. In: Proc. of CAAGET Workshop (2010)
18. Nakov, P., Rosenthal, S., Kozareva, Z., Stoyanov, V., Ritter, A., Wilson, T.: Semeval-2013 task 2: sentiment analysis in twitter. In: Proc. of SemEval. ACL, USA, June 2013
19. Pang, B., Lee, L.: Opinion mining and sentiment analysis. Found. Trends Inf. Retr. **2**(1–2), 1–135 (2008)
20. Rao, D., Ravichandran, D.: Semi-supervised polarity lexicon induction. In: Proc. of the EACL, pp. 675–682. ACL (2009)
21. Rosenthal, S., Ritter, A., Nakov, P., Stoyanov, V.: Semeval-2014 task 9: sentiment analysis in twitter. In: Proc. SemEval. ACL and Dublin City University (2014)
22. Sahlgren, M.: The Word-Space Model. Ph.D. thesis, Stockholm University (2006)
23. Saif, H., Fernandez, M., He, Y., Alani, H.: SentiCircles for contextual and conceptual semantic sentiment analysis of twitter. In: Presutti, V., d'Amato, C., Gandon, F., d'Aquin, M., Staab, S., Tordai, A. (eds.) ESWC 2014. LNCS, vol. 8465, pp. 83–98. Springer, Heidelberg (2014)
24. Shawe-Taylor, J., Cristianini, N.: Kernel Methods for Pattern Analysis. Cambridge University Press (2004)
25. Stone, P.J., Dunphy, D.C., Smith, M.S., Ogilvie, D.M.: The General Inquirer: A Computer Approach to Content Analysis. MIT Press (1966)
26. Suttles, J., Ide, N.: Distant supervision for emotion classification with discrete binary values. In: Gelbukh, A. (ed.) CICLing 2013, Part II. LNCS, vol. 7817, pp. 121–136. Springer, Heidelberg (2013)
27. Turney, P.D., Littman, M.L.: Measuring praise and criticism: Inference of semantic orientation from association. ACM Trans. Inf. Syst. **21**(4), 315–346 (2003)
28. Vanzo, A., Croce, D., Basili, R.: A context-based model for sentiment analysis in twitter. In: Proc. of 25th COLING: Best Paper, pp. 2345–2354. Dublin City University and Association for Computational Linguistics (2014)
29. Vapnik, V.N.: Statistical Learning Theory. Wiley-Interscience (1998)
30. Wilson, T., Wiebe, J., Hoffmann, P.: Recognizing contextual polarity in phrase-level sentiment analysis. In: Proc. of EMNLP. ACL (2005)
31. Zhang, Z., Singh, M.P.: Renew: a semi-supervised framework for generating domain-specific lexicons and sentiment analysis. In: Proc. of 52nd Annual Meeting of the ACL, vol. 1, pp. 542–551. ACL, June 2014 (Long Papers)

Using Semantic Models for Robust Natural Language Human Robot Interaction

Emanuele Bastianelli[2]([⊠]), Danilo Croce[1], Roberto Basili[1], and Daniele Nardi[3]

[1] DII, University of Rome Tor Vergata, Rome, Italy
{croce,basili}@info.uniroma2.it
[2] DICII, University of Rome Tor Vergata, Rome, Italy
bastianelli@ing.uniroma2.it
[3] DIIAG, Sapienza University of Rome, Rome, Italy
nardi@dis.uniroma1.it

Abstract. While robotic platforms are moving from industrial to consumer applications, the need of flexible and intuitive interfaces becomes more critical and the capability of governing the variability of human language a strict requirement. Grounding of lexical expressions, i.e. mapping words of a user utterance to the perceived entities of a robot operational scenario, is particularly critical. Usually, grounding proceeds by learning how to associate objects categorized in discrete classes (e.g. routes or sets of visual patterns) to linguistic expressions. In this work, we discuss how lexical mapping functions that integrate Distributional Semantics representations and phonetic metrics can be adopted to robustly automate the grounding of language expressions into the robotic semantic maps of a house environment. In this way, the pairing between words and objects into a semantic map facilitates the grounding without the need of an explicit categorization. Comparative measures demonstrate the viability of the proposed approach and the achievable robustness, quite crucial in operational robotic settings.

1 Introduction

As robots are moving from industrial environments to consumer markets, their human-like interaction capabilities with people is becoming a key area of interest. Human language is the most natural way of interaction for its expressiveness and flexibility. End-to-end communication processes in natural language are challenging for robots as for the deep interaction of different involved cognitive abilities. Having a robot reacting to a command like *"Take the book on the table"* corresponds to a number of implicit assumptions. First, the environment must contain at least the entities **book** and **table**, and these must be perceived by the interlocutors. Second, the robot must have an inner representation of the objects, e.g. an explicit map of the environment. Third, lexical references to real world entities must be resolved through a *grounding* process [8] linking symbolic knowledge (e.g. words) to the corresponding perceptual information.

Notice that, while a large "natural" vocabulary is available to the user in order to express references to surrounding objects, situations or spatial relations,

© Springer International Publishing Switzerland 2015
M. Gavanelli et al. (Eds.): AI*IA 2015, LNAI 9336, pp. 343–356, 2015.
DOI: 10.1007/978-3-319-24309-2_26

from a robotic perspective grounding particularly depends on both the quality of the robot's lexicon and its ability to link linguistic symbols to its knowledge of the environment. Grounded reasoning seems required here to understand references towards the current state of the world, i.e. the grounded references that symbols exhibit towards objects of the environment, as well as their properties. Failures in such references are crucial. For example, assuming that no book is visually accessible while a table is known, no planning is possible for the robot. Things are even more complicated as the robot may have no entry for *book*, so that such reference fails as for the limited coverage of its internal lexicon. Finally, noise in the environment may lead to mistranscriptions from Automatic Speech Recognition (ASR) so that the sentence may be misunderstood as *"Take the look on the label"*. Such ASR errors, even if mitigated by the adoption of Language Models trained over large corpora, may still arise, e.g. *"Go to the corridor"* vs. *"Go to the coroner"*, as valid transcriptions can be still not consistent with respect to the environment.

In Robotics, grounding is often enabled by exploiting diverse representations of perceptual inputs. Variants are here *ad-hoc* built resources, e.g. handcrafted KBs containing properties of the represented objects, or the outcomes of automatic processes, e.g. visual object recognition or incremental compilation supported by the interaction with the user, as in *symbiotic autonomy* paradigm [18]. Recently, *semantic maps* [17] have been used to represent the perception of the world available to a robot. A semantic map is a knowledge base built over a geometric map used for navigation, containing, *"in addition to spatial information about the environment, assignments of mapped features to entities of known classes"*, i.e. a joint representation of perceptual and symbolic knowledge. Among the different properties about entities, a semantic map may eventually contains linguistic references to them, specified when maps are built with the human assistance, e.g. by pointing and naming the entities in the environment, following the so-called *Human Augmented Mapping* [24] paradigm. In this sense, a map containing entities, such as a book book, characterizes also their linguistic references, e.g. the name *book* for which hasName(b1,*book*) is true.

In this work, we propose a mechanism that allows for linking words in user commands to lexical anchorings representing names of entities, contained in semantic maps. Through this process, perceptual representations of objects and locations are retrieved from the map, thus realizing the grounding process through linguistic information. The proposed approach exploits semantic information derived from Distributional Models (DMs) of the lexicon [21] as well as on phonetic similarity. DMs are broadly used in NLP to acquire semantic relations between words, e.g. quasi-synonymy, through the distributional analysis of large-scale corpora, in order to induce lexical rules such as *"volumes"* are semantically similar to *"books"*. Furthermore, the function also depends on the phonetic distance between words. References to the entities (and their symbolic names into the robot KBs) are thus mapped to distributions able to deal with uncertainty inherent to the grounding. In this way, the grounding capability of the robotic system is able to deal with unseen words as well as to tolerate possible mis-transcriptions coming from the ASR. The original contribution of this work is thus to provide a more robust way of linking words to entities without the need of an explicit categorization, but adopting a linguistic consensus approach, i.e. exploiting lexical usages across large-scale corpora. Grounding is here not restricted to a dictionary available in a set of examples, but enforces properties naturally available to any native speaker.

In the remaining, Section 2 reports a view of the grounding problem in the robotic context and Section 3 discusses the approach we propose. Then, experimental evaluation is reported in Section 4, while Section 5 derives the conclusions.

2   Related Works

The term *grounding* has been firstly defined in the context of cognitive studies by Harnad [8] as the process of mapping symbol meaning in something more than other meaningless symbols. According to Harnad, this is possible only through a perceptual experience, that provides the bridge between the real world a symbolic representation of it. The term grounding has been adopted in Robotics, especially for the problem of linking Natural Language expressions to real world entities, and the work in [22] represents one of the first attempts to face it. Several works suggested some form of supervision to learn the mapping between words and the robot perception of the world. Some of them relied on the multi-modal "show-and-tell" technique, where the user shows or points to an object while naming it, as in [19]. The work in [5] describes a system to give commands to a manipulator. Here the use of new words is allowed, but explicit multi-modal interactions, e.g. dialogues or pointing gesture, are required to link new words and objects in the scene. The new word corresponding to the name is inserted in the grammar used for speech recognition, and is thus learned. In [7], both models for spatial prepositions and for object recognition are trained: words and classes assigned by the object recognition system are linked using the statistical approach described in [6]. However, every previous work requires some form of supervision to learn possible lexical alternatives, as a human directly showing the correspondence between words and classified entities, or by surrogating such experience with artificial training examples. In fact, the vocabulary is closed to the set of words seen in the training phase.

Other works have addressed the problem of grounding language in an environment whose perception is represented in a Semantic Map. The work of [11] parses natural language route instructions in predefined spatial semantic structures, and grounds them in the environment using a probabilistic approach. The work of [4] shows a system that follows navigation instructions in a virtual environment, that is comparable to a Semantic Map. The problem of mapping language to the robot specific language of commands is modeled as a machine translation task. Similarly, in [23] the problem of grounding language in the context of a robotic forklift operation is explored. Here natural language instructions are mapped into instances of probabilistic graphical models that enable a mapping between words and concrete objects, places, paths and events in the real world. Parameters are estimated by looking at human-labeled examples. Again the pairing of words and entities is learned at training time, and unseen words are not treated.

Moreover, some very recent works proposed methodologies to build semantic maps with the help of the user, following the HAM paradigm, as in [1,10,12]. Maps built in this way contain lexical information about how entities have been named during the mapping process. Such information can be exploited to link words in user commands to entities in the environment, through their names. Entities referenced in a vocal command are thus linked to their counterparts in the world representation the robot has, and that has been acquired through perception, realizing a crucial step for grounding.

Previous works seem to rely on different categories to represent perceived entities and then modeling the grounding as the process of pairing words and categories. Our contribution is to facilitate this pairing by providing robust methods for linking words to the names by which entities are defined into the referenced semantic maps, without the need of an explicit categorization. Grounding is thus realized in a two-step process, where the final entity is implicitly linked with a word in a command through a lexical anchor representing its name. This is realized only observing the usage of words in large-scale document collections in an unsupervised fashion. Our method is intended to treat even unseen words, without being restricted to a set of words derived from a set of command examples.

3   Grounding According to Lexical Competence

A general assumption of our work is that grounding corresponds to map entities mentioned in a command against objects of a semantic map, where the latter are located and have an associated name, provided through a Human Augmented Mapping (HAM) process, and specified by the hasName($\cdot, \cdot$) property. According to this, when the map contains an entity e, e.g. a book b1, then a property hasName(e, w_e), such as hasName(b1, *book*), can be associated with it, stating that the word w_e = *book*, represents the name of an entity b1, i.e. a book.

Given a word w occurring in a command used to refer to an entity in the environment, the function expresses a likelihood distribution over all entities e contained in the semantic map. The grounding function corresponds to a fuzzy membership, i.e. a confidence score, to each possible grounding $\langle w, e \rangle$ to entities e corresponding to names w_e. It measures the confidence that w can be grounded to entity e and depends on two major evidences: the **phonetic similarity** between word w and the lexical reference for e, i.e. w_e (e.g. "*look*" vs. "*book*" w.r.t. *book*); the **lexical similarity** that measures a paradigmatic relatedness, e.g. the quasi synonymy, between a w and the lexical reference w_e, e.g. "*volume*" and *book*). Notice that **phonetic** and **lexical similarity** usually act together, as mis-transcriptions, e.g. "*valium*", may well correspond to quasi synonyms, e.g. "*volume*", for a *book*. While lexical similarity could be surrogated by a priori handcrafting the lexicon containing all admissible alternatives for a given w_e, this solution would be applicable only in small or controlled scenarios, as it is very expensive, and existing lexicons, such as WordNet [16], may not be enough versatile, e.g being specialized for domain.

In order to characterize the words admissible for a target w_e, we refer to a Distributional Model (DM) of the lexicon [21]. DMs are intended to acquire semantic relationships between words, mainly by looking at the words usage in large scale corpora. The foundation for these models is the *Distributional Hypothesis* [9], i.e. words that are used and occur in the same contexts tend to share similar meanings. In recent years, DMs have been at the basis of many advances in NLP, and different methods have been proposed to derive them in efficient ways. Main approaches estimate semantic relationships in terms of vector similarity. As an example, a target word w_k is represented through a vector $\boldsymbol{w}_k$ whose dimensions encode all words co-occurring with w_k. As discussed in [20], words in *paradigmatic* relations (e.g. quasi-synonymy) tend to co-occur with the same words and they will be represent through similar vectors. The word representations fostered in this paper are obtained by applying a Recurrent Neural Network architecture [15]. This approach allows to derive a projection function $\Phi(\cdot)$ of target words into a geometrical space, i.e. the vector representation for a word $w_k \in \mathbb{W}$ is obtained through $\boldsymbol{w_k} = \Phi(w_k)$. In such embedded space, targets with similar meaning are represented similarly so that the distance function between their vectors reflects linguistic relatedness. Given two words w_1 and w_2, their similarity relatedness sim can be estimated as the cosine similarity between the corresponding projections $\boldsymbol{w_1}, \boldsymbol{w_2}$ in the space, i.e. $sim(w_1, w_2) = \frac{\boldsymbol{w_1} \cdot \boldsymbol{w_2}}{\|\boldsymbol{w_1}\|\|\boldsymbol{w_2}\|}$. For example, following such approach, given the target word *"book"*, the closest words in the space, i.e. the most related words, are *"booklet"*, *"volume"* *"chapbook"*, *"guidebook"* and *"manuscript"*. Formally, we define

$$S_\tau^e = \{w \mid sim(w, w_e) \geq \tau\} \tag{1}$$

as the set of all the words having a semantic relation with w_e, derived by the DM through the sim function. The threshold τ allows to restrict the set only to the most related words. Given a word w_c in the transcribed vocal command c that references an entity in the environment, the function find the entity e whose w_e maximizes the similarity with w_c. Moreover, as mis-transcriptions may occur, we allow to ground different but phonetically similar words, e.g. *"valium"* vs. *"volume"*, penalizing this grounding proportionally to the phonetic distance. Given a word w_c the final grounding function value will be computed as:

$$g(w_c, e) = \max_{w \in S_\tau^e} \left(ph(w_c, w) * sim(w_e, w) \right) \tag{2}$$

where $ph(\cdot, \cdot)$ is the phonetic similarity between two words. By evaluating the grounding function against all the entities e in the semantic map, we obtain a likelihood distribution of the possible groundings with respect to a pronounced word w_c.

3.1 Phonetic Distances

The factor $ph(\cdot,\cdot)$ in the function $g(\cdot,\cdot)$ is not fixed, and depends on the phonetic distance adopted in the computation. In our investigation, we consider different distance metrics over strings to provide a measure of the phonetic closeness between two words. Such distance has been evaluated over the words transcribed by the ASR.

Four metrics have been selected in this work, as they are of common use in diverse tasks requiring string comparison. Since we expect our function to output a score between 0 and 1, the outcome of some of the adopted metrics has been normalized, as described in the following.

Levenshtein distance [13] is a common edit distance that is used for many purposes, from spell checking to natural language translation. This distance produces a discrete score that is 0 if the two compared strings are exactly equals, or > 0 if the strings differ. The value of $ph(\cdot,\cdot)$ has been then evaluated as $\frac{1}{(dist+1)}$, where *dist* is the Levenshtein distance, to obtain a score in the range [0,1].

Jaro-Winkler [25] is a metric that has been developed for the record-linkage task. It takes into account the number and the order of the common characters between two strings. This metric returns a value in the range [0,1], with 0 meaning no similarity while 1 equates to exact match. No normalization has been thus required for $ph(\cdot,\cdot)$.

Sift3[1] distance is a string distance algorithm in the class of Longest Common Subsequence algorithms. It extends the logic of the Levenshtein distance, by incorporating the sensitiveness with respect to the transpositions from the Jaro-Winkler metric. It is being used for fast string comparison in web applications. Sift3 returns a continuous value between 0 and $+\infty$, and thus $ph(\cdot,\cdot)$ has been evaluated as $\frac{1}{(dist+1)}$, with *dist* the score assigned by the Sift3.

String Kernels [3,14] are similarity measures between texts seen as sequences of symbols (e.g., possible characters) over an alphabet. In general, similarity is assessed by the number of (possibly non-contiguous) matching subsequences shared by two sequences. Noncontiguous occurrences are penalized according to the number of gaps they contain, by a decay factor λ. Normalized String Kernels return a value in the rage [0,1] as for Jaro-Winkler, and thus no normalization has been required for $ph(\cdot,\cdot)$.

3.2 Selection of the Grounding Candidates

The role of the grounding function $g(\cdot,\cdot)$ described in Eq. 2 is to compute a score for each entity in the semantic map with respect to a given word w in a command. It does not directly select which entity may correspond to a correct grounding for that word. To this end, we defined three different policies to select which entity or set of entities represents a possible grounding for a word w. It is worth noting that, from a linguistic point of view, one word may be potentially

---

[1] http://siderite.blogspot.com/2014/11/super-fast-and-accurate-string-distance.html

grounded into more entities in the map. Grounding a word like *picture* in a map containing two entities p1 and p2 with associated names *photograph* and *painting* should result in multiple groundings, as the word *picture* may be used to refer to both entities. Further disambiguation may be necessary to retrieve the final entity, using dialogue schemes or exploiting other perception systems. Such issue is not treated in this paper, and it is left to future works.

Given a word w in a command to be grounded, a semantic map and the set E of entities e_i contained in it, each of them associated with a name w_e^i, each policy select a subset $\Gamma \subseteq E$ of the candidate entities, as described in the following.

- *Argmax* (Γ_{max}): only the entity maximizing the grounding function is returned, according to $\Gamma_{max} = \{e|\arg\max_{e\in E}(g(w,e))\}$;
- Selection based on the *Confidence level* (Γ_n): given μ_w the mean of the grounding function scores over all the entities in E and its standard deviation σ_w, the set of groundings is defined by $\Gamma_n = \{e|g(w,e) > \mu_w + n \cdot \sigma_w\}$, where n defines less, e.g. with $n = 0$, or more restrictive selection policies: with $n > 0$ incrementally fewer entities are selected in Γ_n.

In addition, a simple policy can be assumed as a reference baseline: according to this policy, called *Identity* (Γ_I), only the entity whose name w_e has a perfect match with the input w is returned, i.e. $\Gamma_I = \{e|w_e = w\}$.

Let us consider an example of the grounding mechanism given so far. Given a function $g(\cdot,\cdot)$ that uses the normalization of the Levenshtein distance as phonetic distance and *Argmax* as grounding selection policy. Let us now consider a map with a book b1 and a table t1, with associated names respectively hasName(b1,*book*) and hasName(t1,*table*). Then, consider a command presented to the robot, containing the word w="*volume*" to be grounded. The grounding function is evaluated over the two entities in the map, producing a confidence score for each of them. First, $g($"*volume*"$,$b1$)$ is computed to evaluate the confidence in grounding with respect to b1. The set $S^{b1} = \{$ "*book*", "*booklet*", "*volume*", ...$\}$ is first retrieved: for each $w \in S^{b1}$, according to phonetic similarity against "*volume*" is multiplied by the value of $sim(w_{b1}, w)$, according to Equation 2. With a $sim($b1, "*volume*"$)=0.857$ and a phonetic similarity $ph($"*volume*", "*volume*"$)=1$, the higher score assigned by the grounding function is 0.857. Then, $g($"*volume*"$,$t1$)$ is evaluated with $S^{t1} = \{$ "*table*", "*desk*", ...$\}$. The value of $g($"*volume*"$,$t1$)$ is lower than the value of $g($"*volume*"$,$b1$)$, as the phonetic similarity between w and every word in S^{t1} penalizes the score of $g($"*volume*"$,$t1$)$. Hence, w is grounded in b1.

4 Experimental Evaluation

In order to validate our approach, we need to verify the performance of our linguistic grounding mechanism. Mainly, we want to measure whether such mechanism improves the coverage of correctly grounded terms over a semantic map, with respect to a baseline represented by the *Identity* policy, described in Section 3.2. Starting from a set of vocal commands, transcribed and parsed using off-the-shelf tools, we evaluated the performances of different combinations of phonetic distances and selection policies, by grounding the transcriptions of the afore-mentioned vocal commands. In this context, grounding a transcription means retrieving all the entities in a specific semantic map that are denoted by each noun phrase in the corresponding sentence. As an example, for the sentence *"Take the book on the table"*, both *the book* and *the table* are extracted and mapped into the entities b1 and t1 belonging to the map. Given the status of the map, it can be the case that a noun phrase returns an empty grounding as no entity can been found for its mapping. The final evaluation has thus been carried out in terms of *i*) Precision (P), i.e. the percentage of correct groundings among the ones retrieved by the function; *ii*) Recall (R), i.e. the percentage of correct groundings among the set of *gold* groundings; *iii*) the F-Measure, as the harmonic mean of P and R.

4.1   Experimental Setup

All the commands used during the experiments belong to HuRIC [2], a corpus of spoken commands for robots in the house servicing scenario[2]. It consists of a set audio files paired with their correct transcriptions. All transcriptions are tagged at different linguistic levels, from morphology and syntax, to different semantic formalisms. A total of 538 audio files from the corpus have been used for the experimental evaluation. Every command in the corpus is given with respect to a semantic map representing a house environment, containing all (but not only) the entities that are referred by the command itself.

One of the aspects we wanted to stress is the ability of the grounding function in dealing with mis-transcribed entity names. For this reason, instead of using the correct transcriptions given by the corpus, we used the ones produced using the Google ASR Api[3], that may contain mis-transcriptions. For each command, only the transcription with the lowest Word Error Rate in its n-best list has been considered, for a total of 538 transcriptions. Dependency trees of the resulting sentences have been provided using the Stanford CoreNLP[4].

On the other hand, the grounding function should be able to deal with synonyms. Unfortunately, the commands in HuRIC do not contain enough variations in terms of synonymy to test this feature. To this end, for each semantic map paired with a command, we generated 10 lexical variations of it, by select-ing alternatives for each w_e associated with an entity e in the map from a set defined as follows. For each w_e we built three sets of possible lexical alter-natives: $D = \{w \mid w \in sim(w, w_e)\}$ is the set of words related to w_e accord-ing to the Distributional Model introduced in Section 3 and described below; $W = \{w \mid w \in WN(w_e)\}$ is the set of synonyms of w_e defined in WordNet [16];

---

[2] Available at www.http://sag.art.uniroma2.it/huric
[3] www.google.com/intl/it/chrome/demos/speech.html
[4] nlp.stanford.edu/software/corenlp.shtml

$ND = \{w \mid w \notin sim(w, w_e)\}$ is a control set specifically generated to introduce noise. For each word $w \in D \cup W \cup ND$, we created a pair $\langle w_e, w \rangle$ representing the association between w and the entity name w_e that generated it. We asked three annotators to validate the associations, and for each w_e only those words w being annotated as valid alternatives by all the annotators were considered in the final set of alternatives A_e. To build a variation of a semantic map, for each w_e in it a word from A_e has been randomly chosen and used as a name in the new map. Using such approach to deal with lexical variation, the same command "take the book" is analyzed by considering a map containing the object b1 called "book", but also against maps containing "volume" or "guidebook". Among all the maps, we count a total of 81 types of entities, including artifacts, furnitures and names of locations (e.g. rooms). After the validation process, the average number of lexical alternatives for an entity (i.e. the sets A_e^i) is 3.76 nouns. The Distributional Model introduced in Section 3 has been acquired according to a Skip-gram model [15] through the word2vec tool[5]. In particular, we derive 250 dimensional word vectors, by using the UkWaC corpus[6] a large scale web document collection made of 2-billion words. The resulting model allows to model more than 110,000 words. The grounding function required some parameters to be set. For ease of implementation, and since we did not want to push on the experimentation in order to leave the digression clear, parameters have been tuned on a validation set. The threshold τ (in Equation 1) for the every set S_τ^e has been set to 0. A value of 0.4 has been assigned to the decay factor λ of the String Kernel.

To better study how the versions of the grounding function deal with mistranscriptions and synonymy, the 538 transcriptions have been arranged according to a score that represents the degree of complexity in grounding a transcription over a semantic map. Such complexity score is computed by applying a function that takes as input i) the transcription of a command produced by the ASR, ii) its correct transcription, iii) a semantic map. The score is evaluated considering i) the overall WER of the transcription; ii) the number of entities in the correct transcription that need to be grounded; iii) the number of entities in the transcription that have been correctly transcribed, and that need to be grounded; iv) the number of entities in the correct transcription that are correctly grounded by the Identity function on the semantic map. Indeed, the complexity is directly proportional to the first two factors, while it is inversely proportional to the second ones.

4.2   Results

This Section reports and discusses the results that have been obtained in our investigation. Specifically, we tested the performances of the grounding function in different configurations of phonetic distance and selection policy. Every phonetic distance reported in Section 3.1 has been combined with the *Argmax*

---

[5] word2vec is available at https://code.google.com/p/word2vec/. The settings are: *min-count=50, window=5, iter=10* and *negative=10*.

[6] http://wacky.sslmit.unibo.it/doku.php?id=corpora

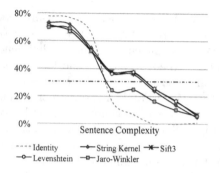

Fig. 1. F1 results for the *Argmax* policy with respect to sentence complexity

(Γ_{max}) and *Confidence Level* selection policies. For the latter, values of $n = 3$ and $n = 4$ have been selected for the discussion (denoted by Γ_3 and Γ_4 respectively) as the others did not produced significant results. Thus, 12 experiments have been run, combining 4 phonetic distances with 3 selection policies. Every combination has been compared with a baseline represented by the *Identity* (Γ_I) selection policy (see Section 3.2). The set of transcriptions have been ordered according to their complexity scores and divided in 8 beans of the same size, each containing transcriptions that are comparable with respect to our notion of complexity.

Our goal is to prove the robust behavior of the proposed approach. Robustness is here intended as the ability of correctly ground terms over semantic maps when the complexity of the transcriptions grows. What we expect is that the Γ_I policy is the most penalized when complexity is high, as it does not use any kind of semantic or phonetic information. On the contrary, using such information represents a solution to this issue, as the significant degradation showed by Γ_I is gently smoothed when other policies are applied. This is noticeable from the plots in Figures 1, 2 and 3, that report the value of F-Measure obtained by Γ_{max},

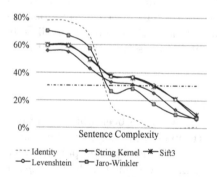

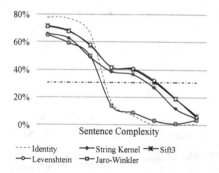

Fig. 2. F1 results for the *Confidence level* policy ($n = 3$) with respect to sentence complexity

Fig. 3. F1 results for the *Confidence level* policy ($n = 4$) with respect to sentence complexity

Table 1. Results, in terms of Precision, Recall and F1, for all the phonetic distances for the *Argmax* selection policy

bean	Identity P	R	F1	String Kernel P	R	F1	Sift3 P	R	F1	Levenshtein P	R	F1	Jaro-Winkler P	R	F1
1	82.8	73.7	**78.0**	72.2	74.6	73.4	70.5	72.8	71.6	69.9	72.2	71.0	68.9	71.2	70.0
2	84.4	70.6	**76.9**	68.6	75.7	71.9	64.2	70.9	67.4	64.2	70.9	67.4	65.9	72.7	69.1
3	81.7	55.9	**66.4**	52.4	58.4	55.2	50.2	55.9	52.9	50.3	56.0	53.0	50.5	56.3	53.2
4	48.4	10.0	16.6	37.5	37.0	37.2	38.1	37.6	**37.9**	36.7	36.2	36.5	24.7	24.4	24.5
5	25.8	4.1	7.1	34.0	38.4	36.0	35.2	39.8	37.4	35.6	40.2	**37.8**	23.4	26.5	24.9
6	0.0	0.0	0.0	22.8	25.2	23.9	24.7	27.2	**25.9**	24.5	27.0	25.7	15.7	17.3	16.4
7	0.0	0.0	0.0	14.3	13.4	13.9	17.2	16.1	16.6	17.6	16.5	**17.0**	10.5	9.9	10.2
8	16.7	0.5	1.0	5.2	4.7	4.9	7.8	7.0	**7.4**	6.6	6.0	6.3	6.4	5.7	6.0

Γ_3 and Γ_4 respectively; here Γ_I behavior is represented by the dashed line, while its mean value is drawn as a dashed-dotted line. Tables 1, 2 and 3 reports the values of Precision, Recall and F-Measure that generated such plots. A drastic drop of Γ_I is noticeable between beans 2 and 6, where it falls from 76.9 to 0, due to its inability of dealing with mistranscriptions and lexical variation. The robustness of the proposed method is significant in this range of complexity: higher values of complexity (e.g. beans 7 and 8) may correspond to not treatable cases, as transcriptions may contain too many errors, e.g. the command *"go to the bathroom"* being transcribed as *"booty butt booty"*. In such cases we may consider a request for clarification from the robot as acceptable, due to the impossibility to being robust against commands that are totally mistranscribed.

In general, the robustness achieved by the proposed policies is noticeable especially from the bean number 3. A drop of 50 points characterizes the Γ_I between beans 3 and 4, while every other policy shows a smoother drop when Sift3, Levenshtein and String Kernels are applied. Going along the beans, the performance stays stable between the beans 4 and 5, and drops down again on the 6*th* bean, but still remaining over the mean value of Γ_I. These values demonstrate how the proposed approach represents a more robust solution.

Looking at the single behaviors of the policies, Γ_4 achieves the best results on average, while Γ_{max} is the one that best compete with the Γ_I when the complexity is low. Here Γ_I shows high value of Precision, as predictable, but

Table 2. Results, in terms of Precision, Recall and F1, for all the phonetic distances for the *Confidence Level* selection policy ($n = 3$)

bean	Identity P	R	F1	String Kernel P	R	F1	Sift3 P	R	F1	Levenshtein P	R	F1	Jaro-Winkler P	R	F1
1	82.8	73.7	**78.0**	40.6	88.3	55.7	45.4	88.9	60.1	45.1	88.9	59.8	72.0	68.8	70.4
2	84.4	70.6	**76.9**	39.9	88.9	55.0	45.2	88.8	59.9	44.8	88.8	59.2	70.8	63.6	67.0
3	81.7	55.9	**66.4**	31.0	69.9	43.0	38.4	69.5	49.5	39.2	69.8	50.2	64.5	51.9	57.5
4	48.4	10.0	16.6	23.7	57.0	33.4	28.2	58.0	**37.9**	27.8	57.0	37.3	35.3	21.3	26.5
5	25.8	4.1	7.1	21.4	60.8	31.7	25.9	60.7	36.3	26.3	61.5	**36.9**	36.2	23.5	28.5
6	0.0	0.0	0.0	17.5	46.8	25.5	22.2	48.1	30.4	23.0	48.4	**31.2**	25.1	13.3	17.3
7	0.0	0.0	0.0	8.9	26.4	13.3	17.5	27.1	**21.2**	17.0	27.2	20.9	16.0	7.1	9.8
8	16.7	0.5	1.0	4.9	9.6	6.5	10.0	9.9	**9.9**	7.8	7.9	7.9	14.6	4.8	7.3

Table 3. Results, in terms of Precision, Recall and F1, for all the phonetic distances for the *Confidence Level* selection policy ($n = 4$)

bean	Identity			String Kernel			Sift3			Levenshtein			Jaro-Winkler		
	P	R	F1	P	R	F1	P	R	F1	P	R	F1	P	R	F1
1	82.8	73.7	**78.0**	54.9	81.7	65.7	63.3	82.4	71.6	63.0	82.4	71.4	87.4	52.0	65.2
2	84.4	70.6	**76.9**	52.7	77.3	62.7	60.6	77.5	68.0	59.7	78.0	67.6	87.1	45.0	59.4
3	81.7	55.9	**66.4**	41.4	59.9	49.0	55.1	60.1	57.5	55.6	60.4	57.9	81.2	36.1	50.0
4	48.4	10.0	16.6	34.8	43.5	38.7	40.1	43.7	**41.8**	40.2	43.4	41.7	52.9	8.0	13.9
5	25.8	4.1	7.1	31.7	43.2	36.6	38.0	42.7	40.2	38.1	43.9	**40.8**	41.5	5.0	9.0
6	0.0	0.0	0.0	22.8	32.7	26.9	30.8	31.6	31.2	31.7	32.8	**32.3**	17.3	1.7	3.1
7	0.0	0.0	0.0	9.1	17.1	11.9	24.3	15.6	19.0	24.4	16.0	**19.3**	5.0	0.5	0.9
8	16.7	0.5	1.0	4.7	5.2	4.9	13.5	4.7	**7.0**	12.1	3.8	5.8	30.2	2.0	3.8

Recall also reaches good values. This is due mainly to the fact that most of the 81 entities in the map have at most 3 lexical alternatives. Such setting facilitated Γ_I in selecting the right grounding when no transcription errors were reported. Sift3, Levenshtein and String Kernels show a similar trend, with the first two sharing the best results on high complexity beans, while the F-Measure of the third one is always slightly lower. Instead, the Jaro-Winkler exposes a behavior that is more conservative with respect to the phonetic similarity. Generally, the values assigned by such metric produce grounding scores that are closer to the mean in the distribution with respect to other metrics. As a consequence, when applied to Γ_n policy, the threshold defined as $\mu + n \cdot \sigma$ discards more candidates for values of n that are lower with respect to other metrics. In fact, for $n = 4$ Precision is privileged on Recall, lowering the F-Measure, resulting in a behavior that is similar to Γ_I.

5   Conclusions

In this work we propose a grounding function that is robust with respect to mis-transcriptions coming form an ASR engine and to the lexical variability of the Natural Language. Such function is based on Distributional Models of Lexical Semantics that allow to acquire semantic relations between words through the automatic analysis of large scale document collections. With respect to existing solutions, our contribution is that the proposed solution allows to ground unseen words without using large scale dictionaries or the need of artificial examples to train supervised models. Such score may be used to support the interpretation process of Natural Language. The first further step to take in this direction is a deeper use of the information contained in semantic maps. In fact, these contain spatial information about the entities that can be fundamental in select-ing the right interpretation of a sentence. Such information could be crucial for the grounding problem, especially in the Semantic Parsing phase, where spatial relations between the arguments of an expressed action can substantially con-tribute. Moreover, semantic maps may contain also images of objects, that can be additionally used in the grounding and disambiguation process. In general, robots are provided with different perception system. The use of information extracted from many sensing apparatuses in Language Understanding may represent an additional source of information for such task. Finally, with regard to the grounding function, a possible improvement may consist in the mixed use of different policies and phonetic distances, that could be activated depending on some variables of the environment or of the spoken commands.

References

1. Bastianelli, E., Bloisi, D.D., Capobianco, R., Cossu, F., Gemignani, G., Iocchi, L., Nardi, D.: On-line semantic mapping. In: ICAR 2013, November 2013
2. Bastianelli, E., Castellucci, G., Croce, D., Basili, R., Nardi, D.: Huric: a human robot interaction corpus. In: Proceedings of LREC 2014, Reykjavik, Iceland, May 2014
3. Bunescu, R., Mooney, R.J.: Subsequence kernels for relation extraction. In: Submitted to the Ninth Conference on Natural Language Learning (CoNLL-2005), July 2006
4. Chen, D.L., Mooney, R.J.: Learning to interpret natural language navigation instructions from observations. In: Proceedings of the 25th AAAI Conference on AI, pp. 859–865 (2011)
5. Connell, J., Marcheret, E., Pankanti, S., Kudoh, M., Nishiyama, R.: An extensible language interface for robot manipulation. In: Bach, J., Goertzel, B., Iklé, M. (eds.) AGI 2012. LNCS, vol. 7716, pp. 21–30. Springer, Heidelberg (2012)
6. Golland, D., Liang, P., Klein, D.: A game-theoretic approach to generating spatial descriptions. In: Proceedings of the EMNLP 2010, Stroudsburg, PA, USA, pp. 410–419 (2010)
7. Guadarrama, S., Riano, L., Golland, D., Gohring, D., Jia, Y., Klein, D., Abbeel, P., Darrell, T.: Grounding spatial relations for human-robot interaction. In: IEEE/RSJ International Conference on Intelligent Robots and Systems, November 2013
8. Harnad, S.: The symbol grounding problem. Physica D: Nonlinear Phenomena **42**(1–3), 335–346 (1990)
9. Harris, Z.: Distributional structure. In: Katz, J.J., Fodor, J.A. (eds.) The Philosophy of Linguistics. Oxford University Press (1964)
10. Hemachandra, S., Kollar, T., Roy, N., Teller, S.: Following and interpreting narrated guided tours. In: Proceedings of the ICRA 2011, Shanghai, China, pp. 2574–2579 (2011)
11. Kollar, T., Tellex, S., Roy, D., Roy, N.: Toward understanding natural language directions. In: Proceedings of the 5th ACM/IEEE, HRI 2010, Piscataway, NJ, USA, pp. 259–266 (2010)
12. Kruijff, G., Zender, H., Jensfelt, P., Christensen, H.: Clarification dialogues in human-augmented mapping. In: Proceedings of the HRI 2006 (2006)
13. Levenshtein, V.: Binary codes capable of correcting deletions. Insertions and eversals. In: Soviet Physics Doklady, vol. 10, p. 707 (1966)
14. Lodhi, H., Saunders, C., Shawe-Taylor, J., Cristianini, N., Watkins, C.: Text classification using string kernels. J. Mach. Learn. Res. **2**, 419–444 (2002)
15. Mikolov, T., Chen, K., Corrado, G., Dean, J.: Efficient estimation of word representations in vector space. CoRR abs/1301.3781 (2013)
16. Miller, G.A.: Wordnet: A lexical database for english. Commun. ACM **38**(11), 39–41 (1995)

17. Nüchter, A., Hertzberg, J.: Towards semantic maps for mobile robots. Robot. Auton. Syst. **56**(11), 915–926 (2008)
18. Rosenthal, S., Biswas, J., Veloso, M.: An effective personal mobile robot agent through symbiotic human-robot interaction. In: Proceedings of AAMAS 2010, vol. 1, May 2010
19. Roy, D., yuh Hsiao, K., Gorniak, P., Mukherjee, N.: Grounding natural spoken language semantics in visual perception and motor control
20. Sahlgren, M.: The Word-Space Model. Ph.D. thesis, Stockholm University (2006)
21. Schütze, H.: Word space. In: Advances in Neural Information Processing Systems, vol. 5, pp. 895–902. Morgan Kaufmann (1993)
22. Steels, L., Vogt, P.: Grounding adaptive language games in robotic agents. In: Proceedings of the Fourth European Conference on Artificial Life, pp. 474–482. MIT Press (1997)
23. Tellex, S., Kollar, T., Dickerson, S., Walter, M., Banerjee, A., Teller, S., Roy, N.: Approaching the symbol grounding problem with probabilistic graphical models. AI Magazine **32**(4) (2011)
24. Topp, E.A.: Human-Robot Interaction and Mapping with a Service Robot: Human Augmented Mapping. Ph.D. thesis, Royal Institute of Technology, School of Computer Science and Communication (2008)
25. Winkler, W.E.: The state of record linkage and current research problems. Tech. rep., Statistical Research Division, U.S. Census Bureau (1999)

Automatic Identification and Disambiguation of Concepts and Named Entities in the Multilingual Wikipedia

Federico Scozzafava[(✉)], Alessandro Raganato, Andrea Moro, and Roberto Navigli

Dipartimento di Informatica, Sapienza Università di Roma,
Viale Regina Elena 295, 00161 Roma, Italy
federico.scozzafava@gmail.com,
{raganato,moro,navigli}@di.uniroma1.it

Abstract. In this paper we present an automatic multilingual annotation of the Wikipedia dumps in two languages, with both word senses (i.e. concepts) and named entities. We use Babelfy 1.0, a state-of-the-art multilingual Word Sense Disambiguation and Entity Linking system. As its reference inventory, Babelfy draws upon BabelNet 3.0, a very large multilingual encyclopedic dictionary and semantic network which connects concepts and named entities in 271 languages from different inventories, such as WordNet, Open Multilingual WordNet, Wikipedia, OmegaWiki, Wiktionary and Wikidata. In addition, we perform both an automatic evaluation of the dataset and a language-specific statistical analysis. In detail, we investigate the word sense distributions by part-of-speech and language, together with the similarity of the annotated entities and concepts for a random sample of interlinked Wikipedia pages in different languages. The annotated corpora are available at http://lcl.uniroma1.it/babelfied-wikipedia/.

Keywords: Semantic annotation · Named entities · Word senses · Disambiguation · Multilinguality · Corpus annotation · Sense annotation · Word sense disambiguation · Entity linking

1 Introduction

The exponential growth of the Web has resulted in an increased number of Internet users of diverse mother-tongues, and textual information available online in a wide variety different languages. This has led to a heightened interest in multilingualism [9], [15]. Over the last decade, collaborative resources like Wikipedia (an online encyclopedia) and Wiktionary (an online dictionary) have grown not only quantitatively, but also in terms of their degree of multilingualism, i.e., the range of different languages in which they are available. For this reason these resources have been exploited in many Natural Language Processing tasks, such as Word Sense Disambiguation (WSD) [16], [22], [26], i.e., the task of determining the sense of a word in a given context, and Entity Linking (EL) [30], i.e.,

© Springer International Publishing Switzerland 2015
M. Gavanelli et al. (Eds.): AI*IA 2015, LNAI 9336, pp. 357–366, 2015.
DOI: 10.1007/978-3-319-24309-2_27

the task of discovering which named entities are mentioned in a text. Although there are knowledge bases that incorporate these different kinds of knowledge [25], i.e. encyclopedic and lexicographic knowledge, currently there are only few datasets that integrate annotations from both kinds of repositories. This is due to the fact that the research community has typically focused its attention on WSD and EL tasks separately. The main difference between WSD and EL lies in the kind of inventory used, i.e., dictionary vs. encyclopedia respectively; however the tasks are pretty similar, as they both involve the disambiguation of textual mentions according to a reference inventory. Recently, work in the direction of joint word sense and named entity disambiguation has been promoted in order to concentrate research efforts on the common aspects of the two tasks, such as identifying the right meaning in context [21]. The system presented by [21], called Babelfy, attains state-of-the art accuracy in both WSD and EL tasks, including in a multilingual setting. As its sense inventory Babelfy draws upon BabelNet, a multilingual encyclopedic dictionary, that has lexicographic and encyclopedic coverage of terms.

Moreover, a first corpus annotated with both concepts and named entities has also been created [20]. However, as this corpus is only in English, we decided to annotate a sample of Wikipedia automatically with both word senses and named entities in two languages.

The paper is organized as follows. In Section 2 we cover related work on annotated text with senses. In Section 3, 4 and 5 we briefly describe Wikipedia, BabelNet and Babelfy respectively. In Section 6 and 7 we provide statistics and evaluations. Finally, Section 8 presents our conclusions.

2   Related Work

Over the years, several datasets annotated either with concepts or with named entities have been created [7], [20], [31]. Moreover, numerous tasks in competitions such as Senseval/Semeval [12–14], [19], [23,24], [27], [29], [32], TAC KBP EL [11], Microposts [1] and ERD [3] have been organized together with the development of frameworks for comparison of entity-annotation systems [4], [34].

As regards WSD, i.e., the task of determining the sense of a word in a given context, many disambiguation competitions resulted in several datasets that were manually annotated with word senses. However, these datasets are pretty small. In fact, the largest dataset manually annotated with word senses is Sem-Cor [18], a subset of the English Brown Corpus, containing 360K words and more than 200K sense-tagged content words according to the WordNet lexical database [28]. However, these datasets contain only lexicographic annotations without considering named entities, and moreover no dataset of comparable size to SemCor exists for languages other than English.

The EL task, i.e., the task of discovering which named entities are mentioned, was introduced more recently [30]. Usually the reference inventory for this task is Wikipedia. In fact, the largest manually annotated dataset is Wikilinks [31], which contains links to Wikipedia pages. Wikilinks consists of web pages, crawled from Google's web index, containing at least one hyperlink that points to English Wikipedia. This dataset consists of roughly 13M documents with 59M annotated mentions. Another well-known corpus is the Freebase Annotations of the ClueWeb Corpora (FACC1) [7], built by researchers at Google, who annotated English-language Web pages from the ClueWeb09 and ClueWeb12 corpora. The corpus consists of an automatic annotation of 800M documents with 11 billion entity annotations. The annotations are generally of high quality with a precision around 80-85% and a recall around 70-85% (as stated by the authors). However, as for the WSD task, these resources contain only named entities without taking into account word senses, and are available only for the English language. Recently, [21] proposed a new unified approach for WSD and EL, called Babelfy, which jointly disambiguates word senses and named entities, reaching the state of the art on both tasks in a multilingual setting. Babelfy was used in the first automatic semantic annotation of both named entities and word senses on the MASC corpus [10], [20]. However, this resource is smaller compared to other resources, containing roughly 200K total annotations available only for the English language. In this paper we perform a high quality automatic annotation with both word senses and named entities of a large corpus of English and Italian.

3 Wikipedia

Wikipedia[1] is a well-known freely available collaborative encyclopedia, containing 35 million pages in over 250 languages. The Wikipedia internal links (see Figure 1) are one of the features that makes Wikipedia a valuable project and resource. In fact it was estimated that the network of internal links offers the opportunity to proceed from any article to any other with an average of 4.5 clicks [5].

Fig. 1. A sample Wikipedia page with links.

The freedom to create and edit pages has a positive impact both qualitatively and quantitatively, matching and overcoming the famous *Encyclopedia Britannica* [8]. It was estimated that the text of the English Wikipedia is currently equivalent to over 2000 volumes of the *Encyclopedia Britannica*.

---

[1] http://www.wikipedia.org

Wikipedia users are free to create new pages following the guidelines provided by the encyclopedia. In fact, each article in Wikipedia is identified by a unique identifier allowing the creation of shortcuts, expressed as: [[ID |anchor text]], where the anchor text is the fragment of text of a page linked to the identified page ID, and [[anchor text]], where the anchor text is linked to the corresponding homonymous page.

For instance, in the following sentence taken from the Wikipedia page Natural Language Processing: *"Natural language processing (NLP) is a field of [[computer science]], [[artificial intelligence]], and [[computational linguistics]] concerned with the interactions between [[computer]]s and [[Natural language|human (natural) languages]]. As such, NLP is related to the area of [[human-computer interaction]]. Many challenges in NLP involve [[natural language understanding]], that is, enabling computers to derive meaning from human or natural language input, and others involve [[natural language generation]]."*, the users decided to link *human (natural) languages* to the Wikipedia page *Natural language*.

In our settings by exploiting the Babelfy disambiguation system we leverage these hand-made connections to improve the quality of our automatic annotation.

4 BabelNet

Our reference inventory is BabelNet[2] [25], version 3.0, a multilingual lexicalised semantic network obtained from the automatic integration of heterogeneous resources such as WordNet [17], Open Multilingual WordNet [2], Wikipedia[3], OmegaWiki[4], Wiktionary[5] and Wikidata[6]. The integration is performed via an automatic mapping between these resources which results in merging equivalent concepts from the different resources. BabelNet covers and links named entities and concepts present in all the aforementioned resources obtaining a wide coverage resource containing both lexicographic and encyclopedic terms.

Fig. 2. An illustrative overview of BabelNet (picture from [25]).

---

[2] http://babelnet.org
[3] http://www.wikipedia.org
[4] http://www.omegawiki.org
[5] http://www.wiktionary.org
[6] http://www.wikidata.org

For instance in Figure 2 the concepts *balloon, wind, hot-air balloon and gas* are defined in both Wikipedia and WordNet while *Montgolfier brothers* and *blow gas* are respectively named entities and concepts retrieved from Wikipedia and WordNet. Each node in BabelNet, called Babel synset, represents a given meaning and contains all the synonyms, glosses and translations harvested from the respective resources. The latest release of BabelNet, i.e., 3.0, provides a full-fledged taxonomy [6], covers 271 languages and it is made up of more than 13M Babel synsets, with 117M senses and 354M lexico-semantic relations (for more statistics see http://babelnet.org/stats). It is also available as SPARQL endpoint and in RDF format containing up to 2 billion RDF triples.

5   Babelfy

To perform the automatic annotation of the considered dataset with both concepts and named entities, we used the latest version of Babelfy[7], i.e., version 1.0. Babelfy is a unified graph-based approach to Entity Linking and Word Sense Disambiguation, a state-of-the-art system in both tasks. A detailed description of the system can be found in [21]. Differently from version 0.9.1, this new release features many parameters among which adding pre-annotated fragments of text to help the disambiguation phase and to enable or disable the most common sense (MCS) backoff strategy that returns the most common sense for the text fragment when the system does not have enough information to select a meaning. Therefore we exploit the links of Wikipedia which are contained in BabelNet as pre-annotated fragments of text.

Babelfy is based on the BabelNet 3.0 semantic network and jointly performs disambiguation and entity linking in three steps. The first step associates with each node of the network a set of semantically relevant vertices, i.e. concepts and named entities, thanks to a notion of semantic signatures. This is a preliminary step which needs to be performed only once, independently of the input text. The second step extracts all the textual mentions from the input text, i.e. substrings of text for which at least one candidate named entity or concept can be found in BabelNet. Consequently, for each extracted mention, it obtains a list of the possible meanings according to the semantic network. The last step consists of connecting the candidate meanings according to the previously-computed semantic signatures. It then extracts a dense sub-graph and selects the best candidate meaning for each fragment.

Therefore our approach is comprised of two main phases: the identification of the semantic context given by the BabelNet synset corresponding to the link in the page (see Figure 3) and the disambiguation of the Wikipedia article through the use of Babelfy. Each Wikipedia page, together with its internal links, corresponds to a Babel synset. Thus providing that information (i.e. the Babel synset) as disambiguation context for the text associated with the link in the page helps the Babelfy algorithm exclude less relevant candidates.

---

[7] http://babelfy.org

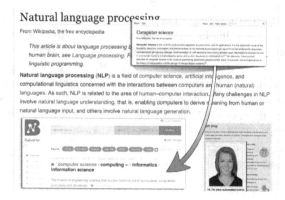

Fig. 3. *Computer science* Wikipedia link with its relative Babel synset.

6 Statistics

In this section we present the statistics of our automatically annotated dataset. We used a sample of 500K articles of English Wikipedia and over 450K articles of Italian Wikipedia POS tagged with the Stanford POS Tagger [33] (for Italian we trained a model using the dataset from the Universal Dependency Treebank Project[8]). The corpora contain respectively 501M and 310M words (see Table 1), among which in both cases 42% are content words (i.e. words POS tagged as noun, adjective, adverb or verb).

Table 1. Statistics of the Wikipedia sample.

	English		Italian	
# Articles		500,000		474,887
# Content Words	209,066,032		133,022,968	
# Non-Content Words	292,796,219		177,786,434	
# Words		501,862,251		310,809,402

In Table 2 and 3, we show the total number of our automatic annotations divided between concepts and named entities with and without the most common sense backoff strategy. As expected we have more annotations with the MCS, while without it we annotated 31% and 21% of the content words, respectively in English and Italian.

7 Evaluation

We performed an evaluation over a restricted sample of annotations to estimate the performance of the system using the accuracy measure, which is defined

---

[8] https://code.google.com/p/uni-dep-tb/

Table 2. Statistics of our automatic annotation of the Wikipedia corpus with MCS.

	English		Italian	
# Adjective Word Senses	14,662,188		5,921,520	
# Adverb Word Senses	3,402,554		2,604,358	
# Noun Word Senses	55,597,241		31,003,356	
# Verb Word Senses	26,072,320		11,942,285	
# Word Senses		99,734,303		51,471,519
# Named Entities		14,162,561		5,503,556
# Total Number of annotations		113,896,864		56,975,075

Table 3. Statistics of our automatic annotation of the Wikipedia corpus without MCS.

	English		Italian	
# Adjective Word Senses	7,816,765		2,848,886	
# Adverb Word Senses	2,450,533		1,385,650	
# Noun Word Senses	32,398,013		14,313,556	
# Verb Word Senses	8,683,852		3,302,068	
# Word Senses		51,349,163		21,850,160
# Named Entities		14,162,220		5,469,766
# Total Number of annotations		65,511,383		27,319,926

Table 4. Annotations with and without using the internal Wikipedia's links.

	English	Italian
# Articles	1,000	1,000
# Annotations with Wikipedia links	72,142	72,597
# Annotations without Wikipedia links	71,354	71,236

Table 5. Annotations in common between comparable Wikipedia pages in two languages.

	English	Italian
# Articles	1,000	1,000
# Annotations	258,273	107,448
# Annotations in common	23,439	

as the number of correct meanings/entities over the whole number of manually annotated mentions.

For evaluation purposes, we also annotated 1K articles for both languages with and without using the internal Wikipedia links as help to the disambiguation phase. As we can see from Table 4, using the Wikipedia page internal links as semantic context for disambiguation, as described in Section 5, the system yields more annotations.

Moreover, we manually evaluated a random sample of 200 concepts and 200 named entities obtaining an estimated accuracy of 77.8% for word senses and 63.2% for named entities for English, and 78.6% and 66% respectively for Italian.

To estimate the similarity of annotated entities and concepts for cross-lingual interlinked Wikipedia pages, we randomly selected 1K English Wikipedia articles and their equivalent in Italian. In Table 5 we show the number of common annotations (i.e. Babel synsets) between corresponding articles in the two languages.

8   Conclusion

In this paper we presented a large sample of the English and Italian Wikipedias disambiguated with both named entities and concepts, thanks to the use of a state-of-the-art disambiguation and entity linking system, i.e., Babelfy [21]. As sense inventory we used BabelNet 3.0, a multilingual encyclopedic dictionary containing lexicographic and encyclopedic terms obtained from the automatic integration of WordNet, Open Multilingual WordNet, Wikipedia, OmegaWiki, Wiktionary and Wikidata. In order to obtain high quality annotations, we exploited the internal links of Wikipedia as an additional aid for the disambiguation phase. We performed a manual evaluation of our automatic annotation, which indicated an estimated accuracy of 77.8% for word senses, 63.2% for named entities in English, and 78.6% and 66%, respectively, in Italian.

The annotated corpora are available at http://lcl.uniroma1.it/babelfied-wikipedia/.

Acknowledgments

 The authors gratefully acknowledge the support of the ERC Starting Grant MultiJEDI No. 259234.

References

1. Basave, A.E.C., Rizzo, G., Varga, A., Rowe, M., Stankovic, M., Dadzie, A.S.: Making sense of microposts (#Microposts2014) named entity extraction & linking challenge. In: 4th Workshop on Making Sense of Microposts (#Microposts2014) (2014)
2. Bond, F., Foster, R.: Linking and extending an open multilingual wordnet. In: ACL (1), pp. 1352–1362 (2013)
3. Carmel, D., Chang, M.W., Gabrilovich, E., Hsu, B.J.P., Wang, K.: ERD'14: entity recognition and disambiguation challenge. In: ACM SIGIR Forum, vol. 48, pp. 63–77. ACM (2014)
4. Cornolti, M., Ferragina, P., Ciaramita, M.: A framework for benchmarking entity-annotation systems. In: Proc. of WWW, pp. 249–260 (2013)
5. Dolan, S.: Six Degrees of Wikipedia (2008). http://mu.netsoc.ie/wiki/

6. Flati, T., Vannella, D., Pasini, T., Navigli, R.: Two is bigger (and better) than one: the wikipedia bitaxonomy project. In: Proc. of ACL, pp. 945–955. Association for Computational Linguistics, Baltimore (2014)

7. Gabrilovich, E., Ringgaard, M., Subramanya, A.: FACC1: Freebase annotation of ClueWeb corpora, Version 1. Release date, pp. 06–26 (2013)

8. Giles, J.: Internet encyclopaedias go head to head. Nature **438**(7070), 900–901 (2005)

9. Gracia, J., Montiel-Ponsoda, E., Cimiano, P., Gómez-Pérez, A., Buitelaar, P., McCrae, J.: Challenges for the multilingual web of data. Web Semantics: Science, Services and Agents on the World Wide Web **11**, 63–71 (2012)

10. Ide, N., Baker, C., Fellbaum, C., Fillmore, C.: MASC: the manually annotated sub-corpus of American English. In: Proc. of LREC (2008)

11. Ji, H., Dang, H., Nothman, J., Hachey, B.: Overview of tac-kbp2014 entity discovery and linking tasks. In: Proc. of TAC (2014)

12. Lefever, E., Hoste, V.: Semeval-2010 task 3: cross-lingual word sense disambiguation. In: Proc. of SemEval, pp. 15–20 (2010)

13. Lefever, E., Hoste, V.: Semeval-2013 task 10: cross-lingual word sense disambiguation. In: Proc. of SemEval, pp. 158–166 (2013)

14. Manandhar, S., Klapaftis, I.P., Dligach, D., Pradhan, S.S.: SemEval-2010 task 14: word sense induction & disambiguation. In: Proc. of SemEval, pp. 63–68 (2010)

15. McDonald, R.T., Nivre, J., Quirmbach-Brundage, Y., Goldberg, Y., Das, D., Ganchev, K., Hall, K.B., Petrov, S., Zhang, H., Täckström, O., et al.: Universal dependency annotation for multilingual parsing. In: ACL (2), pp. 92–97 (2013)

16. Mihalcea, R.: Using wikipedia for automatic word sense disambiguation. In: HLT-NAACL, pp. 196–203 (2007)

17. Miller, G.A.: WordNet: a lexical database for English. Communications of the ACM **38**(11), 39–41 (1995)

18. Miller, G.A., Leacock, C., Tengi, R., Bunker, R.T.: A semantic concordance. In: Proc. of the workshop on Human Language Technology, pp. 303–308 (1993)

19. Moro, A., Navigli, R.: SemEval-2015 task 13: multilingual all-words sense disambiguation and entity linking. In: Proc. of SemEval, pp. 288–297 (2015)

20. Moro, A., Navigli, R., Tucci, F.M., Passonneau, R.J.: Annotating the MASC corpus with BabelNet. In: Proc. of LREC, pp. 4214–4219 (2014)

21. Moro, A., Raganato, A., Navigli, R.: Entity Linking meets Word Sense Disambiguation: A Unified Approach. TACL **2**, 231–244 (2014)

22. Navigli, R.: Word sense disambiguation: A survey. ACM Computing Surveys (CSUR) **41**(2), 10 (2009)

23. Navigli, R., Jurgens, D., Vannella, D.: Semeval-2013 task 12: multilingual word sense disambiguation. In: Proc. of SemEval, vol. 2, pp. 222–231 (2013)

24. Navigli, R., Litkowski, K.C., Hargraves, O.: Semeval-2007 task 07: coarse-grained english all-words task. In: Proc. of SemEval, pp. 30–35 (2007)

25. Navigli, R., Ponzetto, S.P.: BabelNet: The automatic construction, evaluation and application of a wide-coverage multilingual semantic network. Artificial Intelligence **193**, 217–250 (2012)

26. Navigli, R., Ponzetto, S.P.: Joining forces pays off: multilingual joint word sense disambiguation. In: Proc. of EMNLP, pp. 1399–1410 (2012)

27. Palmer, M., Fellbaum, C., Cotton, S., Delfs, L., Dang, H.T.: English tasks: all-words and verb lexical sample. In: Proc. of the Second International Workshop on Evaluating Word Sense Disambiguation Systems, pp. 21–24 (2001)

28. Pilehvar, M.T., Navigli, R.: A large-scale pseudoword-based evaluation framework for state-of-the-art word sense disambiguation. Computational Linguistics **40**(4), 837–881 (2014)
29. Pradhan, S.S., Loper, E., Dligach, D., Palmer, M.: Semeval-2007 task 17: English lexical sample, SRL and all words. In: Proc. of SemEval, pp. 87–92 (2007)
30. Rao, D., McNamee, P., Dredze, M.: Entity linking: finding extracted entities in a knowledge base. In: Multi-source, Multilingual Information Extraction and Summarization, pp. 93–115. Springer (2013)
31. Singh, S., Subramanya, A., Pereira, F., McCallum, A.: Wikilinks: a large-scale cross-document coreference corpus labeled via links to Wikipedia. University of Massachusetts, Amherst, Tech. Rep. UM-CS-2012-015 (2012)
32. Snyder, B., Palmer, M.: The English all-words task. In: Senseval-3: Third International Workshop on the Evaluation of Systems for the Semantic Analysis of Text, pp. 41–43 (2004)
33. Toutanova, K., Klein, D., Manning, C.D., Singer, Y.: Feature-rich part-of-speech tagging with a cyclic dependency network. In: HLT-NAACL, vol. 1, pp. 173–180 (2003)
34. Usbeck, R., Röder, M., Ngonga Ngomo, A.C., Baron, C., Both, A., Brümmer, M., Ceccarelli, D., Cornolti, M., Cherix, D., Eickmann, B., Ferragina, P., Lemke, C., Moro, A., Navigli, R., Piccinno, F., Rizzo, G., Sack, H., Speck, R., Troncy, R., Waitelonis, J., Wesemann, L.: GERBIL - general entity annotation benchmark framework. In: Proc. of WWW, pp. 1133–1143

A Logic-Based Approach to Named-Entity Disambiguation in the Web of Data

Silvia Giannini[1], Simona Colucci[1(✉)], Francesco M. Donini[2], and Eugenio Di Sciascio[1]

[1] DEI, Politecnico di Bari, Bari, Italy
simona.colucci@poliba.it
[2] DISUCOM, Università della Tuscia, Viterbo, Italy

Abstract. Semantic annotation aims at linking parts of rough data (*e.g.*, text, video, or image) to known entities in the Linked Open Data (LOD) space. When several entities could be linked to a given object, a Named-Entity Disambiguation (NED) problem must be solved. While disambiguation has been extensively studied in Natural Language Understanding (NLU), NED is less ambitious—it does not aim to the meaning of a whole phrase, just to correctly link objects to entities—and at the same time more peculiar since the target must be LOD-entities. Inspired by semantic similarity in NLU, this paper illustrates a way to solve disambiguation based on Common Subsumers of pairs of RDF resources related to entities recognized in the text. The inference process proposed for resolving ambiguities leverages on the DBpedia structured semantics. We apply it to a TV-program description enrichment use case, illustrating its potential in correcting errors produced by automatic text annotators (such as errors in assigning entity types and entity URIs), and in extracting a description of the main topics of a text in form of commonalities shared by its entities.

1 Introduction

Web pages represent rich and powerful sources of information collected in form of semi-structured or unstructured text. Semantic technologies offer the possibility to make this knowledge available in a machine processable way, realizing a step forward in the integration and linking of heterogeneous data sources, and enabling different querying mechanisms, involving reasoning and inferences over them. In this context, semantic annotation forms the bridge between textual information and existing ontologies or Semantic Web data sets [29].

Semantic annotation is an Information Extraction (IE) process consisting in applying semantic tags to the information conveyed as free text [2], *i.e.*, it produces a set of Named-Entities (NE) mentioned in the text. It typically requires the recognition of textual spans corresponding to relevant entities, and their classification in a semantic category, possibly assigning a link to real world objects through web identifiers.

© Springer International Publishing Switzerland 2015
M. Gavanelli et al. (Eds.): AI*IA 2015, LNAI 9336, pp. 367–380, 2015.
DOI: 10.1007/978-3-319-24309-2_28

In this paper we focus on a specific task within a semantic annotation process: KB-supported entity disambiguation. In brief, it aims at matching a recognized text entity-mention with the corresponding KB entity, resolving polysemy. Entity disambiguation techniques should also be designed for working in open domains, assuming that there are entities not linkable, *i.e.,* with no match in the reference KB [11]. Disambiguation involves both the identification of the right category of an entity and the choice of the correct entity to link within the KB. In this paper we show how Linked Open Data[1] (LOD) can be successfully used in a disambiguation task, relying on an inference process based on the semantics encoded in the Web of Data. In particular, we present a disambiguation strategy that, starting from a list of DBpedia[2] instances possibly disambiguating an entity, is able to chose the linked web identifier on the basis of a confidence score expressing the specificity level of the information conveyed by that instance when referred to the context defined by other entities mentioned in text. The specificity level is derived with the support of a deductive strategy based on Common Subsumers (CSs) extraction for Resource Description Framework[3] (RDF), for which a definition has been proposed by some of the authors in [8]. CSs are logically computed as RDF descriptions of the features shared by pairs of RDF resources. In this paper we propose a completely novel approach to NE disambiguation, which exploits the informative content embedded in CSs to support the choice—among candidate named entities—of the ones solving the entity linking problem. The adoption of a fully logic-based process—CS extraction— as basis for the proposed strategy makes it logic-based, as well.

The benefit of the proposed approach is illustrated with a possible use-case presented throughout the paper, reporting on the problem of enriching any TV-program description by combining several automatic semantic annotation-tool results.

The paper is organized as follows. Next section gives an overview of the challenges arising in text enrichment through semantic annotations, with particular reference to works related to NE disambiguation. A motivating example is given in Section 3. The formal approach for resolving disambiguation and a proof–of–concept are presented, respectively, in Sections 4 and 5. Conclusions and discussion of future work close the paper.

2   Semantic Annotation: An Overview

The problem of text enrichment refers to all the tasks involved in extracting useful information from unstructured data. In particular, dealing with semantic annotation of texts spanning over heterogeneous topics—such as TV-programs descriptions—is a challenging goal due to difficulties that arise in training a single extractor able to perform well with texts covering different domains. State-of-the-art NE extractors, including both web services APIs and software frameworks

---

[1] http://linkeddata.org/

[2] http://dbpedia.org/About

[3] http://www.w3.org/RDF/

were described by Gangemi *et al.* [14]. They are developed using different algorithms and training data, thus making each of them either targeted for specific NE recognition and classification tasks, or more reliable on particular document types (*e.g.*, newspaper articles, scientific papers, etc.) [28]. Therefore, the enrichment of text with heterogeneous topics and formats could benefit from the integration of annotations provided by several extractors [5,16,30].

In order to make the problem definition self-contained, we give below an overview of all the processes and challenges to be carried on for producing a valid semantic annotation of a text out of the results provided by different NE extraction tools. The discussion is then focused only on NE disambiguation, for which our contribution is proposed in Sec. 4.

2.1 Named-Entity Recognition (NER)

The first goal to be accomplished by an annotator is to recognize all relevant NE present in a document snippet. As stated in past literature [27,31], this task is equivalent to recognize all NE that are also keyphrases, discarding irrelevant ones. NER is a challenging problem due to variation in semantically identical but orthographically different entity names (*e.g.*, abbreviations, alternative spellings, or aliases that give rise to different surface forms for the same entity), and presence of entity names with several possible interpretations, that makes the definition of the relevance of an entity context-dependent. Evaluating agreement and disagreement among different extractors results [5] is particularly effective for this process (*e.g.*, an entity missing in one extractor reference KB can be repaired by the others; or, list of equivalent surface forms for the same entity can be obtained evaluating overlapping text spans).

2.2 Named-Entity Classification (NEC)

Given a recognized NE, the classification task requires to assign it to the right semantic category or type. Different supervised or unsupervised methods have been developed in literature for addressing this process [1,6,7,12,26]. However, an integration strategy of different classifiers outcomes enables the correction of mis-classified entities in terms of category or taxonomy granularity, always selecting the most specific concept representing the class of an entity [5].

2.3 Named-Entity Linking (NEL)

Named-Entity Linking (or Resolution) refers to the association of a NE to the referent entity in a KB. Providing the right resource describing an entity often requires to disambiguate among multiple instances linking to the same entity name or to resolve co-references [10,22]. The integration of different NE extractors results should improve the enrichment output, repairing missing URIs or correcting wrong links.

The main contribution of this paper is aimed at the formal definition of a NE disambiguation strategy using the DBpedia KB. Therefore, now we report more in detail on NE disambiguation related work.

Named-Entity Disambiguation (NED). Early approaches to NED were exclusively based on lexical resources, such as WordNet[4] [25]. Linking with encyclopedic knowledge, like Wikipedia[5], then gained more attention due to the larger coverage that these resources can offer in terms of sense inventory for NE and multi-word expressions [3,20,24]. Machine learning approaches have been used to identify NE in unstructured text and enrich them with links to appropriate Wikipedia articles [21]. This process, also referred to as wikification, embeds a disambiguation phase relying on a comparison between features (*e.g.*, terms, part–of–speech tags) extracted from the text to be annotated and the Wikipedia pages candidates for disambiguation. Nowadays, the Semantic Web community efforts in structuring and classifying real world entities described in web documents, through fine-grained classification schemes and relational facts in which they are involved, encourage the development of disambiguation techniques based on KB such as DBpedia, Freebase[6] or YAGO[7] [18,32].

The problem of entity disambiguation has been widely studied for the automatic construction of KBs [11,23], where entity resolution errors modify the truth value of facts extracted during the KB population process. Hoffart *et al.* ([19]) propose a disambiguation heuristic based on context similarity, where the context is obtained constructing a bag-of-words around the surface name to be linked and each entity the name could be possibly mapped to. The similarity is evaluated considering the word-level overlap between bag-of-words, combined with values expressing the popularity of an entity and the coherence among the context entities. Differently from traditional methods based on bag–of–words for extracting the context, LOD makes implicit relations explicit, thus offering the right structure to exploit pragmatics in disambiguation processes [4], following an idea of resolving interpretation problems with an inference mechanism that is well known in literature [17]. The approach we propose in Section 4 evaluates the most specific information shared by pairs of entities for resolving the disambiguation task, relying on a fully semantic-based exploration of the RDF descriptions of the involved resources.

3   A Motivating Example

Consider the text in Fig.1 representing the description of the first episode of the BBC TV-series *Rococo: Travel, Pleasure, Madness*[8]. As a baseline motivating our work, we report in Tab.2 the annotation results provided by the tool NERD[9]. Each NE is anchored to the corresponding text phrase, labeled with a type of the NERD ontology, and linked to a web page describing the entity (all prefixes introduced throughout the paper are listed in Tab. 1). NERD is

---

[4] http://wordnet.princeton.edu/
[5] http://www.wikipedia.org/
[6] http://www.freebase.com/
[7] https://www.mpi-inf.mpg.de/departments/databases-and-information-systems/
  research/yago-naga/yago/
[8] http://www.bbc.co.uk/programmes/b03sg830#programme
[9] http://nerd.eurecom.fr/analysis#

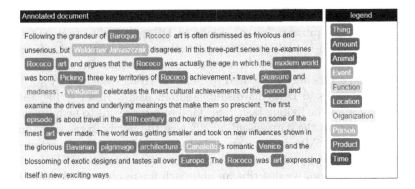

Fig. 1. Annotation output produced by the tool NERD for the first-episode synopsis of the BBC TV-series *Rococo: Travel, Pleasure, Madness*. All recognized NE are highlighted in the annotated document, with different colors depending on the category assigned in NERD. Category colors are explained in the legend on the right. For the list of assigned URIs see Tab. 2.

Table 1. Prefixes used in the examples of this paper

```
dbpedia:      < http : //dbpedia.org/resource/ > .
dbpedia-owl:  < http : //dbpedia.org/ontology/ > .
dc-terms:     < http : //purl.org/dc/terms/ > .
foaf:         < http : //xmlns.com/foaf/0.1/ > .
all-art:      < http : //www.all-art.org/history252_contents_ > .
nerd:         < http : //nerd.eurecom.fr/ontology# > .
wikipedia:    < http : //en.wikipedia.org/wiki/ > .
skos:         < http : //www.w3.org/2004/02/skos/core# >
```

a web framework [30] that provides a set of axioms for aligning taxonomies of several other NE extractors with the one present in NERD, and a machine-learning based strategy for combining different extractor outputs with conflicts resolution. We highlight some kind of errors still contained in the outcome of the aggregation process for the NEL task. In particular, a distinction between mapping errors (the URIs in italic font in Tab. 2) and disambiguation errors (the ones in bold font) can be made. The former are represented by links that are not DBpedia resources (the AlchemyAPI[10], Extractiv[11], OpenCalais[12], TextRa-zor[13], Wikimeta[14], Yahoo! Content Analysis[15], and Zemanta[16] extractors integrated in NERD return links to Wikipedia pages) and are easily repairable (the

---

[10] http://www.alchemyapi.com/
[11] http://www.programmableweb.com/api/extractiv
[12] http://www.opencalais.com/
[13] https://www.textrazor.com/
[14] http://www.wikimeta.com
[15] https://developer.yahoo.com/search/content/V2/contentAnalysis.html
[16] http://developer.zemanta.com/

DBpedia resource corresponding to a Wikipedia page can be identified through the property foaf : primaryTopic). Links to DBpedia redirection pages, consisting of redirection hyperlinks from alternative names to the actual resource describing the entity, can also be treated as mapping errors. On the other hand, disambiguation errors are constituted by links to wrong resources (Rows 11, 13, 17), or by references to DBpedia disambiguation pages (Rows 7, 14), that are not useful URIs for the linking task. This kind of error seems to be influenced by the type assigned to the corresponding NE (see the nerd : Organization type in Row 11 and 17). Moreover, missing links are also present, denoting a bad co-reference resolution process or the recognition of not relevant NE.

In the next section, we describe an approach for resolving ambiguity in NE linking. The approach is based on an inference service specifically defined for RDF.

4 NE Disambiguation: A Deductive Approach

4.1 Problem Definition

Consider a text as a sequence of tokens $T = \langle t_1, \ldots, t_n \rangle$, where n is the number of words in T. A knowledge-based semantic-annotation tool extracts a set of entities $\mathcal{E} = \{e_1, \ldots, e_m\}$ from T, where each entity e_i, $i \in \{1, \ldots, m\}$, is identified by a tuple $e_i = (label_i, offset_i, type_i, URI_i)$[20], where $label_i$ is a sequence $\langle t_j, \ldots, t_k \rangle$ of tokens in T s.t. $1 \leq j < k \leq n$ and $label_i \cap label_{i+1} = \emptyset$, $\forall i \in \{1, \ldots, m-1\}$, and $offset_i$ is the position of the first character of $label_i$ in the text. With URI_i and $type_i$ we indicate a set of resources, and corresponding types, identified in the KB as possible annotation entities for $label_i$. Entity e_i is defined *unambiguous* if it results $|URI_i| \in \{0, 1\}$ and, consequently, $|type_i| \in \{0, 1\}$; otherwise, it is called *ambiguous*. Here, we illustrate a completely novel method for resolving URI conflicts, and possibly the related type conflicts, in presence of ambiguous entities, leveraging on the DBpedia KB. The disambiguation process is grounded on the fully logic–based extraction of commonalities between pairs of RDF resources, through the computation of Common Subsumers (CS). Although adopting an already presented algorithm for such a computation [8], the proposed approach to disambiguation is completely novel.

We argue that the informative content determining the context of an entity e_i is conveyed by the entity itself and a set of neighbor entities, expressible though a sliding window $[e_{i-j}, e_{i+k}]$. Given an ambiguous entity e_i, indices $j < i$ and $k > i$ can be respectively decremented and incremented until meaningful commonalities between the RDF description of a possible URI_i and the URI

---

[17] http://spotlight.dbpedia.org/

[18] https://ner.vse.cz/thd/

[19] http://www.old.ontotext.com/lupedia

[20] Please note that semantic annotation systems often return also a confidence value for the extracted NE denoting the probability for that entity to be correctly annotated by the given type and URI.

Table 2. List of entity label, offset, type, and URI for each NE returned by the NERD framework for the text snippet shown in Fig. 1 (the extractor producing each NE is also reported in the last column).

ID	Label name	Offset	Type	URI	Extractor
1	18th century	533	nerd : Thing	dbpedia : 18th_century	DBpedia Spotlight[17]
2	Architecture	710	nerd : Thing	dbpedia : Architecture	TextRazor
3	Art	176	nerd : Thing	*wikipedia : Art*	TextRazor
4	Art	598	nerd : Thing	*wikipedia : Art*	TextRazor
5	Art	832	nerd : Thing	*wikipedia : Art*	TextRazor
6	Baroque	26	nerd : Thing	dbpedia : Baroque	Wikimeta
7	Bavarian	690	nerd : Thing	dbpedia : Bavarian	THD[18]
8	Canaletto	724	nerd : Person	dbpedia : Canaletto	Wikimeta
9	Episode	502	nerd : Thing	dbpedia : Episode	TextRazor
10	Europe	809	nerd : Location	dbpedia : Europe	DBpedia Spotlight
11	Madness	339	nerd : Organization	dbpedia : Madness_(band)	DBpedia Spotlight
12	Modern world	241	nerd : Thing	dbpedia : Modern_history	DBpedia Spotlight
13	Period	409	nerd : Thing	dbpedia : Period_piece	DBpedia Spotlight
14	Picking	264	nerd : Thing	dbpedia : Picking	THD
15	Pilgrimage	699	nerd : Thing	dbpedia : Pilgrimage	Wikimeta
16	Pleasure	326	nerd : Thing	dbpedia : Pleasure	TextRazor
17	Rococo	35	nerd : Organization	dbpedia : Rococo_(band)	Lupedia[19]
18	Rococo	169	nerd : Thing	dbpedia : Rococo	Wikimeta
19	Rococo	200	nerd : Thing	dbpedia : Rococo	Wikimeta
20	Rococo	297	nerd : Thing	dbpedia : Rococo	Wikimeta
21	Rococo	821	nerd : Thing	dbpedia : Rococo	Wikimeta
22	Venice	745	nerd : Location	dbpedia : Venice	Wikimeta
23	Waldemar	349	nerd : Person		Wikimeta
24	Waldemar Januszczak	97	nerd : Person	dbpedia : Waldemar_Januszczak	Wikimeta

of another entity in the window are not found by the CS algorithm, resolving ambiguity. Moreover, the method we propose extracts a description of the context shared by neighbor entities in terms of their CS, thus paving the way also to other types of IE tasks (*e.g*, topic extraction [13]). We also remark that in this paper we do not deal with the identification of candidate entities, but we rely on results provided by existing semantic-annotation tools.

4.2   Algorithm Description

The adoption of a deductive approach allows for investigating on the informative content hidden in the input resources in order to find out the features shared among them and to ground the disambiguation on the evaluation of such commonalities. In other words, CSs are analyzed to discover in what two RDF resources are similar and consequently to rank pairs of resources according to the significance of the informative content they share. Here we just sketch the distinguishing features of the adopted logic-based approach for computing a CS of two RDF resources. The reader interested in more details about the computation algorithm may refer to Colucci *et al.* [8].

Given an ambiguous entity e_i, for each $uri_i^k \in URI_i$, $k \in \{1, \ldots, |URI_i|\}$, the Common Subsumer (CS) extraction algorithm is run, taking as input the pair $uri_i^k, \overline{uri}_j$, where $\overline{uri}_j$ is the URI of a unambiguous selected entity in the neighborhood of e_i.

The strategy heavily relies on the representation of the resources input to the CS computation. In fact, the meaning of an RDF resource r changes depending which triples r is involved in, since different sets of triples entail (in general) different new triples. The approach by Colucci *et al.* proposes to consider a set of triples T_r for resource r, defining the pair $\langle r, T_r \rangle$ as a rooted-graph (r-graph). The criterion for computing T_r is flexible, allowing a user for setting several selection parameters, such as the exploration level d of the RDF graph rooted in r, the datasets to be used as information sources for triples selection and the properties to be included in the chosen triples.

In brief, given an exploration depth d, the algorithm performs a process similar to a depth-first exploration of the two r-graphs describing resources uri_i^k and $\overline{uri}_j$. Namely, the CS of a pair $(\langle r, T_r \rangle, \langle s, T_s \rangle)$ is $\langle r, T_r \rangle$ itself if $\langle r, T_r \rangle = \langle s, T_s \rangle$; otherwise, it is represented by a pair $\langle \_ : x, T \rangle$, with $\_ : x$ a blank node and T a set of triples entailed (according to Simple Entailment [15]) during the parallel exploration of the two r-graphs. Moreover, the traditional process of depth-first exploration is changed to also explore RDF triples describing predicates encountered in previously investigated triples. Accordingly, the algorithm recursively computes the CS by comparing all resources (both predicates and objects) at the same level in the parallel depth-first search of the two r-graphs. At the end, it results that each triple in T is entailed by a pair of triples encountered in the joint exploration of the two input r-graphs. By example, given a CS $\langle \_ : x, T \rangle$ of a pair of RDF resources $(\langle r, T_r \rangle, \langle s, T_s \rangle)$ and an investigation distance $d = 1$, triples in T are of the form $t = \_ : x \ y \ z$.

If such a triple belongs to T, then the two triples $t_1 = r\ p_1\ q_1$. (with $t_1 \in T_r$), and $t_2 = s\ p_2\ q_2$. (with $t_2 \in T_s$), both exist and both entail t. By looking at t, we notice that _ : x is the root of the CS and y and z may be either URIs or blank nodes. If y (respectively z) is a URI (or a Literal value for z), $y = p_1 = q_1$ (respectively $z = p_2 = q_2$[21]).

4.3   Ranking Common Subsumers

The CS still comes in an r-graph format which includes an RDF description: the result set T. Therefore, we need here to define metrics for identifying the best candidate URI resolving the disambiguation task. Such metrics, which make part of the main contribution of this paper, quantify the amount of information embedded in CS description, and therefore shared by the two initial RDF resources.

In most ambiguous cases, disambiguation involves choosing a *sequence* of entities—two or more, not a single one—for a sequence of labels. In some cases, entities in the "right" disambiguation share a common context—some specific description—that their CS can make explicit. In contrast, wrong choices assign entities which do not share a common context, and this fact can be detected because their CS is some very generic description. We propose to resolve these cases by extracting and comparing the CSs of each candidate sequence of entities—in this paper, we limit to pairs. To this end, we provide a function converting the informative content conveyed by the CS in a score estimating how much specific is the common context of the two resources. Such a score is useful for establishing the best candidate URI resolving the disambiguation task.

Intuitively, the triples $t = \_ : x\ y\ z$. in T which make a CS more significant are the ones in which both y and z are identifiable resources—*i.e.*, they are not blank nodes. As a consequence, the ranking function is designed in order to assign a weight equal to 1 to such triples. Triples t in which only the object is defined (*i.e.*, z is a non-anonymous resource), are ranked with a lower weight, 0.8. Finally, if a triple t has only a definite predicate (*i.e.*, y is not a blank node and the object z is an anonymous resource), the ranking function gives a weight 0.2 to t. Therefore, the triple ranking function is represented by the following relation:

$$rf(t) = rf_p(t) + rf_o(t), \qquad (1)$$

where $rf_p(t)$ and $rf_o(t)$ are, respectively, the functions ranking the predicate and the object of triple t, defined as

$$rf_p(t) = \begin{cases} 0 & \text{if } p \text{ is a blank node} \\ 0.2 & \text{otherwise} \end{cases} ; \qquad (2)$$

$$rf_o(t) = \begin{cases} 0 & \text{if } o \text{ is a blank node} \\ 0.8 & \text{otherwise} \end{cases} . \qquad (3)$$

---

[21] This relation is also valid when q_1 and q_2 are Literal nodes.

Then, the metrics expressing how much r and s have in common on the basis of the significance of their CS $\langle \_ : x, T \rangle$ is given by

$$rank(r, s) = \frac{\sum_{t \in T} rf(t)}{|T|} \tag{4}$$

It is worth noting that the minimum value assumed by $rank(r, s)$ is 0.2 and it is assigned to pair of resources sharing only some predicates of the data model. The maximum value, that is 1, is instead obtained when $r = s$ holds.

5   Proof of Concept

For illustrating the disambiguation methodology, we consider the first wrong entity of the text in Fig. 1, which is (Rococo, 35, nerd : Organization, dbpedia : Rococo_(band)).
In order to obtain a list of candidate URIs for the repairing process, the DBpedia endpoint[22] is queried for the resource dbpedia : Rococo_(disambiguation)[23]. Table 3 shows the CS triples and the value of the CS ranking function for all possible pairs of resources constituted by dbpedia : Baroque as first item and one of the disambiguating URIs listed in dbpedia : Rococo_(disambiguation) as second item (the exploration depth d is set to 1). Resource dbpedia : Rococo, obtaining the highest specificity score, is correctly selected by the disambiguation method. Please note that the second result is still pertaining to the Decorative Art macro-category, but less specific. A side advantage of the proposed disambiguation method is the correction of classification errors for entities with a disambiguation error in the NEL phase (the entity considered in the example was classified as Organization by NERD).

We evaluated also the performance of the approach in the unlucky case that also entity (Baroque, 26, nerd : Thing, dbpedia : Baroque), previously considered unambiguous, is affected by a disambiguation error in the semantic annotation process[24]. Values obtained by the ranking function for the 440 pairs of resources disambiguating entities Baroque and Rococo are reported at http://193. 204.59.20/ned/baroque_rococo.html. The highest value of specificity (0.9) is obtained by pair ⟨dbpedia : Baroque_Architecture, dbpedia : Rococo⟩, while the first result belonging to the Music category obtains a 0.74 value for ⟨dbpedia : Baroque_orchestra, dbpedia : Variations_on_a_Rococo_Theme⟩.
In this case, other neighboring entities can be considered for further defining the context and disambiguate clearly the meaning of the two resources (*e.g.*,

---

[22] http://dbpedia.org/sparql

[23] Alternatively, a query searching for all resources having object literal values of properties rdfs : label or foaf : name similar to the label of the ambiguous entity can be set up.

[24] The interested reader could verify through the Lupedia web service that this extractor annotates Baroque and Rococo respectively with the wrong entities dbpedia : Baroque_(band) and dbpedia : Rococo_(band)

Table 3. CSs enumeration for correcting the annotation of entity (Rococo, 35, nerd : Organization, dbpedia : Rococo_(band)) ordered by values of the ranking function (each evaluated resources pair is reported in bold).

rank	⟨**dbpedia : Baroque, dbpedia : Rococo**⟩	
0.84	_:cs0 dc − terms : subject	dbpedia : Category : 18th_century_in_art, dbpedia : Category : Decorative_arts, dbpedia : Category : Early_Modern_period, _ : z1;
		dbpedia : Category : 18th_century_in_art, dbpedia : Category : Decorative_arts, dbpedia : Category : Early_Modern_period;
	_ : y1	all − art − history : Baroque_Rococo.html, _ : z2;
	dbpedia − owl : wikiPageExternalLink	dbpedia : Category : 18th_century_in_art, dbpedia : Category : Decorative_arts,
	dbpedia − owl : wikiPageWikilink	dbpedia : Category : Early_Modern_period, dbpedia : Category : History_painting,
		dbpedia : Victoria_and_Albert_Museum, dbpedia : Art_history,
		dbpedia : Protestant_Reformation, dbpedia : William_Kent,
		dbpedia : Counter − Reformation, dbpedia : Jean − Philippe_Rameau,
		dbpedia : Neoclassicism, dbpedia : History_of_wood_carving,
		dbpedia : Palladian_architecture, dbpedia : Aleijadinho, _ : z3;
	foaf : isPrimaryTopicOf	_ : z4.

rank	⟨**dbpedia : Baroque, dbpedia : Rococo_Revival**⟩	
0.60	_ : cs0 dc − terms : subject	dbpedia : Category : Decorative_arts, _ : z1;
	_ : y1	dbpedia : Category : Decorative_arts;
	dbpedia − owl : wikiPageWikilink	dbpedia : Category : Decorative_arts, _ : z2;
	foaf : isPrimaryTopicOf	_ : z3.

rank	⟨**dbpedia : Baroque, dbpedia : Variations_on_a_Rococo_Theme**⟩	
0.47	_ : cs0 dc − terms : subject	_ : z1;
	dbpedia − owl : wikiPageExternalLink	_ : z2;
	dbpedia − owl : wikiPageWikilink	dbpedia : Concerto, dbpedia : Rococo, _ : z3;
	foaf : isPrimaryTopicOf	_ : z4.

rank	⟨**dbpedia : Baroque, dbpedia : Rocky_Rococo**⟩	
0.20	_ : cs0 dc − terms : subject	_ : z1;
	dbpedia − owl : wikiPageExternalLink	_ : z2;
	dbpedia − owl : wikiPageWikilink	_ : z3;
	foaf : isPrimaryTopicOf	_ : z4.

rank	⟨**dbpedia : Baroque, dbpedia : Rococo_(band)**⟩	
0.20	_ : cs0 dc − terms : subject	_ : z1;
	dbpedia − owl : wikiPageExternalLink	_ : z2;
	dbpedia − owl : wikiPageWikilink	_ : z3;
	foaf : isPrimaryTopicOf	_ : z4.

rank	⟨**dbpedia : Baroque, dbpedia : Rococo_(club)**⟩	
0.20	_ : cs0 dc − terms : subject	_ : z1;
	dbpedia − owl : wikiPageExternalLink	_ : z2;
	dbpedia − owl : wikiPageWikilink	_ : z3;
	foaf : isPrimaryTopicOf	_ : z4.

through a sequence of skos : broader predicates, it is possible to reach the two categories

dbpedia : Category : Aesthetics and dbpedia : Category : Visual_arts that are also the subject of the recognized linked entity dbpedia : Art). For what concerns the refinement of the entity linking process for dbpedia : Picking, dbpedia : Madness_(band) and dbpedia : Period_piece resources, the computed rank value is 0.2 for the CSs of all possible combination of disambiguation URIs with entities dbpedia : Rococo and dbpedia : Pleasure of the same sentence. It means that the reference KB does not contain any instance representing those entities. Moreover, the reader may agree that the considered ambiguous entities are not relevant, apart from the one labeled with Madness, whose relevance is gained by its occurrence in the title.

Finally, the last unresolved entity is (Bavarian, 690, nerd : Thing, dbpedia : Bavarian). Here, the proposed method returns the highest value, that is 0.57, for the pair ⟨dbpedia : Bavarian, _Iran, dbpedia : Europe⟩, due to the peculiarity of being both of type dbpedia-owl : PopulatedPlace. Future work will be spent on managing these cases, introducing a threshold on the CS ranking values or adopting other NE techniques to support the same task.

6   Conclusions and Future Work

This paper addresses the most challenging aspects in semantic annotation processes and presents the formal definition of a method for resolving NE disambiguation. We show that realizing disambiguation by evaluating commonalities of entities contained in a text powerfully leads to the identification of extensional-based alternative contexts, and the evaluation of the degrees of specificity of the identified contexts drives the disambiguation process. This paper is far from being exhaustive. Extensive studies on the effectiveness of the approach compared with other techniques already proposed in the literature are under analysis. In particular, performance on linking accuracy, in terms of known and unknown KB-entity correctly linked, will be evaluated. Performances after the inclusion of other LOD datasets have also to be studied. Moreover, meaningful CSs descriptions could be used for retrieving other resources conveying the same informative content of some NE in the text [9], providing a support for serendipitous encounters [19,21].

Acknowledgments. We acknowledge support of project "A Knowledge based Holistic Integrated Research Approach" (KHIRA - PON 02_00563_3446857).

References

1. Alfonseca, E., Manandhar, S.: An unsupervised method for general named entity recognition and automated concept discovery. In: Proc. of the 1st Int. Conf. on General WordNet, Mysore, India, pp. 34–43 (2002)

2. Bellot, P., Bonnefoy, L., Bouvier, V., Duvert, F., Kim, Y.M.: Large scale text mining approaches for information retrieval and extraction. In: Innovations in Intelligent Machines-4, pp. 3–45. Springer (2014)

3. Bunescu, R.C., Pasca, M.: Using encyclopedic knowledge for named entity disambiguation. In: Proc. of the 11th Conf. of the European Chapter of the Association for Computational Linguistics (EACL-06), vol. 6, pp. 9–16 (2006)

4. Cambria, E., White, B.: Jumping NLP curves: A review of natural language processing research. IEEE Computational Intelligence Magazine 9(2), 48–57 (2014)

5. Chen, L., Ortona, S., Orsi, G., Benedikt, M.: Aggregating semantic annotators. Proc. of the VLDB Endowment 6(13), 1486–1497 (2013)

6. Chieu, H.L., Ng, H.T.: Named entity recognition: a maximum entropy approach using global information. In: Proc. of the 19th Int. Conf. on Computational linguistics, vol. 1, pp. 1–7. ACL (2002)

7. Cimiano, P., Völker, J.: Towards large-scale, open-domain and ontology-based named entity classification. In: Proc. of the Int. Conf. on Recent Advances in Natural Language Processing (RANLP) (2005)

8. Colucci, S., Donini, F.M., Di Sciascio, E.: Common subsumers in RDF. In: Baldoni, M., Baroglio, C., Boella, G., Micalizio, R. (eds.) AI*IA 2013. LNCS, vol. 8249, pp. 348–359. Springer, Heidelberg (2013)

9. Colucci, S., Giannini, S., Donini, F.M., Di Sciascio, E.: A deductive approach to the identification and description of clusters in Linked Open Data. In: Proc. of the 21th European Conf. on Artificial Intelligence (ECAI 2014). IOS Press (2014)

10. Cucerzan, S.: TAC entity linking by performing full-document entity extraction and disambiguation. In: Proc. of the Text Analysis Conference, vol. 2011 (2011)

11. Dredze, M., McNamee, P., Rao, D., Gerber, A., Finin, T.: Entity disambiguation for knowledge base population. In: Proc. of the 23rd Int. Conf. on Computational Linguistics, pp. 277–285. ACL, Beijing, August 2010

12. Etzioni, O., Cafarella, M., Downey, D., Popescu, A.M., Shaked, T., Soderland, S., Weld, D.S., Yates, A.: Unsupervised named-entity extraction from the web: An experimental study. Artificial Intelligence 165(1), 91–134 (2005)

13. Fetahu, B., Dietze, S., Pereira Nunes, B., Antonio Casanova, M., Taibi, D., Nejdl, W.: What's all the data about?: creating structured profiles of linked data on the web. In: Proc. of the Companion Publication of the 23rd Int. Conf. on World Wide Web Companion, pp. 261–262. International World Wide Web Conferences Steering Committee (2014)

14. Gangemi, A.: A comparison of knowledge extraction tools for the semantic web. In: Cimiano, P., Corcho, O., Presutti, V., Hollink, L., Rudolph, S. (eds.) ESWC 2013. LNCS, vol. 7882, pp. 351–366. Springer, Heidelberg (2013)

15. Hayes, P.: RDF Semantics, W3C Recommendation (2004). http://www.w3.org/TR/2004/REC-rdf-mt-20040210/

16. Hellmann, S., Lehmann, J., Auer, S., Brümmer, M.: Integrating NLP using linked data. In: Alani, H., et al. (eds.) ISWC 2013, Part II. LNCS, vol. 8219, pp. 98–113. Springer, Heidelberg (2013)

17. Hobbs, J.R., Stickel, M., Martin, P., Edwards, D.: Interpretation as abduction. In: Proc. of the 26th Annual Meeting on Association for Computational Linguistics, pp. 95–103. ACL (1988)

18. Hoffart, J., Yosef, M.A., Bordino, Ilaria Fürstenau and H., Pinkal, M., Spaniol, M., Taneva, B., Thater, S., Weikum, G.: Robust disambiguation of named entities in text. In: Proc. of the Conf. on Empirical Methods in Natural Language Processing, pp. 782–792. ACL, Edinburgh, July 2011

19. Maccatrozzo, V.: Burst the filter bubble: using semantic web to enable serendipity. In: Cudré-Mauroux, P., et al. (eds.) ISWC 2012, Part II. LNCS, vol. 7650, pp. 391–398. Springer, Heidelberg (2012)
20. Mihalcea, R., Csomai, A.: Wikify!: linking documents to encyclopedic knowledge. In: Proc. of the 16th ACM Conf. on Information and Knowledge Management, pp. 233–242. ACM (2007)
21. Milne, D., Witten, I.H.: Learning to link with wikipedia. In: Proc. of the 17th ACM Conf. on Information and Knowledge Management, pp. 509–518. ACM (2008)
22. Moro, A., Raganato, A., Navigli, R.: Entity linking meets word sense disambiguation: a unified approach. Transactions of the Association for Computational Linguistics 2, 231–244 (2014)
23. Nakashole, N., Theobald, M., Weikum, G.: Scalable knowledge harvesting with high precision and high recall. In: Proc. of the Fourth ACM Int. Conf. on Web Search and Data Mining, pp. 227–236. ACM, Hong Kong, February 2011
24. Navigli, R., Ponzetto, S.P.: BabelNet: The automatic construction, evaluation and application of a wide-coverage multilingual semantic network. Artificial Intelligence 193, 217–250 (2012)
25. Navigli, R., Velardi, P.: Structural semantic interconnections: a knowledge-based approach to word sense disambiguation. IEEE Transactions on Pattern Analysis and Machine Intelligence 27(7), 1075–1086 (2005)
26. Niu, C., Li, W., Ding, J., Srihari, R.K.: A bootstrapping approach to named entity classification using successive learners. In: Proc. of the 41st Annual Meeting on Association for Computational Linguistics, vol. 1, pp. 335–342. ACL (2003)
27. Prokofyev, R., Demartini, G., Cudré-Mauroux, P.: Effective named entity recognition for idiosyncratic web collections. In: Proc. of the 23rd Int. Conf. on World Wide Web, pp. 397–408. International World Wide Web Conferences Steering Committee (2014)
28. Rizzo, G., Erp, M.V., Troncy, R.: Benchmarking the extraction and disambiguation of named entities on the semantic web. In: Proc. of the 9th Int. Conf. on Language Resources and Evaluation (LREC 2014). European Language Resources Association (ELRA), Reykjavik, May 2014
29. Studer, R., Burghart, C., Stojanovic, N., Thanh, T., Zacharias, V.: New dimensions in semantic knowledge management. In: Towards the Internet of Services: The THESEUS Research Program, pp. 37–50. Springer (2014)
30. Van Erp, M., Rizzo, G., Troncy, R.: Learning with the web: spotting named entities on the intersection of NERD and machine learning. In: # MSM, pp. 27–30. Citeseer (2013)
31. Zhang, L., Pan, Y., Zhang, T.: Focused named entity recognition using machine learning. In: Proc. of the 27th Annual Int. ACM SIGIR Conf. on Research and Development in Information Retrieval, pp. 281–288. ACM (2004)
32. Zheng, Z., Si, X., Li, F., Chang, E.Y., Zhu, X.: Entity disambiguation with freebase. In: Proc. of the The 2012 IEEE/WIC/ACM Int. Joint Conf. on Web Intelligence and Intelligent Agent Technology, vol. 01, pp. 82–89. IEEE Computer Society (2012)

Scheduling, Planning, and Robotics

Efficient Power-Aware Resource Constrained Scheduling and Execution for Planetary Rovers

Daniel Díaz[1], Amedeo Cesta[2], Angelo Oddi[2], Riccardo Rasconi[2(✉)],
and Maria Dolores Rodriguez-Moreno[1]

[1] Universidad de Alcalá, Alcalá de Henares, Madrid, Spain
{ddiaz,malola.rmoreno}@uah.es
[2] CNR, Italian National Research Council, ISTC, Rome, Italy
{amedeo.cesta,angelo.oddi,riccardo.rasconi}@istc.cnr.it

Abstract. This paper presents and evaluates an integrated power-aware, model-based autonomous control architecture for managing the execution of rover actions in the context of planetary mission exploration. The proposed solution is embedded within an application scenario of reference which consists on a rover-based mission concept aimed at collecting Mars samples that may be returned to Earth at a later date for further investigation. This study elaborates on the exploitation of advanced decision-making capabilities within a flexible execution process targeted at generating and safely executing scheduling solutions representing mission plans, seamlessly supporting online plan optimization and dynamic management of new incoming activities. In this work, an experimental analysis on the performance of the control architecture's capabilities is presented, throughout two representative cases of study running upon an integrated test-bed platform built on top of the 3DROV ESA planetary rover simulator.

1 Introduction

The current scenario in space exploration is characterized by the use of automation and robotics. The latest and most outstanding example is the NASA's Mars Science Laboratory (MSL) mission and its rover Curiosity aimed at assessing the planet habitability, i.e., discovering hints about whether Mars ever housed small life forms. Curiosity represents the synthesis of the NASA's vast experience on the deployment of mobile robots on the Mars' surface, gained along many successful missions such as the Mars Pathfinder probe in 1997 [13] or the twin Mars Exploration Rovers (MER) vehicles [12] – Spirit and Opportunity. In the near future, top space agencies' roadmap actually passes through continuing fostering robotics able of maximizing science data return while keeping pace with unexpected events arising during mission execution. For instance, the European Space Agency (ESA), in collaboration with other partners, is envisioning rover-based missions such as the Mars Science Return (MSR) mission concept [1]: a lightweight rover-based mission aimed at acquiring Martian materials from known locations on Mars to be later delivered back to Earth for a further analysis. Future mission baseline requirements, like in MSR, will be surely grounded on

© Springer International Publishing Switzerland 2015
M. Gavanelli et al. (Eds.): AI*IA 2015, LNAI 9336, pp. 383–396, 2015.
DOI: 10.1007/978-3-319-24309-2_29

the investigation of multiple widely-distributed science targets in a single command or working cycle (e.g. one Martian Sol). To successfully attain this end, the solution relies on promoting continuous autonomy *on-board*, since purely manual tele-operation becomes an excessively inefficient option [15], if not infeasible due to the great distances involved between Earth and celestial bodies, as well as the possibly harsh environmental conditions.

There are a few examples of real on-board autonomy applied to space exploration. The validation of the Remote Agent Experiment [16] (RAX) in the Deep Space 1 flight mission or the *Autonomous Science Experiment (ASE)* [5] onboard the Earth Observing One (EO-1) spacecraft [6] are the most representative ones. Related to on-ground automated systems we can mention the *Mixed-initiative Activity Plan GENerator (MAPGEN)* [2] used to support the MER operation or the *Mars-Express Scheduling Architecture (Mexar2)* [3], an AI-based tool currently in use on the Mars-Express mission. With the exception of the previous examples, we have to refer to field demonstrations (on Earth) of advanced autonomy such as the *Limits of Life in Atacama (LITA)* project [21], a NASA and CMU effort where a mobile robot was targeted at characterizing the presence and distribution of microbiological life on the desert of Atacama (Chile). From the European side, we can mention the LAAS autonomous control architecture [11] integrated and tested on-board of the robots *Lama* and *Dala*, two experimental rovers fully equipped with advanced sensing and processing capabilities. In this direction, our work presents a model-based control architecture targeted at generating and executing planetary rover mission plans inspired on the MSR mission concept. In our architecture we propose a constrained-based flexible model able to represent battery power constraints and that integrates plan *re-scheduling* and plan *execution*.

In this paper the rescheduling capabilities of our architecture have been tested by closing the execution loop with the 3DROV planetary rover system simulator [17], an ESA asset providing a realistic behavioral reproduction (i.e., from the temporal, dynamic and energy-related standpoint) of all the rover subsystems (e.g., locomotion, drilling, etc,) necessary to perform the tasks covered in our experimental model. The remainder of the paper is organized as follows. In Section 2 we give a general description of the problem scenario. Section 3 provides a description of the power-aware autonomous control architecture. Section 4 describes the integrated testbed platform built on the top of the ESA's 3DROV planetary rover system simulator [17]. In Section 5 we conduct an empirical analysis on the performance of the basic target capabilities of the controller through the dynamic simulation of two representative and comprehensive cases of study. Finally, some concluding remarks close the paper.

2   Mission Scenario and Problem Definition

We envisioned a martian rover domain inspired on the Mars Sample Return (MRS) concept [1], the second Flagship mission in the ESA Aurora Programme, where an autonomous rover is in charge of traversing long distances to collect soil samples which will be delivered back to Earth to be analyzed. In a conventional day-to-day mission, a set of scientific experiments are scheduled and executed by the rover in a safe and efficient way, as an ordered sequence of actions. A typical experiment basically consists of the execution of the three following activities: (a) traveling to remote locations of scientific interest previously identified from Earth; (b) acquiring Martian soil samples by using a drilling subsystem (s/s) and a sample container to transport them; and (c) delivering the collected samples to a final location where an Ascent Vehicle (AV) will retrieve them from the rover by using a robotic arm.

In a previous work [9] we provided a solution to the static version of the problem of creating such action sequences (referred to as *Power Aware Resource Constrained Mars Rover Scheduling (PARC-MSR)*) where an integrated power-aware decision-making strategy was introduced to synthesize robust plans from a set of high level mission requirements with special attention to the energy constraints. Referring the reader to [9] for a more detailed description of the implemented solution, the current paper aims at broadening the previous work by presenting an on-line experimentation showing the performances of the $CoRe^p$ mission control architecture targeted at providing a reliable and efficient mission execution management under uncertainty, based on advanced decision-making capabilities similar to those presented in [19], while supporting both continuous plan improvement and dynamic management of new incoming activities. Therefore, the problem here addressed can be explained through its decomposition into a threefold scheme that represents both its static and dynamic dimensions.

1. **Plan Synthesis.** The rover exploits advanced decision-making capabilities to both synthesize and repair baseline mission plans. Mission plans represents dispatchable schedules which cover the whole rover activities about the attainment of the mission requirements within a specific horizon, which might span one or more command cycles (i.e., one or more sols or Martian days). More concretely, the rover is in charge of executing a set of experiments $E = \{Exp_1, ..., Exp_n\}$, where each experiment consists of a sequence of activities $Exp_i = \langle Nav_{S,i}, Drill_i, Nav_{i,F}, Rel_i \rangle$. Subindexes $i \in W$ denote waypoint locations: a scientific target position i, the initial location S of the rover (the position from which the rover starts the mission), or the final sample retrieval location F (i.e., the final position where the samples are retrieved by the ascent vehicle). $Nav_{i,j}$ is the *navigation activity* which involves long-range traversals between the two different waypoints i, j to acquire or retrieve a soil sample; $Drill_i$ is the *science acquisition* activity executed at a specific waypoint i in order to collect and store a soil sample; and Rel_i is the *sample release* activity, through which the sample collected at a specific waypoint i is released at the final location F.

2. **Flexible Schedule Execution and Contingency Solving.** The mission execution process starts with the synthesis of a feasible initial solution schedule (baseline schedule). Once the baseline schedule is ready, the rover proceeds to its execution by timely dispatching the set of low-level commands associated to each rover activity. The whole schedule execution is based on

a Sense-Plan-Act (SPA) closed-loop control model [14], which continuously supervises/monitors the execution process so that as soon as a misalignment between the expected and the rover behaviour is detected, a contingency solving strategy is deployed with the aim of recovering the execution to a consistent state. Deliberation for *acting* is becoming a hot topic (see [10]) given the importance of properly and promptly adapting the plan's model despite the domain's uncertainty, which in our work means to re-schedule the plan's operations so as to accommodate the observations received by the rover and always provide consistent information to the executive.

3. **Plan Improvement.** The rover will be also in charge of deploying a continuous plan optimization process running in the background in order to search better quality solution plans (in terms of makespan length), as the current plan (i.e., the most recent feasible solution found) is being executed[1].

3   The Power-Aware Autonomous Control Architecture

In this section we propose an integrated model-based execution control architecture $CoRe^p$ which stands for (Co)ntrol (Re)sources & (p)ower. It enables the rover to: (a) deploy advanced decision-making capabilities involving constraints concerned to both *global path-planning* and pure scheduling like *sequence-dependent setup-times, power production/consumption or multi-capacity resource allocation*; and (b) keep pace with the environmental uncertainty so that possible incoming disturbances threatening the plan under execution can be suitably managed. Figure 1 illustrates the main functional building blocks of the $CoRe^p$ autonomous controller. $CoRe^p$ implements a Sense-Plan-Act (SPA) closed-loop execution scheme which continuously guarantees the consistency of the entire plan execution process by exploiting advanced reasoning capabilities at its core, monitoring and analysing the execution status and deploying reactive strategies in case of misalignments on the internal model updating, i.e., on the face of possible deviations between the expected and the real execution process evolution. In

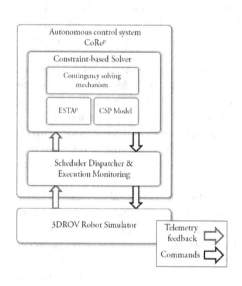

Fig. 1. $CoRe^p$ conceptual schema

---

[1] In the current version of the MRS mission domain, maintaining *solution continuity* (see [20]) during the improvement process is not considered an issue.

addition, $CoRe^p$ allows deploying both continuous (on-line) plan optimization and dynamic management of new incoming mission activities. More concretely, $CoRe^p$ is responsible for generating complete and consistent long-term mission plans covering kilometre-scale distances, and guaranteeing a safe and efficient plan execution through a flexible mission execution management process consisting on the following day-to-day readiness operations:

1. Synthesize feasible and robust baseline schedule solutions, exploiting a constraint based solver called $ESTA^p$ (see [9] for more details).
2. Timely dispatch the nominal schedule and monitoring the execution evolution (Scheduler Dispatcher & Execution Monitoring module) on the basis of the rover telemetry such as position, orientation, command execution status and battery state of charge (SoC). The execution monitor is also aware of the arrival of new experiments (i.e., strings or sequences of rover activities defining a complete scientific experiment execution) externally submitted by the scientific team, or possibly a result of an unexpectedly interesting science observation (*opportunistic science* support).
3. Update the internal mission execution model with the new information provided by the monitor, and detect possible misalignments between the planned and the real rover behaviour, in terms of timing and resource usage. This process is known as "execution monitoring & consistency checking".
4. Provide (on-line) alternative schedule solutions to face with the possible contingent situations (like command execution delays or battery overconsumption), through the execution of reactive strategies to correct possible constraint violations after they have arisen as a result of the internal model updating.
5. Continuously search for better plans (in terms of makespan length) through the execution of an iterative sampling optimization loop which uses $ESTA^p$ in a similar fashion to the Iterative Sampling Earliest Solutions (ISES) [4] strategy for makespan minimization: an efficient multi-pass approach which performs quite well in the face of scheduling problems that involve very large search spaces.

It is worth to mention that the controller reasons at a high abstract level in regards to path planning decisions, and therefore the kind of uncertainties considered must also defined at the same level. The "execution monitoring & consistency checking process" is involved with detecting and assessing temporal, resource usage and energy misalignments, but does not address local navigation issues like route deviations, as it is designed to work on top of a robotic system endowed with local navigation capabilities providing obstacle detection & avoidance, as well as smart, terrain-aware mobility features that mitigate the typical wheel slippage/sinking effects.

4   The Integrated Test-Bed Platform

In this section we provide some more information about the rover simulator integrated with the testbed platform depicted in Figure 1. The platform has the twofold aim of (i) validating the $CoRe^p$ autonomous control architecture; and (ii) performing an experimental analysis on the performance of the control architecture's capabilities throughout two representative cases of study that will be described in the next section.

The rover simulator we chose for our integration is the "Planetary Robot Design, Generic Visualization and Validation Tool" [17] (3DROV), and advanced software simulation platform created to the aim of providing ESA's Automation and Robotics (A&R) section with a supporting tool for the complete development process of planetary robot systems within *realistic mission scenarios*. One of the many interesting features of 3DROV is that it provides the necessary elements to accurately reproduce both the robotic systems and the environmental surrounding aspects involved within a planetary mission operation context. Its modular and distributed scheme design basically consists of the following integrated building blocks:

- The *Generic Controller* encapsulates all the required functionality [2] to operate with all the rover subsystems (i.e., locomotion, drill, sample cache, etc.). It relies on the ESA's SIMSAT simulation framework [1], and assumes the role of the on-board flight software by controlling all the rover operations at low level, and providing standard simulation services such as scheduling of events and time management. The Generic Controller represents the 3DROV's interface with our autonomous control system architecture via the *Functional Layer*, which will be described later.
- The *Rover and Environment models* is a compound of accurate and integrated models representing both: (i) rover physical s/ss like kinematics, power or thermal, sensors and scientific instruments; and (ii) environmental features like the ephemeris and timekeeping, terrain and atmospheric conditions.
- The 3D visualization component is the front-end of the 3DROV simulator and allows us to track (in real-time) the evolution of the simulation execution through a 3D representation of the complete mission scenario.

5   Performance Evaluation

The experimental analysis presented in this section is targeted at both validating the main concepts and capabilities of the autonomous controller $CoRe^p$, and provide a quantitative insight on the performance of the control architecture.

The study cases here considered represent two different simulations of the same problem instance: the problem entails the synthesis and execution of a schedule solution consisting of 5 scientific experiments allocated in a $50m^2$ area of easy traversability. The rover contains a sample cache with a transportation capacity of up to 3 soil samples, hence the rover is forced to reach the ascent

---

[2] 3DROV implements a generic A&R control system based on the ESA's A&R standardised development concepts and guidelines captured within the Control Development Methodology (CDM) [18] framework.

vehicle and release the first soil samples *at least once* before returning to the experiment locations and complete the plan. In the experiment simulations, the time scale was intentionally accelerated by a factor of 60 so that a full working cycle of hours is collapsed into minutes; the aim was to allow to complete a realistic mission simulation execution, which might span several martian sols, into a few hours.

Table 1. Exogenous events description (second simulation MSR_1 execution)

Set of exogenous events injected during MSR_1		
Exogenous Event Type	*Value*	*Affected activity*
Linear speed reduction	-0.5 m/sec.	Plan's 1st Traversal
Consumption rate increment	+0.01%/sec.	Plan's 1st Traversal
Linear speed reduction	-0.3 m/sec.	Plan's 3rd Traversal
Angular speed reduction	-6 deg/sec.	Plan's 2nd Rotation
Consumption rate increment	+0.02%/sec.	Plan's 2nd Rotation

The simulations were executed to the aim of testing the main controller's capabilities, in particular its adaptability under two different scenarios characterized by different levels of uncertainty. In the first simulation (MSR_0) all the information about the scientific experiments to perform is known from the beginning of the execution, and the only source of uncertainty taken into account is the one naturally provided by the simulator itself; in the second simulation (MSR_1), four experiments are scheduled from scratch to constitute the baseline schedule, while the fifth experiment is injected and scheduled *on-the-fly*. In addition, a set of artificially created exogenous events are injected during the simulation execution, in order to further disturb the rover's nominal behaviour in a controllable fashion. Figures 2 and 3 present the preliminary experimental results corresponding to the plan executions obtained in the MSR_0 and MSR_1 simulations, respectively. In the figures, the x axis represents time (simulation time), and

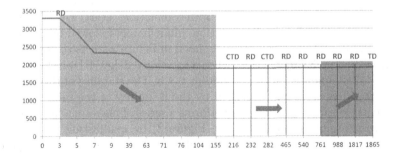

Fig. 2. Set of feasible solutions generated during the first simulation (MSR_0)

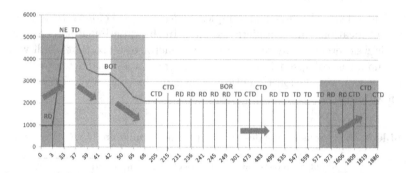

Fig. 3. Set of feasible solutions generated during the second simulation (MSR_1)

the ticks mark the instants (not in scale) at which a new plan has been produced during the execution due to the onset of a conflict making the current solution infeasible; the y axis provides a measurement of the solution makespan (in minutes). The arrows are there to represent the makespan trend (flat, decreasing or decreasing), and are particularly useful in the time intervals where the graphical representation is not precise enough to show a clear trend (especially at the end of the execution). All produced solutions are highlighted with labels characterizing the cause that originated the execution of the related re-scheduling process. The labels have the following meanings: (i) *RD (Rotation Delay)*, (ii) *CTD (Command Termination Delay)*, (iii) *TD (Traversal Delay)*, (iv) *NE (New Experiment)*, (v) *BOT (Battery Overconsumption on Traversal)*, and (vi) *BOR (Battery Overconsumption on Rotation)*. Table 1 presents the set of deviations from the nominal values (i.e., the exogenous events) injected during the second simulation; for each deviation, the table presents the events' type, magnitude and affected plan's activity. As the execution proceeds, all the detected deviations reported in Table 1 are summed up until a pre-defined threshold is reached and a re-scheduling process needs to be triggered. Plan executability is continuously maintained through such re-schedulings, each time the current solution is made infeasible.

For example, the reason behind the re-scheduling executed at the instant 988 in Figure 2 is a rotation delay. Both simulations start with the *off-line* synthesis of a baseline low quality solution, immediately followed by the start of its execution. The MSR_0 simulation depicted in Figure 2 starts the execution of a high makespan (≈ 3300) solution, when a rotational delay (RD) is detected and propagated, and a new solution is found at $t = 3$. The stability of the makespan indicates that the event was absorbed by the flexibility of the solution. At the same time, the optimization process performs a series of makespan improvements on the current solution (green area in Figure 2) by readjusting the execution order of the future experiments' tasks, ultimately allowing the makespan to decrease to the 1906 value by $t = 155$. No need to perform readjustments due to non-nominal behavior on behalf of 3DROV is detected during this time. From the

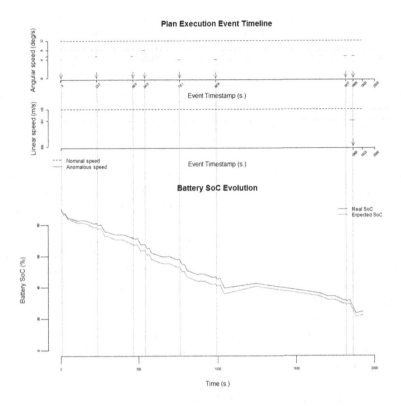

Fig. 4. First simulation telemetry analysis. Relation between: (a) the rover speeds, angular (top) and linear (middle), and the re-scheduling triggers occurring as a consequence of a speed reduction; and (b) the battery state of charge (SoC) and the re-scheduling triggers resulting from a battery overconsumption (bottom)

instant $t = 216$ onward, the telemetry data obtained from the rover determine a number of significant deviations from the rover's nominal performance, hence requiring the production of new solutions. As the figure show, five alternative solutions are synthesized between the instants $t = 216$ and $t = 540$. It should be observed that the solution's makespan remains unchanged despite the re-schedulings (the arrow is horizontal); this is due to two main reasons: (i) the temporal flexibility of the solutions, which allows to absorb temporal delays within a certain extent, in combination with (ii) the re-scheduling policy used, which attempts to maintain a certain degree of *continuity* between the various solutions, allowing a heavy re-shuffling of the activities only when no solution can be found otherwise. During the last part of the execution described in Figure 2 (red area between $t = 761$ and $t = 1865$) four more solutions are computed due to misalignments between the scheduling model's predictions and the real 3DROV's performances. Incidentally, it can be observed that these new solutions do entail an slight increase of the solution's makespan (experimental data provide a makespan increase from 1906 to 1923), due to the fact that the solution's flexibility can no longer accommodate further deviations without affecting the plan's quality. Lastly, it should be noted how, once the solution has reached a good quality level (i.e., $makespan = 1906$ at $t = 155$), the optimization does not find any further improvements for the rest of the execution.

Figure 3 describes the second execution simulation. As previously mentioned, in this case only four of the five experiments to be executed by the rover are known at creation time of the baseline solution; the fifth experiment location is provided at plan execution time. The idea of this second simulation is to demonstrate the capabilities of the reasoner to accept, schedule and optimize new activities in the current plan as the latter is being executed. In order to better highlight the reasoner's online-optimization features, and without loss of generality, it has been chosen to: (i) insert the fifth experiment at an early stage of the plan execution process, so as to maximize the part of the plan that can be readjusted (i.e., the part that has not yet been executed), and (ii) to start the simulation with a low-quality solution in terms of makespan. It should be underscored that online-injected experiments can be accepted and executed regardless of their injection time; in the worst case (and considering no deadlines), new experiment injected at a later stage are added to the end of the plan and executed in due time, possibly requiring extra recharge activities. In fact, this is well within the presented model's features and capabilities, as the reasoner is supposed to delay the execution of each activity until the battery is charged enough to guarantee proper execution. The first remarkable difference that can be appreciated w.r.t. the previous example is the slight makespan increase occurred after the rotational delay (RD) detected at $t = 3$ (from $t = 0$ to $t = 33$). As opposed with the previous case, in the current MSR_1 simulation there is a slight increase in the makespan (see the upward arrow in the picture), due to the fact that the running plan is already very tight, and therefore its inherent flexibility (i.e., the capability to absorb delays) is limited. Seconds later, the fifth experiment is added to the plan and a new solution representing the complete routing for all the five experiments to be executed is synthesized at $t = 33$. Note that this solution entails a significant increase in the makespan ($= 4960$); however, the optimization phase continuously running in the background eventually produces five new solutions in the interval between $t = 37$ and $t = 68$, reducing the makespan to 2099, despite the occurrence of two further exogenous events, i.e., a traversal delay (TD) at $t = 37$, and a battery overconsumption on traversal (BOT) at $t = 42$.

Likewise in the MSR_0 simulation, due to detected deviations from the rover's nominal performance, the solver is called to produce a number of new solutions in the interval between $t = 205$ and $t = 571$, all characterized by the same makespan (i.e., exploiting the solution's flexibility). As the execution proceeds and new re-schedulings are necessary, such flexibility is eventually exhausted, and the makespan starts to increase again (see the climbing arrow in the figure) reaching the value 2134 in the interval between $t = 571$ and $t = 1886$, instant at which the execution terminates. A further objective of our analysis was to test

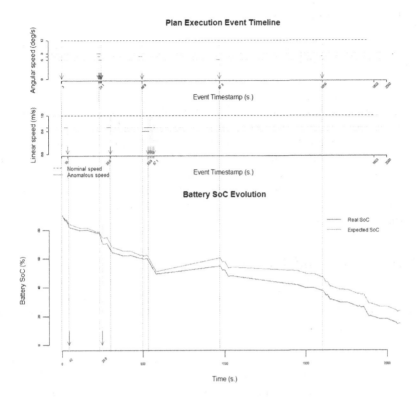

Fig. 5. Second simulation telemetry analysis. Relation between: (a) the rover speeds, angular (top) and linear (middle), and the re-scheduling triggers occurring as a consequence of a speed reduction; and (b) the battery state of charge (SoC) and the re-scheduling triggers resulting from a battery overconsumption (bottom)

the precision of our predictive model with respect to 3DROV's nominal performance. To this aim, we analyzed in both figures the SoC telemetry data captured during the whole execution, and focused on the deviations between the real "battery usage" and the estimations provided by our predictive model. Figures 4 and 5 provide a graphic representation of: (a) the connection between the rover speed evolution, both angular (top) and linear (middle), and the re-scheduling triggers occurring as a consequence of a reduction on the nominal speed; and (b) the connection between the battery usage evolution as the state of charge (SoC) and the re-scheduling triggers resulting from a battery overconsumption (bottom). In particular, Figure 4 shows the rover performance deviations in terms of linear and angular speed during the MSR_0 execution are reflected in the power consumption. It can be observed how the SoC regularly undergoes a steeper consumption trend in correspondence with the occurrence of single as well as "bursts" of exogenous events (e.g., at $t = 3, 232, 465, 540, 761, 988, 1817$ for the angular speed case, and $t = 1865$ for the linear speed case). Outside those bursts, the consumption rate seems to re-gain the ordinary steepness. Also, note the charging operation decided by the scheduler immediately after the consumption peak occurred at $t = 988$, in order to allow the feasible execution of all the remaining tasks, until plan completion.

Figure 5 conveys similar information relative to the MSR_1 simulation, where it is confirmed that the battery consumption rate significantly increases at the occurrences of exogenous events. The difference with respect to the previous case is mainly quantitative, as the execution is soon affected by a larger number of disturbances, thus requiring a re-charge at an earlier stage during plan execution (around $t = 550$). Lastly, it should be observed that in both executions the real and the expected SoC profiles follow the same trend and are characterized by an acceptable deviation (in the second simulation, the predictive model utilized in [9] suffers a slightly larger misalignment in terms of "long run" SoC projection due to the artificially injected disruptions).

6   Conclusions and Future Work

In this paper we presented the power-aware, model-based autonomous control architecture $CoRe^p$ for managing the execution of robot actions in the context of a planetary mission exploration inspired by the Mars Science Return (MSR) mission concept. The main contribution of this work is twofold: (a) the integration of advanced constraint-based reasoning capabilities with flexible contingency solving mechanisms to cope with environmental uncertainty within a Sense-Plan-Act (SPA) closed-loop execution management process; and (b) the demonstration of the basic target capabilities of the controller as well as a performance evaluation on the resolution of contingent situations, through the dynamic simulation of two representative problem examples characterized by different levels of environmental uncertainty, running upon an integrated testbed platform built on top of an existing ESA asset, namely the 3DROV planetary rover system simulator.

In [8], an example resembling a domain similar to that analyzed here is presented, where the authors focus on the executive's temporal problem, describing a labeled temporal network capable of capturing disjunctive information (i.e., pre-compiled alternative plans) with the aim of increasing the executive's reactivity in the face of exogenous events. In future work it would be very interesting to integrate the temporal model adopted in [8] in our resource-based solver, in order to assess possible improvements in execution reactivity. Lastly, our work has also some similarities with the ideas proposed in [7], where the authors present a resource-based planning model expressed as numeric resource flows. A significant difference from our approach is that in [7] the actions are not durative; in the future, it would be desirable to encourage a comparison between the two approaches, at least relatively to the deliberative phase.

Acknowledgments. Daniel Diaz is supported by the European Space Agency (ESA) under the Networking and Partnering Initiative (NPI) *Autonomy for Interplanetary Missions* (ESTEC-No. 2169/08/NI/PA). Thanks for all the support obtained through

ESA-ESTEC, in special to its ESA's technical officer Mr. Michel Van Winnendael. Last author is supported by the Junta de Comunidades de Castilla-La Mancha project PEII-2014-015-A.

References

1. Baglioni, P., Fisackerly, R., Gardini, B., Giafiglio, G., Pradier, A., Santovincenzo, A., Vago, J., Van Winnendael, M.: The Mars Exploration Plans of ESA (The ExoMars Mission and the Preparatory Activities for an International Mars Sample Return Mission). IEEE Robotics & Automation Magazine **13**(2), 83–89 (2006)
2. Bresina, J.L., Jonsson, A.K., Morris, P.H., Rajan, K.: Activity planning for the mars exploration rovers. In: ICAPS, Monterey, California, USA, pp. 40–49. AAAI (2005)
3. Cesta, A., Cortellessa, G., Denis, M., Donati, A., Fratini, S., Oddi, A., Policella, N., Rabenau, E., Schulster, J.: Mexar2: AI Solves Mission Planner Problems. IEEE Intelligent Systems **22**(4), 12–19 (2007)
4. Cesta, A., Oddi, A., Smith, S.F.: A Constraint-Based Method for Project Scheduling with Time Windows. J. Heuristics **8**(1), 109–136 (2002)
5. Chien, S., Sherwood, R., Tran, D., Cichy, B., Rabideau, G., Castaño, R., Davies, A., Mandl, D., Frye, S., Trout, B., D'Agostino, J., Shulman, S., Boyer, D., Hayden, S., Sweet, A., Christa, S.: Lessons learned from autonomous sciencecraft experiment. In: Fourth International Joint Conference on AAMAS, pp. 11–18. ACM, New York (2005)
6. Chien, S., Sherwood, R., Tran, D., Cichy, B., Rabideau, G., Castano, R., Davies, A., Lee, R., M, D., Frye, S., Trout, B., Hengemihle, J., Shulman, S., Ungar, S., Brakke, T.: The EO-1 autonomous science agent. In: AAMAS, New York City, NY, USA (2004)
7. Coles, A., Coles, A., Fox, M., Long, D.: A Hybrid LP-RPG Heuristic for Modelling Numeric Resource Flows in Planning. J. Artif. Int. Res. **46**(1), 343–412 (2013). http://dl.acm.org/citation.cfm?id=2512538.2512548
8. Conrad, P.R., Williams, B.: Drake: An Efficient Executive for Temporal Plans with Choice (January 2014). ArXiv e-prints
9. Diaz, D., Moreno, M.D., Cesta, A., Oddi, A., Rasconi, R.: Efficient Energy Management for Autonomous Control in Rover Missions. IEEE Computational Intelligence Magazine, Special Issue on Computational Intelligence for Space Systems and Operations **8**, 12–24 (2013)
10. Ghallab, M., Nau, D., Traverso, P.: The actor's view of automated planning and acting: A position paper. Artificial Intelligence **208**, 1–17 (2014). http://www.sciencedirect.com/science/article/pii/S0004370213001173
11. Ingrand, F., Lacroix, S., Lemai-Chenevier, S., Py, F.: Decisional autonomy of planetary rovers. J. Field Robotics **24**(7), 559–580 (2007)
12. Maimone, M.W., Leger, P.C., Biesiadecki, J.J.: Overview of the mars exploration rovers autonomous mobility and vision capabilities. In: IEEE International Conference on Robotics and Automation (ICRA), Roma, Italy (2007)
13. Mishkin, A., Morrison, J., Nguyen, T., Stone, H., Cooper, B., Wilcox, B.: Experiences with operations and autonomy of the mars pathfinder microrover. In: IEEE Aerospace Conference, Snowmass, CO, USA (1998)
14. Murphy, R.R.: Introduction to AI Robotics. MIT Press (2000)

15. Muscettola, N., Nayak, P., Pell, B., Williams, B.: Remote Agents: To Boldly Go Where No AI Systems Has Gone Before. Artificial Intelligence 103(1–2), 5–48 (1998)

16. Nayak, P.P., Kurien, J., Dorais, G., Millar, W., Rajan, K., Kanefsky, R., Bernard, E.D., Gamble Jr., E.B., Rouquette, N., Smith, D.B., Tung, Y.W., Muscoletta, N., Taylor, W.: Validating the DS1 remote agent experiment. In: International Conference on Artificial Intelligence, Beijing, China, vol. 440 (1999)

17. Poulakis, P., Joudrier, L., Wailliez, S., Kapellos, K.: 3DROV: a planetary rover system design, simulation and verification tool. In: 10th International Symposium on Artificial Intelligence, Robotics and Automation in Space, Hollywood, USA (2008)

18. Putz, P., Elfving, A.: Control Techniques 2, Automation and Robotics Control Development Methodology Definition Report. Tech. Rep. ESA CT2/CDR/DO (1992)

19. Rasconi, R., Cesta, A., Policella, N.: Validating scheduling approaches against executional uncertainty. Journal of Intelligent Manufacturing 21(1), 49–64 (2010). http://dx.doi.org/10.1007/s10845-008-0172-7

20. Smith, S.F.: Is scheduling a solved problem? In: Kendall, G., Burke, E., Petrovic, S. (eds.) Proceedings of the 1st Multidisciplinary International Conference on Scheduling: Theory and Applications (MISTA 2003), Nottingham, UK, pp. 11–20, August 13–16, 2003

21. Wettergreen, D., Cabrol, N., Teza, J., Tompkins, P., Urmson, C., Verma, V., Wagner, M., Whittaker, W.: First experiments in the robotic investigation of life in the Atacama Desert of Chile. In: International Conerence on Robotics and Automation, Barcelona, Spain, pp. 9–12890 (2005)

Graph-Based Task Libraries for Robots: Generalization and Autocompletion

Steven D. Klee[1], Guglielmo Gemignani[2]([✉]), Daniele Nardi[2], and Manuela Veloso[1]

[1] Computer Science Department, Carnegie Mellon University, 5000 Forbes Ave., Pittsburgh, PA 15213, USA
{sdklee,veloso}@cmu.edu
[2] Department of Computer, Control, and Management Engineering "Antonio Ruberti", Sapienza University of Rome, Rome, Italy
{gemignani,nardi}@dis.uniroma1.it

Abstract. In this paper, we consider an autonomous robot that persists over time performing tasks and the problem of providing one *additional* task to the robot's task library. We present an approach to *generalize tasks*, represented as parameterized graphs with sequences, conditionals, and looping constructs of sensing and actuation primitives. Our approach performs graph-structure task generalization, while maintaining task executability and parameter value distributions. We present an algorithm that, given the initial steps of a new task, proposes an autocompletion based on a recognized past similar task. Our generalization and autocompletion contributions are effective on different real robots. We show concrete examples of the robot primitives and task graphs, as well as results, with Baxter. In experiments with multiple tasks, we show a significant reduction in the number of new task steps to be provided.

1 Introduction

Different mechanisms enable robots to perform *tasks*, including directly programming the task steps; providing actions, state, goals or rewards, and a planning algorithm to generate tasks or policies; and instructing the tasks themselves in some representation. In this work, we consider tasks that are explicitly provided and represented in a graph-based task representation in terms of the robot's sensing and actuation capabilities as *parameterized primitives*.

We address the problem of efficiently giving an *additional* task to a robot that has a task library of previously acquired tasks. We note that just by looking at a robot, one does not know the tasks the robot has already acquired and can perform. The new task may be a repetition, or a different instantiation, or be composed of parts of other known tasks. Our goal is to enable the robot to recognize when the new task is similar to a past task in its library, given some initial steps of the new task. Furthermore, the robot recommends *autocompletions* of

---

S.D. Klee and G. Gemignani—The first two authors have contributed equally to this paper.

© Springer International Publishing Switzerland 2015
M. Gavanelli et al. (Eds.): AI*IA 2015, LNAI 9336, pp. 397–409, 2015.
DOI: 10.1007/978-3-319-24309-2_30

the remaining parts of the recognized task. The robot recommendations include proposed parameter instantiations based on the distribution of the previously seen instantiated tasks.

The contributions of the work consist of the introduction of the graph-based *generalized task* representation; the approach to determine task similarity in order to generate a library of generalized tasks; and the algorithm for recognition of the initial steps of an incrementally provided task, and for proposing a task completion. The task and the generalized tasks are represented as a graph-based structure of robot action primitives, conditionals, and loops.

For generalizing graph-based tasks, and since the general problem of finding labeled subgraph isomorphisms is NP-hard, our approach includes performing subtree mining in a unique spanning tree representation for each graph. The tree patterns, which are executable by the robot and frequent, are added to the library. The saved patterns are generalized over their parameters, while keeping the distribution of values of the parameters of the corresponding instantiated task graphs. The library is then composed of the repeated parts of the given graph-based generalized tasks.

When a new task is given to the robot, the robot uses the library of generalized tasks to propose the next steps with parameters sampled from its distribution, performing *task autocompletion*. The robot performs the proposed completions, which are accepted as a match to the new task or rejected. The process resembles the autocompletion provided by search engines, upon entering initial items of a search. The task autocompletion in our robots can incrementally propose a different task completion as the new task is defined. The result is that not all the steps of the new task need to be provided, and the effort of giving a new task to the robot is reduced.

Our complete generalization and autocompletion contributions are effective in different real robots, including a 2-arm manipulator (Baxter), and a mobile wheeled robot. In the paper, we focus on the illustrations of the robot primitives and task graphs, and results with Baxter. Our approach is general to task-based agents and aims at enabling the desired long-term deployment of robots that accumulate experience.

This introduction section is followed by an overview of related work, sampling the extensive past work on task learning from experience, highlighting our approach for its graph-based generalized representation, and for its relevance and application to autonomous agents. We then introduce the technical components of our approach, present results, and conclude the paper with a review of the contributions and hints of future work.

2   Related Work

Accumulating and reusing tasks has been widely studied in multiple contexts, including variations of case-based planning [4,9,11,22], macro learning [7], and chunk learning at different levels of granularity [1,5,13,14,16]. Tasks, seen as plans, are solved and generated by algorithms that benefit from accumulated experience to significantly reduce the solving effort. Such methods learn based on explanations captured from the dependencies in actions, producing sequential representations. Our approach does not depend on an automated problem solver, as users could be providing the tasks. We represents tasks as graph-based structures with sequences, conditionals, and looping constructs. Generalization and autocompletion use task structures instead of domain-based dependencies.

Multiple techniques address the problem of a robot learning from demonstration [3]. Approaches focus on teaching a single tasks, using varying representations. Tasks have been represented as acyclic graphs composed of nodes representing finite state machines [21] without loops or variables. More recently, complex tasks are represented as Instruction Graphs [15], which only handle instantiated tasks, or Petri Net Plans [8], which support parameterized tasks defined by a user. We introduce Generalized Instruction Graphs, parameterized tasks that are automatically generated from instantiated tasks. Tasks have also been represented as policies that determine state-action mappings [2,6,17,20].

Tasks have been generalized from multiple examples, where each example corresponds to exactly one past task, and the user specifies the generalized class [18]. Our algorithm generalizes and finds parts of tasks patterns that share any graph structure beyond dependency structure [11,22].

To represent conditionals and loops, many task representations are graph-based. Therefore, generalization from examples requires finding common subgraphs between different tasks. The problem of finding labeled subgraph isomorphisms is NP-Hard [12]. However, the problem becomes tractable on trees. To solve it, multiple algorithms have been proposed for mining common frequent subtrees from a set of trees [10]. One, Treeminer, uses equivalence class based extensions to effectively discover frequently embedded subtrees [24]. Instead, GASTON divides the frequent subgraph mining process into path mining, then subtree mining, and finally subgraph mining [19]. We use SLEUTH, an open-source frequent subtree miner, able to efficiently mine frequent, unordered or ordered, embedded or induced subtrees in a library of labeled trees [23]. SLEUTH uses scope-lists to compute the support of the subtrees, while adopting a class-based extension mechanism for candidate generation. Our mined tree patterns are further filtered out to capture executable and parameterized task graphs.

3   Approach

We consider a robot with modular primitives that represent its action and sensing capabilities. We assume the robot has a library of common tasks, where each task is composed of these primitives. Our goal is to identify common frequent subtasks and generalize over them with limited user assistance. In this section, we first give a brief overview of the graph-based task representation used in this work. Then, we present an in-depth description of the generalization and task autocompletion algorithms.

3.1   Instruction Graphs

The task representation we use is based on Instruction Graphs (IG) [15]. In the IG representation, vertices contain robot-primitives, and edges represent possible transitions between vertices. Mathematically, an Instruction Graph is a graph $G = \langle V, E \rangle$ where each vertex v is a tuple:

$$v = \langle id, InstructionType, Action \rangle$$

where the *Action* is itself a tuple:

$$Action = \langle f, P \rangle$$

where f is a function, with parameters P. The function f represents the action and sensing primitives that the agent can perform. We introduce parameter *types* related to their purpose on the robot. For instance, on a manipulator, the function *set_arm_angle* has parameters of type *Arm* and *Angle*, with valid values of $\{left, right\}$ and $[0, 2\pi]$ respectively. The primitives and parameter types are robot-specific.

Each Instruction Graph is executed starting from an initial vertex, until a termination condition is reached. During execution, the *InstructionType* of the vertex describes how the robot should interpret the output of the function f in order to transition to the next vertex. The IG framework defines the following types:

- *Do and DoUntil*: Used for open-loop and closed-loop actuation primitives. The output of f is ignored as there is only one out-edge. For simplificty, we refer to both of these types as Actions.
- *Conditionals*: Used for sensing actions. The output of f is interpreted as a boolean value used to transition to one of two children.
- *Loops*: Used for looping structures. The output of f is interpreted as a boolean value, and actions inside of the loop are repeated while the condition is true.

Additionally, we use a *Reference InstructionType* for specifying hierarchical tasks.

- *References*: Used to execute Instruction Graphs inside of others. The output of f is interpreted as a reference to the other task.

Figure 1 shows an example node with id 4, *InstructionType* Action, function *set_arm_angle*, and parameters *left* and π. For a more detailed overview of Instruction Graphs, we refer the reader to [15]. However, the generalization algorithms we discuss can be applied to other graph-like representations [8,21].

3.2   Generalizing Tasks

In this section we describe our algorithm to extract generalized tasks from a library of Instruction Graphs. We define a general task as a *Generalized Instruction Graph* (GIG). In a GIG, the parameters of some actions are ungrounded.

```
4                      Type:Action

f:set_arm_angle("left", π)
```

Fig. 1. Example of Instruction Graph node with id 4, *Instruction Type* Action, function *set_arm_angle*, and parameters *left* and π.

In such cases, we know the type of these ungrounded parameters, but not their value. So, for each parameter we associate a distribution over all known valid groundings. For instance, in the case of a grounded parameter, the distribution always returns the grounded value. Formally, a GIG is also a graph $GIG = \langle V, E \rangle$ where each vertex v is a tuple:

$$v = \langle id, Instruction Type, GeneralAction \rangle$$

$$GeneralAction = \langle f, P, \Phi \rangle$$

where $\phi_i \in \Phi$ is a distribution over groundings of the parameter $p_i \in P$.

These distributions are learned during task generalization and are used to propose initial parameters during task autocompletion. A GIG can be instantiated as an IG by grounding all of the uninstantiated parameters. This process consists of replacing any unspecified p_i with an actual value.

Our approach generates a library of GIGs from a library of IGs, as shown in Algorithm 1. The general problem of finding labeled subgraph isomorphisms is NP-Hard. However our problem can be reformulated into the problem of finding common labeled subtrees in a forest of trees. To this end, we create a tree representation of each IG. As the first step, we define a mapping from IGs to Trees (T):

$$toTree : IG \to T$$

and its corresponding inverse:

$$toIG : T \to IG$$

The function *to Tree* computes a labeled spanning tree of an input Instruction Graph (line 3). Specifically, *to Tree* creates a spanning tree rooted at the initial vertex of the input IG, by performing a depth first search and by removing back edges in a deterministic manner. This ensures that instances of the same GIG map to the same spanning tree.

Each node in the tree is labeled with the *Instruction Type* and function f of the corresponding node in the IG. In this label, we do not include the parameters because we eventually want to generalize over them.

Next, we use a labeled frequent tree mining algorithm to find frequently occurring tree patterns (line 5). A frequently occurring tree pattern is a subtree that appears more than a threshold σ, called the *support*. A tree-mining algorithm *ftm* takes as input a set of trees and the support. As output, it provides a mapping from each tree pattern to the subset of trees that contain it. Then, since each tree pattern is associated to a set of trees and each tree corresponds

---

Algorithm 1. Task Generalization

1: **procedure** GENERALIZETASKS(IGs, σ, L)
2:    // IG library is converted to trees
3:      $IGTrees \leftarrow \{toTree(g) \mid g \in IGs\}$
4:    // Tree patterns are found by a tree mining algorithm
5:      $tp \leftarrow$ ftm($IGTrees, \sigma$)
6:    // Mapping from tree patterns to IGs is created
7:      $igp \leftarrow \{\langle p, toIG(T)\rangle \mid \langle p, T\rangle \in tp\}$
8:    // Filters remove unwanted tree patterns
9:      $igp \leftarrow$ filter_not_exec(igp)
10:      $igp \leftarrow$ filter_by_length(igp, L)
11:    // Tree patterns of full tasks are reintroduced
12:      $igp \leftarrow$ add_full_igs(IGs, igp)
13:    // Vertices and edges of the GIGs are constructed
14:      $gigs \leftarrow$ create_ugigs(IGs, igp)
15:    // Parameters and distributions are computed
16:      $gigs \leftarrow$ parametrize($IGs, igp, gigs$)
17:      **return** $gigs$
18: **end procedure**

---

to a specific IG, we can create a mapping directly from tree patterns to IGs (line 7). We denote this mapping as *IGP*.

A tree mining algorithm will return many tree patterns. In particular, for any tree pattern, any subtree of it will be returned. This is because each subtree will have a support at least as large as its parent. Rather than keeping all these patterns, we focus on storing those that are the most applicable. There are many possible ways to filter the patterns. We propose several heuristic filters that select patterns based on their *executability, frequency,* and *usefulness*.

- **Executable** patterns are those that the robot can run.
- **Frequent** patterns are statistically likely to appear in the future.
- **Useful** patterns reduce many interactions when correctly proposed.

Each filter is formally defined as a function:

$$filter : IGP \rightarrow IGP$$

We first filter patterns that cannot be executed by the robot. In particular, we remove patterns with incomplete conditionals and loops (line 9).

Then, there is a tradeoff between highly frequent patterns and highly useful patterns. Patterns that occur with a large frequency are typically smaller, so they provide less utility during task autocompletion. Larger patterns provide a lot of utility, but they are usually very specific and occur rarely.

Less frequent patterns are already filtered out by the tree-mining algorithm when we provide it a minimum support σ. To optimize for larger patterns that save more steps during autocompletion, we also remove patterns that are shorter than a threshold length L (line 10). This ensures that we do not keep any pattern that is too small to justify a recommendation to the user.

Finally, since we are dealing with autocompletion for robots, one desirable feature is to be able to propose entire tasks. Even if a full task has a low support, or is below the threshold length, there is value in being able to propose it if a reparameterized copy is being provided to the robot. Consequently, we reintroduce the tree patterns corresponding to full IGs (line 12).

Even with these filters, we still keep some tree patterns that are complete subtrees of another pattern. In practice, many of these patterns provide useful task autocompletion suggestions. However, in memory-limited systems, we suggest also filtering them.

Finally, the algorithm processes the filtered set of tree patterns to create GIGs by creating vertices and edges from the tree pattern and then parameterizing the vertices. First, to create the GIG's vertices and edges, we copy the subgraph corresponding to the tree pattern from any of the IGs containing the pattern (line 14). This gives us a completely unparameterized GIG ($uGIG$), with no parameter distributions. Next, we determine which parameters are grounded in the GIG, and which are left ungrounded. A parameter is instantiated if it occurs with the same value, with a frequency above a given threshold, in all corresponding IGs. Otherwise, the parameter is left ungrounded with an empirical distribution.

This process is repeated for every subtree pattern not removed by our heuristic filters, creating a library of GIGs (line 16). When this algorithm is run incrementally, this library is unioned with the previous library.

Figures 2a and 2c show example IGs for a task that picks up an object, and drops it at one of two locations. Figures 2b and 2d depict their corresponding spanning trees. Finally, Figure 3 shows the general task that is extracted. In this GIG, the parameters in nodes 3 and 4 are kept instantiated, since they were shared by the two original IGs. The others parameters instead are left ungrounded. These parameters have a type id, and a distribution over the landmark ids $\{1,3\}$, which were extracted from the IGs in Figure 2.

3.3   Task Autocompletion

We now consider an agent that is provided a task incrementally through a series of interactions. Each interaction consists of adding a vertex to the graph or modifying an existing vertex. At any step of this process, the agent knows a *partial task*. After each interaction, this partial task is compared against the library of GIGs to measure task similarity and perform autocompletion.

The algorithm performs this comparison by checking if the end of the partial task being provided is similar to a GIG (Algorithm 2). Specifically, we keep a set of candidate proposals, denoted *props*, that match the final part of the partial task. When the partial task changes (line 2), we first update this set to remove any elements that no longer match the task being taught (line 3). Then, we add new elements for every GIG that starts with the new vertex (line 4). When a threshold percentage τ of one or more GIGs in this set matches

404   S.D. Klee et al.

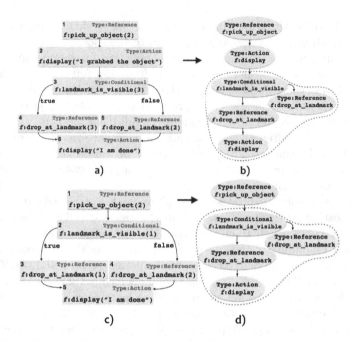

Fig. 2. Example of two Instruction Graphs (a, c) converted into their corresponding spanning trees (b, d). The tree pattern shared between them is circled in red.

the current partial task, the robot proposes the largest GIG and breaks ties randomly (lines 6 and 7).

When a specific proposal is found, the robot displays a representation of the GIG and asks for permission to demonstrate the task. Having previously filtered all the incomplete GIGs, all the proposals can in fact be executed. When granted permission, the agent demonstrates an instance of the GIG, noting when a parameter is ungrounded.

At the end of the demonstration, the agent asks if the partial task should be autocompleted with the demonstrated task. If so, the agent asks for specific values for all of the ungrounded parameters. At this stage, the agent suggests initial values for each ungrounded parameter p_i by sampling from its corresponding distribution ϕ_i.

After all of the parameters are specified, the nodes matching the general task in the partial task are replaced with one *Reference* node. When visited, this node executes the referenced GIG, instantiated with the provided parameters. With this substitution, the length of the task is reduced.

Figure 4 shows a sample task acquisition for a Baxter manipulator interacting with a user. After the first command, the robot finds at least one general task starting with *display_message*. However, none of the GIGs recognized surpass the similarity-threshold τ. When the third instruction is given to the agent, this threshold is surpassed, and the autocompletion procedure is started. First, the

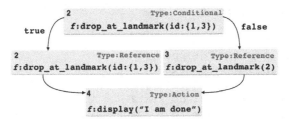

Fig. 3. Example GIG that is extracted from the graphs in Figures 2a and 2c. The parameters in nodes 3 and 4 are instantiated, since they were shared by the two original IGs. Instead, the parameters in nodes 1 and 2 are left ungrounded.

Algorithm 2. Task Autocompletion

1: **procedure** AUTOCOMPLETE($GIGs, ig, props, \tau$)
2:    **if** $hasChanged(ig)$ **then**
3:       $props \leftarrow$ deleteNotMatching($ig, props$)
4:       $props \leftarrow$ addNewMatching($ig, props, GIGs$)
5:       ($best, similarity$) $\leftarrow$ bestMatch($ig, props$)
6:       **if** $similarity \geq \tau$ **then**
7:          propose($best$)
8:       **end if**
9:    **end if**
10: **end procedure**

---

robot asks permission to perform a demonstration of the general task. After completing the demonstration, the robot asks if the autocompletion is correct. If so, it also asks for ungrounded parameters to be specified and suggests values using each parameter's distribution.

4   Experiments

Several of our robots can perform generalization and autocompletion, including a manipulator and a mobile base. In order to demonstrate the value of our approach, we define two sets of tasks. Intuitively, the first set of tasks represents a robot that is still acquiring completely new capabilities. Instead, the second set of tasks represents a robot that is acquiring instances of tasks that it already knows. More formally, in the first set, S_d no tasks are repeated. They share only a small fraction of similar components that can be generalized. To show that generalization takes place, we use a second set, S_r consisting of two repetitions of each of elements in S_d with different parameters. We see that the algorithm recognizes and autocompletes the second instance of each task.

An additional benefit of this approach is that we we can keep one common library for all of our robots. If the robots have different primitives, their tasks are automatically generalized apart. However, if this share primitives, our approach can learn subgraphs common to both robots. One concern is that this library

Example Interaction

```
U: Open Gripper
R: I will open my gripper.
R: What should I do next?
U: Display message "Hello".
R: Ok, what should I do next?
U: Set your left arm to 80 degrees.
R: I think you are teaching me
   something similar to: GIG_14.
R: Can I demonstrate it to you?
U: Yes.
R: First I will display the message "Hello".
R: Then I will set my left arm to 80 degrees.
R: Now I will set my right arm to 90 degrees (open).
R: This is my full suggestion.
R: Would you like to use it?
U: Yes.
[User specifies open parameters]
[User can rename the GIG]
```

Fig. 4. Sample autocompletion interaction during task acquisition.

could be very large. We have accumulated libraries of up to 10,000 tasks, by creating parametric variations of a smaller core of tasks. Even on libraries this large our approach runs in under 4 seconds.

In the rest of this section, we show in detail experiments run on a Baxter manipulator robot.

4.1   Experiments with Baxter

Baxter has two 7 degree of freedom arms, cameras on both arms, and a mounted Microsoft Kinect. The robot-primitives on Baxter manipulate the arms, open their grippers, display messages, and sense landmarks. The frequent subtree mining algorithm we employ is an open source version of SLEUTH [1].

The tasks that Baxter can perform range from waving to making semaphore signs to pointing at landmarks. Many of Baxter's tasks involve picking up an object and moving it to another location. For instance, Figure 5 shows Baxter searching for a landmark to see if a location is unobstructed to drop a block. Without task generalization and autocompletion, a new task must be provided for each starting location and ending location in Baxter's workspace. With generalization and autocompletion, these locations become ungrounded parameters that can be instantiated with any value.

For this experiment, 15 tasks were taught by two users familiar with robots but not the teaching framework. These tasks ranged from waving in a direction, to pointing to visual landmarks, to placing blocks at different positions, to per-

---

[1] www.cs.rpi.edu/~zaki/software/

Fig. 5. Baxter performing an instance of the GIG shown in Figure 3. First, Baxter picks up the orange block (a); then Baxter checks if a location is unobstructed with its left arm camera (b); Since the location is unoccupied, Baxter drops the Block. (c); Finally, Baxter says that it is done (d).

forming a series of semaphore signals to deliver a message. S_d has 15 distinct tasks and S_r has 30 tasks. The average length of a task in both sets is 9.33 nodes.

4.2 Experimental Results

As we accumulate the library incrementally, the order in which tasks are provided affects the generalization. To account for this, we ran 1000 trials where we picked a random ordering to our sets and had a program incrementally provide them to Baxter. The GIG library was updated after each task was provided. At the end of every trial, we measured the number of steps saved using autocompletion compared to providing every step of the task. This measurement includes steps added due to incorrect autocompletion suggestions. For this particular experiment, the support was fixed to 2, the minimum GIG length to 2, and task autocompletes were suggested if $\tau = 30\%$ of a GIG matched the partial task.

We compare our results to an near-upper bound on the number of steps saved by an optimal autocompletion algorithm. For any set of tasks that take n steps to be provided, and are only proposed once τ percent of a task matches a GIG, we have:

$$OPT \leq (1 - \tau) \cdot n$$

This corresponds to perfectly generalizing every task after τ percent of it has been acquired. We note that an algorithm could do slightly better by also making smaller proposals for the first τ percent of each task. For S_d and S_r we have:

$$OPT_d \leq 92$$

$$OPT_r \leq 184$$

Table 1 reports the result of the experiment. Specifically, in this table we report the following measures:

- **Maximum Steps Saved** (%): maximum percentage of steps saved over all permutations, in comparison to the theoretical upper bound.
- **Average Steps Saved** (%): average percentage of steps saved over 1000 permutations, in comparison to the theoretical upper bound.

- **Average Partially Autocompleted:** average number of tasks that were *partially* autocompleted with a GIG.
- **Average Completely Autocompleted:** average number of tasks that were *completely* autocompleted with a GIG.

Table 1. Results obtained for the two sets taught to the Baxter robot.

	1st Set (S_d)	2nd Set (S_r)
Max. Steps Saved	70.65%	100%
Avg. Steps Saved	33.44 ± 14%	81.92 ± 7.05%
Part. Autocompleted	4.72 ± 1.56	4.70 ± 1.58
Compl. Autocompleted	0 ± 0	15 ± 0

As expected, S_r benefits the most from the autocompletion method, saving 82% of the steps compared to OPT_r. For S_d, the savings compared to OPT_d is 33%. In the former case, the robot can leverage the knowledge of the similar tasks it already knows. Indeed, our approach meets the theoretical upper bound when provided tasks from S_r in the optimal ordering. This fact is additionally underlined by the number of tasks in which the robot suggested any correct GIG. In particular, on average the robot proposed a correct autocompletion suggestions for 65% of graphs in the second set, and 30% in the first set.

Also, the 15 IGs added to the second set are all completely autocompleted from their other similar instance. Furthermore, this happens with a statistically insignificant change to the effectiveness of the partial autocompletions.

Finally, the size of the GIG library for the first set was 21 and that the size of the GIG library for the second set was 45. This shows that the heuristic filters we proposed achieves a good balance between saving steps and library size.

5   Conclusion and Future Work

In this paper, we considered autonomous robots that persist over time and the problem of providing them additional tasks incrementally. To this end, we contributed an approach to generalize graph-based tasks, and an algorithm that enables the autocompletion of partially specified tasks.

Our generalization and autocompletion algorithms have been successfully deployed on multiple robots, acquiring large task libraries. Our experiments report in-detail the effectiveness of our contributions on a Baxter manipulator robot for two sets of tasks. With both sets, we found a significant reduction in the number of steps needed for Baxter to acquire the tasks.

In terms of future work, there may be other applicable filters for deciding which tree patterns should be converted to GIGs. Furthermore, structure-based generalization is just one way for a robot to express its capabilities. Future research may look at domain-specific forms of task generalization.

Acknowledgments. This research was partially supported by a research donation from Google, by NSF award number NSF IIS-1012733 and by the FCT INSIDE ERI grant. The views and conclusions contained in this document are those of the authors only.

References

1. Anderson, J.R., Bothell, D., Lebiere, C., Matessa, M.: An integrated theory of list memory. Journal of Memory and Language (1998)
2. Argall, B.D., Browning, B., Veloso, M.: Learning robot motion control with demonstration and advice-operators. In: IROS (2008)
3. Argall, B.D., Chernova, S., Veloso, M., Browning, B.: A survey of robot learning from demonstration. Robotics and autonomous systems (2009)
4. Borrajo, D., Roubíčková, A., Serina, I.: Progress in case-based planning. ACM Computing Surveys (CSUR) 47(2), 35 (2015)
5. Borrajo, D., Veloso, M.: Lazy incremental learning of control knowledge for efficiently obtaining quality plans. In: Lazy Learning (1997)
6. Chernova, S., Veloso, M.: Learning equivalent action choices from demonstration. In: IROS (2008)
7. Fikes, R.E., Hart, P.E., Nilsson, N.J.: Learning and executing generalized robot plans. Artificial intelligence (1972)
8. Gemignani, G., Bastianelli, E., Nardi, D.: Teaching robots parametrized executable plans through spoken interaction. In: Proc. of AAMAS (2015)
9. Hammond, K.J.: Chef: A model of case-based planning. In: AAAI (1986)
10. Jiang, C., Coenen, F., Zito, M.: A survey of frequent subgraph mining algorithms. The Knowledge Engineering Review (2013)
11. Kambhampati, S.: A theory of plan modification. In: AAAI (1990)
12. Kimelfeld, B., Kolaitis, P.G.: The complexity of mining maximal frequent subgraphs. In: Proc. of the 32nd Symp. on Principles of Database Systems (2013)
13. Laird, J.E., Rosenbloom, P.S., Newell, A.: Chunking in soar: The anatomy of a general learning mechanism. Machine learning (1986)
14. Langley, P., McKusick, K.B., Allen, J.A., Iba, W.F., Thompson, K.: A design for the icarus architecture. ACM SIGART Bulletin (1991)
15. Meriçli, Ç., Klee, S.D., Paparian, J., Veloso, M.: An interactive approach for situated task specification through verbal instructions. In: Proc. of AAMAS (2014)
16. Minton, S.: Quantitative results concerning the utility of explanation-based learning. Artificial Intelligence (1990)
17. Ng, A.Y., Russell, S.J., et al.: Algorithms for inverse reinforcement learning. In: ICML (2000)
18. Nicolescu, M.N., Matarić, M.J.: Natural methods for robot task learning: Instructive demonstrations, generalization and practice. In: Proc. of AAMAS (2003)
19. Nijssen, S., Kok, J.N.: The gaston tool for frequent subgraph mining. Electronic Notes in Theoretical Computer Science (2005)
20. Ratliff, N.D., Bagnell, J.A., Zinkevich, M.A.: Maximum margin planning. In: Proc. of the 23rd International Conference on Machine Learning (2006)
21. Rybski, P.E., Yoon, K., Stolarz, J., Veloso, M.: Interactive robot task training through dialog and demonstration. In: Proceedings of the 2th ACM/IEEE International Conference on Human-Robot Interaction (2007)
22. Veloso, M.M.: Planning and learning by analogical reasoning. Springer Science & Business Media (1994)
23. Zaki, M.J.: Efficiently mining frequent embedded unordered trees. Fundamenta Informaticae (2005)
24. Zaki, M.J.: Efficiently mining frequent trees in a forest: Algorithms and applications. IEEE Transactions on Knowledge and Data Engineering (2005)

Enriching a Temporal Planner with Resources and a Hierarchy-Based Heuristic

Alessandro Umbrico[1]([⊠]), Andrea Orlandini[2], and Marta Cialdea Mayer[1]

[1] Dipartimento di Ingegneria, Università degli Studi Roma Tre, Roma, Italy
alessandro.umbrico@uniroma3.it
[2] Istituto di Scienze e Tecnologie della Cognizione, Consiglio Nazionale delle
Ricerche, Roma, Italy

Abstract. A key enabling feature to deploy a plan-based application for solving real world problems is the capability to integrate Planning and Scheduling (P&S) in the solving approach. *Flexible Timeline-based Planning* has been successfully applied in several real contexts to solve P&S problems. In this regard, we developed the *Extensible Planning and Scheduling Library* (Epsl) aiming at supporting the design of P&S applications. This paper describes some recent advancements in extending the Epsl framework by introducing the capability to reason about different types of "components", i.e., *state variables* and *renewable resources*, and allowing a tight integration of Planning and Scheduling techniques. Moreover, we present a domain independent *heuristic* function supporting the solving process by exploiting the hierarchical structure of the set of timelines making up the flexible plan. Some empirical results are reported to show the feasibility of deploying an Epsl-based P&S application in a real-world manufacturing case study.

1 Introduction

The Timeline-based planning approach has been successfully applied in several real world scenarios, especially in space like contexts [1–3]. Besides these applications, several timeline-based Planning and Scheduling (P&S) systems have been deployed to define domain specific applications, see for example Europa [4], IxTeT [5], Apsi-Trf [6]. However, despite their practical success, these systems usually entails the development of applications closely connected to the specific domain they are made for. As a consequence, it is not straightforward to adapt these applications to domains requiring different solving capabilities and thus, it is often necessary to define new solvers somehow loosing "past experiences". To address the above issue, a research initiative has been started to develop a domain independent *Extensible Planning and Scheduling Library* (Epsl) [7]. Epsl aims at defining a modular and extensible software environment to support the development of timeline-based applications. The structure of Epsl allows to preserve "past experiences" by providing a set of ready-to-use algorithms, strategies and heuristics that can be combined together. In this way, it is possible to develop/evaluate several solving *configurations* in order to find the one which best fits the features of the particular domain to be addressed.

© Springer International Publishing Switzerland 2015
M. Gavanelli et al. (Eds.): AI*IA 2015, LNAI 9336, pp. 410–423, 2015.
DOI: 10.1007/978-3-319-24309-2_31

In this paper two enhancements of the EPSL framework are presented. First, the possibility to model and manage renewable resources is introduced in the framework. Briefly, a renewable resource is a shared component having a limited capacity, that is however not consumed by the processes using it: when it is released, it returns to its full capacity. Examples of renewable resources are the memory of a software device or a machine that can process a limited number of pieces at a time. Renewable resources allow the planning framework to model more realistic domains. Secondly, a domain independent heuristic function is defined exploiting the structure of the timelines and the dependencies among them induced by the rules constraining the domain. Such a structure conveys important information that can be used to improve the performances of the planner. It is worth pointing out that domains modeled following a hierarchical approach often exhibit this kind of structure of the system components. An experimental evaluation on problem domains derived from a real-world manufacturing context are finally presented to assess the deployment of the above mentioned heuristic.

2   Timeline-Based Planning in a Nutshell

Timeline-based planning has been introduced in early 90s [1] and several formalizations have been proposed for this approach [8–10]. The timeline-based approach takes inspiration from control theory. It models a complex domain by identifying a set of relevant features that must be controlled over time. Domain features are modeled by means of state variables, a set of rules that "locally" constrains the temporal evolutions of related features and describing the allowed sequence of values/states a feature can assume over time. Domain features are further constrained by means of *synchronization rules*, i.e. a set of "global" constraints that allow one to coordinate different features in order to obtain consistent behaviors of the overall system. A timeline-based planner uses these rules (called *domain theory*) to build *timelines* for domain features. A timeline is a sequence of valued temporal intervals, called *tokens*, each of which specifies the value assumed by the state variable in that interval. So, a timeline describes the temporal evolution/behavior of the related feature over time. The start and end points of the tokens making up a *flexible timeline* are temporal intervals instead of exact time points [8]. A flexible timeline represents an envelope of possible evolutions of the associated feature that can be exploited by an executive system for a robust online execution of the plan [11,12]. *Flexible Timeline-based Planning* usually follows a partial order planning approach starting from a set of partially defined timelines (i.e. the initial planning problem) and building (if possible) a set of completely instantiated timelines (i.e. a solution plan) within a given temporal horizon.

2.1   A Hierarchical Modeling Approach

This section is devoted to briefly present a general methodology, often used when modeling a domain in the timeline-based style. This approach generates a structure in the model which can be exploited during the solving process. The essential of the methodology is a decomposition analysis (like in [13,14]), aiming at identifying the "relevant" features (system's components) that independently evolve over time. A generic component is then described by a set of activities to carry out and the logical states the system can assume coupled with related timing and causal constraints, i.e. temporal durations as well as allowed state transitions. This approach results in a hierarchical model of the domain, with higher level components abstracting away from the internal structure of the system to be controlled. While deepening the analysis into details, the concrete features of the system are represented.

The modeling approach described here usually identifies three relevant classes of components [15]: (i) *functional*, (ii) *primitive* and (iii) *external* components. A *functional* component provides a logical view of the system as a whole in terms of what the system can do notwithstanding its internal composition. It models the high-level functionalities the system is able to perform. A *primitive* component provides a logical view of a particular element composing the system. Usually, values of such a component correspond to concrete states/actions the related element is able to assume/execute in the environment. Finally, an *external* component provides a logical view of elements whose behaviors are not under the control of the system but affect the execution of its functionalities. They model conditions that must hold in order to successfully perform internal activities. In addition to the description of single state variables, their behaviors are to be further constrained by specifying inter-components causal and temporal requirements (called *synchronization rules* in the timeline-based approach) allowing the system to coordinate its sub-elements while safely realizing complex tasks. In this regard, following a hierarchical approach, such rules map the high-level functionalities of the system into a set of activities on primitive and/or external components enforcing operational constraints that guarantee the proper functioning of the overall system and its elements. Namely, synchronizations allow to specify how the high-level functionalities, modeled by means of functional components, are related to the primitive and external components of the domain. The synchronization rules of the domain often reflect the hierarchy of the system components: the values of a higher level (more abstract) state variable are constrained to occur while suitable values are assumed by corresponding lower level ones, modeling its more concrete counterparts.

3   The Extensible Planning and Scheduling Library

EPSL is a layered framework built on top of APSI-TRF[1], it aims at defining a flexible software environment for supporting the design and development of timeline-based applications. The key point of EPSL flexibility is its interpretation of a planner as a "modular" solver which combines together several elements

---

[1] APSI-TRF is a software framework developed for the European Space Agency by the Planning and Scheduling Technology Laboratory at CNR (in Rome, Italy) for supporting the design and deployment of timeline-based P&S applications.

to carry out its solving process. The main components of the EPSL architecture are the following. The *Modeling layer* provides EPSL with timeline-based representation capabilities. It allows to model a planning domain in terms of timelines, state variables, synchronizations and to represent flexible plans. The *Microkernel layer* is the key element which provides the framework with the needed flexibility to "dynamically" extend the framework with new elements. It is responsible to manage the lifecycle of the solving process and the elements composing the application instances (i.e. the planners). The *Search layer* and the *Heuristics layer* are the elements responsible for managing strategies and heuristics a planner can use during the solving process to support the search. The *Engine layer* is the element responsible for managing the portfolio of algorithms, called resolvers, available. Resolvers characterize the expressiveness of EPSL-defined planners. Namely they define what a planner can actually do to solve problems. Finally the *Application layer* is the top-most element which carries out the solving process and finds a solution if any.

3.1   The Epsl Solving Procedure

The solving procedure of a generic EPSL-based planner is described in Algorithm 1. It consists in a plan refinement procedure which iteratively *refines* a plan π by detecting and solving conditions, called *flaws*, that affect the completeness and/or consistency of π. EPSL instantiates the planner solving process over the tuple $\langle \mathcal{P}, \mathcal{S}, \mathcal{H}, \mathcal{E} \rangle$ where $\mathcal{P}$ is the specification of a timeline-based problem to solve, $\mathcal{S}$ is the search strategy the planner uses to expand the search space, $\mathcal{H}$ is the heuristic function the planner uses to select the most promising flaw to solve, and $\mathcal{E}$ is a set of resolvers the planner uses to detect flaws of the plan and compute their solutions.

Algorithm 1. solve($\mathcal{P}, \mathcal{S}, \mathcal{H}, \mathcal{E}$)

1: // initialize search
2: $\pi \leftarrow InitialPlan\,(\mathcal{P})$
3: $fringe \leftarrow \emptyset$
4: // check if plan is complete and flaw-free
5: **while** $\neg IsSolution\,(\pi)$ **do**
6:      $\Phi \leftarrow DetectFlaws(\pi, \mathcal{E})$
7:      // check the set of flaws
8:      **if** $\Phi \neq \emptyset$ **then**
9:          // select the most promising flaw to solve
10:         $\phi \leftarrow SelectFlaw\,(\Phi, \mathcal{H})$
11:         // call resolver to detect flaws and compute solutions
12:         **for** $resv \in \mathcal{E}$ **do**
13:             $nodes \leftarrow HandleFlaw\,(\phi, resv)$
14:             // expand the search with possible plan refinements
15:             $fringe \leftarrow Enqueue\,(nodes, \mathcal{S})$
16:         // check fringe
17:         **if** $fringe = \emptyset$ **then**
18:             // unsolvable flaws
19:             **return** *Failure*
20:     // go on with search - backtracking point
21:     $\pi \leftarrow GetPlan\,(Dequeue\,(fringe))$
22: // get solution plan
23: **return** π

The plan π is initialized on the problem description $\mathcal{P}$ (row 2) and then the procedure iteratively refines the plan until a solution or a failure is detected (rows 5-21). Plan refinement consists in detecting flaws and compute their solutions by means of resolvers $\mathcal{E}$ (rows 6-15). Given a set of flaws Φ of the plan, the most promising flaw ϕ to solve is selected according to heuristic $\mathcal{H}$ (row 10) and then resolvers compute possible solutions. Each solution represents a possible refinement of the current plan so a new branch of the search space is created for each of them and the resulting node is added to the fringe according to $\mathcal{S}$ (row 15). The search goes on until a plan with no flaws is found, i.e. a solution plan (row 5). However if the fringe is empty (rows 17-19) this means that there are unsolvable flaws in the plan, then the procedure returns a failure.

Algorithm 1 depicts a standard search procedure. It is important to point out that the particular set of resolvers $\mathcal{E}$, the strategy $\mathcal{S}$ and the heuristic $\mathcal{H}$ used can strongly affect the behavior and the performance of the solving process. Resolvers are the architectural elements allowing a planner instance to actually build the plan. A resolver encapsulates the logic for detecting and solving specific type of flaws. The greater is the number of available resolvers the greater is the number of flaw types an EPSL-based planner can handle. The set of available resolvers determine the "expressiveness" of the framework, i.e. the type of problems EPSL can solve. Broadly speaking a flaw represents a particular condition to solve for building a consistent and complete plan. It is possible to identify two main classes of flaws: (i) *goals*, flaws affecting the completion of a plan; (ii) *threats*, flaws affecting the consistency of a plan.

At any iteration of Algorithm 1 the refinement procedure detects the flaws of the current plan (row 6) and selects the most "promising" flaw to solve according to the heuristic $\mathcal{H}$ (row 10). Flaw selection is not a backtracking point of the search but it can strongly affect the performance of the solving procedure. A "good" choice of the next flaw to solve, indeed, can prune the search space by cutting off branches that would not bring to solutions. The heuristic $\mathcal{H}$ encapsulates the policy used by the planner to analyze and select plan flaws. In this regards EPSL framework allows an application designer either to develop a heuristic exploiting some domain-specific knowledge of the domain or develop a domain-independent heuristic to address different domains.

These elements represent the main flexibility features of the framework. Here is where users can focus their development efforts in order to customize EPSL-based applications for the particular problem to address. As a matter of fact EPSL architecture allows to easily extend the solving capabilities of the framework by integrating new implementations of the elements described above.

3.2 Integrating Resources

One of the contribution of the paper is the introduction of renewable resources to extend EPSL modeling capabilities. A renewable resource is a shared element of the domain needed during the execution of an action but is not consumed. The resource maximum capacity $\mathcal{C}$ limits the number of activities that can concurrently require the resource. The access to the resource must be properly managed in order to satisfy its capacity constraint. Namely the amount of resource requirements of concurrent activities must not exceed the resource capacity $\mathcal{C}$. EPSL has been extended with the introduction of a new component type for modeling renewable resources. A component of this type specifies the capacity $\mathcal{C}$ of the resource, and requirement activities are temporally qualified by means of tokens specifying the used amount of the resource. The consumption profile of the resource is constituted by the set of tokens making use of it. Obviously, the total amount of resource used by overlapping tokens must not exceed its maximal capacity.

The EPSL framework has been extended with the introduction of a dedicated resolver for detecting and solving flaws concerning the management of renewable resources. The resolver must schedule requirement tokens by detecting and solving peaks on the resource consumption profile. A peak is detected every time the total amount of resource required by a set of (temporally) overlapping tokens is higher then the capacity $\mathcal{C}$. The peak is resolved by posting precedence constraints between particular subsets of activities, called *Minimal Critical Sets* (MCSs). A subset of activities of a peak is an MCS if a precedence constraint between any pair of these activities removes the peak. The resolver has been implemented by adapting to timelines the algorithms described in [16].

3.3   Integrating Heuristics

The flaw selection heuristic $\mathcal{H}$ is the element supporting the flaw selection step (row 10) of the plan refinement procedure. At any iteration of Algorithm 1 the planner must select the next flaw to refine the plan. If no heuristic is given, all the flaws detected are "equivalent" and the planner can only make a random choice. Namely there is no information characterizing the importance of the flaws. However, the flaws of a plan can often have dependencies: solving a flaw can solve or simplify the solution of other related flaws. Therefore it is important to make "good" choices in order to reduce the number of refinement steps needed to build the plan. Bad choices, indeed, may bring to an inefficient solving procedure. A flaw selection heuristic $\mathcal{H}$ provides a criterion the planner can use to identify the most relevant to handle.

The *Hierarchical Flaw-selection Heuristic* (HFH) is an evaluation criterion relying on the hierarchical structure of timeline-based plans. Timelines may be related one to the other by synchronization rules. Given two timelines A and B, a synchronization $S_{A,B}$ from timeline A to timeline B, typically implies a dependency of B from A. Namely, the presence of some token on the timeline B is due to the need of synchronizing with some other token on the timeline A. Therefore, analyzing synchronization rules it is possible to build a *dependency graph* (DG) of the timelines. Figure 1 shows a set of timelines with synchronization rules (the arrows connecting the timelines) and the resulting DG where nodes are timelines and edges are synchronization rules. A hierarchy describing relationships among timelines can be extracted from the obtained DG. An edge from a node A to a node B in the DG represents a dependency between the two corresponding timelines. Thus the hierarchical level of the timeline A is not lower than the hierarchical level of B; if moreover no path connects B to A, A is at a higher level in the hierarchy, i.e. it is more independent than B. If a timeline A depends from B and vice-versa (i.e. A and B are contained in a looping path in the DG), then A and B have the same hierarchical level, and they are said to be hierarchically equivalent. In general, if the DG has a root, i.e. a node with only outgoing edges, it represents the most independent timeline of the hierarchy. For instance, the hierarchy extracted from the DG in Figure 1 is $A \prec B \prec C$ and $B \prec D$, while C and D are hierarchically equivalent, so A is the most independent timeline while C and D are the less independent ones.

Usually, the DG resulting from a planning domain built by applying the modeling approach described in section 2.1 generates a non-flat hierarchy (sometimes even an acyclic graph) that can be successfully exploited by the HFH. Obviously, if a domain has no hierarchical structure, then the HFH heuristic gives no meaningful contribution. The *hierarchy feature* of a flaw corresponds to the "independence" level of the timeline it belongs to. The idea is to exploit this hierarchy and select first flaws belonging to the most independent timeline. The underlying assumption is that the flaws influence is related to the corresponding timeline independence

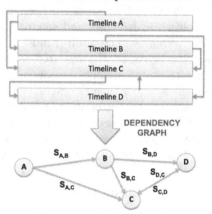

Fig. 1. A Dependency Graph example based on synchronizations.

level. Consequently if "independent" flaws are solved first, i.e. flaws detected on one of the topmost timelines in the hierarchy, then we have a good probability to "automatically" solve or reduce the possible solutions of the "dependent" flaws. Flaw-level reasoning allows the solving process to integrate planning and scheduling steps. This means that at any point in the solving process the planner can make a planning choice by selecting a goal to solve, or a scheduling choice by selecting a scheduling threat to solve, and so on. Similarly to timelines we can give a "structure" to the solving process by assigning a priority to solving steps. Namely we define the *type feature* of flaws and select flaws according to their type. For example we can decide to solve first goals and then scheduling threats in order to force the solving process to take planning decisions before scheduling ones. In addition to the above features we can also specify the *degree feature* of a flaw which characterizes the criticality of a flaw. Similarly to the *fail first principle* in constraint satisfaction problems, the degree of a flaw is a measure of the number of solutions available to solve it. The lower is that degree, the higher is the criticality of the flaw, i.e. few options to solve the flaw. We use this feature to assign an higher priority to flaws with less possible solutions (i.e. more difficult to solve).

The *Hierarchical Flaw-selection Heuristic* (HFH) we have defined, combines together all the features described of the flaw in order to make the "best" choice for selecting the next flaw to solve during the solving process. Given a set of flaws Φ detected on a current plan π, HFH selects the best flaw to solve by applying a *pipeline* of filters that evaluate flaws according by considering the above features as follows:

$$\Phi^0(\pi) \xrightarrow{f_h} \Phi^1(\pi) \xrightarrow{f_t} \Phi^2(\pi) \xrightarrow{f_d} \Phi^3(\pi) \rightarrow \phi^* \in \Phi^3(\pi)$$

Every filter of the pipeline (f_h, f_t and f_d) filters plan flaws by taking into one of flaw feature described. Thus, the initial set of plan flaws $\Phi^0(\pi)$ is filtered by applying the filter f_h which returns the subset of flaws $\Phi^1(\pi) \subseteq \Phi^0(\pi)$ belonging to the most independent timelines. If no hierarchical structure can be found among domain timelines, then the filter f_h returns the initial set of flaws $\Phi^1(\pi) = \Phi^0(\pi)$, i.e. flaws are equivalent w.r.t. hierarchy feature. The filter f_t filters the set of flaws $\Phi^1(\pi)$ by taking into account the type feature of the flaws, e.g. f_t returns the subset of flaws containing only goals $\Phi^2(\pi) \subseteq \Phi^1(\pi)$. Finally the filter f_d filters the set of flaws $\Phi^2(\pi)$ by taking into account the degree feature of flaws and returns the final set $\Phi^3(\pi) \subseteq \Phi^2(\pi)$. The final set of flaws $\Phi^3(\pi) \subseteq \Phi^0(\pi)$ which results from the application of the pipeline, represents equivalent choices w.r.t. the heuristic point of view. Therefore HFH chooses the next flaw to solve ϕ^* by randomly selecting a flaw from the final set $\Phi^3(\pi)$ the next flaw to solve is randomly selected from the final set $\phi^* \in \Phi^3(\pi)$.

4 Applying *Flexible Timeline-based Planning* to a Manufacturing Case Study

As a running example, let us consider a pilot plant from the on-going research project *Generic Evolutionary Control Knowledge-based mOdule* (GECKO): a manufacturing system for Printed Circuit Boards (PCB) recycling [17]. The objective of the system is to analyze defective PCBs, automatically diagnose their faults and, depending on the gravity of the malfunctions, attempt an automatic repair of the PCBs or send them directly to shredding. The pilot plant contains 6 working machines that are connected by means of a Reconfigurable Transportation System (RTS), composed of mechatronic components, i.e., transportation modules. Figure 2(a) provides a picture of a transportation module. Each module combines three transportation units. The units may be unidirectional and bidirectional units; specifically the bidirectional units enable the lateral movement (i.e., cross-transfers) between two transportation modules. Thus, each transportation module can support two main (straight) transfer services and one-to-many cross-transfer services. Figure 2(b) depicts two possible configurations.

Configuration 1 supports the forward (F) and backward (B) transfer capabilities as well as the left (LC1) and right (RC1) cross transfer capabilities. Configuration 2 extends Configuration 1 by integrating a further bidirectional transportation unit with cross transfer capabilities LC2 and RC2. The maximum number of bidirectional units within a module is limited just by its straight length

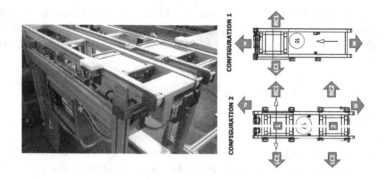

Fig. 2. (a) A transportation module; (b) Their transfer services.

(three, in this particular case). The transportation modules can be connected back to back to form a set of different conveyor layouts. The manufacturing process requires PCBs to be loaded on a fixturing system (pallet) in order to be transported and processed by the machines. The transportation system is to move one or more pallets and each pallet can be either empty or loaded with a PCB to be processed. Transportation modules control systems have to cooperate in order to define the paths the pallets have to follow to reach their destinations.

The description of the distributed architecture and some experimental results regarding the feasibility of the distributed approach w.r.t. the part routing problem can be found in [18,19]. *Transportation Modules* (TMs) rely on P&S technology to synthesize activities for supporting the work flow within the shop floor. Each TM agent is endowed with a Timeline-based planner (build on top of EPSL framework) to build plans for the transportation task.

4.1   The Gecko Timeline-Based Model

Figure 3 shows the timeline-based model of a generic *transportation module* (TM) of the GECKO case study. The timeline-based model has been defined by applying the modeling approach described in 2.1. Namely, a functional state variable *Channel* represents the high level transporting tasks of a TM. Each value of the Channel state variable models a particular transportation task indicating the corresponding input and output port of the module. For instance, *Channel_F_B* models the task of transporting a pallet from *port-F* to *port-B* w.r.t. Figure 2(b).

The *Change-Over* component is a primitive state variable which models the set of internal configurations the transportation module can assume for exchanging pallets with other modules. Namely configurations identify the internal paths a pallet can follow to traverse the module. For instance, *CO_F_B* in Figure 3 represents the configuration needed for transporting a pallet from port F to port B. The *Energy_Consumption* element in Figure 3 is a renewable resource component

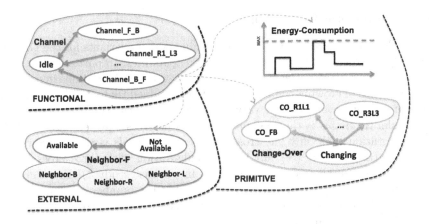

Fig. 3. Timeline-based model for a full instantiated TM

which models the energy consumption profile of a TM of the plant. A system requirement entails that the instant energy consumption of TMs cannot exceed a predefined limit for the physical device. This requirement is modeled by means of synchronization rules between the functional state variable and the renewable resource. These rules specify the energy consumption estimate for each functional activity of the module. For example, the maximum instant energy consumption allowed is 10 units, that *Channel_F_B* activity consumes 9 units of energy during its execution, *Channel_L1_R1*, *Channel_L2_R2*, *Channel_L3_R3* consume 3 units of energy during their execution and so on. Therefore the planner must schedule functional activities satisfying the energy consumption constraint.

A set of external state variables complete the domain by modeling possible states of TM's neighbor modules. Neighbors are modeled by means of external state variables because they are not under the control of the module. Namely, the TM cannot decide the state of its neighbors. However it is important to monitor their status because a TM must cooperate with them in order to successfully carry out its tasks. For instance the TM must cooperate with *Neighbor_F* and *Neighbor_B* to successfully perform a *Channel_F_B* task. Therefore *Neighbor_F* and *Neighbor_B* must be *Available* during task "execution".

Finally a set of synchronization rules specify how a TM implements its channel tasks (see the dotted arrows in Figure 3). According to the modeling approach described in 2.1, the synchronization specification follows a hierarchical decomposition. In this way they allow to specify a set of operative constraints (e.g. temporal constraints) describing the sequence of internal configurations and "external" conditions needed to safely perform channel tasks. For instance, a synchronization rule for the *Channel_F_B* task requires that the module must be set in configuration *CO_F_B* and that neighbor F and neighbor B must be *Available* during the "execution" of the task.

4.2   Experimental Evaluation

We have defined a set of timeline-based planning domain variants for the GECKO case study in order to evaluate EPSL solving capabilities. In particular we have modelled a generic TM in the plant considering several configurations in order to define scenarios of growing complexity. We have assumed that module's neighbor agents are always available and vary the number of cross transfers composing the module: (i) *simple* is the configuration with no cross transfers; (ii) *single* is the configuration with only one cross transfer; (iii) *double* is the configuration with two cross transfer; (iv) *full* is the configuration with three cross transfers (the maximum allowed w.r.t. the plant in our case study 4). The higher is the number of available cross transfers, the higher the number of elements and constraints the planner has to deal with at solving time. Moreover, we defined several EPSL-based planners by varying the *heuristic function* applied during the solving process: (i) The *HFH* planner uses HFH; (ii) The *TFS* planner uses a heuristic based only on the $\mathcal{H}_{ft}$ feature; (iii) The *TLFS* planner uses a heuristic based only on the $\mathcal{H}_{fh}$ feature; (iv) The *DFS* planner uses a heuristic based only on the $\mathcal{H}_{fd}$ feature; (v) The *RFS* planner uses no heuristics at all, i.e. it makes a random selection of the *best* flaw to solve.

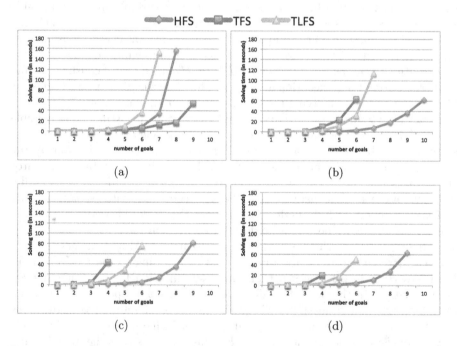

Fig. 4. EPSL-based planners performances on: (a) *simple* configuration; (b) *single* configuration; (c) *double* configuration; (d) *full* configuration

The charts in Figure 4 show the solving time trends of the EPSL-based planners (within a timeout of 180 seconds) w.r.t. the growing dimension of the planning problem (i.e. a growing number of goals) and the growing complexity of the module to control (i.e. the number of available cross transfers). The results show that the *HFH* planner is dominating other planners on the considered planning domains. Comparing the *TFS* planner (the original EPSL planner setting we used before the introduction of the timeline hierarchy) with the *HFS* planner, the deployment of HFH entails a general improvement of performance in terms of both solving times and scalability of the solving capacity of EPSL framework. The results concerning the *DFS* planner and the *RFS* one are not plotted in Figure 4 because these planners could not solve but the simpler cases.

It is worth pointing out that the *TFS* planner outperforms the *HFH* planner only in the *simple* domain, see Figure 4(a). This seems a consequence of the fact that the *TFS* planner maintains a "global" vision of the activities to be performed during the solving process thus reasoning about the overall plan and taking into account all the timelines together. In the *simple* domain, the TM can perform channel tasks only towards two directions (*Channel_F_B* and *Channel_B_F*). Therefore, when building the plan, the planner can more suitably organize the tasks in order to reduce the number of reconfigurations of the module (i.e. change overs). Namely, that planner is able to "group" channel tasks requiring the same configuration of the module (e.g. scheduling all the *Channel_F_B* tasks before *Channel_B_F* tasks). This allows the planner to reduce the number of plan decisions simplifying its construction in the *simple* domain.

Conversely, the *HFS* planner maintains a "local" vision related to the single timelines of the domain because it builds one timeline at a time. Therefore, when that planner must manage the needed reconfiguration of the module (i.e. flaws on *Change Over* timeline), its choices are partially constrained by the channel timeline which has been built before the *Change Over* timeline following the timeline hierarchy. As a consequence, the planner is not able to organize tasks as in the case of *TFS* but it has to manage a larger number of tokens on the timeline at this point. However the *HFS* planner scales better than the *TFS* with the growing complexity of the domain as Figure 4 shows.

5   Conclusions and Future Works

In this paper we have presented some recent advancements in the development of the EPSL framework. In particular, EPSL has been extended in two different aspects by introducing: (i) The capability of modeling and reasoning about renewable resources; (ii) The HFH heuristic to support the solving process. Finally, some experimental results have been reported in order to show the feasibility of EPSL-based planners when deployed to a real world domain. The comparison of EPSL with different timeline-based P&S systems (e.g. EUROPA), the possibility to further extend the framework by integrating different type of resources (i.e., consumable resources) and the definition of new heuristics or improving HFH constitute future works in our research agenda. Moreover, one additional long-term research goal is to identify a set of quality metrics to characterize flexible timeline-based plans and, thus, exploit such metrics in order to generate better plans.

Acknowledgments. Andrea Orlandini is partially supported by the Italian Ministry for University and Research (MIUR) and CNR under the GECKO Project (Progetto Bandiera "La Fabbrica del Futuro").

References

1. Muscettola, N.: HSTS: integrating planning and scheduling. In: Zweben, M., Fox, M.S. (ed.) Intelligent Scheduling. Morgan Kauffmann (1994)
2. Jonsson, A., Morris, P., Muscettola, N., Rajan, K., Smith, B.: Planning in inter-planetary space: theory and practice. In: Proceedings of the Fifth Int. Conf. on AI Planning and Scheduling. AIPS-00, pp. 177–186 (2000)
3. Cesta, A., Cortellessa, G., Denis, M., Donati, A., Fratini, S., Oddi, A., Policella, N., Rabenau, E., Schulster, J.: MEXAR2: AI Solves Mission Planner Problems. IEEE Intelligent Systems **22**(4), 12–19 (2007)
4. Barreiro, J., Boyce, M., Do, M., Frank, J., Iatauro, M., Kichkaylo, T., Morris, P., Ong, J., Remolina, E., Smith, T., Smith, D.: EUROPA: a platform for AI planning, scheduling, constraint programming, and optimization. In: 4th Int. Competition on Knowledge Engineering for P&S (ICKEPS) (2012)
5. Ghallab, M., Laruelle, H.: Representation and control in IxTeT, a temporal plan-ner. In: Proc. of the International Conference on AI Planning Systems (AIPS), pp. 61–67 (1994)
6. Cesta, A., Fratini, S.: The timeline representation framework as a planning and scheduling software development environment. In: Proc. of the 27th Workshop of the UK Planning and Scheduling Special Interest Group (PlanSIG-08) (2008)
7. Cesta, A., Orlandini, A., Umbrico, A.: Toward a general purpose software envi-ronment for timeline-based planning. In: 20th RCRA International Workshop on Experimental Evaluation of Algorithms for Solving Problems with Combinatorial Explosion (2013)
8. Cialdea Mayer, M., Orlandini, A., Umbrico, A.: A formal account of planning with flexible timelines. In: The 21st International Symposium on Temporal Represen-tation and Reasoning (TIME), pp. 37–46. IEEE (2014)
9. Cimatti, A., Micheli, A., Roveri, M.: Timelines with temporal uncertainty. In: Proc. of the 27th AAAI Conference on Artificial Intelligence. AAAI Press (2013)
10. Cialdea Mayer, M., Orlandini, A.: An executable semantics of flexible plans in terms of timed game automata. In: The 22st International Symposium on Temporal Representation and Reasoning (TIME). IEEE (to appear 2015)
11. Orlandini, A., Suriano, M., Cesta, A., Finzi, A.: Controller synthesis for safety criti-cal planning. In: 25th International Conference on Tools with Artificial Intelligence (ICTAI), pp. 306–313. IEEE (2013)
12. Py, F., Rajan, K., McGann, C.: A systematic agent framework for situated autonomous systems. In: Proc. of the 9th Int. Conf. on Autonomous Agents and Multiagent Systems (AAMAS-10) (2010)
13. Bernardini, S.: Constraint-based Temporal Planning: Issues in Domain Modelling and Search Control. Ph.D. thesis, Università degli Studi di Trento (2008)
14. Fratini, S., Pecora, F., Cesta, A.: Unifying Planning and Scheduling as Timelines in a Component-Based Perspective. Archives of Control Sciences **18**(2) (2008)

15. Borgo, S., Cesta, A., Orlandini, A., Umbrico, A.: An ontology-based domain representation for plan-based controllers in a reconfigurable manufacturing system. In: The 28th International FLAIRS Conference. AAAI (2015)
16. Cesta, A., Oddi, A., Smith, S.F.: A Constraint-based method for Project Scheduling with Time Windows. Journal of Heuristics 8(1), 109–136 (2002)
17. Borgo, S., Cesta, A., Orlandini, A., Rasconi, R., Suriano, M., Umbrico, A.: Towards a cooperative -based control architecture for a reconfigurable manufacturing plant. In: 19th IEEE International Conference on Emerging Technologies and Factory Automation (ETFA 2014). IEEE (2014)
18. Carpanzano, E., Cesta, A., Orlandini, A., Rasconi, R., Valente, A.: Intelligent dynamic part routing policies in plug&produce reconfigurable transportation systems. CIRP Annals - Manufacturing Technology 63(1), 425–428 (2014)
19. Carpanzano, E., Cesta, A., Orlandini, A., Rasconi, R., Suriano, M., Umbrico, A., Valente, A.: Design and implementation of a distributed part-routing algorithm for reconfigurable transportation systems. International Journal of Computer Integrated Manufacturing (2015)

Integrating Logic and Constraint Reasoning in a Timeline-Based Planner

Riccardo De Benedictis[✉] and Amedeo Cesta

CNR, Italian National Research Council, ISTC, Rome, Italy
{riccardo.debenedictis,amedeo.cesta}@istc.cnr.it

Abstract. This paper introduces the ongoing work for a novel domain-independent planning system which takes inspiration from both Constraint Programming (CP) and Logic Programming (LP), flavouring it all with Object Oriented features. We will see a specific customization of our environment to the particular kind of automated planning referred to as timeline-based. By allowing for the interesting ability of solving both planning and scheduling problems in a uniform schema, the resulting system is particularly suitable for complex domains arising from real dynamic scenarios. The paper proposes a resolution algorithm and enhances it with some (static and dynamic) heuristics to help the solving process. The system is tested on different benchmark problems from classical planning domains like the Blocks World to more challenging temporally expressive problems like the Temporal Machine Shop and the Cooking Carbonara problems demonstrating how the new planner, named iLoC, compares with respect to other state-of-the-art planners.

1 Introduction

The timeline-based approach to planning [20,23] represents an effective alternative to classical planning [16], especially for complex domains which require the use of both temporal reasoning and scheduling features. The timeline-based approach models a planning and scheduling problem by identifying a set of relevant *features* of the modeled domain which need to be controlled to obtain a desired temporal behaviour (the "desired timelines"). Timelines represent one or more physical (or logical) subsystems which are relevant to a given planning context. The planner/scheduler plays the role of the controller for these entities, and reasons in terms of constraints that bound the values each feature may assume over time and the desired properties of the generated temporal behaviours.

Common timeline-based planners like EUROPA [20], ASPEN [10], IxTeT [18] and APSI-TRF [8] are defined as complex software environments suitable for generating planning applications, but quite heavy to foster research work on specific aspects worth being investigated. Some theoretical work on timeline-based planning like [17] was mostly dedicated to explain details of the CAIP paradigm (and its implementation: EUROPA) and to identify connections with classical planning. The works on IxTeT and APSI-TRF have tried to clarify some key underlying principles but mostly succeeded in underscoring the role of time

© Springer International Publishing Switzerland 2015
M. Gavanelli et al. (Eds.): AI*IA 2015, LNAI 9336, pp. 424–437, 2015.
DOI: 10.1007/978-3-319-24309-2_32

and resource reasoning [9,21]. The search control part of timeline-based planners has always remained significantly under explored with the notable exception of [3].

The current realm is that although these planners capture elements that are very relevant for applications, their theories are often quite challenging from a computational point of view and their performance are rather weak compared with those of state of the art classical planners. Indeed, timeline-based planners are mostly based on the notion of partial order planning [29] and have almost neglected advantages in classical planning triggered from the use of GRAPHPLAN and/or modern heuristic search [4,5,19]. Furthermore, these architectures mostly rely on a clear distinction between temporal reasoning and the use of various forms of constraint reasoning and there is not enough exploration of other forms of reasoning.

This paper introduces the ongoing work for a novel domain-independent planning system called iLoC (Integrated Logic and Constraint reasoner). iLoC extends significantly previous works on timeline-based planning [14] and its fielded applications [7]. The direction we are pursuing consists in proposing a general AI reasoning framework which provides a common glue for the definition of different kinds of AI problems that share constraints among them. In the desiderata of a general purpose cognitive architecture, for example, there might be the possibility of handling both temporal and spatial problems. Although there exist efficient temporal reasoners and spatial reasoners, the definition of constraints among these problems might be cumbersome and highly inefficient to be managed. iLoC aims at general architecture for logic and constraint-based reasoning that brings together some key aspects of intelligent reasoning leaving freedom to specific implementations on both constraint reasoning engines and resolution heuristics. The proposed architecture is characterized by a basic core that allows to define and solve "classical" AI problems (section 2), and includes the possibility to define particular "modules" devoted to the resolution of specific AI sub-problems. We have exploited this basic core for modelling and solving timeline-based planning problems (section 3) and then have evaluated the current results (section 4) to demonstrate the step ahead in the direction of the open problems.

2   The iLoC Reasoning Environment

The basic core of the iLoC architecture provides an object oriented virtual environment for the definition of *objects* and *constraints* among them. The aim here is to provide to the user a minimalistic core that should be both sufficiently expressive as well as easily extensible so as to adapt as much as possible to the most variety of user requirements.

Similar to most object oriented environments, every object in the iLoC environment is an instance of a specific *type*. iLoC distinguishes among *primitive types* (e.g., bools, ints, reals, strings, etc.) and user defined *complex types* (e.g., robots, trucks, locations, etc.) endowed with their *member variables* (variables associated to a specific object of either primitive or complex type), *constructors* (a special type of subroutine called to create an instance of the complex type) and *methods* (subroutines associated with an object of a complex type). Defining a navigation problem, for example, might require the definition of a *Location* complex type having two numeric member variables x and y representing the coordinates of each *Location* instance. In the following, we will address objects and their member variables using a Java style *dot* notation (i.e., given a *Location* instance l, its x-coordinate will be expressed as $l.x$).

Once objects are defined, iLoC allows the definition of constraints among them. For example, in case a robot r should always be more East of a location l, the iLoC user could assert a constraint such $[\![l.x < r.x]\!]$. iLoC considers constraints as logic propositions and, as such, it allows the possibility for negating them (e.g., $\neg[\![l.x \leq 5]\!]$), for expressing conjunctions (e.g., $[\![l.x \leq 10]\!] \wedge [\![l.x \geq 5]\!]$), disjunctions (e.g., $[\![l.x \leq 5]\!] \vee [\![l.x \geq 10]\!]$) and logic implications (e.g., $[\![l.x \geq 10]\!] \rightarrow [\![l.y \geq 10]\!]$). Furthermore, it is possible to impose constraints on existentially (e.g., $\exists l \in Locations : l.x \geq 10$) and universally (e.g., $\forall l \in Locations : l.x \leq 100$) quantified variables. By combining logical quantifiers and object oriented features, iLoC allows to manage all the instances of a given complex type in one shot. It is worth noting that, in case the defined constraints are inconsistent among themselves (e.g., $[\![l.x \leq 10]\!] \wedge [\![l.x \geq 15]\!]$) the system will return a failure.

A rather straightforward method for managing this kind of problems is to translate them into a Satisfiability Modulo Theories (SMT) problem (see, for example, [27]). Among the several available SMT solvers, having different performances and capabilities, we have chosen to use the Java wrapper for the Z3 solver [15]. However, although this basic core allows the definition of quite complex problems (without providing any demonstration, we can state that NP-Complete problems are covered), some of the problems we are interested in are in PSPACE and thus excluded from the possibility of being modelled with this formalism. In order to overcome these limitations, we need something more powerful. Something that, roughly speaking, is able to "decide" the number of involved variables, together with their value. For this purpose, we have chosen to extend the above formalism by allowing first-order Horn clauses[1], i.e., clauses with at most one positive literal, called the *head* of the clause, and any number of negative literals, forming the *body* of the clause. For example, we could use a predicate such as $FirstQuadrant$, with a *Location* l argument, within the clause $FirstQuadrant(Location\ l) \Leftarrow [\![l.x \geq 0]\!] \wedge [\![l.y \geq 0]\!]$, for describing locations in the first quadrant of a Cartesian coordinate system.

2.1 The Resolution Process and the Heuristics

A consequence of what we have seen is that iLoC planning problems can be described by a collection of clauses. There are two types of clauses: *rules* and

---

[1] This means, in general, sacrificing decidability.

requirements. A rule is of the form $Head \Leftarrow Body$. While the head of rules is limited to predicates (i.e., we do no allow constraints in the head of a clause), a rule's body consists of a set of calls to predicates, which are called the rule's *sub-goals*, and a set of constraints, the latter, in any logical combination. Furthermore, we consider rules having the same head as disjunctive. Clauses with an empty body are called requirements and can be either calls to predicates (we distinguish between *facts* or *goals*) or constraints, the latter, in any logical combination. For example, we might use the requirements $goal : FirstQuadrant\,(Location\ l)$, $[\![l.x \geq 5]\!]$ and $\neg[\![l.y < 5]\!]$ for asking the planner to find a location l, among those which are in the first quadrant, having both coordinates greater than or equal to 5.

From an operational point of view, iLoC uses an adaptation of the *resolution principle* [26] for first-order logic, extended for managing constraints in a more general scheme usually known as *constraint logic programming* (CLP) [1]. The basic operations for refining a partial solution π toward a final solution are the following:

1. find the (sub)goals of π.
2. select one such (sub)goals.
3. find ways to resolve it.
4. choose a resolver for the (sub)goals.
5. refine π according to that resolver.

Starting from an initial node corresponding to an empty solution, the search aims at a final node containing a solution that correctly achieves the required goals. For each goal $P\,(t_1^g, \ldots, t_i^g)$, in general, a branch in the search space is created. Resolution, at first, will try to *unify* goals with existing formulas, if any, creating a single branch for all the possible unifications. Specifically, given the existing formulas $P\left(t_1^1, \ldots, t_i^1\right), \ldots, P\left(t_1^j, \ldots, t_i^j\right)$, having the same predicate of the goal, the formula $[\![t_1^g = t_1^1 \wedge \ldots \wedge t_i^g = t_i^1]\!] \vee \ldots \vee [\![t_1^g = t_1^j \wedge \ldots \wedge t_i^g = t_i^j]\!]$ is added to the current solution. Intuitively, the purpose of unification is to avoid considering goals whose (any) rule has already been applied. In addition, a branch is also created for each of the rules whose head unifies with the chosen goal and, whenever such a branch is chosen by the resolution algorithm, the body of the corresponding rule is added to the current solution possibly generating further goals to be managed. Facts, on the contrary, do not trigger further reasoning. Nevertheless their terms are, in general, variables and, therefore, can be constrained during the resolution process whenever they are subject to unification.

Figure 1 shows how this process is applied to the simple problem depicted. At the beginning, the two goals $A\,(w)$ and $B\,(v)$ need to be resolved, so, the former is chosen. Since unification with already existing formulas is not possible (there is no formula yet), the first rule is applied, resulting in the addition of subgoal $B\,(y)$ and constraint $[\![w < y]\!]$ (notice the substitution z/w as a consequence of the unification of the goal $A\,(w)$ with the head of the first rule $A\,(z)$). At this point, goal $B\,(v)$ is chosen. Again, unification is not possible so the second rule is applied, resulting in the addition of constraint $[\![v < 10]\!]$. Last goal to be solved

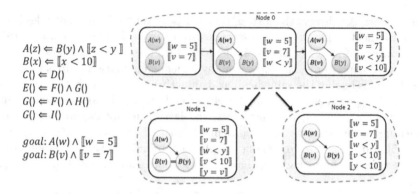

$A(z) \Leftarrow B(y) \wedge [\![z < y]\!]$
$B(x) \Leftarrow [\![x < 10]\!]$
$C() \Leftarrow D()$
$E() \Leftarrow F() \wedge G()$
$G() \Leftarrow F() \wedge H()$
$G() \Leftarrow I()$

$goal: A(w) \wedge [\![w = 5]\!]$
$goal: B(v) \wedge [\![v = 7]\!]$

Fig. 1. An example of resolution in iLoC.

is goal $B(y)$. Since unification is now possible, we create a branch in the search space creating two nodes (i.e., Node 1 and Node 2) and in the former we unify formulas $B(v)$ and $B(y)$ (by adding the constraint $[\![v = y]\!]$) while in the latter we apply the second rule a second time. It is worth noting that by selecting $B(v)$ at the first step, although with different paths, we would have got the same two final nodes. In addition, the first two steps do not require choices so there is no need for creating a branching and the resolution can continue "inside" Node 0. Finally, both the resulting nodes represent a solution for our starting problem and thus the resolution process can terminate. Clearly, the presence of additional (sub)goals would have required further unifications and/or rules applications.

Goal Selection Heuristic. Since all the goals must be solved sooner or later, there is almost no difference in *which* goal is solved first. Selecting the "right" goal, however, impacts heavily with the efficiency of the resolution algorithm. In order to overcome this obstacle we can take advantage of some

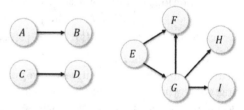

Fig. 2. An example of clauses and the static causal graph originated from them.

heuristics. We can create, for example, a *static causal graph* (we call it static since it doesn't change during the resolution process) and extract information from it to guide the search process. Our graph has a node for each of the predicates that appear in our rules and, for every rule, we add an edge from the head of the rule to each of the predicates that appear in the body of the same rule. We then consider the cost for solving a goal as the number of reachable nodes from the node relative to the predicate associated to the goal. We call such a strategy "All Reachable - Goal Selection Heuristic" (AR-GSH). The idea behind the AR-GSH strategy is to evaluate goals by considering a kind of worst case scenario where none of the formulas unify. Figure 2 shows the static causal graph resulting from the rules of figure 1. As an example, the cost for solving an

$A(w)$ goal, according to AR-GSH, is 1 (since the sole node B is reachable from node A) while the cost for solving an $E(t)$ goal is 4 (since all the nodes F, G, H and I are reachable from node E). Finally, it is worth mentioning that this graph and, consequently, the costs for each of their nodes, solely depend from the rules, therefore, our heuristic is problem independent and thus can be built once and for ever at the beginning of the solving process, allowing constant-time cost retrieval.

Node Selection Heuristic. A further sensible improvement to our resolution process is the choosing of the node from the search space's fringe. By considering the resolution process as an optimization problem, whose objective function is the minimization of the number of unmanaged goals, we can exploit an A* search strategy. The simplest heuristic for choosing such a node could be to select the one having less unresolved (sub)goals. Better than nothing, such a strategy does not appear smart enough since it completely ignores the kind of involved (sub)goals. A better strategy could be to reuse the idea behind AR-GSH by considering the cost for managing a node as the sum of the costs for solving all its goals. We name this strategy "All Reachable - Node Selection Heuristic" (AR-NSH). The estimated cost for managing Node 0 of figure 1, for example, would be equal to the estimated cost for solving goal $A(w)$ (i.e., 1) plus the estimated cost for solving goal $B(v)$ (i.e., 0). It is worth noting that both the AR-GSH and the AR-NSH strategies are not admissible heuristics since they can easily overestimate the cost for solving a goal (e.g., by neglecting some unification), therefore the A* search strategy does not guarantee to find an "optimal" solution in terms of minimum number of steps. We are indeed interested in finding a solution which "simply" guarantees the domain constraints in reasonable time.

Improving the Heuristics. Although these simple heuristics allow us to significantly improve the performance of the resolution process, it is not uncommon that different goals (or different nodes) are estimated alike. Whenever such incident occurs we could choose to further refine the costs. A possible strategy could be to extract information from a different graph that is called *dynamic causal graph* (we call it dynamic since it changes during the resolution process). Specifically, this graph maintains information about all the formulas in a given search space node, together with the reason they are there (i.e., the application of the rules and the unifications). Examples of dynamic causal graphs are those inside nodes in figure 1. A simple way for extracting information from these graphs is to consider the number of possible unifications as a complexity estimation for a (sub)goal. We call such a strategy "Less Merges - Goal Selection Heuristic" (LM-GSH). At the end of Node 0 of figure 1, for example, AR-GSH would estimate the cost of a goal $C()$ alike the cost of a goal $A(u)$ (i.e., 1). Nevertheless LM-GSH would have preferred the goal $C()$ having less possible unifications with respect to $A(u)$ (i.e., 0 vs 1). As a reinforcement of this strategy, it is worth noting that the goal $C()$ would have been chosen anyway since, being not possible to unify it with any other existing formula, there was no other choice than applying the rule. Finally, similar to what we have done for AR-GSH, we adapt the LM-GSH strategy to the node case producing a fourth heuristic that we call "Less Merges - Node Selection Heuristic" (LM-NSH).

How can such a generic framework be related with automated planning? Is it possible to extend the framework to manage other forms of reasoning? In the following, after some basic introductory information on timeline-based planning, we will try to convince the reader that it is possible to give an affirmative answer to these questions and, moreover, performances of the resulting planner are not bad compared with other state-of-the-art planners.

3   Timeline-Based Planning within iLoC

The search space of a timeline-based planner has typically *partially specified plans* as nodes and *plan refinement operations* as arcs. Plan refinement operations are intended to further complete a partial solution, i.e., to achieve an open goal or to remove some possible inconsistency. Intuitively, these refinement operations avoid adding to the partial plan any constraint that is not strictly needed for addressing the refinement purpose (this is called the *least commitment principle*). The solving procedure starts from an initial node corresponding to an empty solution and the search aims at a final node containing a solution that correctly achieves the required goals.

A possible approach to the resolution of timeline-based planning problems is to provide the predicates described in the previous sections with numerical arguments in order to represent their starting times, their ending times and their durations. Also, it will be required to define some specific complex types, whose instances will be called *timelines*, in order to add further "implicit" constraints among the formulas defined "over" their instances. This will also result in a slight adaptation of the resolution procedure in order to check the consistency for every object in the current partial solution so as to make explicit the just mentioned implicit constraints. What does it mean to define a formula "over" a timeline? We simply add a parameter having the same type as the timeline to the predicates and call such a parameter *scope*. It is worth noting that most timeline-based planners like EUROPA, or APSI-TRF, indeed, consider timelines as a sort of "containers" for formulas. In our approach since the core reasoning element are the formulas, and consistently with a classical logical approach, we choose to incorporate the timelines "inside" the formulas. In other words, the type of our scope variables will be a "distinguisher for triggering some further reasoning". Furthermore the resulting scope variables are, to all effects, *variables* and, therefore, could be subject to constraints. We now introduce the the minimal set of the complex types commonly used in timeline-based planning.

State Variables. They are used to describe the "state" of a dynamical system as, for example, the position of a specific object at a given time or a simple manufacturing tool that might be operating or not. The semantics of a state variable (and thus the implicit constraints we need to make explicit) is simply that, for each time instant $t \in \mathbb{T}$, the timeline can assume only one value.

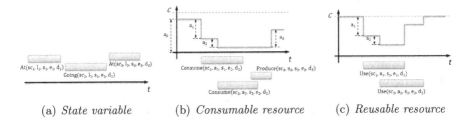

(a) *State variable*   (b) *Consumable resource*   (c) *Reusable resource*

Fig. 3. Different kinds of timelines with predicate instances and resource profiles.

Figure 3(a) represents an example of state variable with three atomic formulas (parameter types are omitted for sake of space). The example shows a robot r_0, a state variable of type *Robot*, which might be *At* a given location or might be *Going* to another location. We thus have the two predicates $At(sc, l, s, e, d)$ and $Going(sc, l, s, e, d)$ each having a parameter sc of type *Robot*, describing the scope of the formulas, and parameters l, s, e and d respectively for the location, the start, the end and the duration. The planner will take care of adding the proper constraints for avoiding the temporal overlapping of the incompatible states (i.e., all the formulas which have the same scope and do not unify) or for "moving" the states on other instances of type *Robot* (i.e., choosing another value, for example r_1, for the scope of the formula).

Resources. They are entities characterized by a *resource level* $\mathcal{L} : \mathbb{T} \to \mathbb{R}$, representing the amount of available resource at any given time, and by a *resource capacity* $\mathcal{C} \in \mathbb{R}$, representing the physical limit of the available resource. We can identify several types of resources depending on how the resource level can be increased or decreased in time. A *consumable resource* is a resource whose level is increased or decreased by some activities in the system. An example of consumable resource is a reservoir which is produced when a plan activity "fills" it (i.e., a tank refueling task) as well as consumed if a plan activity "empties" it (i.e., driving a car uses gas). Consumable resources have two predefined rules, each having an empty body, and a predicate $Produce(sc, id, a, s, e, d)$ ($Consume(sc, id, a, s, e, d)$) as head, so as to represent a resource production (consumption) on the consumable resource sc of amount a from time s to time e with duration d (we use an id parameter to prevent unification among these formulas). In addition, the consumable resource complex type has four member variables representing the initial and the final amount of the resource, the min and the max value for the profile. Quite popular in the scheduling literature, *reusable resources* are similar to consumable resources where productions and consumptions go in tandem at the start and at the end of the activities. Reusable resources can be used for modelling, for example, the number of programmers employed on a given project for a given time interval. Reusable resources have one predefined rule having an empty body and a predicate $Use(sc, id, a, s, e, d)$ as head so as to represent an instantaneous production of resource sc of amount

a at time s and an instantaneous consumption of the same resource sc of the same amount a at time e. In addition, the reusable resource type has a member variable for representing the capacity of the resource. Figures 3(b) and 3(c) represent, respectively, an example of consumable resource and an example of reusable resource with some associated formulas.

By introducing these complex types, we require the reasoner to add further constraints so as to avoid object inconsistencies (e.g., different states overlapping on some state variable; resource levels $\mathcal{L}$ exceeding resource capacity $\mathcal{C}$ or going lower than min, etc.). We chose to refine our resolution process by introducing a step for detecting such inconsistencies and for adding required constraints which would remove them. The resulting basic operations for refining a partial solution π toward a final solution are thus the following:

1. find the (sub)goals of π.
2. select one such (sub)goals.
3. find ways to resolve it.
4. choose a resolver for the (sub)goals.
5. refine π according to that resolver.
6. check for any object inconsistency and remove it.

Similar to [9], we use a lazy approach for detecting inconsistencies. Namely, we let the underlying SMT solver to extract a solution given the current constraints and, in case some inconsistency is detected we add further constraints so as to remove the inconsistency. A simple example should clarify the idea. Let us suppose in a given partial solution there are two formulas describing a state variable sv_k having two overlapping states s_i and s_j, we solve the inconsistency by adding the constraint $[\![s_i.start \geq s_j.end]\!] \vee [\![s_j.start \geq s_i.end]\!] \vee [\![s_i.scope \neq s_j.scope]\!]$ preventing further overlapping of these states on the same state variable. The core idea for solving resource inconsistencies follows a very similar schema.

4 Evaluating iLoC

To assess the value of our planner we discuss two aspects: (a) a direct comparison with other planners with respect of representative problems; (b) a discussion of one of the properties we are pursuing in designing the planner: the flexibility in its use.

4.1 Solving Planning Problems

In order to evaluate our approach we have tried to compare it with different planners on different benchmarking problems. We have had difficulty in finding source code and/or executables of many of the Planning Competition planners so we do not present a complete comparison. Nevertheless we selected three planners that are interesting for their features and compared them with iLoC:

VHPOP [28] shares with our planner the partial ordering approach, OPTIC [2] and COLIN [11] are both based on a classic FF-style forward chaining search [19]. Since there are randomness elements, we ran each problem instance one hundred times and we noted the average execution time.

We start the comparison by solving the Blocks World domain, a workhorse for the planning community. As known, in this domain a set of cubes (blocks) are initially placed on a table. The goal is to build one or more vertical stacks of blocks. The catch is that only one block may be moved at a time: it may either be placed on the table or placed atop another block. Because of this, any blocks that are, at a given time, under another block cannot be moved. As expected, experimental results (see figure 4(a)) show that planners that take advantage from "classical heuristics" perform significantly better than our approach. This shows how domains where causality is a dominant feature still represent a direction for improving iLoC.

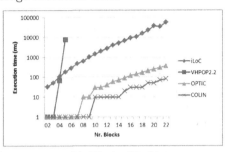

(a) *Blocks World*

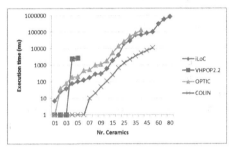

(b) *Temporal Machine Shop*

Without being discouraged, we have checked our system with a second problem named Temporal Machine Shop [13]. The relevance of this problem is twofold. This problem is the only *temporally expressive* problem (i.e., requires concurrency for being solved, see [12]) of the International Planning Competition (IPC) and, within the same competition, it is solved by the sole ITSAT planner (see [25]). The problem models a bak-

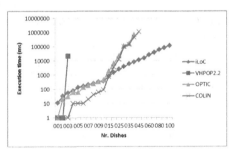

(c) *Cooking Carbonara*

Fig. 4. Experimental results.

ing ceramic domain in which ceramics can be baked while a kiln is firing. Different ceramic types require a different baking time. While a kiln can fire for at most 20 minutes at a time (and then it must be made ready again), baking a ceramic takes, in general, less time, therefore we can save costs by baking them altogether. Additionally, similar to [25], we have slightly complicated the domain by considering the possibility for ceramics to be assembled, so as to produce different structures which should be baked again to obtain the final product. In the experiments, we change both the number of ceramics to be baked and the number of kilns, allowing more parallelism. Figure 4(b) shows that performances of iLoC are comparable with those of state-of-the-art planners. It is worth observing that COLIN performs better than iLoC but is not able to solve problems with more than 50 ceramics since it runs out of memory.

iLoC differences with respect to other approaches become more interesting while solving the Cooking Carbonara domain introduced in [22]. This is another temporally expressive problem in which the aim is the preparation of a meal, as well as its consumption by respecting constraints of warmth. Problems called cooking-carbonara-n require to plan the preparation of n dishes of pasta. The concurrency of actions is required to obtain the goal because it is necessary that the electric hot plates work so that water and oil are hot enough to cook pasta and bacon cubes. It is also necessary to perform this baking in parallel to serve a dish that is still hot during its consumption. It is worth observing that with respect to the previous domain, we do not have a constraint of maximum duration of plate firing, but this problem strongly resembles a basic scheduling problem with renewable resources. This is confirmed by results in Figure 4(c) that show iLoC outperforming the compared solvers.

Although heuristics should be further improved, iLoC seems to show an interesting behaviour when numeric variables (e.g., those representing time) play an important role in the domain. In the current iLoC, the scheduling abilities seem to behave better than the causal reasoning ones, confirming a space for improvements. A separate discussion is worth doing and concerns the expressiveness of iLoC. All the competing planners use the PDDL language (see [16]) for modeling their planning problems and, in general, it is quite cumbersome to impose temporal constraints among plain PDDL actions. In the Cooking Carbonara domain it is important that the cooking happens before the eating but eating should not start too late to avoid that food becomes cold. In [22] a PDDL extension is proposed to overcome this issue and to model properly the domain, however, none of the available planners supports this extension and thus they have been evaluated in a simplified domain in which the warmth constraint decays and dishes can be served anytime after they have been cooked. It is worth noting how this constraint is naturally captured in the iLoC modelling language by creating a rule having as head an action and as body a second action in conjunction with a constraint among the temporal parameters of the two actions.

4.2   Managing Execution Uncertainty with iLoC

We now address a different perspective for evaluating iLoC features. When tackling with plans, indeed, a key aspect to consider is that they have to be executed and, when executed, failures are not uncommon. When dealing with dynamic environments, further constraints might become available at execution time requiring the adaptation of the plan to some real needs or, in some cases, a complete replanning step. Since replanning is often an expensive task, requiring solving a PSPACE problem, generating flexible plans that might be slightly adapted at execution time is highly desirable.

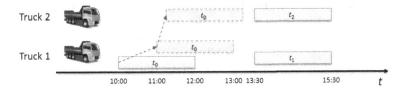

Fig. 5. Plan flexibility for plan adaptation.

Consider, for example, a simple logistic company having a fleet of two trucks (see figure 5). The two tasks t_0 and t_1 are scheduled for Truck 1 respectively from 10:00 to 12:00 and from 13:30 to 15:30. Although task t_0 can be slightly postponed, we have a further constraint asserting it must end strictly before 17:00. However, in order to avoid penalties with its customer, task t_1 must be executed exactly at the scheduled time. Now suppose that at 9:55 our logistic company discovers that Truck 1 has some problem at the engine. Since estimated repair time is one hour, we could temporally adapt the plan for starting t_0 at 11:00 (we achieve this by adding the constraint $[t_0.start >= 11]$). This kind of temporal adaptation is quite common for any planning and execution environment (see, for example, [24] and [6]) up to be considered mandatory. Now imagine at 10:55 we discover that repair time was underestimated and a further hour is required. By moving t_0's start time at 12:00 we would interfere with task t_1, scheduled for time 13:30. Clearly, given these constraints, a replanning phase would assign task t_0 to Truck 2. However, by exploiting the scope parameter introduced earlier, we can adapt the plan by reassigning task t_0 and thus avoiding a possibly time consuming replanning phase. We recall that the scope parameter is indeed a variable, and therefore we can let the reasoner choose a value for it according to involved constraints. In other words, similar to what happens for temporal adaptation, we can perform also more complex forms of adaptation, including this kind of "causal" adaptation. It could be argued that such an adaptation could be expensive as well and, indeed, it is exponential in the worst case. Nevertheless the underlying SMT solver allows efficient constraint propagation, making the adaptation process negligible, with respect to a complete replanning process, on the simple problems that we have used for testing this kind of flexibility. We are currently working to synthesize some benchmark problems for demonstrating the effectiveness of iLoC when such problems arise.

5   Conclusions

This paper has introduced a new planner called iLoC that evolves from our previous work [14] but addresses the solver design from a more general perspective. We have first presented a general framework that integrates logic and constraint based reasoning and a general resolution algorithm for problem solving together with some heuristic reasoning. We have then considered how such a framework could be applied for solving timeline-based planning problems and we have evaluated the framework on three different planning problems (i.e., the blocks world, the temporal machine shop and the cooking carbonara problems). Initial results seem to be encouraging and the flexibility capabilities could be useful in many cases where the ability of managing uncertainty, during plan execution, plays an important role.

It is worth observing how temporally expressive problems, in general, seem to receive little benefit from enhanced state-based heuristics (which, on the contrary, have led to substantial improvements, in recent years, within the classical planning approaches). We believe, indeed, that a domain-independent planner should be able to efficiently solve all kinds of planning problems. For this reason, orthogonal approaches might be valid and useful for the resolution of such problems. Nevertheless, our heuristics need to be further improved, especially for those domains (e.g., the blocks world) in which causal aspects are predominant to the temporal ones. In the future, we also plan to define a formalization of the input language and to include an automatic translation technique from PDDL domains.

Acknowledgments. Authors work is partially funded by the Ambient Assisted Living Joint Program under the SpONSOR project (AAL-2013-6-118). Authors would like to thank Prof. Marta Cialdea Mayer for the time spent reading a previous version of this work and for her precious suggestions. Thanks to the AI*IA reviewers for several comments.

References

1. Apt, K.R., Wallace, M.G.: Constraint Logic Programming Using ECL^{i}PS^e. Cambridge University Press, New York (2007)
2. Benton, J., Coles, A., Coles, A.: Temporal planning with preferences and time-dependent continuous costs. In: Twenty-Second International Conference on Automated Planning and Scheduling (ICAPS) (2012)
3. Bernardini, S., Smith, D.: Developing domain-independent search control for europa2. In: Proceedings of the Workshop on Heuristics for Domain-independent Planning at ICAPS 2007 (2007)
4. Blum, A., Furst, M.L.: Fast planning through planning graph analysis. In: IJCAI, pp. 1636–1642. Morgan Kaufmann (1995)
5. Bonet, B., Geffner, H.: Planning as Heuristic Search. Artificial Intelligence **129**(1–2), 5–33 (2001)
6. Cashmore, M., Fox, M., Larkworthy, T., Long, D., Magazzeni, D.: AUV mission control via temporal planning. In: 2014 IEEE International Conference on Robotics and Automation (ICRA) (2014)
7. Cesta, A., Cortellessa, G., De Benedictis, R.: Training for Crisis Decision Making - An Approach Based on Plan Adaptation. Knowledge-Based Systems **58**, 98–112 (2014)
8. Cesta, A., Cortellessa, G., Fratini, S., Oddi, A.: Developing an end-to-end planning application from a timeline representation framework. In: Proceedings of the 21st Innovative Applications of Artificial Intelligence Conference (IAAI) (2009)
9. Cesta, A., Oddi, A., Smith, S.F.: A Constraint-based Method for Project Scheduling with Time Windows. Journal of Heuristics **8**(1), 109–136 (2002)

10. Chien, S., Tran, D., Rabideau, G., Schaffer, S., Mandl, D., Frye, S.: Timeline-based space operations scheduling with external constraints. In: Proc. of the 20th Int. Conf. on Automated Planning and Scheduling (ICAPS) (2010)
11. Coles, A.J., Coles, A.I., Fox, M., Long, D.: COLIN: Planning with Continuous Linear Numeric Change. Journal of Artificial Intelligence Research 44, 1–96 (2012)
12. Cushing, W., Kambhampati, S., Mausam, Weld, D.S.: When is temporal planning really temporal? In: Proceedings of the 20th International Joint Conference on Artificial Intelligence (IJCAI) (2007)
13. Cushing, W., Weld, D.S., Kambhampati, S., Mausam, Talamadupula, K.: Evaluating temporal planning domains. In: Proceedings of the 17th International Conference on Automated Planning and Scheduling (ICAPS) (2007)
14. De Benedictis, R., Cesta, A.: Timeline planning in the J-TRE environment. In: Filipe, J., Fred, A. (eds.) ICAART 2012. CCIS, vol. 358, pp. 218–233. Springer, Heidelberg (2013)
15. de Moura, L., Bjørner, N.S.: Z3: An efficient SMT solver. In: Ramakrishnan, C.R., Rehof, J. (eds.) TACAS 2008. LNCS, vol. 4963, pp. 337–340. Springer, Heidelberg (2008)
16. Fox, M., Long, D.: PDDL2.1: An Extension to PDDL for Expressing Temporal Planning Domains. Journal of Artificial Intelligence Research 20, 61–124 (2003)
17. Frank, J., Jónsson, A.K.: Constraint-Based Attribute and Interval Planning. Constraints 8(4), 339–364 (2003)
18. Ghallab, M., Laruelle, H.: Representation and control in IxTeT, a temporal planner. In: Proceedings of the 2nd Int. Conf. on AI Planning and Scheduling (AIPS) (1994)
19. Hoffmann, J.: FF: The Fast-Forward Planning System. AI Magazine 22(3), 57–62 (2001)
20. Jonsson, A., Morris, P., Muscettola, N., Rajan, K., Smith, B.: Planning in interplanetary space: theory and practice. In: Proceedings of the Fifth Int. Conf. on AI Planning and Scheduling (AIPS) (2000)
21. Laborie, P.: Algorithms for propagating resource constraints in AI planning and scheduling: existing approaches and new results. Artificial Intelligence 143, 151–188 (2003)
22. Maris, F., Régnier, P.: Planification temporellement expressive. TLP-GP, un planificateur pour la résolution de problèmes temporellement expressifs. Revue d'Intelligence Artificielle 24(4), 445–464 (2010)
23. Muscettola, N.: HSTS: Integrating planning and scheduling. In: Zweben, M. and Fox, M.S. (ed.) Intelligent Scheduling. Morgan Kauffmann (1994)
24. Py, F., Rajan, K., McGann, C.: A systematic agent framework for situated autonomous systems. In: 9th International Conference on Autonomous Agents and Multiagent Systems (AAMAS) (2010)
25. Rankooh, M.F., Mahjoob, A., Ghassem-Sani, G.: Using satisfiability for non-optimal temporal planning. In: del Cerro, L.F., Herzig, A., Mengin, J. (eds.) JELIA 2012. LNCS, vol. 7519, pp. 176–188. Springer, Heidelberg (2012)
26. Robinson, J.A.: A Machine-Oriented Logic Based on the Resolution Principle. Journal of the Association for Computing Machinery 12(1), 23–41 (1965)
27. Sebastiani, R.: Lazy Satisfiability Modulo Theories. JSAT 3, 141–224 (2007)
28. Simmons, R.G., Younes, H.L.S.: VHPOP: Versatile Heuristic Partial Order Planner. J. Artif. Intell. Res. 20, 405–430 (2003)
29. Weld, D.S.: An Introduction to Least Commitment Planning. AI Magazine 15(4), 27–61 (1994)

ASCoL: A Tool for Improving Automatic Planning Domain Model Acquisition

Rabia Jilani[✉], Andrew Crampton, Diane Kitchin, and Mauro Vallati

School of Computing and Engineering, University of Huddersfield, Huddersfield, UK
{rabia.jilani,a.crampton,d.kitchin,m.vallati}@hud.ac.uk

Abstract. Intelligent agents solving problems in the real world require domain models containing widespread knowledge of the world.

AI Planning requires domain models. Synthesising operator descriptions and domain specific constraints by hand for AI planning domain models is time intense, error-prone and challenging. To alleviate this, automatic domain model acquisition techniques have been introduced. Amongst others, the LOCM and LOCM2 systems require as input some plan traces only, and are effectively able to automatically encode a large part of the domain knowledge. In particular, LOCM effectively determines the dynamic part of the domain model. On the other hand, the static part of the domain – i.e., the underlying structure of the domain that can not be dynamically changed, but that affects the way in which actions can be performed – is usually missed, since it can hardly be derived by observing transitions only.

In this paper we introduce ASCoL, a tool that exploits graph analysis for automatically identifying static relations, in order to enhance planning domain models. ASCoL has been evaluated on domain models generated by LOCM for international planning competition domains, and has been shown to be effective.

Keywords: Automated planning · Knowledge engineering · Domain model acquisition

1 Introduction

AI Planning is the process of finding a plan, i.e. a sequence of actions that applied in an initial state reach the desired goals. Planning is a pivotal task that has to be performed by every autonomous system. Planning techniques require domain models as input. Domain models encode the knowledge of the domains in terms of actions that can be executed and relevant properties. Creating such models is a difficult task, that is usually done manually; it requires planning experts and it is time-consuming and error-prone.

The domain model acquisition problem has mainly been tackled by exploiting two approaches. On the one hand, knowledge engineering tools for planning have been introduced over time, for supporting human experts in modelling the knowledge. Two particular examples are itSIMPLE [15] and GIPO [14].

© Springer International Publishing Switzerland 2015
M. Gavanelli et al. (Eds.): AI*IA 2015, LNAI 9336, pp. 438–451, 2015.
DOI: 10.1007/978-3-319-24309-2_33

A review of the state of the art is provided by Shah et al. [13]. Recently, also crowdsourcing has been exploited for acquiring planning domain models [17]. On the other hand, a number of techniques are currently available for automatic domain model acquisition; they rely on example data for deriving domain models. Significant differences can be found in terms of the quantity and quality of the required inputs. For a detailed overview, the interested reader is referred to [10].

The LOCM systems [2,3] perform automated generation of the dynamic aspects of a planning domain model, i.e. changes in the state of the world occurring due to action execution, by considering a set of plan traces, only. A plan trace is a sequence of actions that when applied in an initial state, reach the desired goals. No additional knowledge about initial, goal or intermediate states is needed by LOCM. In comparison with other systems, LOCM approaches require a minimal amount of information; other systems also require at least partial state information. One drawback of the LOCM process is it can induce only a partial domain model which represents the dynamic aspects of objects and not the static aspects. Static aspects can be seen as relations that appear in the preconditions of operators only, and not in the effects. Therefore, static facts never change in the world, but are essential for modelling the correct action execution. This is problematic since most domains require static predicates to both restrict the number of possible actions and correctly encode real-world constraints. This is the case in well-known benchmark domains like *Driverlog*, in which static predicates represent the connections of roads; the level of floors in *Miconic*, or the fixed stacking relationships between specific cards in *Freecell*.

Any missing static relations are manually introduced into the domain models provided by the LOCM systems. Recently, the LOP approach has been proposed [7]. In order to extract static relations for extending LOCM-generated domain models, LOP exploits optimal goal-oriented plan traces. Specifically, they compare the optimal input plans with the optimal plans found by using the extended domain models; if the latter are shorter, some static relations are missing. This approach has drawbacks, as LOP strongly depends on the availability of optimal plans; they are usually hard to obtain for non-trivially solvable problems. It should be noted that trivial problems usually lead to extremely short plans, which tend not to be very informative for extracting knowledge. Moreover, LOP identifies the need for a static predicate between a set of action parameters, but does not provide information about the type of relationship that connects the involved predicates.

In this paper we present ASCoL (Automated Static Constraint Learner), a method that can effectively identify static relations between predicates by considering any type of plan traces: they can be either optimal / suboptimal goal-oriented or random walks. By exploiting a directed graph representation of operator arguments' relations, ASCoL is also able to identify the type of relation that pairs of predicates show. Relations between pairs can be extended for deriving more complex n-arity static predicates. A preliminary version of ASCoL has been presented in [9]; this version was able to identify inequality constraints only. In a large experimental analysis, we demonstrate the ability of

ASCoL in finding static relations for enhancing domain models automatically acquired by LOCM. Remarkable results have been achieved in complex, well-known benchmark domains, with regards to the number of static relations, like *TPP* (Travelling and Purchase Problem) and *Freecell*. It should be noted that ASCoL can also be used without LOCM. In particular, we observed that it can be useful for debugging domain models, in order to identify missing static relations, or to further constrain the search space by pruning useless actions.

The remainder of this paper is organised as follows. Section 2 provides the necessary background on AI Planning and describes the addressed learning problem based on Knowledge Engineering (KE) for AI Planning. Section 3 gives a detailed description of ASCoL. Section 4 presents the results of the empirical evaluation. Finally, Section 5 concludes the paper and gives directions for future research.

2 Background

In this Section we firstly provide the necessary background on AI Planning, and then introduce the learning problem addressed by ASCoL.

2.1 AI Planning

In this work we consider classical planning [5] domain models. Classical planning deals with finding a (partially or totally ordered) sequence of actions transforming the static, deterministic and fully observable environment from some initial state to a desired goal state. In the classical representation *atoms* are predicates. *States* are defined as sets of ground (positive) atoms. A *planning operator* $op = (name(o), pre(o), eff^-(o), eff^+(o))$ is specified such that $name(o) = op\_name(x_1, \ldots, x_k)$ (op_name is a unique operator name and $x_1, \ldots x_k$ are variable symbols (arguments of certain types) appearing in the operator), $pre(o)$ is a set of predicates representing the operator's preconditions, $eff^-(o)$ and $eff^+(o)$ are sets of predicates representing the operator's negative and positive effects. *Actions* are ground instances of planning operators. An action $A = (pre(A), eff^-(A), eff^+(A))$ is *applicable* in a state s if and only if $pre(A) \subseteq s$. Application of A in s (if possible) results in a state $(s \setminus eff^-(A)) \cup eff^+(A)$.

A *planning domain model* is specified via sets of predicates and planning operators. A *planning problem* is specified via a planning domain, initial state and set of goal atoms. A *solution plan* is a sequence of actions such that a consecutive application of the actions in the plan (starting in the initial state) results in a state that satisfies the goal.

Specifically, the *planning domain model* (hereinafter called domain description, action model or operator schema) has two major elements:

1. Dynamic Knowledge: a set of parametrised action schema representing generic actions and resulting effects in the domain under study.

2. Static Knowledge: relationships/constraints that are implicit in the set of operators and are not directly expressed in the plans. These can be seen as predicates that appear in the preconditions of operators only, and not in the effects. Therefore, static facts never change in the world, but are essential for modelling the correct action execution. According to Wickler [16], let $Op = \{op_1, op_2, \ldots, op_n\}$ be a set of operators and let $Pr = \{Pr_1, Pr_2, \ldots, Pr_n\}$ be a set of all the predicate symbols that occur in these operators. A predicate $Pr_i \in Pr$ is fluent *iff* there is an operator $op_j \in Op$ that has an effect that changes the truth of the predicate Pr_i. Otherwise the predicate is static.

For instance, in the drive operator of the *TPP* [1] domain model used in the International Planning Competitions (IPC) [2], the *connected* predicate is a static relation between two places. There is no operator in the TPP domain model that can change that predicate, but it is fundamental for modelling the domain.

```
(:action drive-ipc
 :parameters
 (?t - truck ?from ?to - place)
 :precondition
 (and
     (at ?t ?from)
     (connected ?from ?to))
 :effect
 (and
     (not (at ?t ?from))
     (at ?t ?to))
)
```

ASCoL accepts input plans (plan traces) in the same text-based format supported by LOCM. i.e. a training sequence of N actions in order of occurrence, which all have the form:

$$A_i(O_{i1}, \ldots, O_{ij}) \quad for \quad i = 1, \ldots, N$$

Where A is the action name and O is the action's object name. Each action (A) in the plan is stated as a name and a list of arguments. In each action (A_i), there are j arguments where each argument is an object (O) of the problem.

2.2   The Learning Problem

We define the learning problem that ASCoL addresses as follows. Given the knowledge about object types, operators and predicates, and a set of plan traces, how can we automatically identify the static relation predicates that are needed by operators' preconditions? We base our methodology on the assumption that

---

[1] http://www.plg.inf.uc3m.es/ipc2011-learning/attachments/Domains/tpp.pddl
[2] http://ipc.icaps-conference.org/

the input plan traces contain tacit knowledge about constraints validation and acquisition.

Specifically, a *learning problem description* is a tuple *(P, T)*, where *P* is a set of plan traces and *T* is a set of types of action arguments in *P*. The *output* for a learning problem is a *constraint repository R* that stores all admissible constraints on the arguments of each action *A* in plan traces *P*.

3   The ASCoL Method

We now present the ASCoL method that has been developed for identifying useful static relations. ASCoL requires as input a set of plan traces and a partial domain model. The provided output is an extended domain model including static relations. The process can be summarised as follows:

1. Read the partial domain model and the plan traces.
2. Identify, for all operators, all the pairs of arguments involving the same object types.
3. For each of the pairs, generate a directed graph by considering the objects involved in the matching actions from the plan traces.
4. Analyse the directed graphs and extract hidden static relations between arguments.
5. Run inequality check.
6. Return the extended domain model that includes the identified static relations.

It should be noted that ASCoL relies on two assumptions:

– The dynamic part of the domain model provided as input is correct, and includes the type of each operator argument.
– The plan traces have been generated by considering problems that share the same objects, and object names.

With regards to the first assumption, LOCM systems are able to effectively model the dynamic side of domain models. In terms of plan traces, the ASCoL assumption is particularly reasonable when traces come from real-world application observations; in that scenario sensors are identifying and naming objects – they usually exploit fixed names. Moreover, it is acceptable to assume that the relevant objects of a real-world scenario will not quickly change. For example, machines in a factory or trucks and depots for logistics companies.

The main information available for ASCoL comes from the input plan traces. While reading plan traces as a first step, we remove from the plan traces all the actions that refer to operators that do not contain at least two arguments of the same type.

This condition is due to the reason that, although, theoretically, static relations can hold between objects of different types, they mostly arise between same-typed objects. For example in the *Freecell* domain, where static relations define allowed sequential arrangement between cards but not between cards and cells/columns. Moreover, considering only same-typed object pairs can reduce the computational time required for identifying relations. It is also worth noting that, in most of the cases where static relations involve objects of different types, this is due to a non-optimal modelling process. Furthermore, such relations can be easily identified by naively checking the objects involved in actions; whenever some objects of different type always appear together, they are likely to be statically related.

For the purpose of illustrating the ASCoL process, we consider as a domain model example *Freecell* as our running example, used in the IPC3 planning competition [11]. It is a STRIPS [4] encoding of a card game (similar to Solitaire) that comes free with Microsoft Windows. Starting from an initial state, with a random configuration of cards across eight columns, the user can move cards in a specified order onto four home cells, following typical card stacking rules, and using a number of free cells as a resource. The domain-specific static constraints of the Freecell domain are the allowed sequential arrangement of cards in the free cells, the home cells and among the card columns using actions such as `colfromfreecell` and `sendtohomecell`.

The Freecell domain provides a suitable framework to integrate our approach. We choose this domain as an illustrative example because all of its ten operators satisfy the above mentioned condition of same types and it is rich in terms of a challenge to learn static facts, e.g. *Can-Stack* and *Successor* constraints. Freecell plan traces only contain three types, i.e. *card*, *suit* and *num*. This domain is made more complex by the fact that one single type *num* is used to represent three things and this in turn removes the boundary between three different behaviours. *num* is used to represent the face value of cards and to count free cells and free columns. Let us consider the `sendtofree-b` operator. The type signature of the operator is as follows:

`sendtofree-b(?card -card ?cells ?ncells ?cols ?ncols -num)`

This operator sends the *?card* to a free cell, considering this card as the last one in the column. *?cells* and *?ncells* represent the number of available free cells while *?cols* and *?ncols* represent the number of empty columns after sending the last card of the column to a free cell. Here type *num* is used to represent substitute of free cell count as well as empty column count.

3.1 Generation of Directed Graphs

Step 2 of ASCoL scans the partial domain model provided. This is done in order to identify all the directed graphs that have to be generated. For each operator satisfying the aforementioned condition of at least two same-typed arguments, a directed graph is generated between each ordered couple of same-typed arguments. At Step 3, each of the previously identified directed graphs $G = (IDs, Conn)$ is generated. G is generated by considering the objects that appear in actions of plan traces, corresponding to instantiation of the operators' arguments. Specifically, *IDs* is the set of vertices of the graph G, which includes all the objects observed from plan traces; *Conn* is the finite set of arcs in G. We decide to generate an arc that goes from the object used in the first considered argument, to the second. Therefore, an arc (o_1, o_2) is directed from o_1 to o_2, and it is added to *Conn* if the objects o_1 and o_2 appear, in this order, in the place of the considered arguments of one action of the plan trace.

To explain the structure of G better, we continue with our example. Considering the sendtofree-b action, the following pairs of arguments are identified at Step 2: pair1 (?cell, ?ncells), pair2 (?cell, ?cols), pair3 (?cell, ?ncols), pair4 (?ncells, ?cols), pair5 (?ncells, ?ncols), pair6 (?cols, ?ncols). For each of the identified pair of arguments, a directed graph is generated, by considering the objects used in plan traces. In order to exemplify how directed graphs are generated, let us suppose that from a plan trace we collected the following two instances of the sendtofree-b action.

```
sendtofree-b(cardX N1 N0 N3 N4)
sendtofree-b(cardY N2 N1 N2 N3)
```

In order to generate the directed graph for the pair1 arguments, ASCoL considers all the objects used as third and fourth arguments of the sendtofree-b action. Given our example, $IDs=\{N1, N0, N2\}$. The *Conn* set includes the following arcs: $\{(N1, N0), (N2, N1)\}$. This is because, in the first instance of sendtofree-b, $N1$ is before $N0$, and in the second $N2$ appears before $N1$. Hence, the directed graph G of our example has the following structure:

$$N2 \to N1 \to N0$$

3.2   Analysis of Directed Graphs

In step 4, the generated directed graphs are investigated. In particular, we are interested in identifying their structure, for determining whether there is a specific ordering among the involved objects, and consequently among the arguments of corresponding operators. The directed graphs can have three different shapes:

Totally Ordered Graph. Given the directed graph G, for each pair of vertices a, b we say that $a > b$ if there exists a directed route that connects a to b. Moreover, we say that $a = b$ only if a and b are the same vertex. This ordered relation between vertices allows checking for total ordering among them. Total ordering is assessed by checking the trichotomy property for each pair of vertices [1]. The trichotomy property states that given two objects a and b, exactly one relation between $a < b$, $a = b$, $a > b$ holds. In addition to totality in relationships, also the following properties can hold:

1. Transitivity: $a > b$ and $b > c$ implies $a > c$.
2. Antisimmetry: $a > b$ and $b > a$ implies $a = b$.

Cyclic Graph. In a cyclic graph, there exists at least one route that can return to its starting vertex.

Acyclic Graph. A directed acyclic graph is a graph with no directed cycles. Starting from any vertex, there is no directed route to return to the starting point.

ASCoL examines the structure of all the directed graphs, in order to identify static relations between the corresponding pairs of operator arguments. If the graph G is proved to be totally ordered, this indicates a strong static relation between the arguments. In particular, the arguments are used as a scale of values, i.e. it is always possible to identify, given two values, an ordering between them. This sort of structure is usually exploited in STRIPS for modelling levels or quantities. For instance, in the well-known Zenotravel domain, fuel levels of aircraft are modelled by using `next` static-predicates, which are used for providing an ordering between the different `flevel`.

In case of a directed cyclic graph, ASCoL tests if G is fully connected or not. In the former case, a static relation of type "connection" is added to the preconditions of the operator. This is because, mainly in transport-like domains, these sort of graphs indicate the presence of an underlying map of strongly connected locations.

Finally, if the graph is acyclic, the current version of ASCoL can not derive any information. An acyclic graph can either indicate that no static relation is in place between the considered arguments, or there exists some partial ordering between them. For instance, the presence of a not-strongly connected map of locations; this is particularly true if plan traces are not generated by random walks.

3.3 Inequality Check

For each graph G that has been proven to be cyclic, ASCoL checks the presence of self-loops on vertices. If no loops are observed, then an inequality precondition (i.e., a precondition that forces the two arguments to be different) is added to the operator.

This is the case of the `sail` operator from the *Ferry* domain. The two locations shall never be equal. The following shows the enhanced operator:

```
(:action sail
    :parameters (?from - location ?to - location)
    :precondition (and
                (not (= ?from ?to))
                (at-ferry ?from))
    :effect (and
                (at-ferry ?to)
                (not (at-ferry ?from)))) 
```

ASCoL forces inequality by using *negative precondition* and *equality* features of PDDL [12]. Currently, many planners support such features, also due to the use of them in recent IPCs.

4   Experimental Evaluation

The aim of this experimental analysis is to assess the ability of the ASCoL technique in identifying static relationships between arguments.

15 domain models have been considered, taken either from IPCs or from the FF domain collection (FFd)[3]: Barman (IPC7), Driverlog (IPC3), Ferry (FFd), Freecell (IPC3), Gold-miner (IPC6), Gripper (IPC7), Hanoi (FFd), Logistics (IPC2), Miconic (IPC2), Mprime (IPC1), TPP (IPC5), Trucks (IPC5), Spanner (IPC6), Storage (IPC5) and ZenoTravel (IPC3). These domains have been selected because they are encoded using different models and modelling strategies, and their operators include more than one argument per object type. All domains but Gripper, Logistics and Hanoi, exploit static relations. Plans have been generated by using Metric-FF planner [8] on randomly generated problems, sharing the same objects. ASCoL has been implemented in Java, and run on a Core 2 Duo/8GB processor.

Table 1 shows the results of the experimental analysis. Results are shown in terms of the number of identified static relationships (Learnt SR) and number of additional static relations provided (Additional SR) that were not included in the original domain model. We also provide information about the number of plans required for ASCoL to converge – i.e. adding more plans does not change the number of identified static relations –, the average length of plans, and the CPU time required by ASCoL. Interestingly, we observe that ASCoL is usually able to identify all the static relations of the considered domains. Moreover, in some domains it is providing additional static relations, which are not included in the original domain model. Remarkably, such additional relations do not reduce the solvability of problems, but reduce the size of the search space by pruning useless instantiations of operators. We also compared our results with results of LOP system for a few prominent domains.

With regards to the number of plans required by ASCoL to converge, we observe that in many cases, a few "long" plans are enough. Apparently, there is not a strong relation between the number of required plans and the overall number of static relations. Instead, a larger number of plans is needed when some operators are rarely used, or if the number of arguments of the same type are high, per operator.

Considering a classification terminology, we can divide the relations identified by ASCoL in to four classes: true positive, true negative, false positive and false negative.

True Positive. These are correctly identified static relations. Relations identified by ASCoL are almost always static relations which are included in the original domain models.

True Negative. Dynamic relations that are correctly not encoded as static relations. ASCoL did not identify a static relation between arguments that are actually connected by a dynamic relation in any of the considered domains.

---

[3] https://fai.cs.uni-saarland.de/hoffmann/ff-domains.html

Table 1. Overall results on considered domains. For each original domain, the number of operators (# Operators), and the total number of static relations (# SR) are presented. ASCoL results are shown in terms of the number of identified static relationships (Learnt SR) and number of additional static relations provided (Additional SR) that were not included in the original domain model. Such relations do not compromise the solvability of problems, but prune the search space. The seventh and eighth columns indicate respectively the number of plans provided in input to ASCoL, that allows it to converge, and the average number of actions per plan (A/P). The last column shows the CPU-time in milliseconds

Domain	# Operators	# SR	Learnt SR	Additional SR	# Plans	Avg. A/P	CPU-time
TPP	4	7	7	0	7	28	171
Zenotravel	5	4	6	2	4	24	109
Miconic	4	2	2	0	1	177	143
Storage	5	5	5	0	24	15	175
Freecell	10	19	13	0	20	60	320
Hanoi	1	0	1	1	1	60	140
Logistics	6	0	1	1	3	12	98
Driverlog	6	2	2	0	3	12	35
Mprime	4	7	7	0	10	30	190
Spanner	3	1	1	0	1	8	144
Gripper	3	0	1	1	1	14	10
Ferry	3	1	2	1	1	18	130
Barman	12	3	3	0	1	150	158
Gold-miner	7	3	1	0	13	20	128
Trucks	4	3	3	0	6	25	158

False Positive. In some domains ASCoL infers one or two additional relations facts that are not included in the original domain model. From the Knowledge Engineering point of view, and considering the fact that such additional preconditions do not reduce the solvability of problems, such inferred relations can add value to the original model in terms of effectiveness of plan generation.

False Negative. In Freecell and Gold-miner domains ASCoL does not identify all the static relations.

The ability of ASCoL to correctly identify static relations that should be included as preconditions of specific operators, depends on the number of times the particular operator appears in the provided plan traces. The higher the number of instances of the operator in the plan, the higher the probability that ASCoL will correctly identify all the static relations. For example, in TPP domain only 2 plans of average length 200 actions each yields the learning of 85% of the static relations. With the addition of another plan of length 200 actions in input, 100% are identified. TPP domain is one of the more complex considered domains, in terms of having more same-type arguments (4 same typed) in 3 out of 4 operator headings. We now discuss some of the most interesting results.

4.1   Freecell

In this domain there are six cases out of nineteen where ASCoL can not identify static relations. We observe that such relations are somehow peculiar to the domain. They describe stacking rule relations between cards of different suits. In particular, those relations model, at the same time, the fact that one card $c1$ can be stacked over another card $c2$ if: (i) the colour of the suit of $c1$ is red, and the colour of the suit of $c2$ is black (or vice-versa), and (ii) the nominal value of $c1$ is smaller (higher if stacking at home) then the value of $c2$.

Apart from the aforementioned six cases, which compress two static relations in a single predicate, all the others are correctly identified by ASCoL. Remarkably, even though `sendtofree-b` and `newcolfromfreecell` are operators that appear rarely in plan traces (and are used rarely in the card game as well as they model the fact that one card is sent to a free cell or to a free column), ASCoL correctly identifies all their static preconditions.

Due to the rare occurrence of the above mentioned two operators, it prevents ASCoL from learning the corresponding relations from just a few plans. A large number of plans are required; at least 20. However, only 8 plan traces are required for correctly identifying the static relations of the remaining operators. The high number of plans needed is also due to the fact that num type models different aspects of the game. Empirically, we observed that for reducing the number of plans, splitting the num type is useful. num is used for modelling three different aspects of the game, therefore we split it as follows: $freeCol$ to keep a record of free cells available, col to keep record of empty columns and $home$ to record arrangement of cards in home cell. This way ASCoL only requires 8 plans to learn all the relations.

4.2   TPP Domain

TPP is an example of one of the most complex domains for the ASCoL approach. Out of 4 operators, 3 contain 7 arguments each. Out of the 7, 4 arguments are of the same type $level$. The following shows the mentioned operators; argument names indicate the corresponding type.

```
buy (goods truck market level1 level2 level3 level4)
unload (goods truck depot level1 level2 level3 level4)
load (truck goods market level1 level2 level3 level4)
```

The large number of arguments of the same type gives rise to six pairs per operator, with the corresponding graphs. Out of six pairs, ASCoL correctly identifies the two static binary relations as the benchmark domain also contains only two static facts in each i.e. *(next level2 level1) (next level4 level3)*. In contrast, LOP system is not able to identify these facts as it depends on a LOCM2 encoded domain model for its input. LOCM2 fails to induce complete dynamics of the TPP domain.

4.3   Miconic Domain

Miconic domain is a STRIPS encoding of a system to transport a number of passengers with an elevator from their origin to their destination floors. It has two static predicates: one in *up* and the other in *down* actions. Such static relations are used for indicating the ordering of arguments of type *floor*, i.e. the order between the different floors which are connected by the elevator.

In this domain, the predicate above is used for modelling the fact that one floor is above the lower one. Such predicate is used both for up and down movements of the elevator, but in a different way.

ASCoL successfully detects the presence of a link between floors, but generates two different static relations, one per operator. Although different from the original model, such an encoding is an alternative way for modelling the relation between floors. In contrast, LOP system is not able to identify these relationships as due to more groundings for up and down actions in plans, this never leads to a shorter plan, and even an empty static fact does not preserve the optimality of input plans [7].

4.4   Gold-Miner

Gold-miner domain is a STRIPS encoding of a scenario where a robot is in a mine (grid) and has the goal of reaching a location that contains gold by shooting through the rocks. It exploits a static relation (namely, *connected*), which indicates that two cells are connected, in three operators: *move*, *detonate-bomb* and *fire-laser*.

In our experimental analysis, we generated plans using the problem set provided for IPC6. By considering the corresponding plans, ASCoL learns easily the connection fact used in the *move* operator. Interestingly, ASCoL is not able to correctly identify the same static relationship in the remaining two operators. This is due to the fact that both *detonate-bomb* and *fire-laser* are rarely used, usually once per plan. Moreover, in the considered IPC6 problems, the structure of problems forces the robot to fire the laser (or detonate bombs) within the same row of the grid, but not across different columns. Changing the structure of problems slightly, so that the robot can detonate bombs (or fire laser) also between cells of different columns, allows ASCoL to learn all the static relations of the gold-miner domain model. It should be noted that changing the problems does not modify the overall structure of the domain model. Moreover, the very peculiar shape of the considered problems is possibly due to a bug in the random problem generator.

5   Conclusion and Future Goals

Planning is a fundamental task for autonomous agents. In order to perform planning, it is critical to have a model of the domain in which the agent has to interact and operate. Encoding domain knowledge can be done manually, but it requires both planning and domain experts. Moreover, it is time-consuming and error-prone. To overcome such issues, techniques for automatic domain acquisition have been proposed over time. Most of them require a large amount of knowledge to be provided as input. Currently, LOCM systems can effectively encode the dynamic part of domain models by analysing plan traces, only. Static knowledge, that is represented as preconditions in PDDL operators, can not be identified by LOCM or similar approaches.

In this paper we introduced ASCoL, an efficient and effective method for identifying static knowledge missed by domain models automatically acquired. The proposed approach generates a directed graph for each pair of same-type arguments of operators and, by analysing some properties of the graphs, identifies relevant relations between arguments. Remarkably, the contributions of ASCoL, as demonstrated by our large experimental analysis, are: (i) the ability to effectively identify static relations for a wide range of problems, by exploiting graph analysis; (ii) ASCoL can work with both optimal and suboptimal plan traces; (iii) considering pairs of same-typed objects allow the identification of all the static relations considered in the benchmark models, and (iv) it can be a useful debugging tool for improving existing models, which can indicate hidden static relations helpful for pruning the search space.

ASCoL can be exploited both by domain and planning experts. Domain experts can use ASCoL for improving their PDDL knowledge, and generating more robust and more optimised domain models. Planning experts can exploit the proposed method for identifying hidden static relations between different aspects of the domain, thus improving their understanding of the specific application.

We see several avenues for future work. We plan to thoroughly test the differences between using different kinds of plan traces, e.g., random walks and goal-oriented generated by different planners. Grant, in [6], discusses the limitations of using plan traces as the source of input information. ASCoL faces similar difficulties as the only input source to verify constraints are sequences of plans. We are also interested in extending our approach for considering static relations that involve more than two arguments; although they are not usually exploited in domain models, and can be broken down to a set of static relations between pairs, they can generally be useful. In particular, we aim to extend the approach for merging graphs of different couples of arguments. Finally, we plan to identify heuristics for extracting useful information also from acyclic graphs.

References

1. Apostol, T.M.: Calculus, vol. I. John Wiley & Sons (2007)
2. Cresswell, S., McCluskey, T.L., West, M.M.: Acquisition of object-centred domain models from planning examples. In: Proceedings of the International Conference on Automated Planning and Scheduling (ICAPS) (2009)
3. Cresswell, S.N., McCluskey, T.L., West, M.M.: Acquiring planning domain models using LOCM. The Knowledge Engineering Review 28(02), 195–213 (2013)
4. Fikes, R.E., Nilsson, N.J.: STRIPS: A new approach to the application of theorem proving to problem solving. Artificial Intelligence 2(3), 189–208 (1972)
5. Ghallab, M., Nau, D., Traverso, P.: Automated planning: theory & practice (2004)
6. Grant, T.: Identifying Domain Invariants from an Object-Relationship Model. PlanSIG2010, 57 (2010)

7. Gregory, P., Cresswell, S.: Domain model acquisition in the presence of static relations in the LOP system. In: Proceedings of the International Conference on Automated Planning and Scheduling (ICAPS) (2015)
8. Hoffmann, J.: The Metric-FF Planning System: Translating "gnoring Delete Lists" to Numeric State Variables **20**, 291–341 (2003)
9. Jilani, R., Crampton, A., Kitchin, D.E., Vallati, M.: ASCoL: automated acquisition of domain specific static constraints from plan traces. In: The UK Planning and Scheduling Special Interest Group (UK PlanSIG) 2014 (2014)
10. Jilani, R., Crampton, A., Kitchin, D.E., Vallati, M.: Automated knowledge engineering tools in planning: state-of-the-art and future challenges. In: The Knowledge Engineering for Planning and Scheduling workshop (KEPS) (2014)
11. Long, D., Fox, M.: The 3rd International Planning Competition: Results and analysis. J. Artif. Intell. Res. (JAIR) **20**, 1–59 (2003)
12. Mcdermott, D., Ghallab, M., Howe, A., Knoblock, C., Ram, A., Veloso, M., Weld, D., Wilkins, D.: PDDL - The Planning Domain Definition Language. Tech. rep., CVC TR-98-003/DCS TR-1165, Yale Center for Computational Vision and Control (1998)
13. Shah, M., Chrpa, L., Jimoh, F., Kitchin, D., McCluskey, T., Parkinson, S., Vallati, M.: Knowledge engineering tools in planning: state-of-the-art and future challenges. In: Proceedings of the Workshop on Knowledge Engineering for Planning and Scheduling (2013)
14. Simpson, R.M., Kitchin, D.E., McCluskey, T.: Planning domain definition using gipo. The Knowledge Engineering Review **22**(02), 117–134 (2007)
15. Vaquero, T.S., Romero, V., Tonidandel, F., Silva, J.R.: itSIMPLE 2.0: An integrated tool for designing planning domains. In: Proceedings of the International Conference on Automated Planning and Scheduling (ICAPS), pp. 336–343 (2007)
16. Wickler, G.: Using planning domain features to facilitate knowledge engineering. In: KEPS 2011 (2011)
17. Zhuo, H.H.: Crowdsourced action-model acquisition for planning. In: Proceedings of the AAAI Conference on Artificial Intelligence (2015)

Approaching Qualitative Spatial Reasoning About Distances and Directions in Robotics

Guglielmo Gemignani[✉], Roberto Capobianco, and Daniele Nardi

Department of Computer, Control, and Management Engineering "Antonio Ruberti",
Sapienza University of Rome, Rome, Italy
{gemignani,capobianco,nardi}@dis.uniroma1.it

Abstract. One of the long-term goals of our society is to build robots able to live side by side with humans. In order to do so, robots need to be able to reason in a qualitative way. To this end, over the last years, the Artificial Intelligence research community has developed a considerable amount of qualitative reasoners. The majority of such approaches, however, has been developed under the assumption that suitable representations of the world were available. In this paper, we propose a method for performing qualitative spatial reasoning in robotics on abstract representations of environments, automatically extracted from metric maps. Both the representation and the reasoner are used to perform the grounding of commands vocally given by the user. The approach has been verified on a real robot interacting with several non-expert users.

1   Introduction

One of the long-term goals of our society is to build robots able to live side by side with humans, interacting with them in order to understand the surrounding world. However, to reach such a goal, robots need to be able to understand what humans communicate (Fig. 1). The conventional numeric approach used in robotics is in fact deeply different from natural language interaction between people. The former is based on precise metric information about already well known environments, in which each element is uniquely specified only through its coordinates. The latter can deal, instead, with ambiguities and spatial uncertainties, which are solved by referring to purely qualitative properties of objects, or relations among them: correct grounding of spoken information and places of the world can be obtained even with incomplete spatial knowledge. Many difficulties arise when trying to move from a numeric representation of the world to a qualitative one. Especially if the commands to be executed by the robot are given through speech.

Many theories have been developed in the field of Artificial Intelligence, demonstrating that Qualitative Spatial Reasoning [1] can overcome the problems arising from indeterminacy, by allowing inference from incomplete spatial knowledge. Implementations of this kind of reasoners enable disambiguation between objects through spatial relations like directions or distances, thus allowing to improve symbol grounding on robots equipped with a speech recognition

© Springer International Publishing Switzerland 2015
M. Gavanelli et al. (Eds.): AI*IA 2015, LNAI 9336, pp. 452–464, 2015.
DOI: 10.1007/978-3-319-24309-2_34

system. However, the majority of theories on qualitative spatial reasoning have been developed under the assumption that discretized representations of the world were available. Usually, this is not true for robots, that need an abstraction of the environment strictly depending on its underlying structure, in order to reason about actions which can be executed.

In this paper, we propose a method for performing qualitative spatial reasoning on robots. In detail, we applied the *cone-based* approach, presented by [2], to an abstraction of the environment specifically built for a consistent integration of high-level reasoning and numeric representation. Both the representation and the reasoner are used in order to perform grounding of commands vocally given by the user to the robot. The proposed approach has been subsequently validated. In particular, by analysing the number of grounded commands in different settings, we pointed out multiple relations between such commands and the amount of knowledge available to the robot. Moreover, we performed additional experiments to identify the major issues that occur during the grounding process, by analyzing the user expectations with respect to the system outputs.

Fig. 1: User interacting with a robot through natural language interaction.

The key contributions of our work are the following. First, we introduce an abstract representation of the environment useful for the deployment of many theories of AI on real robotic applications. Second, we present a method for automatically adding, on such a representation, a high-level description of the objects through the interaction with the user: each object is represented as a composition of rectangles, easily enabling for the computation of spatial properties and relationships. Finally, we describe how, by exploiting the easiness of use of both the object and the environment representation, reasoning on areas can be effectively accomplished, according to well founded theories about Qualitative Spatial Reasoning.

The remainder of the paper is organized as follows. Related work is illustrated in Section 2, followed by a description of how the environmental representation is built, given in Section 3. Section 4 will show, instead, how the representation can be used to ground the commands given by a user, by exploiting a specifically implemented qualitative spatial reasoner (QSR). Finally, the experiments undertaken to validate this approach are reported in Section 5, while Section 6 will discuss the work presented and the future developments.

2 Related Work

In order to understand commands that use qualitative spatial references for distinguishing objects in the environment, first of all, a robot needs to be able of performing symbol grounding. The problem of symbol grounding, namely the process of matching natural language expression, with entities of the world and their corresponding representation internal to the robot, has been addressed by many authors. For example, in [3] the authors present a system able to follow natural language directions. In this work, the process of grounding the user commands is divided in three steps: Extracting linguistic structures related to spatial references; Grounding symbols to the corresponding physical objects within the environment; Reasoning on the knowledge acquired to extract a feasible path. In [4], instead, the authors describe a robot able to learn and use word meanings in three kind of tasks: indoor navigation, spatial language video retrieval, and mobile manipulation. They propose an approach for robustly interpreting natural language commands, based on the extraction of shallow linguistic structures. In particular they introduce a Generalized Grounding Graph able to handle multiple arguments or sentences nested into the commands. Finally, in [5] a sophisticated robot is described, equipped with a symbolic high-level spoken dialogue system that uses Discourse Representation Structures [6] to represent the meaning of the dialogues that occur with user.

The second key aspect needed to understand commands of the type "go in front of the closet next to the emergency door", is the ability of reasoning about spatial directions in a qualitative manner. In other words, the robot needs to be able of reasoning about an object with respect to another object in a given reference frame. In the literature, spatial relations are studied and used in various research fields. For example, in the "CogX"[1] project [7] the spatial relations "in" and "on" have been used to define object targets for indirect object search. Kunze et al. [8] have enhanced this work by using more restrictive spatial models to provide more tightly defined viewing probabilities. In particular, by using information about landmark objects and their spatial relationship to the target object, the authors show how a searching task can be improved by directing the robot towards the most likely object locations. The authors of [9] use, instead, an extension of the double cross calculus, introduced by [10], to express robot navigation objectives that include spatial relations in a Mars-like environment. This work, however, lacks an intermediate layer between the metric map and the high-level representation used for reasoning. Finally, Loutfi et al. [11] look

---

[1] http://cogx.eu/

at this problem in the context of perceptual anchoring to provide qualitative relations inferred from observed metric relations.

As well known in the research community of Qualitative Spatial Reasoning [12] the representation of spatial knowledge is usually divided in "propositional" and "pictorial" representations. The former cannot easily express structural properties, being focused on formal properties of the representation itself. The latter, even preserving structural properties, provide a low level representation which is not suitable for fast computations. Hernández suggests, in his work, the use of a hybrid representation, interfacing these two categories as separate representations. Inspired by [12] and adopting a similar approach, we propose a method for performing qualitative spatial reasoning in robotics, where the interface between the metric information and the symbolic knowledge is represented by a grid-based structure automatically built by our robot (briefly described in Section 3). On top of the representation, we adopt a reasoning approach that exploits shapes for distance and orientation qualitative calculus. Specifically, we decompose spaces and objects in rectangles, adopting an intrinsic reference frame. For grounding the command received by the user, instead, we follow the approach recently proposed by [13]. In our system, the output of an automatic speech recognition module (ASR) is matched with one of the frames representing the commands executable by the robots, later grounded using definite clause grammars [14]. A more detailed description of the processing chain from user utterance to task execution will be described in Section 4.

3   Representing the Environment

Starting from a low-level representation of the environment (Metric Map), rich of structural details, we have devised a method for automatically extracting a hybrid representation of the environment able to easily interface the symbolic and the structural layers of knowledge. Such a representation has been obtained by capturing the 2D structure of the environment through a grid, on top of which knowledge can be added. Such a grid, in fact, has the key property of having cells of different size, in order to capture the spatial similarity of close locations, while being uniformly accessible as a matrix. Each element of the matrix, therefore, can be considered independently, being its spatial relations kept by the structure of the matrix itself.

Since this work has been developed for domestic robots, operating in indoor locations, the first step performed in order to create a grid structurally coherent with the environment is the detection of the edges of the map through the Canny Edge Detector (Fig. 2a). Then, we find the segments corresponding to its walls, since they are the basic elements to form the grid itself. This process is performed by extracting the lines in the image through the Hough Transform (Fig. 2b) and then applying a filtering process (Fig. 2c) based on the length and distance of the segments. Starting from those, the grid (called Grid Map) is produced in an adaptive manner: 1) The segments generated through the wall detection are extended to the whole map image; 2) The minimum horizontal and vertical distances (x_{min}, y_{min}) are computed; 3) each pair of parallel lines is considered, and a new one is inserted between them only if their distance is at least twice x_{min} or y_{min}, depending on their angle. The resulting grid, therefore, has cells of a size ranging between $x_{min} \cdot y_{min}$ and $2x_{min} \cdot 2y_{min}$.

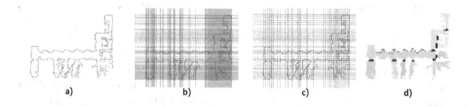

a)   b)   c)   d)

Fig. 2: a) Edges of the Metric Map. b) Segments extracted through the Hough Transform. c) Segments obtained after the filtering process in the wall detection. d) Representation of the Cell Map in which the areas of the environment are represented with different colors, the doors in blue and the objects in red.

On top of such a discretized representation, knowledge is added online by acquiring information through the natural language interaction between the robot and the human, to perform semantic mapping. Specifically, three types of knowledge can be acquired: Areas (e.g., corridors, rooms, etc.); Structural elements (e.g, windows, doors, etc.); Objects in the environment (e.g., tables, closets etc.). Using the Metric Map for the autonomous navigation of the robot in the environment, the positions of the different objects of interest are registered. The user can tag a specific object by naming it and specifying its position through a commercial laser pointer, thanks to which the orientation of the object is obtained by extracting its normal. The robot, in fact, is endowed with a speech recognition system and a Kinect sensor for detecting the laser point and extracting properties and shapes of objects, in addition to the laser range sensor used for the localization process. Once knowledge is acquired, it is first inserted in a relational database and, then, it is processed in order to be reported on the Grid Map, thus obtaining the Semantic Map, composed by the Cell Map and the Topological Graph. The Cell Map (Fig. 2d) contains a high-level description about the regions, structural elements, and objects contained in the environment. The Topological Graph, instead, is created in order to represent the information needed by the robot for navigating and acting in the environment, associating each node of the graph to a cell of the Cell Map. A final representation of the robot's knowledge is shown in Fig. 3, while a more detailed description about the representation, its building algorithms and the knowledge acquisition process can be found in [15].

Given the various kinds of knowledge represented in the Cell Map, various forms of reasoning can be performed. To this end, all the knowledge included in the Cell Map and in the Topological Graph is automatically translated into Prolog assertion predicates. In particular, each element of the Cell Map is represented with a predicate

```
cellIsPartOf(XCoord, YCoord, AreaTag)
```

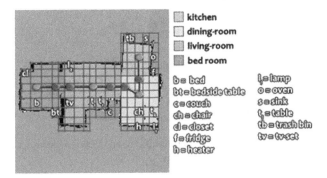

Fig. 3: Semantic map of a domestic environment. The Topological Graph is depicted on top of the Cell Map and the objects in it. The metric map is also depicted in the background.

and each object is represented with the two predicates

```
object(Id, XCoord, YCoord, Properties),
Type(Id, Type).
```

For example, the knowledge of a white plug located in the cell with grid map coordinates 45, 67 belonging to the corridor area will be represented with the three predicates

```
cellIsPartOf(45, 67, corridor),
object(plug1, 45, 67, color-white),
objectType(plug1, plug).
```

The knowledge stored in the Topological Graph is instead represented as an acyclic graph in Prolog with the predicates

```
node(Id, XCoord, YCoord),
arc(Id1, Id2),
```

respectively for the nodes and the arcs of the graph. Having translated in Prolog all the knowledge stored in the representation built with the previously described method, performing certain kind of inference on it becomes straight forward, as it is shown in the next section.

4   Reasoning on the Representation

In this section, we show how the commands uttered by a user are understood by our robot, by qualitatively reasoning about orientations and distances of objects in the environment. In particular, we describe the processing chain applied to the input command (Fig. 4), first describing the natural language processing operations performed on it, while later showing the qualitative reasoning method adopted.

Fig. 4: Processing chain applied to the commands uttered by a user, starting from the ASR and ending with the command execution.

Starting from the sentence uttered by the user, the processing chain for spoken commands follows a well-established [16]: based on a language model built using $SRGS^2$, the system tries to match the output of the ASR within a knowledge base of frames, which represent the commands that can be executed by the robot. According to the results of such a matching, either the processed command is sent to the grounding engine, or an interaction with the user is started, in order to ask for clarifications or for a new command. When a frame is correctly instantiated and an output can be passed to the grounder, the Sentence Analyzer serializes it into a command keyword and a list of tokens representing the specification of the action to be performed. For example, if the command uttered by the user is "go in front of the socket on the right of the closet", the output received by the grounder will be the keyword "GO_FRONT", in addition to the list of tokens "[to, the, socket, on, the, right of, the, closet]".

The output of the Sentence Analyzer is then passed to the Grounding Engine. By using definite clause grammars implemented in *Prolog*, we parse the input tokens in order to extract the located object, the reference object and the spatial relation that relates them. Having extracted these three elements from the user command, the system tries to ground them by querying the knowledge base and the spatial reasoner: two objects that fall in the categories of the located object and the reference object are searched, filtering the results by requiring their positions to agree with the relation specified by the user. To better explain this process, if the list "[to, the, socket, on, the, right of, the, closet]" is received as an input, the tokens "socket", "right of" and "closet" are identified as the located object, the spatial relation and the reference object respectively. The knowledge base is then queried for all the sockets and closets known in the environment with their positions. Finally, the known objects are filtered by the spatial reasoner that discards all the sockets that are not on the right of the closet.

In order to perform this latter operation, by exploiting the representation previously described, we built a spatial reasoner with an intrinsic reference frame. Three vicinity relations (near, next to and nearest), their three opposite relations (far, not next to and furthest) and four orientations (behind, in front, on the right and on the left) have been implemented. In particular, by defining C_{Loc} and C_{Ref}

---

^{2} http://www.w3.org/TR/speech-grammar/

the set of cells belonging to the cell map that include a portion of the objects Loc and Ref respectively, we say that Loc has a vicinity relation with Ref if and only if:

$$d(centroid(C_{Loc}), centroid(C_{Ref})) < t$$

where d is the euclidean distance, t is a threshold constant and $centroid(x)$ is a function that takes a set of cells x in input and returns the coordinates of its centroid in the metric map coordinate system. By specifying a threshold constant for both the relations "near" and "next to", respectively t_{near} and t_{next}, we therefore define the six distance relations (the nearest attribute is computed finding the object that minimizes the above defined distance). In order to define the orientation relations, instead, analogously to [12], we exploited the "intrinsic front side" of the objects (identified with the normal ñ of the surface tagged by the user during the knowledge acquisition phase previously described). Specifically, we have used it to define a forward orientation, later deriving, by rotating clock-wise, respectively the concept of left, backwards, and right regions, as shown in Fig. 5. By defining the general concept of directions, we adopted the *cone-based* approach [2] to explicate the four directional relations, starting from the centroid of the reference object, as shown in Fig. 5. As for the definition of vicinity, in order for two object Loc and Ref to be related from a relation R, by defining A_R^{Ref} the area corresponding to a region in the direction R with respect to the reference object Ref (e.g. A_{right}^{closet} is the area on the right of the closet), we require that:

$$centroid(C_{Loc}) \in A_R^{Ref}$$

where, again, the $centroid(x)$ is a function that takes a set of cells x in input and returns the coordinates of its centroid in the metric map coordinate system. Note that, if the centroid of the located object corresponds to the centroid of the reference object, we consider them connected with all the four directional relations, since our representation is an abstraction of the physical world. After applying the cone-based approach to our representation, by collapsing the represented objects in their centroid, all the properties derived for this approach from the theory are automatically inherited by our system. Finally, by exploiting the representation of the environment automatically built two advantages can be identified: The objects can be automatically inserted in the environment representation through the interaction with the user; The qualitative spatial reasoning can be performed on points (the centroid of the cells representing the object in the Cell Map), allowing for an easy and straightforward approach.

5   Experimental Evaluation

Several tests have been conducted in order to demonstrate the improvements that qualitative spatial reasoning can determine in grounding the commands given by the users to a robot, as well as the efficacy of implementing such an approach on a real robot. Our validation work has been therefore focused on two different kinds of experiments.

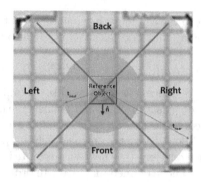

Fig. 5: Reference frame adopted for the implemented spatial reasoner. The image shows how the concept of "near" and "next to" have been implemented as well as how regions in the different direction have been identified through the cone-based approach.

The purpose of the first experiment was evaluating the impact of a qualitative spatial reasoner on an agent whose amount of knowledge continuously grows, as well as the influence of the already available knowledge on such a reasoning. The goal of this work was enabling a real robot to disambiguate the instructions given by a human-being on the basis of the relations between objects in the environment. Such an evaluation has been carried out considering the number of unambiguous and ambiguous commands (i.e., commands referring to more than one object with a specific spatial property, see Fig. 6) grounded by the agent. Indeed, when full knowledge about the environment is available, grounding ambiguous commands would mostly lead to the execution of the wrong action with respect to the user expectation, while all the unambiguous commands are supposed to be correctly grounded.

Fig. 6: Setting in which the command *"go to the socket near the closet"* has an ambiguous meaning. The command, in fact, could be grounded with either the socket on the left or the one on the right of the image, since they are both near the closet.

We therefore analyzed first the impact of the presence or absence of the qualitative spatial reasoner (QSR) and then the amount of knowledge available to the agent. In detail, we first asked each member of a group of 26 students from the First Örebro Winter School in Artificial Intelligence and Robotics[3] to provide a set of 3 commands containing spatial relations between objects, by looking at pictures of the test environment. Then, from the 78 acquired commands, we extracted two types of tasks: 28 ambiguous and 50 unambiguous. By gradually adding knowledge about the objects inside the knowledge base of the agent, we therefore measured how many commands were grounded. We repeated the experiment for both categories of commands, with or without the qualitative spatial reasoner. Since the curves depend on the order of the objects inserted in the knowledge base, the experiment has been performed five times in order to obtain its average trend (Fig. 7). In case the QSR was not present (first curve), only the objects in the environment, whose category has a unique member, were correctly identified. For example, since we had two closets in the test environment, there was no way of distinguish them without exploiting spatial relations. By comparing the first and the second curve in the image, it can be noticed that the presence of the QSR does not greatly affect their trend, when a little amount of knowledge is available, due to the absence of exploitable spatial relations between objects. On the contrary this is not true when substantial environmental information is accessible. In this case, Curve 2, 3 and 4 show that the QSR is essential for grounding all the unambiguous commands, lowering and eventually zeroing the errors that derive from the grounding of ambiguous ones (which should not have been grounded). In order to better understand this point, suppose you have a test environment where two trash bins are in front of two different windows: by not knowing the existence of one of the two trash bins, if the ambiguous command *"go to the trash near the window"* is given, the robot will erroneously ground the command with the only trash known. Differently, if both trashes are known the robot, it will correctly ground both objects, warning the user of such an ambiguity.

The goal of the second experiment was, instead, to understand the limitations of the proposed approach rather than to perform a usability study. In detail, we do not want to analyze in a quantitative manner the obtained results but, our intention is to identify the kind of errors perceived by non-expert users during the interaction and the grounding process. To this end, we implemented our method on a robot, able to interact with a user through natural language: in this setting we measured the agreement between the user expectations and the grounding performed by the robot. In particular, we first produced a Cell Map by carrying the robot on a tour of the environment and tagging 23 objects within the environment, as well as the doors and the functional areas in it, through an online augmentation of the Map. Then, we asked 10 different non-expert users to assign 10 distinct tasks to the robot, asking them to evaluate if the robot correctly grounded their commands, meeting their expectations. The commands have been directly acquired through a Graphic User Interface, in order to avoid possible errors due to misunderstandings from the speech recognition system. In detail, the users had the

---

[3] http://aass.oru.se/Agora/Lucia2013/

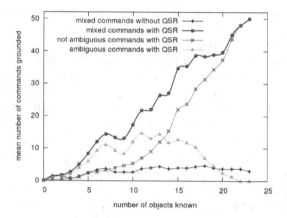

Fig. 7: Mean number of grounded commands with respect to the number of objects known in the environment, added in a random order. Three different curves ("mixed commands without QSR", "mixed commands with QSR", "not ambiguous commands with QSR" and "ambiguous commands with QSR"), respectively, report the results obtained by giving to the robot the complete (mixed) set of commands, only unambiguous commands or only ambiguous commands. As expected, with a qualitative spatial reasoner and a complete knowledge about the relevant elements of the environment, the robot correctly grounds only the not ambiguous commands.

possibility to choose the action to be executed by specifying the located object, the reference object and one of the 10 spatial relations implemented in our reasoner. Table 1 shows that approximately 80% of the uttered commands have been correctly grounded. The remaining 20% of the wrongly grounded commands where due to two different phenomena:

- The command given was ambiguous, requiring other proprieties, in addition to direction and distance, to identify the object;
- The users did not behave coherently during the interaction with the robot, by varying their concept of vicinity or by adopting different reference frames.

While the first issue is intrinsic to the nature of the command and it can be solved by exploiting other proprieties (e.g., the color), the second one could be addressed by using adaptive parameters, learnt over time through the interaction with the user. For example we noticed that the concept of vicinity for a reference object varies with the number of objects around it. By keeping track of the feedbacks given by the user when the system wrongly grounds a command, the change of the vicinity concept (represented in our system with the two thresholds t_{next} and t_{near}) over different settings could be modelled. Moreover, the well established concept that the reference frame adopted for spatially relating objects changes with respect of where the user is standing ([17] and [18]), could be addressed by dynamically changing the robot's reference frame based on the position of the human. Such solutions, however, go beyond the scope of this paper.

Table 1: Results of the second experiment. Ten different users have been asked to give ten different tasks to the robot, using spatial relations about distances and directions. The table shows the number of correctly and wrongly grounded commands with respect to the expectations of the users.

User	Correctly Grounded Commands	Wrongly Grounded Commands
1st	7	3
2nd	8	2
3rd	10	0
4th	6	4
5th	8	2
6th	8	2
7th	10	0
8th	7	3
9th	9	1
10th	8	2
Total	81	19

6 Conclusion

In this paper we have presented a method for applying qualitative reasoning about directions and distances on real robots. In particular, we have shown how a suitable representation of the environment can be automatically extracted from the Metric Map, by creating a grid-based abstraction of the world with the aid of the user. By embedding in such a representation a high-level description of the objects, qualitative spatial reasoning can be performed by the robot to accomplish tasks in real scenarios. Indeed, this is an important task, for performing a further step in the direction of implementing effective human-robot interaction. The proposed approach has been validated by considering the number of grounded commands with respect to different amounts of knowledge available to the robot, as well as with the presence and the absence of a qualitative spatial reasoner. From such an analysis the essential role of a qualitative spatial reasoner for grounding spoken commands has been pointed out. Finally, we have performed a second experiment to identify the major issues that occur during the grounding process. Specifically, several non-expert users have been required to give specific commands to the robots, comparing their expectations with the output of the system. Two issues intrinsically embedded in the human-robot interaction have been identified: the ambiguity of certain commands and the incoherence of reference frames adopted by the users. Solving these issues, as well as using the proposed representation for a further improvement of the reasoner (e.g., considering different kind of commands that exploit other properties to identify objects), will be the focus of our future work. Moreover, we are planning to extend our approach in a 3D representation.

References

1. Knauff, M.: The cognitive adequacy of allen's interval calculus for qualitative spatial representation and reasoning (1999)
2. Frank, A.U.: Qualitative spatial reasoning with cardinal directions. In: Seventh Austrian Conference on Artificial Intelligence (1991)
3. Tellex, S., Kollar, T., Dickerson, S., Walter, M.R., Banerjee, A.G., Teller, S.J., Roy, N.: Understanding natural language commands for robotic navigation and mobile manipulation. In: AAAI (2011)
4. Tellex, S., Kollar, T., Dickerson, S., Walter, M.R., Banerjee, A.G., Teller, S., Roy, N.: Approaching the symbol grounding problem with probabilistic graphical models (2011)
5. Theobalt, C., Bos, J., Chapman, T., Espinosa-Romero, A., Fraser, M., Hayes, G., Klein, E., Oka, T., Reeve, R.: Talking to godot: dialogue with a mobile robot. In: Proceedings of IEEE/RSJ International Conference on Intelligent Robots and Systems (IROS 2002) (2002)
6. Kamp, H., Reyle, U.: From discourse to logic: Introduction to modeltheoretic semantics of natural language, formal logic and discourse representation theory (1993)
7. Sjöö, K., Aydemir, A., Jensfelt, P.: Topological spatial relations for active visual search (2012)
8. Kunze, L., Doreswamy, K.K., Hawes, N.: Using qualitative spatial relations for indirect object search. In: IEEE International Conference on Robotics and Automation (ICRA) (2014)
9. McClelland, M., Campbell, M., Estlin, T.: Qualitative relational mapping for planetary rovers. In: Workshops at the Twenty-Seventh AAAI Conference on Artificial Intelligence (2013)
10. Zimmermann, K., Freksa, C.: Qualitative spatial reasoning using orientation, distance, and path knowledge (1996)
11. Loutfi, A., Coradeschi, S., Daoutis, M., Melchert, J.: Using knowledge representation for perceptual anchoring in a robotic system (2008)
12. Hernández, D.: Diagrammtical aspects of qualitative representations of space (1992)
13. Bastianelli, E., Croce, D., Basili, R., Nardi, D.: Unitor-hmm-tk: structured kernel-based learning for spatial role labeling. In: Proceedings of the Seventh International Workshop on Semantic Evaluation (SemEval 2013). Association for Computational Linguistics (2013)
14. Pereira, F.C., Warren, D.H.: Definite clause grammars for language analysis – a survey of the formalism and a comparison with augmented transition networks (1980)
15. Bastianelli, E., Bloisi, D.D., Capobianco, R., Cossu, F., Gemignani, G., Iocchi, L., Nardi, D.: On-line semantic mapping. In: 2013 16th International Conference on Advanced Robotics (ICAR) (2013)
16. Aiello, L.C., Bastianelli, E., Iocchi, L., Nardi, D., Perera, V., Randelli, G.: Knowledgeable talking robots. In: Kühnberger, K.-U., Rudolph, S., Wang, P. (eds.) AGI 2013. LNCS, vol. 7999, pp. 182–191. Springer, Heidelberg (2013)
17. Baruah, R., Hazarika, S.M.: Qualitative directions in egocentric and allocentric spatial reference frames (2014)
18. Tversky, B.: Cognitive maps, cognitive collages, and spatial mental models. In: Campari, I., Frank, A.U. (eds.) COSIT 1993. LNCS, vol. 716, pp. 14–24. Springer, Heidelberg (1993)

COACHES Cooperative Autonomous Robots in Complex and Human Populated Environments

Luca Iocchi[1]([✉]), M.T. Lázaro[1], Laurent Jeanpierre[2], Abdel-Illah Mouaddib[2], Esra Erdem[3], and Hichem Sahli[4]

[1] DIAG, Sapienza University of Rome, Rome, Italy
iocchi@dis.uniroma1.it
[2] GREYC, University of Caen Lower-Normandy, Caen, France
[3] Sabanci University, Istanbul, Turkey
[4] Vrije Universiteit Brussel, Brussels, Belgium

Abstract. The deployment of robots in dynamic, complex and uncertain environments populated by people is gaining more and more attention, from both research and application perspectives. The new challenge for the near future is to deploy intelligent social robots in public spaces to make easier and safer the use of these spaces. In this paper, we provide an overview of the COACHES project which addresses fundamental issues related to the design and development of autonomous robots to be deployed in public spaces. In particular, we describe the main components in which Artificial Intelligence techniques are used and integrated with the robotic system, as well as implementation details and some preliminary tests of these components.

1 Introduction

Public spaces in large cities are becoming increasingly complex and unwelcoming environments because of the overcrowding and complex information in signboards. It is in the interest of cities to make their public spaces easier to use, friendlier to visitors and safer to increasing elderly population and to the people with disabilities. In the last decade, we observe tremendous progress in the development of robots in dynamic environments populated by people. There are thus big expectations in the deployment of robots in public areas (malls, touristic sites, parks, etc.) to offer services to welcome people in the environment and improve its usability by visitors, elderly or disabled people.

Such application domains require robots with new capabilities leading to new scientific challenges: robots should assess the situation, estimate the needs of people, socially interact in a dynamic way and in a short time, with many people, the navigation should be safe and respects the social norms. These capabilities require new skills including robust and safe navigation, robust image and video processing, short-term human-robot interaction models, human need estimation techniques and distributed and scalable multi-agent planning.

The main goal of the COACHES project (October 2014 - September 2017) is to develop robots that can suitably interact with users in a complex large public

© Springer International Publishing Switzerland 2015
M. Gavanelli et al. (Eds.): AI*IA 2015, LNAI 9336, pp. 465–477, 2015.
DOI: 10.1007/978-3-319-24309-2_35

Fig. 1. COACHES environment and robot.

environment, like a shopping mall. Figure 1 shows the *Rive de l'orne* shooping mall in Caen (France) where the experimental activities of the project will be carried out, as well as a prototype of the robot that will be used.

Previous work on social robotics and human-robot interaction mostly focused on one-to-one human-robot interaction, including elderly assistance (e.g., Giraff-Plus project [5]) and interaction with children (e.g., MOnarCH project [6]). Robots acting as museum tour-guides have also been successfully experimented. One of the first robot interacting with many non-expert users was RHINO deployed at the "Deutsches Museum" in Bonn, Germany [2]. In this work, the main focus was in the mapping, localization and navigation abilities in crowded environment, while human-robot interaction was limited to buttons on the robot, a remote Web interface and pre-recorded sentences issued by the robot.

As shown in the figure, in contrast with previous work, the COACHES environment is very challenging, because populated by many people. Moreover, we aim at a more sophisticated interaction using multiple modalities (speech, gesture, touch user interfaces) and dialog generated on-line according to the current situation and the robot's goals. Consequently, the required level of "intelligence" of the COACHES robots in order to adequately perform complex and effective tasks in this environment in presence of people is much higher than in previous projects.

The proposed methodology to reach the project goals is based on integration of Artificial Intelligence and Robotics. In this paper, we describe the main overall architecture of the system (Section 2) and the components related to Artificial Intelligence and Robotics (Section 3): 1) knowledge representation, 2) planning under uncertainty, 3) hierarchical plan execution and monitoring. Section 4 provides some examples and Section 5 concludes the paper.

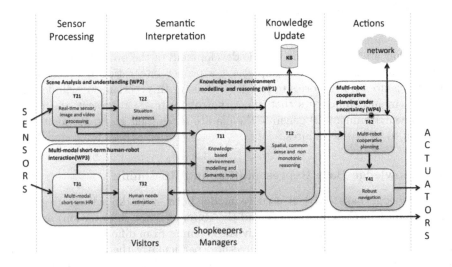

Fig. 2. COACHES software architecture

2   Software Architecture

The software architecture of the COACHES robots is shown in Figure 2). An open architecture (hard/soft) and standard technologies available will be used, so that it will be easy to extend and/or adapt the capabilities of the system during the whole length of the project (especially to integrate and test various algorithms and/or sensors). Such an open architecture will also simplify and optimize integration efficiency as well as re-use of assets in other projects or products.

The main software components that will be developed for control, reasoning and interaction functionalities of the system are listed below.

- *Scene analysis*, including sensor processing procedures for both on-board robot devices and static sensors in order to determine the current situation and understand events that are of interest for the system.
- *Multi-modal HRI*, defining a set of modalities for human-robot interaction, including speech recognition and synthesis, touch interaction, graphical interface on a screen mounted on the robot and Web interfaces.
- *Knowledge-based representation and reasoning*, defining the formalism and the procedure to represent and reason about the environment and the task of the robots.
- *Planning and execution monitoring*, for generating the plans to achieve the desired goals and monitor their execution for robust behaviors.
- *Safe navigation*, for guaranteeing safety operations of the robot in a populated environment.

3   Artificial Intelligence Components

While the overall software architecture described before integrates several components that are all important for the development of the project, in this paper we focus on the modules that implement a proper integration between Artificial Intelligence and Robotics techniques. Thus, in this section, we will describe the three main components that allow the robots to: 1) represent and reason about the environment, 2) generate the plan to reach their goals; 3) monitor the execution to overcome failures.

3.1   Knowledge Base Representation and Reasoning

The knowledge base (KB) is used to model both static knowledge (e.g., the semantic map of the environment and the common sense information) and the dynamic knowledge (e.g., human activities) coming from different units, such as the perception modules of the architecture, particularly the multi-modal HRI interface and the image processing modules. From these information, the reasoning module is able to infer the list of possible tasks to accomplish. This list of tasks is then sent to the decision module (described in the following section) that will compute the policy to accomplish them.

Although there are many existing approaches to semantic representations of the environment (see [7] for a survey), a standard formalism does not exist. In this section, we thus define the main features of the knowledge base used in the project, based on experience in previous work [1,3]. We first introduce the semantic labels used to describe elements of the world, then predicates that determine relations among these labels, and finally its application to the use case in the project.

Semantic Labels. In order to refer to objects and classes of objects in the knowledge base, we introduce a set of labels that will be associated to semantic meanings.

A first set of these labels are called *Concepts*. Concept labels are associated to classes of objects and in this paper they are denoted with an uppercase initial letter. For example, *Restaurant* is a concept used in the semantic map to denote the class of restaurants in the shopping mall. These concepts are organized in a hierarchical way according to the "is-a" relation. In this way, we can express, for example, that the concept *FrenchRestaurant* is a *Restaurant*.

A second set of labels will be used to denote objects. Each object belongs to a concept implementing the relation "instance-of". Object labels are denoted with lowercase letters. Thus, a particular restaurant in the mall will be denoted with an object label *caféMarcel* that will be an instance of the concept *FrenchRestaurant*.

Predicates. Predicates are used to describe relations among the semantic labels. For example, the "is-a" and the "instance-of" relations can be represented by the corresponding predicates **is-a** and **instance-of**. Predicates are also used to denote properties of objects or locations (e.g., the status of a door or the presence of air-conditioned in a shop). For representing the *Rive de l'orne* shopping mall, we consider different types of areas: shops, restaurants, halls, corridors, rest areas, offices, toilettes, etc. For shops, services and restaurants we consider different categories:

- *Shop categories*: dress shop, women dress shop, kid dress shop, men dress shop, makeup store, store perfume, sport store, etc.
- *Restaurant categories*: French, Japanese, Chinese, Italian, Oriental, African, fast-food, etc.
- *Service categories*: security, information, health-care, etc.

All these areas are represented as concepts that are grouped in a more general concept *Area*. The hierarchy of these areas will be defined through the "is-a" relation of the semantic labels described before.

Some examples of predicates for representing the shopping mall are:

is-a(FrenchRestaurant, Restaurant)
instance-of(caféMarcel, FrenchRestaurant)
connect(door12, hall, caféMarcel)
open(door12)
airconditioned(caféMarcel)

Reasoning. The KB is used by reasoning processes that define the goals for the COACHES robots. To this end, the reasoning engine takes into account the available information in the KB related to: semantic map, common-sense knowledge, and dynamic knowledge coming from the scene analysis and HRI modules. With this input, this module determines which goals for the system are consistent with the current situation. Then these goals are passed to the Planning module described below.

A further function of reasoning on the KB is to determine conditions for plan execution that are derived from direct perception. In this way, plan execution can consider properties not directly observable from perception, but coming from a reasoning process that interpret perception with common-sense reasoning.

3.2 Planning under Uncertainty

In this section we describe the Markov Decision Process (MDP) used to model the COACHES planning domain and the algorithm implemented for computing the optimal policy.

Task Structure. The result of the reasoning module (KB module) is a set of goals $G = \{g_1, g_2, \ldots, g_k\}$ concerning advertisement, patrolling, assisting and escorting. We note also that advertising goals could be performed in parallel with the moving ones. Consequently, the task structure is a hierarchy of modules to execute. This structure is inspired by progressive processing units [4], that we name PRU+. In our application, we define four PRU+. Each PRU+ is composed of levels where the first level concerns the execution of the subtask GOTO SITE X, the second level concerns the advertisement at a location (x, y) and the third level consists of DO TASK X where X could be the assistance, the patrolling, the escorting or the surveillance. With such task structures we can also define some joint goals requiring joint PRU+. For example, escorting a people from one location in a building to another location in the other building requires a cooperation between robots. Indeed, the first robot executes a policy of PRU+ for escorting a user to the exit of the first building, provide him/her information to reach the other building and then send information to the other robots in the other building to continue the escorting task at the second building. The structure of tasks we propose for single robot tasks is {GOTO x, ADVERTISEMENT, DO x}, while for the joint task is {GOTO x, ADVERTISEMENT, INFORM PEOPLE, SEND MESSAGE TO THE OTHER ROBOTS}. The task DO x concerns different tasks of assistance.

More formally, a PRU+ is defined by a set of levels $\{l_1, l_2, \ldots, l_k\}$, where each level l_i is composed by a set of modules $\{m_i^1, m_i^2, \ldots, m_i^{p_i}\}$ and each module m_i^j is defined by different execution outcomes that we name options $\{\alpha_i^j, \beta_i^j, \ldots\}$.

MDP Definition and Planning. The planning procedure consists of formalizing the robot activities as an MDP using the PRU+ task definition. This procedure is based on two steps: 1) generating an MDP from a PRU+, 2) compute the optimal policy for the generated MDP. In the following we define the $MDP = < S, A, R, T >$ where :

- S is a set of states defined by $x = [l, m, o, v]$ where l is the level of the PRU, m is a module of the level l, o is an option of module m and v are state variables defining the execution context representing the subset of variables to be considered for the option o.
- A is the set of actions consisting of execution of one module of the next levels E or skipping the level S.
- T is the transition function defined as follows :
  - $Pr([l + 1, m', o', v'] | [l, m, o, v], E) = p(o')$, this means when execution module m' at state $[l, m, o, v]$ we move to state $[l', m', o', v']$ with probability p(o') representing the probability to get the outcome o'.
  - $Pr([l + 2, m', o', v'] | [l, m, o, v], S) = 1$, this transition is deterministic because we skip level $l + 1$ and we move to level $l + 2$.
- R is the reward function related to the options assessing the benefit to get the outcome;

From this definition, the Bellman equation for our model becomes

$$V(x) = R(o) + max_{E,S} \sum_{x'} Pr(x'|x, a)V(x')$$

The optimal policy π is computed by a standard MDP solving algorithm based on value-iteration. Moreover, in this algorithm, a tabu-list of actions is used to choose or drop actions to be inserted in the policy. This tabu-list is built and updated by the Model updater module, described below in this section, representing the actual feedback coming from the execution layer.

3.3 Plan Execution and Monitoring

Plan execution monitoring and interleaving planning and execution are crucial features for an autonomous robot acting in a real environment, specially when human interaction is involved, as for the COACHES robots. Indeed, in complex scenarios, it is not possible to model and foresee all the possible situations that may occur, consequently plans generated off-line (i.e., before the actual execution of the task), when several information about the real environment are not known, may not be optimal or feasible at execution time.

It is thus necessary to explicitly model and consider possible plan failures and to devise a mechanism that is able to properly react to these failures. Moreover, on-line replanning (i.e., planning after plan failures) may not be feasible when the model itself is inaccurate, since the same cause of the plan failure (typically a non-modeled feature of the environment) will likely occur also in next executions.

To this end, we have defined a plan execution and monitoring framework composed by three modules: a planner (as described in the previous section), an executor, and a model updater. The three modules cooperate during the execution of a complex task for a robot and provide for a feedback mechanism from execution to planning. More specifically, the following interactions are devised: 1) the planner notifies on-line to the executor the best plan (policy) to be executed according to the current model of the world; 2) the executor executes this plan (policy) and determines success or failures of the actions; 3) each failure is reported to the model updater that will follow some rules (either automatic domain dependent or manual domain dependent) to modify the current model, so that the planner can generate a new plan that is more suitable for the current situation as detected by the executor.

The execution module is based on the Petri Net Plan (PNP) formalism[1] [8]. PNP is a formalism to represent high-level plans for robot and multi-robot systems. Being based on Petri Nets, it is very expressive and can thus represent durative ordinary and sensing actions, and many constructs such as sequence, loop, interrupt, fork/join, and several multi-robot synchronization operators. PNPs are used to model the behavior (i.e., the policy) that is generated by the planner module and to execute it using the PNP-ROS implementation that allows ROS actions[2] to be executed under the control of a PNP.

The two main components of this process will be described in the rest of this section: 1) Policy to PNP transformation; 2) Model updater.

Policy to PNP Transformation. The policy generated by the MDP planner is automatically transformed in a PNP. For this process, the MDP planner

---

[1] pnp.dis.uniroma1.it
[2] wiki.ros.org/actionlib

produces the following information: the initial state, one or more goal states, state-action pairs implementing the policy and the conditions to be checked when non-deterministic actions are executed. States, actions and conditions are represented just as unique labels. With this input, the algorithm for generating the corresponding PNP is based on a recursive procedure for building the graph corresponding to the policy, starting from the initial state to the goal states, applying the state-action pairs for each state and adding a sensing operator for every non-deterministic effect of an action.

The labels in the policy and in the PNP referring to actions correspond to implemented actions, while labels referring to conditions correspond to sensor processing procedures that evaluate their truth based on the current information available to the system.

The PNP generated with this process does not contain a representation of action failures. Action failures are considered by defining a set of execution conditions for each action and by automatically adding action interrupts when these conditions are not valid. In this way the new PNP will be able to actually check execution conditions of actions and to interrupt the plan whenever these conditions are not valid. For example, an execution condition of a communication action is that a person is in front of the robot. While executing the action, the condition of a person being in front of the robot is checked and, if it becomes false, the action is interrupted.

When an interrupt is activated, the flow of execution of the plan can follow one of the two following lines: 1) *internal recovery procedure*[3], when the current plan itself contains a recovery behavior (i.e., a sub-plan or portion of the plan) for dealing with this failure; 2) *plan failure*, when the current plan is not able to deal with this kind of failure.

In the latter case, the executor sends to the Model updater module the following information: 1) action failed, 2) condition that was checked to determine action failure, 3) status of the plan (that can contain additional conditions useful for diagnosis of the failure). Given this input, the Model updater module (described in the next paragraph) modifies the MDP model of the domain and activates a new planning procedure to generate a new plan (policy) that aims at avoiding at least the failure cause just occurred.

Model Update. The problem of updating a planning model, given the feedback of the execution of the plan, is very relevant for actual application of planning techniques to real problems, but, to the best of our knowledge, a general solution suitable for our needs does not exists.

At this moment, we have implemented a simple method that builds and maintains a tabu list of actions to be selected in the MDP planning process. More specifically, whenever an action fails, the action is inserted in the tabu list and thus it will not be selected in the next generation of the policy. This mechanism is also tied to a time decay mechanism, so that the presence of an

---

[3] At this moment, the *internal recovery procedures* are manually written, while some automatic technique could be devised.

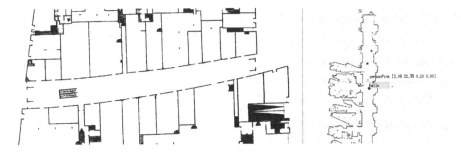

Fig. 3. 2D map of the *Rive de l'orne* shopping center and Stage simulator snapshot of the DIAG example.

action in the tabu list decreases over time, making the action available some time after the action failed, in order to try it again in the future.

For example, if the action of moving to a particular shop fails because there are too many people in that area, the robot will avoid to generate a new plan that will include going to that shop for a while, avoiding thus the main cause of the current failure.

Although not optimal, this strategy allows the robot to generate new plans that will possibly avoid the causes of failure of the previous plans.

4   Implementation and Tests

Before experimenting the robots in the actual environment, it is necessary to develop and test the solutions in a simulator and in a more controlled environment. To this end, we report here the development of a simulated environment for the project and some preliminary tests made with the robot in the DIAG Department.

COACHES architecture is implemented within the ROS framework[4]. ROS includes several ready-to-use modules for basic functionalities of the robots, hardware drivers, simulation, and debugging. Moreover, the ROS environment guarantees an easy porting from simulation to real robots and in particular, our software architecture is implemented in such a way to remain unchanged when passing from simulation to robots.

4.1   2D Simulator Environment

The simulation environment in COACHES is 2D and is based on Stage, in particular on its ROS version[5]. The choice of a 2D simulator (instead of a 3D one) is motivated by: 1) the need of modeling and testing high-level behaviors of the robots that do not involve 3D perception, 2) the possibility of using the simulator for multiple robots and other moving elements representing people in the

---

[4] www.ros.org
[5] wiki.ros.org/stage

environment, 3) the possibility of using the simulator on standard laptops, thus not requiring advanced graphical cards for running 3D simulations.

We have extended the original Stage simulator by adding a characterization of people in the environment and simple forms of HRI: i) the words spoken by the robot appears in the simulation window; ii) a GUI can be used by an operator to simulate human-robot inputs.

In the Stage simulator maps of the *Rive de l'orne* shopping center (Fig. 3 left) and of the DIAG Department (Fig. 3 right) have been realized. The Stage environment models one or more robots that have the same 2D sensor and actuator configurations as the real ones and some additional mobile obstacles that represent people moving in the environment. Several behaviors can be tested in this simulated environment such as: 2D perception of human behaviors, human-robot social navigation (e.g., following a person or guiding a person), safe navigation in the environment.

Several tests have been performed on the simulator, showing that it is a suitable tool for developing high-level robot behaviors.

4.2   Preliminary Tests at DIAG

In order to test the developed modules on a robot, we have defined a task (similar to the COACHES use cases) that can be run in an office environment. We consider a robot assisting users in an office. The robot welcomes people at the entrance and tell them about the latest news. It may also offer assistance for the printer: bringing the printed document to some other person, or informing technicians about printer troubles.

The mission is described in the PRU+ depicted in Figure 4. It has 4 layers: 1) waiting for people, 2) welcoming people and offering assistance, 3) bringing

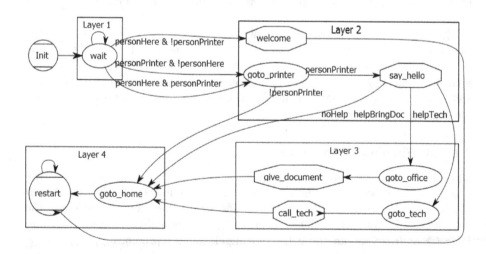

Fig. 4. PRU+ for printer-assistance and welcome

documents and fetching for technicians, and 4) returning to home position. The expected behavior is the following: from action 'wait' in Layer 1 four outcomes are possible: nobody is there, somebody has been detected near the entrance, near the printer, or both. When the 'wait' action completes, the robot might decide to wait again. If somebody is close to it, the robot can welcome and announce news. If somebody is at the printer, the robot can go there. Tasks 'welcome', 'say_hello', 'call_tech' and 'give_document' are also granting the robot a reward. They are represented by octagons in Figure 4.

From this PRU+, a 16-states MDP is built. Once solved, it produces the following policy:

- (Init): wait
- (1,wait,both): goto_printer
- (1,wait,entry): welcome
- (1,wait,print): goto_printer
- (2,goto_printer,err): goto_home
- (2,goto_printer,ok): say_hello
- (2,say_hello,bring): goto_office
- (2,say_hello,done): goto_home

- (2,say_hello,help): goto_tech
- (2,welcome,done): restart
- (3,goto_office,done): give_document
- (3,goto_tech,done): call_tech
- (3,give_document,done): goto_home
- (3,call_tech,done): goto_home
- (4,goto_home,done): restart
- (4,restart,done): restart

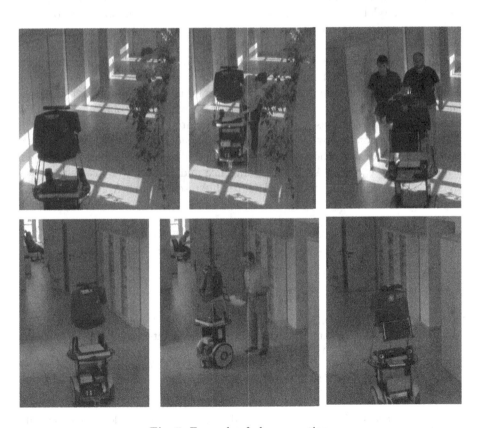

Fig. 5. Example of plan execution.

The policy is denoted by state-action pairs, where states are represented as $[l, m, v]$ (i.e., level, module and state variables, as described in the previous section) and actions correspond to the tasks defined in the PRU+. This policy is then translated into a PNP and executed by the robot.

Figure 5 shows some snapshots of plan execution, in the situation where the robot is asked to bring a document to a person. The interaction with the two persons involved and a few intermediate snapshots are reported. Notice that, although in a simplified setting[6], with these tests we have verified suitability and effectiveness of most of the components of our software architecture and their interconnection.

5   Conclusions

In this paper, we have described the main concepts of the COACHES project and in particular its Artificial Intelligence and Robotics components and their integration. More specifically, we have described a framework for integrating knowledge representation and reasoning, MDP planning and PNP execution, allowing a feedback from execution to reasoning in order to update and improve the current model of the world. Implementation and preliminary tests of such an integration have been performed to assess the suitability of the proposed architecture.

Many interesting results are expected from the COACHES project, since the environment and the challenges considered here are very ambitious. Among the several performance evaluation procedures, we aim at including extensive user studies that will be used to validate the effective development of intelligent social robots performing complex tasks in public populated areas.

We believe that deploying robots in public spaces populated by non-expert users is a fundamental process for the actual design, development and validation of integrated research in Artificial Intelligence and Robotics. Consequently, we envision many significant contributions to this research area from the COACHES project.

Acknowledgments. COACHES is funded within the CHIST-ERA 4th Call for Research projects, 2013, Adaptive Machines in Complex Environments (AMCE) Section.

References

1. Bastianelli, E., Bloisi, D.D., Capobianco, R., Cossu, F., Gemignani, G., Iocchi, L., Nardi, D.: On-line semantic mapping. In: 2013 16th International Conference on Advanced Robotics (ICAR), pp. 1–6, November 2013
2. Burgard, W., Cremers, A.B., Fox, D., Hhnel, D., Lakemeyer, G., Schulz, D., Steiner, W., Thrun, S.: The interactive museum tour-guide robot. In: Proc. of AAAI (1998)

---

[6] At this moment HRI and perceptions modules are not fully implemented and thus we replaced them with the remote control of an operator.

3. Capobianco, R., Serafin, J., Dichtl, J., Grisetti, G., Iocchi, L., Nardi, D.: A proposal for semantic map representation and evaluation. In: Proc. of the European Conference on Mobile Robots (ECMR) (2015)
4. Cardon, S., Mouaddib, A.-I., Zilberstein, S., Washington, R.: Adaptive control of acyclic progressive processing task structures. In: Proc. of IJCAI, pp. 701–706 (2001)
5. Coradeschi, S., Cesta, A., Cortellessa, G., Coraci, L., Galindo, C., Gonzalez, J., Karlsson, L., Forsberg, A., Frennert, S., Furfari, F., Loutfi, A., Orlandini, A., Palumbo, F., Pecora, F., von Rump, S., Štimec, A., Ullberg, J., Ötslund, B.: GiraffPlus: a system for monitoring activities and physiological parameters and promoting social interaction for elderly. In: Hippe, Z.S., Kulikowski, J.L., Mroczek, T., Wtorek, J. (eds.) Human-Computer Systems Interaction: Backgrounds and Applications 3. AISC, vol. 300, pp. 261–271. Springer, Heidelberg (2014)
6. Ferreira, I., Sequeira, J.: Designing a Social [Robot] for Children and Teens: Some Guidelines to the Design of Social Robots. International Journal of Signs and Semiotic Systems 3(2) (2014). Special Issue on The Semiosis of Cognition
7. Kostavelis, I. Gasteratos, A.: Semantic mapping for mobile robotics tasks: A survey. Robotics and Autonomous Systems (2014)
8. Ziparo, V.A., Iocchi, L., Lima, P.U., Nardi, D., Palamara, P.F.: Petri net plans - A framework for collaboration and coordination in multi-robot systems. Autonomous Agents and Multi-Agent Systems 23(3), 344–383 (2011)

Author Index

Printed in the United States
By Bookmasters